BOSNIA AND THE DESTRUCTION OF CULTURAL HERITAGE

The massive intentional destruction of cultural heritage during the 1992–1995 Bosnian War targeting a historically diverse identity provoked global condemnation and became a seminal marker in the discourse on cultural heritage. It prompted an urgent reassessment of how cultural property could be protected in times of conflict and led to a more definitive recognition in international humanitarian law that destruction of a people's cultural heritage is an aspect of genocide. Yet surprisingly little has been published on the subject.

This wide-ranging book provides the first comprehensive overview and critical analysis of the destruction of Bosnia-Herzegovina's cultural heritage and its far-reaching impact. Scrutinizing the responses of the international community during the war (including bodies such as UNESCO and the Council of Europe), the volume also analyses how, after the conflict ended, external agendas impinged on heritage reconstruction to the detriment of the broader peace process and refugee return. It assesses implementation of Annex 8 of the Dayton Peace Agreement, a unique attempt to address the devastation to Bosnia's cultural heritage, and examines the treatment of war crimes involving cultural property at the International Criminal Tribunal for the former Yugoslavia (ICTY).

With numerous case studies and plentiful illustrations, this important volume considers questions which have moved to the foreground with the inclusion of cultural heritage preservation in discussions of the right to culture in human rights discourse and as a vital element of post-conflict and development aid.

Heritage, Culture and Identity

Series Editor: Brian Graham,
School of Environmental Sciences, University of Ulster, UK

Other titles in this series

World Heritage in Iran: Perspectives on Pasargadae
Edited by Ali Mozaffari
ISBN 978 1 4094 4844 0

Edible Identities: Food as Cultural Heritage
Edited by Ronda L. Brulotte and Michael A. Di Giovine
ISBN 978 1 4094 42639

Who Needs Experts?
Counter-mapping Cultural Heritage
Edited by John Schofield
ISBN 978 1 4094 3934 9

The Making of a Cultural Landscape
The English Lake District as Tourist Destination, 1750–2010
Edited by John K. Walton and Jason Wood
ISBN 978 1 4094 2368 3

Cultural Heritage of the Great War in Britain
Ross J. Wilson
ISBN 978 1 4094 4573 9

Many Voices, One Vision: The Early Years of the World Heritage Convention
Christina Cameron and Mechtild Rössler
ISBN 978 1 4094 3765 9

Partitioned Lives: The Irish Borderlands
Catherine Nash, Bryonie Reid and Brian Graham
ISBN 978 1 4094 6672 7

Ireland's 1916 Uprising
Explorations of History-Making, Commemoration & Heritage in Modern Times
Mark McCarthy
ISBN 978 1 4094 3623 2

Bosnia and the Destruction of Cultural Heritage

HELEN WALASEK

With contributions by

Richard Carlton, Amra Hadžimuhamedović, Valery Perry and Tina Wik

Routledge
Taylor & Francis Group
LONDON AND NEW YORK

First published 2015 by Ashgate Publishing

Published 2016 by Routledge
2 Park Square, Milton Park, Abingdon, Oxon OX14 4RN
711 Third Avenue, New York, NY 10017, USA

First issued in paperback 2017

Routledge is an imprint of the Taylor & Francis Group, an informa business

British Library Cataloguing in Publication Data
A catalogue record for this book is available from the British Library

The Library of Congress Cataloging-in-Publication Data
A catalog record for this book is available from the Library of Congress

ISBN 13: 978-1-138-30842-8 (pbk)
ISBN 13: 978-1-4094-3704-8 (hbk)

Dedicated to the Zemaljski Muzej, the Commission to Preserve National Monuments, and all those people and institutions who have tried to preserve the cultural heritage of Bosnia-Herzegovina.

Contents

List of Figures

List of Maps

List of Tables

Notes on Contributors

Richard Carlton is a director of the Archaeological Practice Ltd, Newcastle and is a Visiting Fellow in Archaeology at University of Newcastle. He is an archaeologist working primarily in the field of ethno-archaeology who has carried out extensive research on the pottery traditions of Bosnia-Herzegovina and the Western Balkans before, during and after the 1992–1995 war. He is joint coordinator of the Krajina Post-Mediaeval Frontier Project in collaboration with the National Museum of Bosnia-Herzegovina (Zemaljski Muzej). With Helen Walasek he carried out two field trips in 2000 and 2001 making the first extensive assessment of destroyed historic monuments across Bosnia-Herzegovina for which there was then little accurate information.

Amra Hadžimuhamedović is a member of the Commission to Preserve National Monuments of Bosnia-Herzegovina, serving since 2002 to the present, and is a leading authority on the architectural heritage of Bosnia-Herzegovina. She is an architect who teaches at the International University of Sarajevo and has published and lectured widely on the destruction and recovery of Bosnia's architectural heritage. She is Programme Coordinator for Bosnia-Herzegovina for the Council of Europe's Regional Programme on Cultural and Natural Heritage in South East Europe (*RPSEE*) and was a consultant expert in drafting the new European Convention on Cultural Heritage. She was born in Stolac and is actively involved in restoration and human rights programmes there.

Valery Perry has worked in Bosnia-Herzegovina since 1999 for organizations including the European Centre for Minority Issues (ECMI), the NATO Stabilization Force (SFOR), the Organization for Security and Cooperation in Europe (OSCE), and several local and international NGOs. She has a PhD from George Mason University's Institute for Conflict Analysis and Resolution (ICAR). Between 2001–2002 she implemented the ECMI Civil Society project *The Role of Annex 8 in Bridging Communities, Promoting Reconciliation, and Developing a Unique Identity in Bosnia and Herzegovina*. Most recently Perry was Chief of Party for a Public International Law and Policy Group (PILPG) project in Bosnia-Herzegovina.

Helen Walasek was an Associate of the Bosnian Institute, London and was Deputy Director of Bosnia-Herzegovina Heritage Rescue (BHHR) for which she worked 1994–1998. She has a Diploma in Museum Studies from University of Leicester. She has made many working visits to Bosnia during and after the war and was an Expert Consultant for the Council of Europe reporting on museums in Bosnia-Herzegovina in 1995 and 1996. She has lectured widely on heritage in conflict in academic settings. With Richard Carlton she carried out two field trips in 2000 and 2001 making the first extensive assessment of destroyed historic monuments across Bosnia-Herzegovina for which there was then little accurate information.

Tina Wik is an architect who was Project Manager in Bosnia-Herzegovina for the Swedish NGO Cultural Heritage without Borders (CHwB) from 2000–2007. Her restoration projects included the Zemaljski Muzej (the National Museum of Bosnia-Herzegovina, Sarajevo) and many other historic structures across the country. She was one of the two international members appointed to the Commission to Preserve National Monuments of Bosnia-Herzegovina, serving from 2002–2008. She is currently Visiting Professor of Structural Engineering at University of Dalarna, and has been Professor of Restoration at the Royal Institute of Art (KKH), Stockholm. She is architect in charge of Örebro Castle and runs her own architectural practice.

Note on Terminology

Both Bosnia and Bosnia-Herzegovina will be used throughout the book to mean the internationally-recognized state of Bosnia-Herzegovina, except where a distinction is being made between the two regions of Bosnia and Herzegovina. Muslim was a term of 'nationality' (rather than religion) in the former Yugoslavia. Many Bosnian Muslims now choose to call themselves Bosniak as a national group. Bosniaks, Serbs and Croats are legally Bosnia-Herzegovina's three main 'constituent people' according to the country's constitution as outlined in the Dayton Peace Agreement. For brevity Muslim/Bosniak, Serb and Croat will be used on occasion to mean Bosnian Muslims, Bosnian Serbs and Bosnian Croats except where differentiation between Serbs from Serbia and Croats from Croatia is required.

Note on Pronunciation of Names

Pronunciation is always regular and only letters or combinations of letters that are different or are pronounced differently than in English (such as c, ć, đ) or are similar to English but pronounced in a particular way (such as vowels) are listed.

a	as in p**a**rt
c = ts	as in pe**ts**
č = ch	as in ma**tch**
ć = ch	as in chur**ch**
dž	as in we**dg**e
đ/dj	as in **g**in
e	as in m**e**t
g	as in **g**one
i	as in m**ee**t
j	as in **y**oung
lj	as in mil**li**on

nj	as in o**ni**on
o	as in p**o**t
š = sh	as in **s**ugar
u	as in l**oo**k
ž = zh	as in plea**s**ure

Acronyms

AFCP	US Ambassadors Fund for Cultural Preservation
AKTC	Aga Khan Trust for Culture
ARBiH	*Armija Republike Bosne i Hercegovine* (Army of the Republic of Bosnia-Herzegovina – the Bosnian Army)
Armija	Another term for the Bosnian Army
AVNOJ	*Antifašističko Vijeće Narodnog Oslobođenja Jugoslavije* (Anti-Fascist Council of the People's Liberation of Yugoslavia)
BBC	British Broadcasting Corporation
BHHR	Bosnia-Herzegovina Heritage Rescue
BAM	Bosnian Convertible Marka (see also KM)
BSA	Bosnian Serb Army (see also VRS)
CBH	Court of Bosnia and Herzegovina
CIA	Central Intelligence Agency
CCBiH	Constitutional Court of Bosnia and Herzegovina
CHwB	Cultural Heritage without Borders
COE	Council of Europe
CPIC	Coalition Press Information Center
CSCE	Conference on Security and Co-operation in Europe (precursor of the OSCE to 5–6 December 1994)
CSCE	Commission on Security and Cooperation in Europe (US Helsinki Commission)
DELBIH	Delegation of the European Union to Bosnia and Herzegovina
EC	European Commission
ECMM	European Community Monitor Mission
EU	European Union
EUAM	European Union Administration in Mostar
FRY	Federal Republic of Yugoslavia (the former Yugoslav Republics of Serbia and Montenegro)
HRC	Human Rights Chamber for Bosnia and Herzegovina
HRW	Human Rights Watch
HV	*Hrvatska vojska* (Croatian Army)
HVO	*Hrvatsko vijeće obrane* (Croatian Defence Council: the military formation of the Bosnian Croat para-state of Herceg-Bosna)
ICA	International Council on Archives
ICCROM	International Centre for the Study of the Preservation and Restoration of Cultural Property
ICG	International Crisis Group
ICJ	International Court of Justice
ICOM	International Council of Museums
ICOMOS	International Council on Monuments and Sites

ICRC	International Committee of the Red Cross
ICTY	International Criminal Tribunal for the former Yugoslavia
IFLA	International Federation of Library Associations and Institutions
IFOR	Implementation Force
IPTF	International Police Task Force
IRC	International Rescue Committee
IRCICA	Research Center for Islamic History, Art and Culture
IRPP/SAAH	Integrated Rehabilitation Project Plan/Survey on the Architectural and Archaeological Heritage (a COE/EU programme)
ITN	Independent Television News
IZ	*Islamska Zajednica* (The Islamic Community)
JNA	*Jugoslovenska Narodna Armija* (Yugoslav People's Army)
KFOR	Kosovo Force
KLA	Kosovo Liberation Army
KM	*Konvertibilna marka* (see also BAM)
MDGF	Millennium Development Goals Achievement Fund
NATO	North Atlantic Treaty Organization
NGO	Non-governmental organization
ODA	Overseas Development Administration
OIC	Organisation of the Islamic Conference
OHR	Office of the High Representative
OSCE	Organization for Security and Co-operation in Europe
OSF/OSI	Open Society Fund/Open Society Institute
PACE	Parliamentary Assembly of the Council of Europe
PfP	Partnership for Peace (a NATO programme)
PRDU	Post-war Reconstruction and Development Unit
RFERL	Radio Free Europe/Radio Liberty
RS	*Republika Srpska* (one of the two entities of Bosnia-Herzegovina)
RTRS	*Radio Televizija Republike Srpske*
SFOR	Stabilization Force
SFRY	Socialist Federal Republic of Yugoslavia
Sida	Swedish International Development Cooperation Agency
UN	United Nations
UNDP	United Nations Development Programme
UNESCO	United Nations Educational, Scientific and Cultural Organization
UNHCR	United Nations High Commissioner for Refugees
UNMIBH	United Nations Mission in Bosnia-Herzegovina
UNOPS	United Nations Office for Project Services
UNPROFOR	United Nations Protection Force
USAID	United States Agency for International Development
USCRI	United States Committee for Refugees and Immigrants
USDS	United States Department of State
USGAO	United States General Accounting Office
USHMM	United States Holocaust Memorial Museum
VOA	Voice of America/*Glas Amerike*
VRS	*Vojska Republike Srpske* (Army of Republika Srpska/Bosnian Serb Army)
WHC	World Heritage Committee

WMF	World Monuments Fund
ZAVNOBiH	*Zemaljsko antifašističko vijeće narodnog oslobođenja Bosne i Hercegovine* (National Anti-Fascist Council of the People's Liberation of Bosnia and Herzegovina)

Map I.1 Bosnia-Herzegovina showing its principal internal borders after the Dayton Peace Agreement of December 1995 and places discussed in the volume. © Richard Carlton

Introduction

Helen Walasek

The massive and intentional destruction of cultural and religious property in Bosnia-Herzegovina[1] during the 1992–1995 Bosnian War became a seminal marker in the discourse on cultural heritage in both the professional and non-professional spheres. The destruction (carried out as part of widespread campaigns of ethnic cleansing and one of the defining features of the conflict) provoked worldwide condemnation, particularly the attacks on iconic structures like the National Library[2] (*Vijećnica*) in Sarajevo and the Old Bridge (*Stari Most*) at Mostar. It stimulated a variety of strong responses from international and supranational organizations, national governments, international and national cultural heritage bodies, individual professionals and members of the general public. At the war's end addressing the devastation to Bosnia's cultural heritage was considered so crucial to the success of the peace process that Annex 8 of the 11 Annexes to the Dayton Peace Agreement provided for the formation of a Commission to Preserve National Monuments. Furthermore, the concerted attacks on Bosnia's cultural heritage crystallized a more definitive discussion and recognition in international humanitarian law that destruction of a people's cultural heritage was an aspect of genocide and were to feature in indictments for war crimes issued by the International Criminal Tribunal for the former Yugoslavia (ICTY) and its subsequent prosecutions. It brought to the fore, as well, the question of how to deal with the protection and preservation of cultural property in conflict and post-conflict scenarios, particularly in the context of multinational military peacekeeping and peace enforcing/peace implementing operations.

Yet almost two decades after the end of the war there has yet to appear a comprehensive overview and analysis of the destruction of Bosnia's cultural heritage, its significance, and wide-ranging impact, in part because the attention of the heritage community moved to assessing the disastrous effect of the 2003 invasion and occupation of Iraq on that country's patrimony. Thus there has been little critical analysis of responses to the destruction, with the exception of the case of Mostar and the reconstruction of the Old Bridge, on which there is a surfeit of publications – in itself symptomatic of the focus of post-conflict reconstruction efforts by the international community.

Despite the widespread and widely-publicized censure, what active intervention did the international community take to protect and preserve Bosnia's heritage during the conflict? Was the extensive worldwide condemnation followed by equally extensive international assistance in restoring Bosnia's cultural heritage after the conflict ended? What form did any assistance take and who were the major actors? How did varying external agendas

1 Both Bosnia-Herzegovina and Bosnia will be used throughout to mean the internationally-recognized state of Bosnia-Herzegovina.

2 The full name of the library is the National and University Library. However, as most sources refer to both the building and the institution as the 'National Library,' this designation will be used throughout.

impinge on the reconstruction process? What types of structures were restored and by whom? What were the roles of UNESCO and other intergovernmental/international organizations, national governments and NGOs in the restoration scene? Did Annex 8 and the Commission to Preserve National Monuments fulfill their mandate?

As cultural heritage preservation/restoration and protection are slowly coming to be seen as a significant element in post-conflict and development aid and a crucial part of the recovery process in post-conflict societies, this volume seeks to assess these questions. It gives the first comprehensive overview and analysis of the destruction to the cultural heritage in Bosnia-Herzegovina during the 1992–1995 war. Both a case study and source book, it provides an account of the first significant destruction of European cultural heritage during conflict since World War Two and the international community's responses to that destruction during the conflict and over the first decade of post-war recovery.

While the narrative moves at times close to the present, it will not examine such questions as the construction of new mosques (and churches), particularly those mosques in the Federation built with funds from Saudi Arabian, Indonesian, Malaysian or other Islamic sources.[3] Nor will it comment on the contentious issue of the construction of memorials to those (civilians and soldiers) who died in the 1992–1995 war – or the equally contentious *prohibition* of the construction of such memorials.[4] It does not seek to describe the evolution of more recent heritage/identity politics and the ongoing story of the restoration and preservation of Bosnia's cultural patrimony – one of the frontlines in the continuing battle to defend the country's historically diverse identity, and one with its own problems and dilemmas.

The War in Bosnia

The 1992–1995 Bosnian War was the most violent and destructive of the Wars of Yugoslav Succession that accompanied the breakup of the federal socialist state of Yugoslavia into its constituent republics after the rise of Slobodan Milošević and what became known as the Greater Serbia project.[5] The wars, which most analysts now agree were essentially the outcome of this contemporary political project, were driven by an aggressive ethno-national exclusivism and a race for territory. While the Wars of Yugoslav Succession (including the Bosnian War) are not considered religious wars, those promoting the conflicts extensively mobilized the past and ethno-national/ethno-religious differences in pursuit of their aims.

---

3 See, for instance, Aksamija, A. 2010. Contested identities: Identity politics and contemporary mosques in Bosnia and Herzegovina, in *Mosques in Europe: Why a Solution has Become a Problem*, edited by S. Allievi in cooperation with Ethnobarometer, Network of European Foundations Initiative on Religion and Democracy in Europe. London: Alliance Publishing Trust, 325–32; Ford, P. 2001. Mosques face new danger, *Bosnia Report* (New Series) 23/24/25 (June – October 2001), London: The Bosnian Institute, 46.

4 Such as at Prijedor municipality where the local authorities have banned the construction of memorials to non-Serb victims of ethnic cleansing. See also, for instance, Irwin, R. and Šarić, V. 2010. Calls for war memorials divide Bosnia, *Institute for War & Peace Reporting* (*iwpr.net*), 6 December; Sito-Sucic, D. 2013. Bosnian activists erect 'guerrilla memorials' to war crimes victims, *Chicago Tribune* [online], 26 October.

5 See, for instance, Ramet, S.P. 2006. *The Three Yugoslavias: State-Building and Legitimation, 1918–2005*. Bloomington, IN: Indiana University Press on Greater Serbia.

The continual blocking of attempts (led by Slovenia and Croatia) to realign political relations between the constituent republics of federal Yugoslavia by Milošević's Serbia and a series of aggressive moves by Serbia and the Serb-dominated Yugoslav People's Army (JNA), including its support for breakaway nationalist Serbs in Croatia prompted the secession of the republics of Slovenia and Croatia (led by its president Franjo Tuđman)[6] from Yugoslavia in June 1991. The Wars of Yugoslav Succession began in earnest as the JNA swiftly attacked both countries in response.[7] Ethnically homogenous Slovenia suffered only a ten-day war, but Croatia, historically home to a large Serb minority with many living along its borders with Serbia and Bosnia and where the self-proclaimed Serb para-state of Republika Srpska Krajina had already been established,[8] came under sustained assault.

While the conflict in Croatia followed a very different trajectory to the Bosnian War, it was here that cultural property was first targeted – initially Croat/Catholic and later Serb/Orthodox heritage. Most notorious in terms of heritage destruction by JNA/Yugoslav Navy/Serb paramilitary forces was the bombardment of the non-military target of Dubrovnik (a World Heritage Site) and the devastation of the Baroque city of Vukovar in 1991; a substantial part of Vukovar's museum and gallery collections were subsequently removed to Serbia. But destructive as the attacks on cultural and religious property in Croatia were, they could not compare with the devastation Bosnia-Herzegovina's cultural heritage was to face.

Unlike Slovenia and Croatia, Bosnia-Herzegovina was home to a demographic patchwork of three principal ethno-national groups – Bosnian Muslims (Bosniaks), Bosnian Serbs and Bosnian Croats[9] – and separation from Yugoslavia was always going to be problematic. By 1990 the '...neighbouring nationalisms of Serbia and Croatia had become intimidating presences, with the ultimate ambitions of Milošević and Tuđman [towards Bosnia-Herzegovina's territory] barely concealed.'[10] A stream of (predominately anti-Muslim) propaganda and the alleged threat to Serbs in Bosnia streamed from Belgrade, while Tuđman made no secret of his desire to incorporate Bosnia-Herzegovina into a Greater Croatia.[11] In this menacing environment the first multi-party elections Bosnia in December 1990 brought three ethno-nationalist parties into power: the 'Party of Democratic Action' (SDA)[12] led by Alija Izetbegović, the 'Serb Democratic Party' (SDS)[13] led by

6 Sometimes spelled Franjo Tudjman.

7 For a good brief overview of the war and the events leading up to it see Hoare, M.A. 2010. The War of Yugoslav Succession, in *Central and Southeast European Politics since 1989*, edited by S.P. Ramet. Cambridge: Cambridge University Press, 111–35.

8 Republika Srpska Krajina (RSK) = Republic of the Serbian Krajina.

9 Muslim was a term of 'nationality' (rather than religion) in Yugoslavia. Many Bosnian Muslims now choose to call themselves as Bosniak as a national group; Bosniaks, Serbs and Croats are legally Bosnia-Herzegovina's three main 'constituent people' according to the country's constitution as outlined in the Dayton Peace Agreement. For brevity Muslim / Bosniak, Serb and Croat will occasionally be used to mean Bosnian Muslims, Bosnian Serbs and Bosnian Croats except where differentiation between Serbs from Serbia and Croats from Croatia is required. According to the 1991 census held in Bosnia-Herzegovina Muslims formed 43.47 per cent of the population, Serbs 31.21 per cent, Croats 17.38 per cent, Yugoslavs 5.54 per cent and Others 2.4 per cent.

10 Malcolm, N. 2002. *A Short History of Bosnia*. London: Pan Books, 217.

11 See: Ramet, S.P. 2006. *The Three Yugoslavias: State-Building and Legitimation, 1918–2004*. Bloomington, IN: Indiana University Press, 434.

12 *Stranka demokratske akcije*.

13 *Srpska demokratska stranka*.

Radovan Karadžić, and the 'Croatian Democratic Union' (HDZ),[14] essentially an off-shoot of Franjo Tuđman's party of the same name in Croatia. With the SDA holding the largest number of seats, Alija Izetbegović led a coalition government formed of the three parties.

Following the events in Croatia, in May 1991 Karadžić's SDS called for the secession of parts of northern and western Bosnia with the intention of uniting them with the Republika Srpska Krajina in Croatia and began receiving deliveries of arms from Serbia.[15] The SDS secretly formed so-called 'Crisis Committees' (*krizni štab/krizni štabovi*) in a number of municipalities across the country – parallel local political structures that could assume control and provide assistance to Serb-led forces when the time came.[16] In September 1991 JNA units withdrew from Croatia into Bosnia-Herzegovina and many regard the date of the real start of the war in that country as 1 October 1991 when JNA troops destroyed the town of Ravno in southern Herzegovina on their way to attack Dubrovnik. Over the winter of 1991–1992 the JNA constructed heavy artillery positions around Bosnia's principal cities, including Sarajevo.[17]

Thus, by early 1992 it was clear that the question of Bosnia-Herzegovina's sovereignty needed to be decided and a referendum on independence was held between 29 February and 1 March. Although boycotted by many Bosnian Serbs on the urging of Karadžic and the SDS,[18] 63.4 per cent of the electorate voted overwhelmingly in favour of secession.[19] The Republic of Bosnia-Herzegovina, with Alija Izetbegović as its president, declared independence on 3 March 1992 and received international recognition as a state soon afterwards. Bosnia-Herzegovina was swiftly admitted to the Conference on Security and Co-operation in Europe (CSCE) on 30 April 1992 and became a member state of the United Nations on 22 May 1992.[20]

War broke out almost immediately, a war that was to be characterized above all by the practice known as *ethnic cleansing*. Secessionist Bosnian Serbs, led by Radovan Karadžić and aided by Biljana Plavšić, Nikola Koljević and Momčilo Krajišnik, established the breakaway para-state of Republika Srpska with the support of the JNA and Serbian paramilitary units which moved into Bosnia from Serbia, carrying out a systematic and well-planned programme of ethnic cleansing of non-Serb populations (targeting both Muslims

14 *Hrvatska demokratska zajednica*. The party was first led by Stjepan Kljuić but when Kljuić insisted on supporting the elected government of Bosnia-Herzegovina under Alija Izetbegović and the country's territorial integrity, he was forced out of power under pressure from Tuđman in Zagreb in January 1992 and replaced with the ardent Croat nationalist Mate Boban who was keen to see 'Croat' territories in Bosnia become part of a Great Croatia. See: Ramet, S.P. 2006. *The Three Yugoslavias: State-Building and Legitimation*, 434.

15 Malcolm, N. 2002, 225.

16 For more on the Crisis Committees, see Hanson, D. 2004. *Bosnian Serb Crisis Staffs*, expert report (updated) prepared for the Office of the Prosecutor, the International Criminal Tribunal for the former Yugoslavia (ICTY), Exhibit P528, 26 November 2004.

17 Malcolm, N. 2002, 230.

18 Bosnian Serbs were actively urged to boycott the referendum by Radovan Karadžić and his Serbian Democratic Party (SDS) often through intimidation and the refusal of local officials to participate in the proceedings. See the Commission on Security and Cooperation in Europe (CSCE) 1992 report, *The Referendum on Independence in Bosnia-Herzegovina February 29–March 1, 1992* for a full account.

19 The number voting in favour of independence was 99.7 per cent.

20 The CSCE was the precursor of the OSCE until 5–6 December 1994. The former Yugoslav and now independent republics of Croatia and Slovenia became new UN member states on the same day as Bosnia-Herzegovina. Bosnia-Herzegovina became a member state of the Council of Europe on 23 April 2002.

and Croats) in an attempt to create a contiguous mono-ethnic territory. By autumn 1992 the Bosnian Serbs controlled over 70 per cent of Bosnia-Herzegovina's territory.[21]

During these campaigns of ethnic cleansing many atrocities and human rights abuses considered war crimes and breaches of international humanitarian law took place, including the destruction of the religious and cultural symbols of the expelled populations. The most notorious paramilitary leaders were Arkan (Željko Ražnatović) and Vojislav Šešelj. At the same time, the three and a half-year siege of Bosnia's capital, Sarajevo, by first JNA, then Bosnian Serb forces (VRS/BSA)[22] led by Ratko Mladić began in April 1992. It was during the early months of the siege that the *Vijećnica* was bombarded and set alight, one of the many attacks on the city's cultural institutions.

A separate 'war within a war' followed in January 1993, propelled by the proposals of the Vance-Owen Peace Plan (VOPP) to divide Bosnia-Herzegovina into ethnically-dominated cantons.[23] Now Bosnian Croat forces (HVO),[24] which had been fighting up till then in a brittle alliance with Bosnian government troops (ARBiH),[25] turned on their allies in an attempt to gain territory and establish Mostar as the capital of an ethnically homogenous Croat para-state of Herceg-Bosna,[26] during which they received substantial support from Croatia and the Croatian Army. The removal of Bosnian Muslim populations from Croat-held areas was a key aim of the attacks.[27] Once again, ethnic cleansing took place, accompanied by the wide devastation of religious and cultural heritage, including the total destruction of the *Stari Most* (Old Bridge) in Mostar. With intensive fighting between HVO and ARBiH forces in central Bosnia, there was also a large-scale HVO-assisted transfer of Croat populations from there to Croat-held territory in Herzegovina.[28] This distinct phase of the conflict ended in March 1994 with the Washington Agreement and the establishment of the Federation of Bosnia-Herzegovina, the so-called Muslim-Croat Federation.

In 1995, though the siege of Sarajevo worsened, the UN safe area of Srebrenica fell to the Bosnian Serb Army in July and the Sarajevo marketplace massacre by a BSA mortar attack took place on 28 August, the trajectory of the war was changing as Serb-held territory began to be retaken in the west, first in Croatia and then in western Bosnia-Herzegovina. In May 1995 the Croatian Army retook the Serb-held enclave of Western Slavonia resulting in a flood of Serb refugees into Republika Srpska. The Split Agreement of 22 July 1995 between Bosnia-Herzegovina and Croatia allowed for the intervention of the Croatian

21 On 28 August 1995 Bosnian Serb General Zdravko Tolimir summed up the situation succinctly when he told the RS Assembly how, with material and technical support from the former Yugoslavia, Republika Srpska had acquired the means 'with which 35 per cent of the population succeeded for four years in holding over 70 per cent of the territory on which lived over 65 per cent of the inhabitants of our enemies' (that is, Muslims and Croats); *Assembly Minute Summaries*, ICTY, Exhibit P537.4.

22 VRS = *Vojska Republike Srpske* (Army of Republika Srpska), or BSA = Bosnian Serb Army.

23 Named after its proposers, joint chairs of the International Conference on the Former Yugoslavia, United Nations Special Envoy, former US Secretary of State Cyrus Vance and Lord [David] Owen, representing the European Union.

24 HVO = *Hrvatsko vijeće obrane* or Croatian Defence Council.

25 ARBiH = *Armija Republike Bosne i Hercegovine* / Army of the Republic of Bosnia-Herzegovina. Sometimes referred to in sources as the 'Armija' or ARBH.

26 Herceg-Bosna was an unrecognized and illegal ethno-national para-state formed by extremist Bosnian Croats that came into being in 1991 and ceased to exist with the signing of the Washington Agreement. Sometimes referred to as the Croatian Community (later Republic) of Herzeg-Bosnia.

27 The Bosnian Serb population by this time had largely fled or been expelled.

28 Called by some an attempt at 'reverse ethnic cleansing.' See: Memorable Solidarity or Ethnic Cleansing, *Sense Tribunal* [online], 26 August 2008.

Army in Bosnia and the first operations in western Bosnia-Herzegovina cut Serb supply routes into the Republika Srpska Krajina. In August the Croatian Army swiftly retook the Krajina in Operation Storm; once again thousands of Serb refugees fled to Serb-held northern Bosnia.

Now combined offensives of the Bosnian Army, HVO and Croatian Army moved swiftly eastwards reclaiming territory, helped by the NATO bombing of Bosnian Serb military installations that followed General Ratko Mladić's failure to abide by an ultimatum issued by the international community after the Sarajevo marketplace massacre. Bosnian and Croatian troops came ever nearer to Banja Luka and there was a real prospect of a Serb military defeat.

However, US-led diplomatic efforts were beginning to bear fruit and a Serb defeat was not a prospect the international community was prepared to envisage. Negotiations were stepped up and on 12 October 1995 the final ceasefire of the Bosnian War began. Intensive talks at a US Airforce base in Dayton, Ohio throughout November led to the signing of the Dayton Agreement in Paris on 14 December 1995.[29]

Over three and a half years of war, 2.2 million people were internally displaced or became refugees, just under 100,000 were killed,[30] tens of thousands were raped and tortured, and thousands of monuments of cultural or religious importance were destroyed or badly damaged until the ceasefire was declared in October 1995. While the Bosnian War is even now occasionally characterized as entirely a civil war, the international nature of the conflict and the involvement and material support of Serbia and Croatia have been documented decisively by the investigations and case law of the International Criminal Tribunal for the former Yugoslavia, although it did incorporate aspects of civil conflict.[31]

Ethnic Cleansing

The ethnic cleansing campaigns of the Bosnian War featured forcible mass expulsion, mass killings, mass rape, internment in concentration camps, torture and intentional destruction of the religious and cultural symbols of the expelled populations. There were two explicitly voiced aims in this destruction: first, to eliminate any evidence of the expelled population's historic existence on the territory, and second, by the removal of these markers of community identity, to discourage those who survived ethnic cleansing from ever returning.

Responsibility for the campaigns of ethnic cleansing that took place in Bosnia-Herzegovina has been well-established by investigations such as those of the United Nations Commission of Experts and the International Criminal Tribunal for the former Yugoslavia (ICTY), and by assessments like the 1995 Central Intelligence Agency (CIA) Report.[32] The

29 The Dayton Peace Agreement (or Dayton Accord) is formally known as the General Framework Agreement for Peace, or GFAP. For the full text and Annexes see http://www.ohr.int/dpa/default.asp?content_id=380.

30 The now most commonly accepted figures and statistics on the conflict are produced by the Research and Documentation Center (*Istraživačko dokumentacioni centar*), Sarajevo.

31 For the ICTY's case law on the international nature of the conflict see the Human Rights Watch 2006 report, *Genocide, War Crimes, and Crimes Against Humanity: A Topical Digest of the Case Law the International Criminal Tribunal for the former Yugoslavia*, 22–9. See also: Hoare, M.A. 2010. The War of Yugoslav Succession, 112.

32 *Sanitized [ ] Version of Ethnic Cleansing Paper*, Central Intelligence Agency (CIA), 5 January 1995, available at http://www.foia.cia.gov/browse_docs.asp?doc_no=0001074874. See also

Commission of Experts' Final Report of 28 December 1994 (Annex IV) found that the majority of systematic and well-planned ethnic cleansing operations (and consequently of the destruction of cultural and religious property) were carried out by Bosnian Serb forces, together with Yugoslav People's Army (JNA) forces and paramilitary units from Serbia, mainly against Bosnian Muslim populations, but also against Bosnian Croats. The report further determined that Bosnian Croat forces had also carried out policies of ethnic cleansing targeted principally at Muslims, but also Bosnian Serb populations, but on a significantly lesser scale than the Bosnian Serbs and their allies. Finally, it found while Bosnian government forces (largely Muslim) had engaged in practices that constituted 'grave breaches' of the Geneva Conventions, they had no policy of ethnic cleansing and did not carry out such operations.[33]

The International Presence

The war in Bosnia was not a war entirely carried out far from the gaze of the global public. It was a war in which the international community was deeply involved on the ground, and as the conflict played out thousands of external actors were present in the country as military peacekeepers and observers, humanitarian aid workers, journalists and other members of the media. The headquarters of UNPROFOR,[34] the multinational United Nations peacekeeping force formed to monitor the ceasefire in the Croatian war in February 1992, was based in Sarajevo itself when the Bosnian conflict began in earnest in April 1992. By June 1992 UNPROFOR's mandate was extended to Bosnia-Herzegovina, but it began with narrow focus: to ensure the functioning of Sarajevo airport so that aid could be delivered, and in September to support the humanitarian activities of the United Nations High Commissioner for Refugees (UNHCR) and the International Committee of the Red Cross (ICRC). These two organizations worked across Bosnia-Herzegovina during the war and its employees were often the sole international witnesses to human rights abuses, including the destruction of cultural property and its immediate aftermath.[35] Both UNHCR and ICRC frequently found themselves in effect accomplices in campaigns of ethnic cleansing when it seemed the only way to protect targeted populations was to evacuate them.[36]

Observers from the European Community Monitor Mission (ECMM) which had been operating in Yugoslavia since July 1991 were also deployed throughout Bosnia, though they had limited access to Bosnian Serb-held areas and were intermittently blocked from movement elsewhere. Though not all operated in all parts of the country, hundreds of other

---

Ethnic Cleansing and Atrocities in Bosnia, Statement by CIA Deputy Director for Intelligence, John Gannon, Joint SSCI SFRC Open Hearing, 9 August 1995.

33 *Final report of the United Nations Commission of Experts established pursuant to Security Council resolution 780 (1992), Annex IV The policy of ethnic cleansing.* S/1994/674/Add.2 (Vol. I), 28 December 1994, Introduction.

34 UNPROFOR = United Nations Protection Force.

35 See Young, K. 2001. UNHCR and ICRC in the former Yugoslavia: Bosnia-Herzegovina, *International Review of the Red Cross*, September 2001, Vol. 83, No. 843, 784–6 on bearing witness. On destruction of cultural property see, for instance, Gentile, L. 1994. In Banja Luka, terror seems uncannily normal, *The New York Times*, Opinion, New York, 14 January. See also Helen Walasek, *Destruction of the Cultural Heritage* in this volume for more on this subject.

36 Young, K. 2001, 796–9. See also Gušić, B. 1995 Život i smrt u okupiranoj banjaluci, *Oslobođenje*, 15–23 March 1995; English translations of excerpts from these articles 'Testimony of Bedrudin Gusic,' parts 1 and 2 available online at https://groups.google.com.

humanitarian organizations were present too, from the extensive operations of the United Nations and its agencies, the International Rescue Committee (IRC), state international aid departments like Britain's Overseas Development Administration (ODA), NGOs (secular and faith-based) like the Danish Refugee Council, the Methodist agency UMCOR and Islamic Relief, civil society organizations like George Soros' Open Society Fund, to small voluntary and activist efforts such as SOS Bosnia and Workers' Aid which, while they only travelled to the country sporadically with humanitarian relief, were often both witnesses and targets.[37]

What is Bosnia? International Perceptions of Bosnia and the Bosnian War

The Bosnian War has been called one of the most written about events in modern history,[38] and the responses of the international community to the conflict have been comprehensively documented.[39] Moreover, it was a war not only written about after the event, but played out daily on television screens to global audiences as it happened, reported and analysed at length in the media week after week. It was a war where the views of many quasi-experts, a few real experts, historians, anthropologists, journalists, travel writers and other opinion-formers were to loom unusually large and have a far-reaching impact.

For despite the extensive media coverage, the roots of the conflicts that accompanied the dissolution of Yugoslavia were all but unknown to policy-makers who existed in a 'fog of historical ignorance'[40] as the war began – and for long afterwards. The history of Bosnia-Herzegovina itself was largely unfamiliar outside the former Yugoslavia. Few western academics had made a study of the complexities of its political, religious and cultural history and the interrelationships between Bosnia's three principal ethno-national/ethno-religious groups: Bosnian Muslims (Bosniaks), Bosnian Serbs and Bosnian Croats.[41] And unlike Iraq, site of the long-researched ancient civilization of Mesopotamia from which artefacts were on display at major museums around the world and had the benefit of a corps of well-informed international experts ready to speak out on its behalf, even fewer knew anything of Bosnia's cultural heritage.[42]

---

37 The numbers of NGOs operating in Bosnia have been estimated to be between 300 (Eyre, Lt Col D. 1998 Working with NGOs: What Every SOF Soldier Should Know, *Special Warfare*, Spring 1998, 15, cited in Baumann, R F., Gawrych, G.W., Kretchik, W.E. 2004. *Armed Peacekeepers in Bosnia*. Fort Leavenworth, Kansas: Combat Studies Institute Press, 194) and 530 (Wentz, L. 1998. *Lessons from Bosnia: The IFOR Experience*. Washington, DC: Command and Control Research Program (CCRP), 135. Wentz reports an estimated 530 NGOs operating in Bosnia-Herzegovina when IFOR troops arrived in Bosnia in December 1995).

38 Hoare, Q. and Malcolm, N. (eds) 1999. *Books on Bosnia*. London: The Bosnian Institute, ix.

39 For good brief overviews of the Bosnian War see Hoare, M.A. 2010 and the last two chapters of Hoare, M.A. 2007 *The History of Bosnia: From the Middle Ages to the Present Day*. London: Saqi Books. For a longer account see Silber, L. and Little, A. 1996. *The Death of Yugoslavia*. London: Penguin Books.

40 Malcolm, N. 2002. *A Short History of Bosnia*. London: Pan Books, 14.

41 Bosnian Muslims (Bosniaks), Bosnian Serbs and Bosnian Croats are all Slavs. Historically the great majority of Serbs were Eastern Orthodox by religion, Croats Roman Catholic, while Muslims/Bosniaks are descended from local converts to Islam (variously Orthodox, Catholic and followers of the local schismatic Bosnian Church) after Bosnia became part of the Ottoman Empire in 1463.

42 Chief among them were historian John Fine and art historian Marian Wenzel.

Washington Post commentator Amy E. Schwartz wrote in September 1993 on this lack of knowledge in the United States, a lack that led to a cultural disconnect of little help to Bosnia:

> Even though Americans in large numbers are agonizing over the slaughter in Bosnia, there's a dimension of the destruction … that outsiders can't grasp. News reports last week described the gradual decimation of the Ottoman bridge at Mostar – the most famous of the many gorgeous stone bridges that used to be a Bosnian trademark, built by Ottoman architects in the 1500s to splice the mountainous realm together. Post reporter John Pomfret quoted a local soldier who told him, "I'm fighting for the bridge," not for his family or nationality or ethnic group.[43] But not many people this far from the Balkans have read the literature and poetry that return over and over throughout centuries to the beauty of the bridge at Mostar. To those who have, hearing of the bridge's destruction would be something like a person in the West hearing of the flattening of Notre Dame or Canterbury.[44]

Schwartz goes on to compare a booklet on Bosnia's destroyed heritage with its grainy photos[45] and the glossy productions brought out by UNESCO to support the rebuilding of the tourist destination of Dubrovnik in Croatia: 'No contrast could be as depressing, nor as obliquely illustrative,' she writes, 'of the distance between Bosnian history and our "canon"' as the difference between the publications. Despite the evidence of a staggering and ongoing destruction, Schwartz wonders why the publication cataloguing Bosnia's devastated heritage has 'so different and pathetic an effect' compared with those on Dubrovnik: 'In innumerable subtle ways it lies beyond a veil and a dividing line, out of the reach of tourism authorities and familiar literary and historical references.' Schwartz quotes Shreve Simpson, Islamic curator at the Freer and Sackler galleries who says 'In North America there's just no sense that these places with funny names can be repositories of vital heritage. It's all at a remove somehow. It must be for UN officials too … because otherwise how could they let this happen?"'

Thus pundits and decision-makers on both sides of the Atlantic turned to travelogues old and new to learn the 'history' of the region and a handful of books were to have an astonishing influence. Two of them, Rebecca West's *Black Lamb and Grey Falcon*, an account of her travels in Yugoslavia, and Ivo Andrić's fictional *The Bridge on the Drina* – often held up as an accurate and telling account of 'Balkan' history going back to the Ottoman era – had been written not within the recent past, but during the years of the Second World War,[46] an era repeatedly invoked as a paradigm for understanding

43 Pomfret, J. 1993. As Mostar Bridge crumbles, so does Bosnian dream of ethnic unity, *The Washington Post*, 30 August. The soldier was a Croat Catholic fighting in the Bosnian Army (ARBiH).

44 Schwartz, A.E. 1993. What outsiders can't grasp, *The Washington Post*, 9 September 1993.

45 The publication was: Ihsanoglu, E. (ed.) 1993. Bosnia-Hercegovina: History, culture, heritage. *Research Center for Islamic History, Art and Culture Newsletter*, 31, April 1993. Istanbul: IRCICA.

46 West's book, based on her travels in Yugoslavia in 1937 was written and first published in wartime Britain in 1941; Andrić's book was written in Belgrade between 1942–1943 and first published in 1945. For more on the impact of West's book in during the 1990s see also: Ramos, R. 2003. Imagining Peace, *Trickster's Way*: Vol. 2, Iss. 2, Article 4, and Hall, B. 1996. Rebecca West's War, *The New Yorker*, 15 April 1996, 74–83. See Antić on the role of The Bridge on the Drina in Western academic and political discourse: Antić, M. 2003. Living in the Shadow of the Bridge: Ivo Andric's 'The Bridge on the Drina' and Western Imaginings of Bosnia, spacesofidentity.net, The Canadian

events and mentalities in the former Yugoslavia (particularly how the Serbs/Bosnian Serbs would behave militarily) by the corps of experts and advisors, among them the elderly Fitzroy Maclean (who had fought with Tito's Partisans) and whose *Eastern Approaches* also became an essential reference book.

Most notorious in its impact on policy-making was the effect of reading Robert Kaplan's *Balkan Ghosts* on US President Clinton when making his decision not to lift the arms embargo against the countries of the former Yugoslavia – a decision which in reality only affected the beleaguered Bosnian government.[47] Kaplan, who declared: 'I would rather have lost my passport and money than my heavily thumbed and annotated copy of *Black Lamb and Grey Falcon*,'[48] was heavy in his emphasis on 'ancient ethnic hatreds.'

Thus an image of Bosnia was constructed which was to have a critical impact on Euro-Atlantic decision-making. It was an image mediated by the activities and influence of an energetic Serb lobby which sought to present the conflict on their terms,[49] combined with a carefully-nurtured confusion as to whether the war was a civil war or international in nature. Viewed through the lens of racist representations and pejorative stereotypes of 'the Balkans' as a cauldron of 'ancient ethnic hatreds,'[50] this image was infused, too, with the growing Islamophobia of the early 1990s which had a negative effect on international relationships (including relationships with UNPROFOR) for Bosnia-Herzegovina's legitimate and legally recognized government – usually characterized as 'Muslim' and as one of the 'warring factions.' An additional ingredient in this representation was a relativist insistence of the equivalence of the three main parties in the conflict in respect of atrocities, human rights abuses and destruction of religious and cultural property committed, despite considerable evidence to the contrary.[51]

Thomas Cushman has written on such relativistic thinking during the Bosnian War, seeing this 'moral equivalence' as 'an independent cultural force, which affected the interpretation of concrete, objective historical events' and 'concrete policies of western

---

Center for German and European Studies, Vol. 3, No. 2 (2003). The author has herself heard this fictional text written in mid-twentieth-century Yugoslavia cited by academics as an accurate document of the behaviour of the Turks in seventeenth-century Ottoman Bosnia.

47 For an account see Ramos, R. 2003; Kaplan, R.D. 1993 *Balkan Ghosts: A Journey Through History*, New York: St. Martin's. Described originally in Drew, E. 1994 *On the Edge: The Clinton Presidency*, New York: Simon & Schuster. The Bosnian Serbs had the huge resources of the JNA (Yugoslav People's Army) to draw on; the Republic of Bosnia-Herzegovina had no army at the start of the war.

48 Quoted in Ramos, R. 2003, 4.

49 On the Serb lobby see Hodge, C. 1999 *The Serb Lobby in the United Kingdom*. Seattle: University of Washington, Donald Treadgold Papers, No. 22, September 1999, and Conversi, D. 1996 Moral relativism and equidistance in British attitudes to the war in the former Yugoslavia, in T. Cushman and S. Mestrovic (eds) *This Time We Knew: Western Responses to Genocide in Bosnia*. New York and London: New York. See Campbell, D. 1998. *National Deconstruction: Violence, Identity and Justice in Bosnia*. Minneapolis: University of Minnesota Press, 49 on the dominance of Serb representations of the conflict and 50, on US Secretary of State Warren Christopher's attempt to attribute atrocities to all three sides in the conflict (particularly Muslims) despite being given evidence to the contrary by his staff. See also Gordon, Michael R. 1993. US memo reveals dispute on Bosnia, *The New York Times*, 25 June.

50 Despite that, outside the countries of the former Yugoslavia, no open conflict between different ethno-religious groups followed the collapse of Communism in other parts of the Balkans. Much has been written on this characterization of the Balkans, but see, for instance Campbell, D. 1998, 90–91 on the subject in relation to Bosnia.

51 See Campbell, D. 1998, 33–82 on representations of Bosnia.

political elites.'[52] While looking particularly at the work of anthropologists, his remarks might apply equally to all those who took on the status of experts during the conflict when he noted that:

> ' … in many cases they advised political leaders or their accounts have been quoted in justifications for particular political policies. The articles and their authors thus form part of the core of the expert system, which drove much of western understanding of the Balkan conflict and political action in the region … ' and ' … how supposedly objective, dispassionate, academic analysis of the conflict worked at another level as propaganda which both legitimated Serbian military aggression in Croatia and Bosnia and helped to justify the lack of response to stop genocide by western political powers.'[53]

Brendan Simms in *Unfinest Hour*, his work on British foreign policy towards Bosnia, has also written extensively on the detrimental role of 'experts' during the conflict.[54] The central role Britain played in driving international policy towards Bosnia until the very end of the war has been widely acknowledged.[55] Reviewing *Unfinest Hour*, Noel Malcolm described how in Britain:

> An official doctrine reigned supreme. It was promulgated at the highest levels of government, never challenged by the opposition, and accepted by pundits, newspaper editors and leader writers throughout the land. Those who did criticize it were dismissed as unreliable or extreme. And yet every element of it was simply and obviously false.[56]

The starting point for this 'official doctrine' was that the war in Bosnia was 'an incomprehensibly complex "ethnic" war, of a sort that had raged there for centuries.'[57] As Simms himself noted later, though now it was widely accepted as being 'essentially a Serbian war of aggression waged by externally supported proxies,' at the time the conflict was presented as a civil war ('a three-sided quagmire'), and describes how,

> British official sources tried to head off demands for military intervention by suggesting a rough moral equivalence between aggressor and victim; that this equivalence was reinforced by a rhetoric of 'warring factions,' which failed to distinguish between rebel Serb – and later Croat – perpetrators and their victim, the internationally recognised government of Bosnia-Herzegovina; that the 'orientalist' mindset of much of the military profoundly affected perceptions on the ground.[58]

---

52 Cushman, T. 2004. Anthropology and genocide in the Balkans: An analysis of conceptual practices of power. *Anthropological Theory*, March 2004, Vol. 4, 9.

53 Cushman, T. 2004, 22.

54 Simms, B. 2002. *Unfinest Hour: Britain and the Destruction of Bosnia*. London: Penguin.

55 Britain is one of the five permanent members of the UN Security Council, it held the Presidency of the European Union during the crucial period July–December 1992, and key negotiators during conflict were the British peers Lord Carrington and Lord Owen. Owen was co-author of the failed Vance-Owen and Owen-Stoltenberg Plans.

56 Malcolm, N. 2001. Britain's fatal foreign policy, *The Sunday Telegraph*, 9 December.

57 Malcolm, N. 2001.

58 Simms, B. 2003. The end of the 'official doctrine': The new consensus on Britain and Bosnia, *Civil Wars*, Vol. 6, Iss. 2, 53. He was writing on the impact of his book *Unfinest Hour*.

James Pettifer attributed this 'official doctrine' more pungently to 'profoundly Serbophile key decision-makers in the Government, the Foreign and Commonwealth Office and Army swimming in a murky sea of sentimentality and bogus history about Tito and the Second World War, Germanophobia after the Croatian conflict, appalling anti-Americanism, and Orientalizing ideology about the Balkans,' noting 'the entrenched anti-Islamic assumptions in much of the SIS, Joint Intelligence Committee, and military intelligence worlds, and the still largely unknown levers Belgrade can pull in a few British elite circles.'[59]

Observers like the journalist Maggie O'Kane described these depictions and characterizations as a cynical political ploy, part of a calculated process of fostering disengagement and distance in domestic audiences by western (particularly British, French and US) governments 'in order to justify their inaction and political cowardice,' citing as an example British Foreign Secretary Douglas Hurd's continued insistence on referring to the Bosnian government – the legally recognized government of a UN member state – as the 'Muslim' government.[60]

Gregory Kent's work on media 'framing' of the Bosnian War, has demonstrated that, despite unambiguous evidence of responsibility, with a few notable exceptions, the greater part of the media colluded with this 'official version' of the conflict, showing how 'the manner in which the problem of Bosnia had been framed and, therefore, defined meant there were only "shades of grey", "no heroes, only villains" ….'[61]

Added to this, Bosnia was described repeatedly (and erroneously) in the media as never having been a nation or state, or even a defined territorial unit – and therefore, seemingly not worth taking international action in support of its integrity – and that a distinctive Bosnian cultural identity did not exist. An outspoken commentator in this respect was former US Secretary of State Henry Kissinger whose views were given wide coverage in the media. Kissinger declared on more than one occasion throughout the war that 'Bosnia has never been a nation; there is no specifically Bosnian cultural identity.'[62]

Despite the efforts of a group of activists, academics, analysts, journalists, aid workers, officials and politicians who sought a more accurate representation of Bosnia and the conflict, this dominant 'framing' of the Bosnian War undoubtedly had its effect. As we shall see, these perceptions of Bosnia and the nature of the conflict and of the moral equivalence of the warring parties was to have an impact on international responses towards preserving and protecting Bosnia's cultural heritage during the war and on restoration and rebuilding in the years following the final ceasefire.

---

59 Pettifer, J. 2002. How the Serbs were appeased, *Times Literary Supplement*, 10 May, 29, reviewing Simms' *Unfinest Hour*. Pettifer had reason to know as an advisor to the UK government.

60 O'Kane, M. 1996. *The Role of the Media in Bosnia*. Balkan Papers/Interview Series, Australia Region Media (armedia.net.au), May 1996.

61 Kent, G. 2004. Framing and problem definition: British responses to the war against Bosnia, in *Media in Security and Governance: The Role of the News Media in Security*, edited by M. Caparini, Bonn: Nomos / Bonn International Center for Conversion (BICC) and Geneva: Geneva Centre for the Democratic Control of Armed Forces (DCAF), 171.

62 Kissinger, H.A. 1993. Bosnia has never been a nation and has no specific cultural identity. Why are we intent on preserving this Balkan no man's land? *Los Angeles Times*, 16 May, M2.

The Peace in Bosnia

The signing of the Dayton Peace Agreement (DPA) on 14 December 1995 marked the legal conclusion of the Bosnian War. Yet despite bringing the fighting to an end, there was widespread dissatisfaction with Dayton and a belief that it was a peace without justice for anyone, victims and aggressors (who often felt themselves victims) alike. Regardless of the peace agreement's aim to reverse the effects of ethnic cleansing, the DPA effectively formalized the territorial gains achieved through the process. Bosnia-Herzegovina was divided into two entities: the Federation of Bosnia-Herzegovina (the so-called Muslim-Croat Federation) formed of ten cantons, with either Bosniak (Bosnian Muslim) or Bosnian Croat majority populations, and Republika Srpska, a unitary authority dominated by a Bosnian Serb majority. The treaty's endorsement of Bosnia's three 'constituent peoples'[63] (Bosniaks/Muslims, Croats and Serbs) worked towards the reinforcement of ethno-national/ethno-religious identities and the entrenchment of powerful ethnocracies who came to dominate the post-war realm.[64] Broad political powers were devolved to the entities and new institutions were formed at entity-level. What central state structures remained were fragile and constantly undermined by the activities of (typically) Serb and Croat ethno-nationalist politicians, while many former state institutions (including cultural institutions such as the Zemaljski Muzej, Bosnia-Herzegovina's national museum) struggled for legal recognition and consequently a place in central state budgets and funding streams.[65]

The host of international organizations responsible for implementing and monitoring the peace agreement was led by the Office of the High Representative (OHR) headed by an appointed High Representative that oversaw civilian aspects of the treaty,[66] while its military aspects were supervised by the NATO-led multinational peacekeeping force IFOR (becoming SFOR a year later).[67] An International Police Task Force (IPTF) was also created under the DPA to oversee local law enforcement agencies.[68] Most important and visible of the attendant intergovernmental organizations were the Organization for Security and Cooperation in Europe (OSCE),[69] the United Nations High Commissioner for Refugees (UNHCR), the International Committee of the Red Cross (ICRC), the UN Mission in Bosnia-Herzegovina (UNMIBH), other UN agencies such as UNDP and UNESCO,[70] the World Bank, and the Council of Europe (COE). Humanitarian aid and development agencies/NGOs came in their hundreds to implement recovery and reconstruction programmes.

But unlike those who had spent the war in the country, many in the post-conflict aid and development community arrived in Bosnia-Herzegovina for the first time and had little knowledge or understanding of the recent history of the environment in which they

63 See DPA, Annex 4, Constitution of Bosnia and Herzegovina.

64 Toal, G. and Dahlman, C.T. 2011. *Bosnia Remade: Ethnic Cleansing and its Reversal*, New York: Oxford University Press, 184.

65 At the time of writing in October 2012, the Zemaljski Muzej closed its doors to the public for the first time in its 124-year history due to the continued lack of funding stemming from recognition of its legal status.

66 See Annex 10 of the DPA.

67 See Annex 1-A of the DPA. IFOR = Implementation Force, SFOR = Stabilisation Force from 20 December 1996. SFOR was followed by EUFOR = European Union Force from 2 December 2004.

68 The IPTF was created under Annex 11 of the DPA.

69 See Annex 1-B of the DPA.

70 UNDP = United Nations Development Programme, UNESCO = United Nations Educational, Scientific and Cultural Organization.

were operating. Human rights expert and former aid worker Diane Paul, reporting to the Commission on Security and Cooperation in Europe (CSCE) in June 1996 on a recent mission to Banja Luka (where she had been a frequent visitor during the war) observed:

> My last visit in April 1996 was deeply disturbing … for I found a city newly populated with internationals who seemed to have little concept of the horrors which had happened there. It was as if the past had never happened, as if the Dayton Agreement was the beginning of time.[71]

All too frequently this mindset grew into empathy with the majority population among whom international workers were living – the last group prepared to discuss the human rights abuses which had taken place in their communities during the war and where many of the perpetrators still held public office. Not only that, some national contingents within IFOR/SFOR were believed to have a positive bias towards particular ethno-national groups (often co-religionists or historic former allies), as for instance, Spanish troops based in Croat (Catholic) dominated municipalities in Herzegovina were regarded as pro-Croat and French troops as actively pro-Serb.[72]

This was to have a negative impact on (largely Muslim) displaced populations as they began to return to their pre-war towns and villages and rebuild not only homes and workplaces, but more contentiously, their places of worship and other cultural monuments. When returnees began to exert their rights in such spheres as education, the practice of religion and equality in the public space – particularly through the reconstruction of the physical markers of community identity – they were too often seen by members of the international community as demanding, obstructive and negative.[73] That great numbers were categorized as so-called 'minority returns' in spite of the fact that the returned community had been in the majority before they had been ethnically cleansed did not help.

Yet, paradoxically, within months of the end of the conflict there was mounting pressure from the international community for the so-recently warring ethno-national groups to 'reconcile' – regardless of who had been the victim and who the aggressor. The restoration of the built heritage came to be linked by these external actors with this drive towards reconciliation as conveniently visible symbols of the process.

The Impact of the Wars of Yugoslav Succession on Cultural Property Protection

The deliberate and widespread destruction of cultural heritage in Bosnia and Croatia galvanized considerable activity in the sphere of cultural property protection though, as will be discussed elsewhere in this volume, few of these activities were in, or even directed

71 *Banja Luka. Ethnic Cleansing Paradigm or Counterpoint to a Radical Future*, Briefing of the Commission on Security and Cooperation in Europe (US Helsinki Commission), Washington DC: CSCE, 11 June 1996, 10.

72 This bias towards historic allies could extend to British troops. In 1996 the author was told by a British Army sergeant based in Bijeljina, Republika Srpska: 'We have to be pro-Serb because we're British.'

73 See Helen Walasek, *Domains of Restoration*, and Amra Hadžimuhamedović in this volume. See also for instance Kolind, T. 2008. *Post-war Identification: Everyday Muslim Counterdiscourse in Bosnia Herzegovina*. Aarhus: Aarhus University Press, 248–52 on negotiations over education in Stolac and attitudes of international officials.

towards, Bosnia itself. But in the wider domain of heritage preservation, the impact of the devastation of the Wars of Yugoslav Succession was such that 'Yugoslavia' was to become the paradigm of cultural property destruction during the so-called non-traditional conflicts that followed World War Two, with the Old Bridge at Mostar as the instantly recognizable visual signifier that inevitably accompanies discussions on the subject and adorns the covers of scores of books.

On the level of international humanitarian law and policy-making the catastrophe in Bosnia and Croatia brought forward an urgent review of the 1954 *Hague Convention for the Protection of Cultural Property in the Event of Armed Conflict* by UNESCO, prompted by a proposal from the Netherlands during the 140th session of the organization's Executive Board in October 1992,[74] which led ultimately to the drafting and adoption of the *Second Protocol* to the convention in 1999.[75] The *Statute of the International Criminal Tribunal for the former Yugoslavia (ICTY)* of 1993 was an important addition to international criminal law on the destruction of cultural property. In October 1993 the heads of state and government of the member states of the Council of Europe met in Vienna for the first time in the organization's history, marking the end of the division of Europe and the integration of new member states; their *Vienna Declaration* explicitly referred to the events in the former Yugoslavia in reference to the declaration's focus on democratic values and respect for human rights and the protection of minorities (stressing their right to culture), naming cultural co-operation (including protection and enhancement of cultural heritage) as essential for creating a 'cohesive diverse Europe.'[76]

The events in Bosnia were to prompt an eruption of symposia and workshops on disaster preparedness and damage to cultural heritage during armed conflict, such as the series of round tables organized by ICOMOS and UNESCO beginning in October 1992 in conjunction with the review of the Hague Convention[77] from which a *Risk Preparedness Scheme* and an *Inter-Agency Task Force (IATF) on Cultural Heritage at Risk* consisting of ICOMOS, ICCROM, UNESCO, ICOM (among others) emerged.[78] The meetings also saw the formation of the *Blue Shield movement* that resulted in the establishment of the International Committee of the Blue Shield (ICBS) in July 1996;[79] while characterizing

74 Hladík, J. 1999. The 1954 Hague Convention for the Protection of Cultural Property in the Event of Armed Conflict and the notion of military necessity, *International Review of the Red Cross* No. 835, 30 September 1999, 621–35. Boylan, P.J. 1993. *Review of the Convention for the Protection of Cultural Property in the Event of Armed Conflict (The Hague Convention of 1954)*, UNESCO Doc.CLT-93/WS/12.

75 *Second Protocol to the Hague Convention of 1954 for the Protection of Cultural Property in the Event of Armed Conflict*, The Hague, 26 March 1999. On the impact of the cultural destruction in Bosnia and Croatia on drafting the Second Protocol, see E. Roethlisberger and Y. Sandoz in Dutli, M.T. (ed.) 2002. *Protection of Cultural Property in the Event of Armed Conflict*, 18 and 24; and O'Keefe, R. 2006. *The Protection of Cultural Property in Armed Conflict*, Cambridge: Cambridge University Press, 238–9.

76 *Vienna Declaration*, the Council of Europe, Vienna, 9 October 1993.

77 *Second Information Report on war damage to the cultural heritage in Croatia and Bosnia-Herzegovina*, Parliamentary Assembly of the Council of Europe (COE), Doc. 6869, Strasbourg, 17 June 1993, 21; ICOMOS' Risk Preparedness Scheme, *ICOMOS Newsletter, No. 2 1994*, July 1994, 18. The organizations represented at the meeting included UNESCO, ICCROM, ICOMOS, the Getty Institute and Patrimoine sans Frontières.

78 Stovel, H. 1998. Risk Preparedness: A Management Manual for World Cultural Heritage, Rome: ICCROM/UNESCO/ICOMOS/WHC, 2.

79 See the Blue Shield website at www.ancbs.org/. The ICBS brings together representatives of the Co-ordinating Council of Audiovisual Archives Associations (CCAAA), International Council on

itself as 'the cultural equivalent of the Red Cross,' this description unfortunately proved very far from true, and ICBS became notable for its inaction. Another gathering prompted by the destruction in Bosnia and Croatia was the seminar held in Stockholm in June 1994 on *Information as an Instrument for Protection against War Damages to the Cultural Heritage* which brought together not only heritage professionals from international organizations and NGOs, but a former Commander of UNPROFOR as well.[80]

For the destruction of the Bosnian War undoubtedly raised an awareness in the military, and more particularly in NATO and NATO member states, of the importance of cultural property protection in conflict and post-conflict scenarios (with the Netherlands and Austria showing a special interest). In June 1996, not long after the NATO-led multinational force IFOR had deployed in Bosnia on its peace-implementing operation, NATO held a conference on *Cultural Heritage Protection in Wartime and in State of Emergency* in Krakow as part of its Partnership for Peace (PfP) programme during which participants expressed their support for the *Appeal for International Aid for Croatian and Bosnia-Herzegovina Monuments Destroyed During the War* made at the IVth European Conference of Ministers Responsible for the Cultural Heritage at Helsinki on 31 May 1996.[81]

The concept of CIMIC (civil military cooperation) evolved in concrete form during NATO's post-conflict IFOR/SFOR operations in Bosnia and came to be seen by the organization as an essential element of its peacekeeping and peace-implementing operations.[82] NATO's Strategic Concept of April 1999 (SC-99)[83] recognized the new environment in which the Alliance's forces were operating, mentioning its role in Bosnia, as well as stressing the importance of military cooperation with civilian actors in such circumstances. This led to an initial interest in incorporating cultural property protection within the framework of CIMIC operations. However as Joris Kila, a Dutch art historian and archaeologist involved at the early stages of the NATO/CIMIC project in 1999, has traced, interest gradually waned apparently in large part due to the same perceptions as in the wider sphere of post-conflict aid (described below) of the low status of cultural property protection, a preference for 'high-visibility, quick impact' projects (like refurbishing schools), and a reluctance to be seen as competing with state international development departments and the NGOs that traditionally provided humanitarian assistance.[84]

While its response to the destruction of cultural property in Bosnia and Croatia undoubtedly spurred some soul-searching at UNESCO which had legal responsibility for application of the 1954 Hague Convention, the International Committee of the Red Cross (ICRC), too, now had to take stock. For the ICRC also had unambiguous responsibilities towards protection of cultural and religious property under Additional Protocols I

Archives (ICA), International Council of Museums (ICOM), International Council on Monuments and Sites (ICOMOS) and International Federation of Library Associations and Institutions (IFLA).

80 The seminar was held by the Swedish Central Board of Antiquities, ICOMOS Sweden and the Swedish National Commission for UNESCO in Stockholm in June 1994, *Information as an Instrument for Protection against War Damages to the Cultural Heritage*.

81 See: Final Communiqué of the NATO-Partnership for Peace conference on *Cultural Heritage Protection in Wartime and in State of Emergency*, Krakow, 18–21 June 1996.

82 See for instance: Phillips, Col. W.R. Civil-Military Cooperation: Vital to peace implementation in Bosnia, NATO Review [online], Vol. 46, No. 1, Spring 1998, 22–5.

83 *The Alliance's Strategic Concept*, agreed at Washington DC, April 1999. http://www.nato.int/cps/en/natolive/official_texts_27433.htm.

84 Kila, J.D. 2012. *Heritage under Siege: Military Implementation of Cultural Property Protection Following the 1954 Hague Convention*. Leiden: Brill, xix–xxiv, 46–50. Dr Kila went on to found the International Military Cultural Resources Work Group (IMCuRWG).

(Article 53) and II (Article 16) to the Geneva Conventions,[85] as well under international humanitarian law applicable in armed conflicts more generally, which the ICRC's own statutes undertook to uphold and seek to apply.[86] Not only that, the ICRC's Legal Division was an active contributor to the drafting of the Second Protocol to the Hague Convention.[87] It is difficult to discover evidence of any action the ICRC took in respect of cultural property protection during the Bosnian War.

After the war was over, however, the ICRC began to work more actively in promoting and disseminating information on international humanitarian law on the issue, through forums such as the *Meeting of Experts on National Implementation of the Rules for the Protection of Cultural Property in the Event of Armed Conflict* held by its Advisory Service on International Humanitarian Law in Geneva 5–6 October 2000[88] or *Heritage Under Fire: The Protection of Cultural Property in Wartime*, a conference held by the British Red Cross in London on 26 June 2001. On 14 November 2001 the Council of Delegates of the International Red Cross and Red Crescent Movement, issued a *Resolution on Protection of Cultural Property in the Event of Armed Conflict* when it met in Geneva.[89] UNESCO and the ICRC began to work more frequently together on the issue of the protection of cultural property during armed conflict, mostly through the joint organization of regional seminars and workshops.[90]

Heritage and Warfare

As will be shown, responses to the destruction of Bosnia's heritage during the 1992–1995 war plainly illustrate the paradox of cultural property protection in times of conflict. Though international humanitarian law clearly mandates protection of a people's cultural property, concerns over being seen to be privileging buildings over people remained problematic for external actors – even for those engaged in heritage protection – which without doubt led to a reluctance to take action. Yves Sandoz, in 2000 a Legal Advisor at the International

---

85 Additional Protocols I and II of 1977 to the Geneva Conventions under the articles on Protection of cultural objects and of places of worship: Article 53 of Additional Protocol I (Protection of Victims of International Armed Conflicts) and Article 16 of Additional Protocol II (Protection of Victims of Non-International Armed Conflicts) of the Geneva Conventions of 12 August 1949, 8 June 1977. See also Bugnion, F. 2004. The origins and development of the legal protection of cultural property in the event of armed conflict, *International Review of the Red Cross, Special Issue: Protection of Cultural Property in Armed Conflict*, 854-2004, 11 November 2004.

86 The ICRC Statutes assert its role to 'undertake the tasks incumbent upon it under the Geneva Conventions, to work for the faithful application of international humanitarian law applicable in armed conflicts and to take cognizance of any complaints based on alleged breaches of that law' (Article 5.2c), and 'to work for the understanding and dissemination of knowledge of international humanitarian law applicable in armed conflicts and to prepare any development thereof' (Article 5.2g).

87 Henckaerts, J-M. 1999. New rules for the protection of cultural property in armed conflict, *International Review of the Red Cross*, 835, 30 September 1999.

88 For the proceedings of that meeting see Dutli, M.T. (ed.) 2002. *Protection of Cultural Property in the Event of Armed Conflict.*

89 *Resolution 11. Protection of Cultural Property in the Event of Armed Conflict, Resolutions of the Council of Delegates – 2001, adopted at its session of 11–14 November 2001.* Geneva: Council of Delegates of the International Red Cross and Red Crescent Movement.

90 *Report on the Implementation of the 1954 Hague Convention for the Protection of Cultural Property in the Event of Armed Conflict and its Two 1954 and 1999 Protocols, Report on the Activities from 1995 to 2004.* UNESCO, (Clt-2005/WS/6), 4.

Committee of the Red Cross, noted the dilemma of trying to implement international laws on cultural property protection when faced with human suffering in the field, but emphasized:

> Deliberate attacks on cultural property are marks of contempt, and contempt can serve as an excuse or a pretext for the worst outrages, of which it is often the precursor. The struggle to defend the cultural property of a population, and hence respect for its dignity, is therefore an integral part of the humanitarian operation aimed at protecting that population.[91]

However, the Bosnian War raised the question, too, of whether it is even possible to protect cultural heritage during conflict where warring parties/aggressors are determined on destruction. This question was brought to the fore less than three years after the end of the war in Bosnia, as well as that of the role of multinational peacekeeping forces in cultural property protection, with the outbreak of war in nearby Kosovo, then still part of Serbia and the Federal Republic of Yugoslavia (FRY).[92] When open conflict broke out in the province between Serbian police units and the Kosovo Liberation Army (KLA) over spring 1998, what appeared to be a repeat of the Bosnian War unfolded. Serbian forces once more engaged in ethnic cleansing, this time of the majority Kosovar Albanian population, engaging in the widespread destruction of their cultural and religious heritage.

As in Bosnia, claims over the damaged and destroyed heritage became highly politicized. Assertions by the FRY/Serb authorities of extensive damage to Kosovo's important medieval Serbian Orthodox heritage following a NATO bombing campaign proved to be largely without foundation.[93] However, despite the arrival of a multinational NATO-led UN peacekeeping force (KFOR) in Kosovo in June 1999, many minor Orthodox churches did suffer revenge attacks.[94] UNESCO and ICRC responded to these increasing attacks on cultural property in 2000 with the preparation of a leaflet outlining the basic principles of cultural property protection in Albanian, Serbian and English aimed at both the local population and international peacekeepers.[95]

But worse was to follow in March 2004, however, when notwithstanding the presence of KFOR troops guarding Orthodox sites, an eruption of interethnic violence resulted in destruction and extensive damage to 35 Serbian Orthodox sacral structures (many of them important historic monuments) by Kosovar Albanian rioters.[96] Lessons (in theory) had been

---

91 Sandoz, Y. 2002. Competing priorities: Placing cultural property on the humanitarian law agenda, in Dutli, M.T. (ed.) 2002. *Protection of Cultural Property in the Event of Armed Conflict; Report on the Meeting of Experts (Geneva, 5–6 October 2000)*, Geneva: International Committee of the Red Cross, Advisory Service on International Humanitarian Law, 22.

92 The dynamics of the conflict in Kosovo where the majority ethnic Albanian population had long endured repression on the part of Serbia was very different than that in Bosnia-Herzegovina. The Federal Republic of Yugoslavia (FRY), consisting of the former Yugoslav republics of Serbia and Montenegro, was formed on 28 April 1992. It became the state union of Serbia and Montenegro in 2003 and in 2006 the union was dissolved and Serbia and Montenegro became independent states.

93 Herscher, A. and Riedlmayer, A. 2000. Architectural Heritage in Kosovo: A Post-War Report, first published in *US/ICOMOS Newsletter* 4 (July–August 2000). The survey was carried out in October 1999.

94 Herscher, A. and Riedlmayer, A., 2000. KFOR = Kosovo Force.

95 *Fifth Meeting Of The High Contracting Parties To The Convention For The Protection Of Cultural Property In The Event Of Armed Conflict (The Hague, 1954)*, UNESCO Headquarters, Paris, 5 November 2001, Final Report, CLT-01/CONF/204/4, 4.6.

96 See *Activity Report of the Reconstruction Implementation Commission (RIC) for Orthodox Religious Sites in Kosovo 2005* for information on the sites affected.

learned from the Bosnian experience, and KFOR troops had a mandate to intervene with the use of force in police/military grey areas such as had occurred at the 2001 riots at Banja Luka and Trebinje during cornerstone-laying ceremonies to begin the rebuilding of destroyed historic mosques.[97]

Yet the KFOR responses to the 2004 attacks in Kosovo differed from contingent to contingent, with each following their own national rules of engagement.[98] Whereas Italian troops resisted the attack on the monastery of Visoki Dećani, at Devič Danish troops evacuated the residents of the monastery and then withdrew, leaving it to be set alight. In Prizren German troops guarding the Orthodox cathedral of St George, whose orders only permitted the use of force in self-defence, returned to barracks, leaving the church to be destroyed. The German forces' behaviour was widely condemned, compelling the commander of the German KFOR contingent in the city to defend their actions, saying: 'We acted exactly according to our regulations.'[99]

Heritage and post-Conflict Recovery

In 2007 Sultan Barakat, director of the Post-war Reconstruction and Development Unit (PRDU), writing as someone who worked extensively in the field,[100] noted that even with a dramatic step change in the recognition of the importance of post-war reconstruction in development and humanitarian fields, cultural heritage reconstruction was still perceived as an unaffordable luxury in post-war scenarios and received little attention either in academic discourse or in international policies or practice.[101] Indeed, in the scores of analyses of post-war Bosnia-Herzegovina, Dayton implementation and refugee return published over the nearly two decades following the end of the war there is scarcely a mention of Annex 8 or the reconstruction of cultural property.

In his approach to post-war heritage reconstruction, Barakat argues for (among other things) the 'integration of cultural heritage into the wider physical, economic and social responses' to post-conflict recovery.[102] As will be seen, despite the widely acknowledged role destruction of cultural and religious property had played as a key war aim, and despite the importance attached to preservation of Bosnia's heritage by the inclusion of Annex 8 in the Dayton Peace Agreement, in the years following the end of the war, implementation of the annex was almost completely neglected by the international community until forced to

97 See Friesendorf, C. 2012. *International Intervention and the Use of Force: Military and Police Roles*, SSR Paper 4, Geneva: The Geneva Centre for the Democratic Control of Armed Forces (DCAF), 39–58. For more on the 2001 riots at Banja Luka see Valery Perry and Helen Walasek, *Domains of Restoration*, in this volume.

98 See Friesendorf, C. 2012. *International Intervention and the Use of Force*; Serbenco, E. 2005. The Protection of Cultural Property and Post-Conflict Kosovo, *Revue québécoise de droit international*, (2005) 18.2, 121.

99 Von Flottau, R., Ihlau, O., Szandar, A., Ulrich, A. 2004. Deutsche Soldaten: Die Hasen vom Amselfeld, *Der Spiegel* [online], 3 May. See also: Paterson, T. 2004. German troops 'hid like rabbits' in Kosovo riots, *The Telegraph* [online], 9 May.

100 Barakat and PRDU had worked in Bosnia. See Helen Walasek, *Destruction of the Cultural Heritage* in this volume for more on the work of PRDU in Bosnia.

101 Barakat, S. 2007. Postwar reconstruction and the recovery of cultural heritage: Critical lessons from the last fifteen years, in *Cultural Heritage in Postwar Recovery, papers from the ICCROM Forum held in Rome 4–6 October 2005*, edited by N. Stanley-Price. Rome: ICCROM, 26–7.

102 Barakat, S. 2007, 27.

do so by the violent events of mid-2001 over the rebuilding of historic mosques at Banja Luka and Trebinje in Republika Srpska. It was to take almost a decade after Dayton for restoration of cultural heritage to be seen as an important part of the recovery process in Bosnia-Herzegovina.

Bosnia and the Destruction of Cultural Heritage

As this book will demonstrate, the catastrophic attacks on Bosnia-Herzegovina's cultural heritage were almost entirely intentional and systematic. In Chapter 1, Helen Walasek gives a detailed account of this devastation, looks at patterns of destruction, offers an analysis of losses to the built heritage and assesses the main perpetrators of the destruction and their motivation and justification for their acts. The fate of movable heritage from museums, libraries, archives and other collections is examined, including the looting and other movement of cultural property as a result of the conflict. She then turns to examine international responses to these deliberate attacks which she shows to have been (with a few exceptions) lacking in concrete action during the war and far from meeting the expectations raised after the conflict's end.

As noted above, the war in Bosnia has been one of the most written about events in modern history, yet despite the worldwide censure, was the destruction of Bosnia's heritage so extensively analysed? In Chapter 2, Richard Carlton and Helen Walasek provide an appraisal of documentation of the damage to Bosnia-Herzegovina's cultural and religious property, noting that, 'Given the scale and significance of the cultural destruction in Bosnia-Herzegovina during the 1992–1995 war … the surveys or scholarly works on the subject are remarkably few.' Their assessment reveals that although at first glance the destruction appears to be extensively documented, yet precise figures of destruction or damage remain elusive. They note as well how wartime damage reports became a politicized tool in the propaganda battle being waged by competing interest groups regarding the pattern of destruction and accusations of culpability. They examine both wartime and post-war reports produced by local and regional sources, especially those coming from the different religious communities and heritage preservation bodies, turning finally to surveys carried out by international bodies and organizations.

The lack of independently verified information and doubts about the accuracy of the reports coming from local and regional sources led – as in other spheres – to early (and erroneous) assumptions of an equivalence of destruction of cultural and religious heritage by all three main warring parties in the conflict, an assumption which Carlton and Walasek note 'continues to have a long afterlife in academia.' Even after the end of the war, with the dismantling of Bosnia's pre-war heritage system and a lack of resources (and political will) for local bodies to complete a comprehensive survey and the inability to physically access large areas where the most widespread devastation had occurred, even as late as 2000–2001, the true extent of the destruction was hard to gauge. With the conclusion of hostilities the international community had little interest in making such an assessment.

Carlton and Walasek attempted to address this partial, often inaccurate record and the absence of a comprehensive assessment by carrying out targeted surveys to validate and enhance existing data. Chapter 3 describes the authors' fieldtrips across Bosnia-Herzegovina in 2000–2001 and subsequently, which in addition to verifying the nature and scale of monumental destruction of all types, allowed a critical examination of the validity of the highly charged and politicized wartime claims and counter claims of heritage destruction, including scrutiny of assertions made during the war by Bosnian government authorities that Christian sacral monuments were protected by Bosnian Army forces, as well as challenging the commonly-held view of the equivalence of destruction by all three main warring parties in the conflict.

With the end of the war, the issue of how to deal with the damage to Bosnia's cultural and religious heritage was one among the many enormous human and material problems that faced the ravaged country – one that was to a large extent ducked by the international community which had been so vociferous over its destruction during the conflict. Yet that damage to the cultural heritage had been perceived as one of the peace settlement's critical concerns by those who framed the Dayton Peace Agreement with its ambition to reverse the effects of ethnic cleansing. Arguably the most overlooked of Dayton's 11 annexes, Annex 8 uniquely attempted to address the issue of Bosnia-Herzegovina's destroyed and damaged cultural heritage, providing for the creation of the Commission to Preserve National Monuments of Bosnia-Herzegovina mandated to receive petitions on monuments and sites for designation as National Monuments.

With Chapter 4, Valery Perry provides an overview of Annex 8 and the evolution of the Commission during its first five years under the aegis of UNESCO. She moves on to assess the lack of attention to the annex by those implementing Dayton until the crisis of mid-2001 when violent demonstrations over mosque reconstruction in Republika Srpska highlighted its importance in underpinning refugee return. Coming at a time when the mandate of the UNESCO-led Commission had ended, the violence prompted a new focus on Annex 8 and the body's re-establishment as a state-level institution of Bosnia-Herzegovina under the terms of the Dayton Agreement. Perry traces the reconstituted Commission's emergence as a key actor in the battle to preserve and reconstruct Bosnia-Herzegovina's heritage and its historic diversity. Yet as the only heritage body in the country with a state-wide remit, the limited mandate that Annex 8 gave the Commission and its status as a Dayton institution came to be a straightjacket in the following years as it increasingly became the focus of attacks by Serb and Croat ethno-nationalist politicians and power structures.

With the Dayton Agreement focused on reversing the effects of ethnic cleansing and the global condemnation of monument destruction during the war, restoration and reconstruction of cultural heritage should have been high on the international agenda. Yet, influenced by the negative 'framing' of Bosnia and the aid and development community's perceptions of the low importance of restoring historic monuments among the many challenges facing post-war reconstruction, international heritage restoration efforts converged on the iconic site of Mostar and its Old Bridge with its easily-grasped value as a symbol of reconciliation, almost to the exclusion of other sites. In Chapter 5, Helen Walasek examines international and domestic responses in respect of post-war restoration including lack of/obstructions towards restoration (including the mobilization of archaeology), the politicization of restoration, the linkage of restoration with reconciliation, and restoration as form of justice and a human right. She shows how the moral equivalence of the war years was to evolve into an often fruitless search for equivalence in restoring historic structures of importance to each of Bosnia's three principal ethno-national groups. Yet community-led restoration of sacral structures began, despite the lack of international or local support.

The first of two case studies that continue the assessment of post-war restoration looks in greater depth at one such community-led effort. In Chapter 6, Amra Hadžimuhamedović describes the mobilization of displaced Muslims (and some Serbs) who returned to their home town of Stolac and in the face of determined and prolonged resistance from the dominant Croat ethno-nationalist power structures, began to restore its destroyed heritage in one of the most difficult settings in Bosnia-Herzegovina, a process in which she was deeply involved. A second appraisal of the restoration scene is offered in Chapter 7 by Tina Wik, who gives an overview of the work of the Swedish organization Cultural Heritage without Borders (CHwB), virtually the only international NGO carrying out heritage restoration and preservation across Bosnia-Herzegovina in the years following the end of the war. Wik outlines the many issues and challenges she confronted during her time implementing conservation/restoration projects as CHwB's Bosnia project leader, among them how to apply international conservation charters (like the Venice Charter) in this post-conflict setting where destruction had been so catastrophic and where local praxis or political events demanded contradictory solutions.

Finally, Helen Walasek examines how the search for justice for victims of the 1992–1995 Bosnian War became a testing ground of international humanitarian and human rights law in respect of the protection and preservation of cultural and religious property, the right to a people's enjoyment of their cultural heritage and the development of concepts of cultural heritage and identity. In Chapter 8, Walasek outlines the ground-breaking legal precedents of the International Criminal Tribunal for the former Yugoslavia (ICTY) in their treatment of the destruction of cultural property not only as a crime in itself, but also as a manifestation of persecution and genocide, as well as summarizing influential decisions of the International Court of Justice (ICJ), the Human Rights Chamber for Bosnia and Herzegovina (HRC) and domestic courts in Bosnia-Herzegovina.

Chapter 1
Destruction of the Cultural Heritage in Bosnia-Herzegovina: An Overview

Helen Walasek

Patterns of Destruction

The scale of the devastation of the cultural and religious heritage of Bosnia-Herzegovina during the 1992–1995 Bosnian War was extensive. The exceptional defining feature of the devastation was the systematic and deliberate nature of the attacks. Thus, the greater part of the destruction occurred as the result of the intentional targeting of cultural property, rather than from collateral damage inflicted during the course of battle. These premeditated attacks were a fundamental part of violent attempts to create mono-ethnic territories by the process known as *ethnic cleansing*,[1] while in cities like Sarajevo and Mostar structures and institutions that symbolized or contained material proofs of Bosnia's historic pluralist identity were targeted. Thus the destruction could be seen as assaults on the physical evidence of co-existence and heterogeneity.

There were two principal phases of destruction of cultural and religious property. The first followed the outbreak of war in Bosnia-Herzegovina[2] in March 1992 as part of a campaign of ethnic cleansing carried out by secessionist Bosnian Serbs supported by the Yugoslav People's Army (JNA) and Serbian paramilitary units against non-Serb populations in their efforts to consolidate the territory of an ethnically 'pure' Serb para-state of Republika Srpska.[3] However, sporadic attacks on non-Serb religious property by the JNA had begun even earlier. Probably the first such attack was the dynamiting of the eighteenth-century Ljubović Mosque in Odžak near Nevesinje by JNA troops on the night of 23–24 September 1991.

By autumn 1992 the Bosnian Serbs held 70 per cent of the territory of Bosnia-Herzegovina and the targets of their ethnic cleansing operations were Bosnian Muslims/Bosniaks (demographically the most numerous) and Bosnian Croats (who traditionally formed the majority in areas like the Bosanska Posavina along the south bank of the Sava River).[4]

1 See *Introduction* for more on the trajectory of the Bosnian War and ethnic cleansing.

2 See *Introduction* for an overview of the course of the war. March 1992 was the 'official' start of the war in Bosnia-Herzegovina; however, religious property was being attacked by JNA troops as early as 1991 and the start of 1992, such as, for instance, the sixteenth century Catholic church in Ravno in Herzegovina was badly damaged on 3 October 1991.

3 For a detailed account, see *Application of the Convention on the Prevention and Punishment of the Crime of Genocide (Bosnia and Herzegovina v. Serbia and Montenegro)* at the International Court of Justice (ICJ), pleading by Laura Dauban (CR 2006/5, 1 March 2006) and expert witness evidence of András Riedlmayer (CR 2006/22, 17 March 2006) on cultural heritage destruction in Bosnia-Herzegovina.

4 See *Introduction* Note 4 for an explanation of Bosnia-Herzegovina's main ethno-national groups.

Attacks on Islamic (Muslim) and Roman Catholic (Croat) sacral structures were frequently tied with the murders of clerics and parishioners, as well as other atrocities and human rights abuses.[5]

The 43-month siege of Bosnia's capital Sarajevo by JNA/Bosnian Serb (VRS)[6] forces began, a notable feature of which were the attacks on the city's major cultural institutions and historic monuments. The most notorious was the bombardment of the *Vijećnica* (National Library)[7] with incendiary shells over the night of 25–26 August 1992. Mostar was subjected to intensive JNA/VRS shelling between mid-April to mid-May 1992 causing considerable damage to its built heritage, particularly in the historic core. There were reprisal attacks on Orthodox/Serb religious property, principally by Bosnian Croat/HVO forces,[8] in the Posavina to the north and in Herzegovina, where in mid-June the important monuments of Žitomislić monastery and the Serbian Orthodox Cathedral (*Saborna Crkva*) in Mostar were destroyed.

The second phase of significant destruction of cultural property followed the announcement of the Vance-Owen Peace Plan in January 1993 which proposed to divide Bosnia-Herzegovina into mono-ethnic cantons, encouraging separatist Bosnian Croat forces (HVO) to begin attacking their Bosnian government allies in an attempt to gain territory for an ethnically 'pure' Croat para-state of Herceg-Bosna with Mostar as its capital. Here again, ethnic cleansing of Bosnian Muslim populations[9] by HVO/Bosnian Croat forces was accompanied by the wide-scale devastation chiefly of Ottoman/Islamic, but also of Orthodox/Serb religious and cultural heritage. It was during this phase that the historic Ottoman urban complex of Stolac in Herzegovina was intentionally destroyed by HVO/Bosnian Croat extremists beginning in July 1993 after the forcible internment of the entire male Muslim population of military age and the expulsion of the remainder. This so-called Croat-Muslim War was epitomized by the deliberate assaults on the historic Ottoman core of Mostar (especially on the government-held east bank of the Neretva River) during the HVO siege of the town that began in May 1993. The zenith of this devastation was the premeditated shelling of the *Stari Most* (Old Bridge) which finally collapsed into the Neretva River on 9 November 1993.

Meanwhile, in Bosnian Serb-controlled territories, destruction continued into 1993 with the systematic dynamiting of Bijeljina's and Banja Luka's mosques and Islamic/Ottoman heritage by Bosnian Serb authorities/VRS forces, as well as, for instance, the demolition of such important monuments as the Osman-paša Mosque in Trebinje, the Careva/Obradovića Mosque in Bileća, the Avdić Mosque in Plana and the Esma Sultana/Čaršija Mosque in Jajce. Thus the greater part of cultural heritage destruction (though not all) took place during 1992–1993.

There was relatively little destruction in 1994, but 1995 saw a further spate of attacks on religious property. Some were prompted by the retaking of Serb-held territory in both

5 See the evidence of expert witness András Riedlmayer at the ICJ for examples (CR 2006/22, 17 March 2006, Nos. 48–50).

6 JNA = *Jugoslovenska Narodna Armija*; VRS = *Vojska Republike Srpske* (Army of Republika Srpska).

7 The full name of the institution is the National and University Library. However, as most sources refer to the building and the institution as the 'National Library,' this designation will be used throughout. A few sources refer to the library as the 'university library' or 'town library'.

8 HVO = *Hrvatsko vijeće obrane*, or Croatian Defence Council, the military formation of the unrecognized Bosnian Croat para-state of Herceg-Bosna).

9 Mostar's Serb population had largely departed or been expelled by then.

Croatia and Bosnia-Herzegovina and were focused on Roman Catholic religious structures in Republika Srpska (by this time almost all Islamic structures in the territory had been completely destroyed or were in a ruinous state). In July mosques in the UN-protected 'safe areas' of Srebrenica and Žepa were destroyed by Bosnian Serb forces after they overran the two enclaves and after the towns were empty of their Muslim inhabitants. In the final months of the war as the forces of the HVO, Croatian and Bosnian Armies advanced, a number of Orthodox religious structures were attacked and damaged. But all the destruction of 1995 amounts to a fraction of the destruction which took place over 1992 and 1993.

Analysis of the built heritage destroyed or badly damaged during the conflict shows that the types of monuments attacked were overwhelmingly religious,[10] and of those, structures were overwhelmingly Muslim – many from the Ottoman and Austro-Hungarian periods. Of non-religious structures, similarly, those destroyed or damaged were largely Ottoman period, associated with Muslims, or were structures or institutions which represented Bosnia-Herzegovina's historic pluralism or held records of centuries of co-existence. Scores of secular buildings without ethno-national/ethno-religious affiliation were targeted (especially in Sarajevo and Mostar), particularly structures from the Austro-Hungarian period, where many institutions and official bodies (cultural and otherwise), archives, libraries, museums and headquarters of infrastructure bodies were housed. However, although Muslim/Ottoman sacral and secular monuments particularly suffered attack, there was also widespread destruction and damage to the Catholic (Croat) and Orthodox (Serb) religious heritage.

A proportion of the intentional targeting of cultural and religious property by artillery shelling took place in the context of long-running sieges, the best known being the attack on the *Vijećnica* (National Library, 1896) and the destruction of the *Stari Most* (Old Bridge, 1566) mentioned above. Other examples of this type of intentional targeting (often with incendiary shells) were the Oriental Institute (1891) and the Olympic Museum (Villa Mandić, 1903) in Sarajevo and the Franciscan church of Ss Peter and Paul in Mostar (1866). The famous dervish tekke (*tekija*) at Blagaj which stands at the base of the sheer cliff from which the Buna river emerges was targeted from HVO positions, apparently with rockets; the missiles struck the cliff face, causing rocks to fall onto the roof of the building below.[11]

However, by far the greatest percentage of deliberate targeting had a distinct feature: it did not occur during military operations, but rather took place in locales either far from the frontlines or where fighting had ceased and were in secure control of the perpetrators of the destruction. Thus, in Bosnian Serb-controlled Banja Luka, where there were no hostilities at any time, 15 mosques[12] along with other Ottoman/Islamic structures were destroyed between April and December 1993 – a year and more after the conflict had started.

The destruction began on 9 April 1993 with the demolition of the small Bosnian style Sefer-beg (Pećinska) Mosque with its wooden minaret.[13] Twelve of Banja Luka's 15

---

10 Of course, many religious structures of no architectural or historic merit were also destroyed as part of the same processes.

11 See *Prosecutor v. Jadranko Prlić et al*, ICTY Case No. (IT-04-74), testimony Ratko Pejanović, 4 May 2006, 1281–4.

12 See *The Islamic Community in Bosnia and Herzegovina v. The Republika Srpska, (Banja Luka Mosques), Decision on the Admissibility and Merits, (Delivered on 11 June 1999)*, Human Rights Chamber (HRC), Case No. CH/96/29. Some lists include the modern mosque in suburban Vrbanja, bringing the total of mosques destroyed to sixteen.

13 See Husedžinović, S. 2002. *Zlocinacko unistavanje: spomenika islamske arhitekture u Banjoj Luci*, http://bhmedia.se/ [online] for a detailed account.

mosques were listed national monuments dating from the Ottoman period and included the domed Ferhadija (1579) and Arnaudija (1594/95) mosques, both of which were blown up during the early hours of 7 May 1993.[14] The previous day, Đurđevdan (St George's Day), an important marker of Serb ethno-national identity, had been celebrated and Bedrudin Gušić, Chairman of the Islamic Community in the city during the war, has related how Banja Luka's remaining Muslims, headed by their Mufti Ibrahim Halilović who had remained in the city to help and try to protect his parishioners, feared some kind of attack.[15] Late that night at around 23.00, police cordoned off roads near the historic mosques; some living in nearby houses and apartments, apparently alerted to the coming blast, had left their windows opened wide. The huge explosions which rocked the city just after 03.00 were so great some residents believed a munitions depot had blown up.[16]

Yet the minaret of the stone-built Ferhadija, the city's most beautiful and celebrated mosque, remained standing – probably as Sabira Husedžinović (then an expert working at the Banja Luka Institute for the Protection of Cultural, Historical and Natural Heritage) believed, because of the internal steel frame added to the minaret during rehabilitation work in 1989.[17] Late on the night of 7 May in an attempt to legitimize their actions, the Bosnian Serb authorities tried to persuade Bedrudin Gušić to sign a document drawn up by a hastily-formed official 'Commission' stating that in its semi-destroyed state the minaret of the Ferhadija posed a danger to public safety and needed to be demolished.[18] But Gušić refused to sign. Back at home, at about 23.30, he saw spotlights trained on the ruins of the Ferhadija and knew the inevitable was about to happen. At half past midnight on 8 May amid cheers from nearby apartment blocks where former JNA (Yugoslav People's Army) officers lived, the minaret of the Ferhadija Mosque was mined and demolished.[19]

The destruction of the Ferhadija in two stages allowed Banja Luka's citizens to brave the heavy police presence on the mornings of 7 and 8 May to stare at its ruins, many – not only Muslims – in a state of shock and disbelief.[20] Indeed, politician (and later convicted war criminal) Radoslav Brđanin clearly felt too many Serbs were protesting about the demolition of the mosque, commenting in the Assembly of Republika Srpska on 9 May:

> I propose to ban the Satanization of Serbs by Serbs. The main topic in Banja Luka day and night is about the mosque. I don't say we should thank those who destroyed it, but you mustn't complain so much, at least not in our media. For all those who complain I'll enlarge a post card in colour and they can carry it with them.[21]

---

14 Kinzer, S. 1993. 2 Major mosques blown up by Serbs, *The New York Times*, 8 May.

15 Bedrudin Gušić was chairman of the Islamic Community in Banja Luka from May 1992 to November 1994. Gušić wrote a series of long articles giving a detailed account of this period in Banja Luka published in Sarajevo in 1995: Gušić, B. 1995. Život i smrt u okupiranoj banjaluci, *Oslobođenje*, 15–23 March; English translations of excerpts from these articles, 'Testimony of Bedrudin Gusic,' parts 1 and 2, are available online at https://groups.google.com.

16 Selman, M. 2006. *Banjaluka – Uspomena i Stvarnost sa Mejdana*, www.bosnafolk.com, October 2006, 109; ICTY Exh. P444A, *1st Krajina Corps Report on Destruction of Ferhadija and Arnaudija Mosques*, 7 May 1993. site postwar, clock tower site postwar

17 Husedžinović, S. 2002. *Zlocinacko unistavanje: spomenika islamske arhitekture u Banjoj Luci*,

18 Gušić, B. 2012. In memoriam: 'Banjalučka Ferhadija – ljepotica koju su ubili' (7 maj 1993 – 7 maj 2012) Zločin bez kazne, *http://bedrudingusic.wordpress.com*, 7 May.

19 Gušić, B. 1995. *Testimony of Bedrudin Gusic*.

20 Gušić, B. 1995. *Testimony of Bedrudin Gusic*; see also Selman, M. 2006.

21 *Republika Srpska Assembly Minutes Extracts*, ICTY Exh. P537.4, 32.

Fig. 1.1 (a) A prewar photo of the sixteenth-century Ferhadija Mosque, Banja Luka, and two mausolea (before 1981). The Muftijstvo (Islamic Community offices) can be seen, left, behind the mosque. © Ekrem Hakki Ayverdi / ICTY

Fig. 1.1 (b) Citizens of Banja Luka stare at the remains of the minaret of the Ferhadija Mosque in Banja Luka. The mosque was deliberately blown up with explosives in the early hours of 7 May 1993. The minaret survived the first attempt at demolition but was blown up the following day. The *turbe* (mausoleum) of Ferhad-paša, the mosque's founder, still stands at the far right; it was blown up later in the year on 15 December. The offices of the Islamic Community can be seen in the background. © Estate of Aleksander Aco Ravlić

Fig. 1.2 The remains of the Ferhadija Mosque in Banja Luka after being deliberately blown up with explosives over 7 and 8 May 1993 before they were bulldozed and removed. The sixteenth-century Ottoman clock tower, or *sahat kula*, still stands; it was to be blown up later in the year on 15 December. © Estate of Aleksander Aco Ravlić

Among those who went to witness the aftermath of the destruction were Bedrudin Gušić, as well as local journalist and historian Aleksander Aco Ravlić (himself of Bosnian Serb origin), engineer Aida Krajišnik and Sabira Husedžinović, who all dared to photograph the devastation at the site.[22] Ravlić was forced into exile in Croatia because of his views on the destruction, and went on to publish a book on the Ferhadija reproducing his and Krajišnik's photographs of the outrage.[23] Despite the frantic efforts of the Islamic Community led by Mufti Halilović to try to preserve the remains of the mosques and prevent them from being removed, the sites were cleared and levelled by the Banja Luka authorities, the fragments of the two buildings taken to be dumped at the municipal waste disposal site and elsewhere. As the bulldozers did their work, Husedžinović was to risk arrest while supervising the rescue of architectural fragments from the rubble of the Ferhadija by members of the Islamic Community.[24]

22 Husarska, A. 1998. Without A Prayer? Rebuilding a Muslim Temple is a Test of Bosnia's Peace, *The Washington Post*, 10 May, C01.

23 Ravlić, A.A. 1996. *Banjalučka Ferhadija: ljepotica koju su ubili*. Rijeka: AARis.

24 Author interview with Sabira Husedžinović June 2001; Gušić, B. 2012. In memoriam; Selman, M. 2006. *Banjaluka – Uspomena i Stvarnost sa Mejdana*, www.bosnafolk.com.

While mosques in the urban centre of Banja Luka were demolished to their foundations, many in suburban areas survived as shells, although heavily damaged and all with their minarets torn down. British journalist Tim Judah had made his way to the city shortly after the destruction of the Ferhadija and watched as an earth mover clawed at the remains of its solid masonry and lorries 'carted the rubble of one of the finest examples of Ottoman architecture in Bosnia to the municipal rubbish dump.'[25] One Muslim onlooker told him: 'It is as though they have torn our heart out. They wanted us to understand that we have no place here.' Others Muslims were more defiant: 'Just because of the mosques we will stay,' said one woman, 'If they put us in convoys, we'll go' said another, 'otherwise we won't go. This is our country.'[26]

But after the destruction of the city's last mosques in September 1993 the sizeable number of Muslims who still lived in Banja Luka were so intimidated that most wanted to leave. Larry Hollingworth, head of UNHCR operations in Bosnia commented then: 'This is the end of an era. There is no question that this has frightened the community so much they all want to go, so it could be the end of their history here.'[27] But there was to be a final assault on Banja Luka's Ottoman heritage on 15 December when the city's sixteenth-century clock tower (*sahat kula*) and important tombs and mausolea (*turbe*) such as those of Ferhad-paša (builder of the Ferhadija Mosque) and Admiral Halil-paša were dynamited in a final eruption of violence against these emblematic markers of the Muslim community and its long historic presence in the city.

The Croat community in the Banja Luka region was also under attack. Like Mufti Halilović, the Roman Catholic Bishop of Banja Luka, Dr. Franjo Komarica, remained in his diocese throughout the war, actively speaking out against the ethnic cleansing that was taking place.[28] Although the city's Catholic cathedral of St Bonaventure survived, by the beginning of 1994 over 50 per cent of Catholic churches and ecclesiastical buildings in the diocese had been destroyed, 21 per cent heavily damaged and 25 per cent partially damaged.[29]

In an interview on 23 March 1995 Bishop Komarica observed of Banja Luka's Muslim community: 'Their situation, if that is possible, is even worse than ours. While "only" half of our churches have been pulled down, all six Banja Luka mosques, some of which were exceptionally beautiful and old, have been levelled to the ground and covered up.'[30]

Two months later, at the beginning of May, as the Croatian Army retook Serb-held Western Slavonia in neighbouring Croatia and thousands of ethnic Serbs fled to the Banja Luka region, reprisals were launched against the area's remaining Bosnian Croats. Six Catholic churches and a monastery in the diocese were dynamited or set alight within a few weeks of each other. Two priests and a nun died during the destruction and scores

---

25 Judah, T. 1993a. Razing of mosques gives new resolve to Muslims. *The Times*, 14 May.

26 Judah, T. 1993a.

27 Judah, T. 1993b. Banja Luka loses last mosque, *The Times*, 10 September. Almost uniquely amongst the larger towns of Republika Srpska, Banja Luka retained a significant, though ever declining and constantly persecuted Muslim minority population throughout the war years.

28 For some documents and photographs, see Komarica, F. 2000. *In Defence of the Rightless/U Obrani Obespravljenih*. Zagreb: Targa. Available at http://www.biskupija-banjaluka.org/izdanja_/obespravljeni_/english_/index.htm

29 Human Rights Watch 1994. *War Crimes in Bosnia-Hercegovina: UN Cease-Fire Won't Help Banja Luka*, D608, 1 June 1994.

30 Vujnović, B. 1995. The Fetish of Nationalism is Unworthy of Man. *AIM Press* [online], Alternativna Informativna Mreža (Alternative Information Network), 3 April. The total number of mosques destroyed should, of course, be fifteen (sixteen including the mosque at Vrbanja), not six.

Fig. 1.3 All that remained of Banja Luka's sixteenth-century Ottoman clock tower in 2001. © Commission to Preserve National Monuments of Bosnia & Herzegovina

of Croats (and Muslims) were expelled.[31] The destruction continued into the summer and on 19 August 1995, the Church of the Assumption of the Blessed Virgin Mary in Sanski Most was mined and totally destroyed. By the war's end of the 91 Catholic churches which had existed in the diocese at the start of the war, 45 were destroyed and only six remained untouched.[32]

Elsewhere, the historic Ottoman cores of small towns like Foča (in Bosnian Serb-controlled territory) and Stolac (in a Bosnian Croat-controlled area) were devastated by the well-organized and methodical demolition of structures in the absence of military operations in the area. Other major incidences of this type of intentional destruction include many structures in the exquisite and well-preserved Ottoman town of Počitelj (in Bosnian Croat-held territory), and Ottoman/Muslim heritage in Trebinje, Bijeljina, Zvornik and Prijedor (all in Bosnian Serb-controlled territory).

The gradual takeover of Foča by a Bosnian Serb 'Crisis Committee' in the spring of 1992 and the targeting of the town's Muslim population with the aim of eliminating their presence in the locality has been well-documented by numerous reports and during prosecutions of the International Criminal Tribunal for the former Yugoslavia (ICTY).[33] Once with a Muslim population of almost 52 per cent, by 1995 there were virtually no

31 See: Gutman, R./Newsday 1995. Bosnia priest, nun killed, church says – Serbs blamed for terror against Catholics, Muslims, *The Seattle Times*, 18 May; Chronicle News Services 1995. Serb homes, Croatian church leveled in Bosnia – Orthodox Church condemns violence against Catholics, *The San Francisco Chronicle*, 20 May. See also: *Human Rights Watch 1995. The Croatian Army Offensive in Western Slavonia and its Aftermath, Ensuing Abuses in Bosnian Serb-Held Areas*, 7:11 (July), 13–15; the report attributes the destruction to heavily armed Serb refugees from Western Slavonia and Bosnian Serb Army forces.

32 Roane, K.R. 1996. Church in ruins, 27 Croats invoke the spirit of 6,000. *Chicago Tribune*, 2 January.

33 Human Rights Watch 1998. *'A Dark and Closed Place': Past & Present Human Rights Abuses in Foca*, 10:6 (D), 1 July 1998; ICTY Outreach 2005. View from The Hague: Facts about Foča, in *Justice in Transition* (*Pravda u tranziciji*), Vol. 2, December 2005, Belgrade: Office of the War Crimes Prosecutor of Serbia; McDowall, L (ed.) 2009. *Bridging the Gap Between the ICTY and Communities in Bosnia and Herzegovina, Conference Proceedings, Foča, 9 October 2004*, The Hague: ICTY; ICTY Cases: *Prosecutor v. Dragoljub Kunarac, Radomir Kovač and Zoran Vuković*, IT-96-23-T and IT-96-23/1-T and *Prosecutor v. Milorad Krnojelac (Foča)*, Case No. IT-97-25-T.

Muslims left in Foča and entire Muslim neighbourhoods and all the town's mosques had been intentionally destroyed, the majority razed to the ground and their remains removed. In January 1994 the Assembly of Republika Srpska declared Foča's name changed to Srbinje, or 'Serb-town.'

Though there was some initial Muslim resistance, they were no match for local Bosnian Serbs, well-armed with weapons from the JNA depot in nearby Livade and assisted by paramilitary units from Serbia and JNA reservists called in from Montenegro. The final takeover culminated in an attack on the town on 8 April 1992. By 17 April the Serb forces had established complete control over Foča and fighting there had ceased. Over the following months some of the worst atrocities of the war were committed against Foča's Muslim population, ranging from expulsion to the murder, detention and torture of that part of the population which remained, and, most notoriously, the systematic rape of scores of women and girls. Entire Muslim neighbourhoods such as the historic Prijeka Čaršija were systematically attacked and the destruction and demolition of all 11 of Foča's mosques began.[34]

Among the first to be destroyed was the Careva, or Imperial Mosque (1501) which had stood on an elevation at the town's historic centre; the structure was set alight, its minaret blown up and the ruins bulldozed and removed.[35] A witness described how as he saw a mosque burning in Donje Polje on 15 May 1992, the local fire brigade stood by and watched, turning their attentions instead to a haystack standing beside the mosque to prevent it from catching fire.[36] On 28 May 1992 the eighteenth-century Mehmed-paša Kukavica Mosque and its minaret, at the heart of the Prijeka Čaršija, was dynamited and its adjoining medresa, *abdesthana* and muderis' apartment set alight.[37] On 1 June the historic Šehova or Kadi Osman-efendijina Mosque was destroyed.[38] But it was not until 2–3 August 1992 that the stone-built Aladža Mosque (1550/1) was blown up in a huge explosion.[39] The domed Aladža Mosque, one of the most important examples of classical Ottoman architecture, not only in Bosnia-Herzegovina but in all of South East Europe, was renowned for its fine stone work and exquisitely painted interior; it was the last intact mosque standing in Foča. A US satellite photograph of Foča taken one week later on 10 August 1992 vividly shows empty spaces where the Aladža, Careva, Šehova and other mosques had once stood, the dome of the Mehmed-paša Kukavica Mosque, whose central structure had somehow survived the attacks, and the roofless shell of the sixteenth-century Musluk (Atik Ali-paša) Mosque.

---

34 See *Prosecutor v. Milorad Krnojelac (Foča), Judgement*, Case No. IT-97-25-T, 15 March 2002, Nos. 12–50.

35 Also known as the Imperial Mosque or the Sultan Bajezid Velija II Mosque. Although the dated of its destruction was stated to be 22–23 April 1992, its minaret at least was still in existence on 1 May when photographer David Brauchli recorded the destruction in the Prijeka Čaršija (see illustration).

36 *Prosecutor v. Dragoljub Kunarac, Radomir Kovač and Zoran Vuković*, IT-96-23-T, testimony Safet Avdić, 22 March 2000, 665.

37 *Mosque and Medresa of Mehmed-paša Kukavica, the Architectural Ensemble*, designation as an National Monument, Commission to Preserve National Monuments of Bosnia and Herzegovina, 2 March 2004.

38 *Sheikh's Mosque (Kadi Osman-effendi's Mosque)*, designation as an National Monument, Commission to Preserve National Monuments of Bosnia and Herzegovina, 10 November 2005.

39 *Prosecutor v. Dragoljub Kunarac, Radomir Kovač and Zoran Vuković (Trial Judgment)*, IT-96-23-T and IT-96-23/1-T, ICTY, 22 February 2001, No. 46; McDowall, L (ed.) 2009 *Bridging the Gap Between the ICTY and Communities in Bosnia and Herzegovina, Conference Proceedings, Foča, 9 October 2004*, presentation by Peter Mitford-Burgess, Investigator, Office of the Prosecutor, ICTY, 12.

Fig. 1.4 **(a)** Residents of Foča walk through the devastated Prijeka Čaršija, a Muslim neighbourhood, on 1 May 1992. The minaret of the Careva Mosque still stands in the background; it was later destroyed and the mosque site cleared. © David Brauchli / Sygma / Corbis.

(b) The Careva Mosque, Foča, before its destruction in 1992. © Gazi Husrev-beg Library, Sarajevo

(c) The rubbish-strewn site of the mosque in October 2000. © Richard Carlton

Fig. 1.5 **(a)** The sixteenth-century Aladža Mosque in Foča before its destruction in 1992. © Dragan Resner / ICTY

Fig. 1.5 **(b)** The site of the Aladža Mosque, Foča, in June 1996. The foundations of the mosque and the ablutions fountain (*šadrvan*) can clearly be seen. © Lucas Kello / ICTY

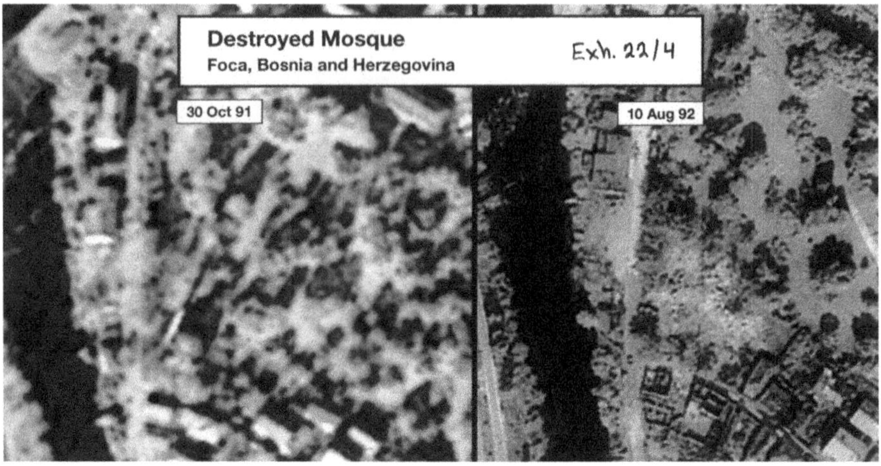

Fig. 1.6 (a) Satellite photographs of Foča taken by US Reconnaissance Systems on 31 October 1991 and 10 August 1992. In the image from October 1991 the dome of the Aladža Mosque and its minaret are plainly visible bottom left; the image from August 1992 shows the same site as an empty space. © United States Reconnaissance Systems / ICTY

The city of Bijeljina in the far north-east of Bosnia and strategically located on the main road to Serbia, was one of the first towns in Bosnia to be attacked by Serb paramilitary units on 1 April 1992 and was under the control of Bosnian Serb authorities from that time. Yet Bijeljina's five mosques, including the historic Atik Mosque, were not destroyed until almost a year later, all blown up with explosives in the early hours of 13 March 1993.[40] The Atik Mosque, facing the city's central square, was one of Bosnia's finer single-space mosques and the town's oldest.[41]

Notable cases of the destruction and severe damage to Catholic cultural heritage far from the frontlines were the Franciscan monastery of St Mark the Evangelist at Plehan in northern Bosnia (church 1902) and the church of St John at Podmilačje near Jajce, both destroyed by Bosnian Serb forces. St John incorporated the remains of a fifteenth-century church in its fabric – the sole surviving medieval church in Bosnia (of which there are only a handful) to remain in continuous and unchanged use.[42] Before the war the shrine at St John had been a shared space of multiconfessional pilgrimage visited by Muslims and Orthodox Christians,

40 Mlakar, M. 1993. Bijeljina dynamiters: The night the minarets fell, *Vreme News Digest Agency*, No. 78, 22 March; Bosnia Herzegovina: Report from the town of Bijeljina, *Channel 4 News*, 17 March 1993, Gaby Rado reporting, ITN BSP170393007.

41 The Atik mosque was first constructed c. 1520–1566, but rebuilt a number of times, particularly after the Treaty of Belgrade in 1739 ended a period of Austrian occupation of Bijeljina; there were later additions in the nineteenth and early twentieth centuries. See: *Atik Mosque (Old Mosque, Sultan Sulayman's Mosque), Harem and Turbe, the Site and Remains of the Architectural Ensemble*, http://kons.gov.ba/main.php?id_struct=50&lang=4&action=view&id=872.

42 All others which survived did so because they were converted into mosques, such the Fethija Mosque at Bihać, formerly the medieval church of St Anthony (probably 14th century).

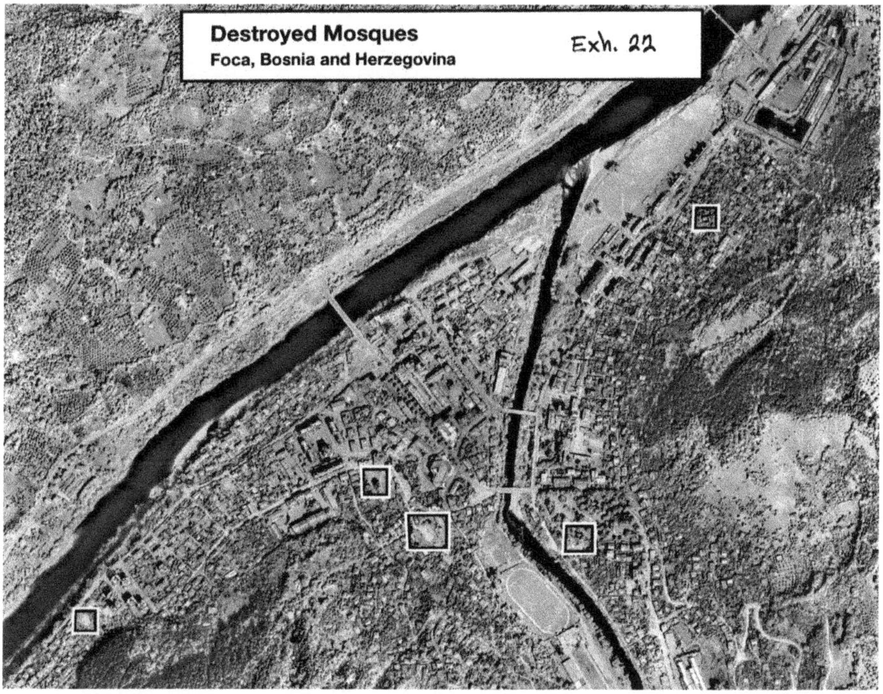

Fig. 1.6 (b) A second satellite photograph taken on 10 August 1992 shows the sites of destroyed mosques in Foča. Squared sites are from bottom left: 1. Tabačka Mosque (by right bank of the Drina river), 2. Mehmed-paša Kukavica Mosque (centre left), 3. Careva Mosque (centre). 4. Aladža Mosque (centre right, on right bank of Ćehotina river), 5. Musluk / Ali-paša Mosque Mosque (top right). Both photographs were used as prosecution evidence at the ICTY. © United States Reconnaissance Systems / ICTY

as well as Catholics.[43] At Plehan Bosnian Serb forces shelled the monastery repeatedly from 24 April 1992 and on 2 July 1992 drove a truck carrying two tons of explosives at the monastery, leaving it in ruins.[44] Among the Serb/Orthodox structures destroyed in this way (by Bosnian Croat forces) were the renowned sixteenth-century Žitomislić Monastery near Mostar that was dynamited and burned in mid-June 1992 and the church of St Nicholas at Trijebanj near Stolac with its important frescoes, as well as more modest structures like the sixteenth-century Church of the Transfiguration of Christ at Klepci, near Čapljina, where the first Serb school in the region had been built nearby in 1857.

Where destruction took place in the absence of military operations the usual methods used were the controlled blasting of structures by dynamite, tank mines or other explosives

43 Particularly on St John's Eve (24th June).

44 Živković, I. (ed.) 1997. *Raspeta Crkva u Bosni i Hercegovini: Uništavanje Katoličkih Sakralnih Objekata u Bosni i Hercegovini (1991–1996)*, Banja Luka, Mostar, Sarajevo: Biskupska Konferencija Bosne i Hercegovine and Zagreb: Hrvatska Matica Iseljenika Bosne i Hercegovine and Hrvatski Informativni Centar, 270.

Fig. 1.7 The ruins of Plehan Franciscan monastery in northern Bosnia after being shelled and finally dynamited by the Bosnian Serb Army on 2 July 1992. © Biskupska Konferencija Bosne i Hercegovine / ICTY

Fig. 1.8 **(a)** The church of the sixteenth-century Serbian Orthodox monastery of Žitomislić near Mostar in 1989 and **(b)** still in ruins, October 2001. © Peck Yee Tan, 1989 / Aga Khan Visual Archive and © Tina Wik

(pointing to the involvement of military units), by a combination of burning with explosives used to complete demolition of the building, or by simply setting the structure alight.[45] One of the largest structures totally demolished in this way was the neo-Baroque Serbian Orthodox Cathedral of the Holy Trinity or *Saborna Crkva* (1863) which had dominated the skyline

45 For methods used to damage/destroy cultural and sacral buildings, see, for instance, the testimony of Colin Kaiser at the International Criminal Tribunal for the former Yugoslavia (ICTY), *Prosecutor v Radovan Karadžić and Ratko Mladić*, Case Nos IT-95-18-R61 and IT-95-5-R61, *Rule 61 Hearing*, 2 July 1996, 438–9; and his report for the Office of the Prosecutor at the ICTY: Kaiser, C. 2002. *Report on Destruction of Cultural Property* 09-Jul-02, Exhibit P906, *Krajišnik* Case No. IT-00-39.

of Mostar. After being shelled on 7–8 June 1992, the structure was set alight and its bell tower destroyed on 15 June and its remains later dynamited until reduced to heaps of rubble; responsibility for the act is still unclear, although Bosnian Croat (HVO) forces are suspected.[46]

In many instances, particularly in the case of mosques (and churches) in city centre locations, the remains of destroyed structures were bulldozed and debris removed completely from the site which was subsequently levelled to remove any trace of its former presence. Among the most notorious incidences of this practice were those of the Ferhadija Mosque in Banja Luka, the Aladža Mosque in Foča and the Sinan-beg Boljanić Mosque (1570) in Čajniče with their ensembles of fountains, mausolea and other monuments; all were among the most important single-space domed mosques in Bosnia-Herzegovina. Others included the Osman-paša Mosque (1726) in Trebinje and the Čaršijska Mosque in Prijedor with its characteristically Bosnian wooden minaret. The Roman Catholic Church of the Assumption of the Blessed Virgin Mary in Nevesinje, a handsome stone-built neo-Romanesque structure dating from the Austro-Hungarian period (1901) designed, like the Bishop's ordinariate in Mostar by the prominent architect Max David, suffered a similar fate, its remains dumped at a landfill site together with those of the Careva Mosque, one of the two mosques in this small Herzegovinian town that were also completely destroyed and razed.[47] The earliest parts of the Careva Mosque[48] dated to the fifteenth century and it was regarded as one of the oldest mosques in the country. All the above were in the territory of Republika Srpska.

The removal of the remains of destroyed buildings (usually on the grounds of 'public safety') was typically carried out by contractors authorized by the local authorities, or in the case of the mosques in Bijeljina, by bulldozers of the Bosnian Serb Army (VRS).[49] Much of the remains of the Ferhadija and Arnaudija Mosques were taken to the municipal waste dump at Ramići where they were identified and excavated long after the end of the war. Similarly, after an anonymous tip-off, other large fragments of the Ferhadija were located at the bottom of an artificial lake formed from a gravel pit near the village of Karanovac; the blocks were later brought to the surface with the help of divers.[50]

In a number of cases, like that of the Savska Mosque (1739) in Brčko, mosque rubble was used to cover the mass graves of Muslims massacred during ethnic cleansing.[51] On 18 September 2000 a team from the Federation Commission on Missing Persons was

46 See Kaiser, C. 1993. *War Damage to the Cultural Heritage in Croatia and Bosnia-Herzegovina*, in COE *Information Report*, Doc. 6756, 2 February, 46.

47 See website of the website of the Commission to Preserve National Monuments, monument designations for Nevesinje: http://kons.gov.ba.

48 Also known as the Imperial or Sultan Bajezid Velija Mosque.

49 *The Islamic Community in Bosnia and Herzegovina v. The Republika Srpska (Bijeljina Mosques), Decision on the Admissibility and Merits (Delivered on 6 December 2000)*, Human Rights Chamber (HRC) Case No. CH/99/2656, VI.A.3, 11.

50 Personal knowledge and see Riedlmayer, A. 2008. Banja Luka's Ferhadija mosque rises again, *Bosnian Institute news* [online], 3 March; Hadžihasanović-Katana, M. 2012. The reappearance of Ferhadija Mosque in Banja Luka, in *Protecting Cultural Heritage in Times of Conflict: Contributions from the Participants of the International Course on First Aid to Cultural Heritage in Times of Conflict*, edited by S. Lambert and C. Rockwell. Rome: ICCROM, 69. See also the Ferhadija website for more information: http://www.ferhadija.ba.

51 For the Savska Mosque, see Riedlmayer, A. 2002a. *Destruction of Cultural Heritage in Bosnia-Herzegovina, 1992–1996: A Post-war Survey of Selected Municipalities*, (*Milošević*, ICTY Case No. IT-02-54, Exhibit P486, Date: 08/07/2003, and *Krajišnik*, ICTY Case No. IT-00-39, Exhibit P732, Date: 23/05/2005), 12.

Fig. 1.9 **(a)** The Čaršijska Mosque in Prijedor with its characteristically Bosnian wooden minaret before the war. © Gazi Husrev-beg Library, Sarajevo **(b)** The site of the Čaršijska Mosque in October 2000. The same apartment block as in the prewar photograph can be seen at the rear. © Richard Carlton

Fig. 1.10 **(a)** The Roman Catholic Church of the Assumption of the Blessed Virgin Mary, Nevesinje, designed by Max David. Constructed 1901–1903, destroyed 1992. © Ćiril Ćiro Raič / ICTY **(b)** The levelled site of the church in 2002. © András Riedlmayer

excavating a suspected grave site near KP Dom (Kazneno-Popravni Dom) in Foča, a former prison and the principal location where male Muslim civilians had been detained, mistreated and tortured and from where many disappeared. During the course of the excavations what appeared to be rubble from the Aladža Mosque was uncovered, including parts of columns, a discovery recorded by ICTY representatives observing the excavations.[52] The fragments were subsequently definitively identified as coming from the mosque, many identified by

52 The ICTY personnel were OTP Forensic Project Manager Brian Strongman and investigator Don King. See *Prosecutor v. Milorad Krnojelac (Foča)*, Case No. IT-97-25-T, testimony Racine Manas, 18 January 2001, 1854–67.

Fig. 1.11 A team from the Federation Commission on Missing Persons and an ICTY investigator on 18 September 2000 as a column from the Aladža Mosque is uncovered during the excavation of a suspected mass grave near KP Dom, Foča. The photograph was used as prosecution evidence at the ICTY. © Brian Strongman / Don King / ICTY

their distinctive decoration and coloured paintwork.[53] Four years later in the summer of 2004 easily identifiable fragments of the Aladža Mosque, among them sections of columns and parts of the mihrab and minaret, were again discovered by another exhumations team from the Federation Commission for Missing Persons excavating suspected mass grave sites on the banks of the River Drina near the Iron Bridge, here at a depth of 7 metres under a covering of hospital waste, garbage and soil.[54]

Where monuments were levelled, new structures were sometimes built on the sites as, for example, at Divič near Zvornik. Here 'Serbianization' of the once predominately Muslim village by Bosnian Serb authorities, who renamed it Sveti Stefan, was accompanied by the erection of an Orthodox church of the same name on the site of the village's demolished mosque.[55] More typically, levelled sites of destroyed monuments

53 *Prosecutor v. Milorad Krnojelac*, Case No. IT-97-25-T, testimony Racine Manas, 18 January 2001, 1861–4.

54 See the Commission to Preserve National Monuments website: *Exhumations Team Discovers Fragments of the Aladža Mosque in Foča*, at http://kons.gov.ba/main.php?mod=vijesti&extra=107546 0534&action=view&id_vijesti=307&lang=4. Pictures of the fragments recovered from the suspected mass grave sites can be seen at http://www.rijaset.ba/index.php?option=com_content&view=article&id =16239:fotogalerija-ostaci-porusene-aladze-dzamije&catid=259:godina-obnove-aladze&Itemid=457.

55 Riedlmayer, A. 2002a and Kovac, E. 2009. Controversial church relocated, *Southeast European Times* [online], Sarajevo 8 June 2009. http://www.setimes.com/cocoon/setimes/xhtml/en_ GB/features/setimes/features/2009/06/08/feature-02

Fig. 1.12 Typical uses for the levelled site of a destroyed mosque: parked cars, communal garbage containers and small kiosks. The site of the Krpića Mosque in Bijeljina in 2000. © Richard Carlton

were used as parking lots, for small markets, for storage, as locations for large communal garbage containers or were left as empty (frequently rubbish-strewn) spaces.

Blaming the Victim for the Crime

A recurring ploy of the destroyers of cultural and religious property was to blame the victim for the crime. These accusations were usually made by the perpetrators against 'the Muslims.' Thus, for instance, the destruction of the National Library (*Vijećnica*) in Sarajevo, the Old Bridge (*Stari Most*) in Mostar, and the Ferhadija Mosque and other Islamic structures in Banja Luka were all blamed by those who carried out or approved of the destruction on 'the Muslims.' This could be because the attacks on some structures attracted wide negative international, regional and even domestic condemnation, or be part of the propaganda efforts of ethno-nationalist elites.

The cases of the National Library and the Old Bridge are discussed in more detail below. Another was the case of Ahmići in central Bosnia, where on 16 April 1993 approximately 120 Bosniaks (Muslims) were massacred, hundreds of houses burned and two mosques deliberately dynamited and destroyed by HVO troops. Yet Colonel Tihomir Blaškić, the HVO commander held ultimately responsible for what happened at Ahmići, regularly maintained that the acts had been carried out by Muslim infiltrators, or Muslims dressed in HVO uniforms,[56] claiming in one interview that the whole crime had been

56 See testimony Alastair Duncan, *Blaškić* ICTY Case No. IT-95-14, 3 June 1998, 9055; Blaškić also suggested Serb extremists might be responsible.

carefully staged with the assistance of local British UNPROFOR commander Colonel Bob Stewart for the benefit of foreign journalists and European Community Monitor Mission (ECMM) observers to throw the blame onto the HVO.[57]

With regard to Banja Luka, after the blowing up of the Ferhadija and Arnaudija Mosques on 7 May 1993, Radio Banja Luka reported that 'Islamic fundamentalists threw explosive devices at two famous mosques;'[58] similar reporting followed the demolition of other Islamic structures in the city, such as that of the Turbe of Halil-paša on 15 December 1993.[59] The local newspaper *Glas Srpski* asserted that the mosques had been destroyed by 'the Muslims' themselves.[60] A member of the Committee of the Islamic Community in Banja Luka was threatened and beaten by local police who attempted to make him sign a statement that the Ferhadija and Arnaudija Mosques had been destroyed by the Muslims in order to put blame on the Serbs.[61]

Losses to the Built Heritage

Reporting to the Parliamentary Assembly of the Council of Europe on war damage to the cultural heritage in Bosnia-Herzegovina in early 1993, consultant Colin Kaiser described in detail the devastation of Mostar's historic centre after the JNA/Bosnian Serb siege of the city '*beside which,*' he emphasized, '*the damage in the Old Town of Dubrovnik pales in comparison.*'[62] Yet at that time the devastating Bosnian Croat/HVO attacks on the city were still to come.

Mostar was by far the most extensively ravaged of Bosnia-Herzegovina's historic cities and towns in two waves of destruction that left its central core with its many important Ottoman and Austro-Hungarian buildings in ruins. During the JNA/Bosnian Serb Army attacks of April–May 1992, mosques, Catholic churches, cultural institutions and other important structures were targeted, some reduced to a sequence of crumbling walls, such as the row of bazaar buildings along Kujundžiluk on the approach to the Old Bridge, or as unstable burned out shells like the neo-Moorish fantasy of the Hotel Neretva (1892). The Catholic Bishop's ordinariate (palace), designed by Max David in 1906, came under bombardment for a day and its library totalling 60,000 books destroyed.[63] In unambiguous reprisal attacks, the magnificent Serbian Orthodox cathedral (Saborna Crkva) was totally destroyed in June 1992 and the Old Orthodox church (1832) burned to a shell on 6 January 1993 on Orthodox Christmas Eve.

---

57 See transcipt of an interview with Tihomir Blaškić in the Croatian magazine *Danas* on 5 October 1993, *Blaškić*, Exh. 380.

58 Tabaković, E. 1999. Ostalo je malo pravih Banjalucana koji vole svoj grad, *Dnevni Avaz*, 24 October: ' … islamski fundamentalisti bacili eksplozivne naprave na dvije cuvene dzamije … .'

59 Reported to the author in 2000 by a Muslim resident of Banja Luka who had witnessed the destruction of the turbe and the subsequent reporting on television.

60 Gušić, B. 1995. Život i smrt u okupiranoj banjaluci, *Oslobođenje*, 15–23 March.

61 Gušić, B. 1995. Život i smrt u okupiranoj banjaluci, *Oslobođenje*, 15–23 March.

62 Kaiser, C. *War Damage to the Cultural Heritage in Croatia and Bosnia-Herzegovina*, in COE *Information Report*, Doc. 6756, 2 February 1993, No. 129; emphasis in the original.

63 See Kaiser, C. *VII. War Damage to the Cultural Heritage of the City of Mostar*, in *Information Report on the Destruction by War of the Cultural heritage of Croatia and Bosnia-Herzegovina*, Parliamentary Assembly of the Council of Europe (COE), Doc. 6756, Strasbourg, 2 February 1993, Nos 125–95.

(a)

(b)

(c)

Fig. 1.13 (*facing page*) **(a)** A view of Mostar in 1986. The Kujundžiluk in the foreground runs along the east bank of the Neretva River towards the Old Bridge. The minaret of the Nesuh-aga Vučijaković Mosque is seen bottom left, while the neo-Baroque Serbian Orthodox cathedral (Saborna Crkva) rises above the town. © Reha Gunay / Aga Khan Trust for Culture (AKTC) **(b)** Destroyed bazaar buildings along Kujundžiluk on the way to the Old Bridge destroyed by JNA/Bosnian Serb shelling in 1992. © Institute for the Protection of the Cultural, Historical and Natural Heritage of Bosnia-Herzegovina **(c)** Mostar, June 2001. The rubble of the Serbian Orthodox Cathedral of the Holy Trinity (Saborna Crkva), dynamited in June 1992. © András Riedlmayer

But as the HVO began to lay siege to Mostar in May 1993 it was the city's Ottoman/Islamic heritage that was the target: a report by the European Community Monitor Mission (ECMM) of 3 June 1993 observed how Mostar's Old Town was beginning to resemble Beirut, and that all Muslim monuments were being intentionally and systematically destroyed.[64]

In the historic core of Sarajevo, with its rich blend of Ottoman and Austro-Hungarian architecture, where a functioning sixteenth-century *bezistan* (covered market) faced Secessionist-style shop fronts and where the Gazi Husrev-beg Mosque (1531), the Old Orthodox church (sixteenth–twentieth centuries), the Orthodox cathedral (1868) the Austro-Hungarian period Roman Catholic cathedral (1889) and the historic Old Synagogue (1581) stood within a few hundred metres of each other, nearly every building sustained some damage. The famous Sephardic Jewish cemetery on the heights of Kovačići (founded 1630) was used by the Bosnian Serb Army as an artillery emplacement from where they shelled the city below, inevitably attracting return fire that damaged graves and the cemetery's small chapel; before withdrawing, the BSA left the cemetery heavily mined.

The range and scale of monuments destroyed or substantially damaged across the country during the conflict cannot be overestimated. Yet definitive statistics for the destruction and damage to Bosnia-Herzegovina's cultural heritage remain hard to pin down due to the patchy nature of existing surveys (many published by, or on behalf of, the different religious communities) and the lack of a comprehensive independent post-war assessment and evaluation of monuments destroyed and damaged. Many wartime damage lists were regarded as suspect and lacking in both accuracy and credibility as they became weapons in highly politicized and well publicized propaganda battles regarding levels of heritage destruction, particularly of religious property.[65]

However, as noted above, Muslim or Ottoman sacral and secular monuments formed by far the greatest part of structures targeted. Of religious buildings in general, Riedlmayer has noted the disproportionate targeting of structures of historic importance.[66] While attacks on Catholic/Croat and Orthodox/Serb monuments generally were attacks on individual structures, it was the destruction of the still-surviving complexes of what Balić has called harmonious 'living examples' of Ottoman urban planning[67] that caused the greatest devastation to the historic cores of towns and villages across the country. Despite

64 *Special Report: Situation on the Provisional Implementation of the Peace Plan in Middle Bosnia-Herzegovina*, ECMM, HRC Zenica, 3 June 1993, ICTY Exh. DH 1870.

65 See Richard Carlton and Helen Walasek, *Documenting the Destruction* in this volume for a detailed discussion.

66 Riedlmayer, A. 2002a, 15.

67 Balić, S. 1994. Destruction of the Bosnian architectural heritage: An interim report, in *Journal of Islamic Studies*, 5 (1994), 2, 268–73.

Fig. 1.14 **(a)** View of the historic core of Sarajevo in 1981 with its characteristic mix of Ottoman and Austro-Hungarian buildings. The surrounding steep hills would be used as positions by the Bosnian Serb Army to shell the city below from 1992 to 1995. The Vijećnica (National Library) is seen centre, with the small domes of the Brusa Bezistan below. © William Remsen, 1981 / Aga Khan Visual Archive **(b)** An Austro-Hungarian period building on Titova (Marshal Tito Street) in the centre of Sarajevo damaged by shelling 1992–1993. © Christian Maréchal / Wikimedia Commons

the damage, the central core of Sarajevo retained its characteristic identity, but the historic centres of smaller towns like Foča and Stolac where this intricate complex of structures had endured best became unrecognizable, with devastated areas of crumbling ruins broken by blank empty spaces where major monuments had formerly stood.

The lengths to which the determined obliteration of Muslim heritage extended is illustrated by the demolition of the tiny mosque of Sultan Mehmed the Conquerer at Kušlat near Zvornik (fifteenth-century, restored eighteenth century), an ancient traditional Bosnian style mosque with steeply pitched roof and wooden minaret. Perched high on a steep crag requiring a two hour climb up an unpaved road, it was nevertheless completely destroyed in February 1993 by Bosnian Serb forces and its remaining ruins later removed *after* the signing of the Dayton Peace Agreement.[68]

68 See Riedlmayer, A. 2002a. monument records, on the ICTY database and his testimony at the trial of Vojislav Šešelj, *Prosecutor v. Vojislav Šešelj*, Case No IT-03-67, 22 May 2008.

Fig. 1.15 The war-damaged tower of Husein-Kapetan Gradaščević in Gradačac, full of shell holes c.1995–1996. © Kemal Zukić / Centra za Islamsku Arhitekturu, Sarajevo

The Commission to Preserve National Monuments of Bosnia-Herzegovina[69] estimates that Foča lost 60 per cent of its historic urban fabric during the war.[70] The Commission further estimates that of 60 valuable urban nuclei in Bosnia-Herzegovina, 49 were destroyed or very badly damaged. Among them were the towns of Počitelj, Maglaj, Jajce, Prusac and Gradačac, where the tower of Husein-kapetan Gradaščević (1802–1834), one of Bosnia's greatest national heroes, was badly damaged by Bosnian Serb shelling.

With its circling defensive walls, Jajce, seat of Stjepan Tomašević, the last king of Bosnia, still retains features of the type of fortified medieval urban settlement that existed before the Ottoman conquest, possessing a unique array of monuments that reflect Bosnia's multifaceted history. Jajce's steep hill, dominated by a massive fortress and dotted with steeply-roofed traditional houses, became a landscape of roofless ruins and rubble-strewn spaces, its historic mosques badly damaged or totally destroyed, as were the town's Orthodox and Catholic churches and structures such as the early nineteenth century Ottoman *musafirhana* and *tekija*. Prusac, an important Ottoman religious and cultural centre in the sixteenth and seventeenth centuries, was subjected to bombardment by Bosnian Serb forces that badly damaged its rich collection of Islamic monuments, such as the complex of religious buildings associated with its most famous citizen, the sixteenth-century scholar Hasan Kafi Pruščak, and the Handanija Mosque (1617), starting point for the pilgrimage to Ajvatovica, the most important Islamic pilgrimage in Europe.[71]

But secular buildings unaffiliated with any particular group were targeted, particularly Austro-Hungarian buildings where many institutions (cultural and otherwise), and headquarters of infrastructure bodies were housed. Thus, the splendid Central Post Office (1913) in Sarajevo, purpose-built by Bosnia's Austro-Hungarian administration was so attacked.[72] While an element of spurious military legitimacy might be raised in respect of this assault (though there was no legitimacy attached to the conflict), none could be offered for the shelling of the Oriental Institute on 17 May 1992. The Institute was intentionally targeted by the Bosnian Serb artillery with phosphorus shells which burned and completely

69 See Valery Perry elsewhere in this volume for more on the Commission to Preserve National Monuments.

70 Figures prepared for the author by the Commission to Preserve National Monuments, compiled in September 2010: *National Monuments Destroyed or Damaged in the War*. The author would like to thank the Commission for these statistics.

71 See Tina Wik in this volume for more on Jajce and Prusac.

72 The Military Post Office and Telegraph or PTT Building.

Fig. 1.16 The gutted interior of the Oriental Institute in Sarajevo, deliberately targeted with phosphorus shells by the Bosnian Serb Army on 17 May 1992. The Institute was home to the largest collection of Islamic and Jewish manuscripts and Ottoman documents in South East Europe. © Orijentalni institut u Sarajevu / ICTY

destroyed its large and important collection of Islamic and Jewish manuscripts in Arabic, Persian, Ottoman Turkish, Hebrew and *aljamiado* (Bosnian written in Arabic script), as well as the Ottoman provincial archives and cadastral registers – primary sources for Bosnia's history and culture. It was only one of the many important libraries and archives and institutions holding documentation on Bosnia's history and heritage that were attacked.[73] Among other such institutions intentionally targeted were the Archive of Herzegovina and the Institute for the Protection of Monuments in Mostar.

The Bosnian Serb bombardment of the National Library in Sarajevo (*Vijećnica*) with incendiary shells over the night of 25/26 August 1992 on the eve of the important London Conference on Bosnia has been extensively described.[74] As flames erupted, windows exploded and Serb sniper fire raked the area, firemen attempted to extinguish the blaze but

73 For a selection of the many articles by András Riedlmayer on the subject, see bibliography. A number are available at http://bmip.info/articles/articles.html.

74 The London Conference took place 26–27 August 1992. It was sponsored by the United Nations and the European Community; among those attending were Alija Izetbegović, Radovan Karadžić, Slobodan Milošević and Franjo Tuđman.

For two of the best eyewitness accounts from the international print media, see Schork, K. 1992. Jewel of the city destroyed by fire, *The Times*, 27 August; Pomfret, J. 1992. Battles for Sarajevo intensify as Bosnian peace conference opens, *Associated Press*, 26 August.

Fig. 1.17 The Vijećnica, the National Library of Bosnia-Herzegovina, Sarajevo, in flames after being targeted by heavy artillery and incendiary shells on the night of 25–26 August 1992. © Kemal Hadžić

Fig. 1.18 Smoke pours from the Vijećnica, the National Library of Bosnia-Herzegovina after the shelling on the night of 25–26 August 1992. The photograph is a prosecution exhibit for the ICTY case against Radovan Karadžić and Ratko Mladić. © ICTY

were hampered by low water pressure and the continuing shelling. A chain of volunteers tried to rescue what they could from the building. The burning continued for days while ashes from the burned books drifted over Sarajevo, what Bosnian writer Valerijan Žujo has called the 'black birds.'[75] In an interview shortly afterwards Bosnian Serb leader Radovan Karadžić characteristically blamed the 'Muslims' themselves for the fire – because of the library's Austro-Hungarian heritage.[76] He expounded in more detail to the visiting Elie Wiesel, telling the Nobel Peace Prize laureate that the Muslims had removed 'their' books from the library and then 'burned it because they didn't like its Austro-Hungarian architecture in the midst of their favorite Turkish buildings.'[77]

75 Huseinovic, S. and Arbutina, Z. 2012. Burned library symbolizes multiethnic Sarajevo, *Deutsche Welle* [online], 25 August 2012.

76 Interview with Radovan Karadžić, *New Perspectives Quarterly*, 9:4, Fall 1992, 50–51. In reply to the question: 'During the London conference (last month), the historic National Library burned down during fierce fighting. Who is responsible?' Karadžić answered: 'Muslims. They burned down the Sarajevo library because they saw it as the heritage of the Austro-Hungarians. We know it was arson and not shelling because the fire came from the lower floor.'

77 Firestone, D. 1992. Peace rebuff 'This is not a simple war,' Wiesel is told in Bosnia, *Newsday*, 30 November; and Wiesel, E. 2008. Nobel laureate Elie Wiesel remembers Radovan Karadzic, the Serbian murderer, *NY Daily News* [online], 26 July.

But perhaps the most infamous of all the attacks on cultural property during the Bosnian War was the destruction of the sixteenth-century *Stari Most* (Old Bridge) at Mostar.[78] The bridge had been targeted from the beginning of the conflict, though it was noted by observers that while it had been struck during the JNA's two and a half month siege of Mostar (who had blown up all of the city's bridges with the exception of the Old Bridge),[79] it was hit many more times during the HVO siege which began in May 1993.[80] Two months into this new phase of the war the Bosnian government was alarmed enough about the state of the bridge and the intentions of the HVO to send an urgent request on 9 July to the UN Security Council, UNESCO and the Council of Europe asking that they prevent the destruction of the Old Bridge in Mostar and urgently send a UNESCO team to ensure its protection.[81]

On 8 November an HVO tank positioned on Stotina hill under Mount Hum began a sustained shelling of the bridge and the mayor of Mostar made a desperate radio appeal to the international community to condemn the attack and called for urgent assistance in saving it.[82] But the assault continued into the following day, and at approximately 10.15 on the morning of 9 November 1993 the *Stari Most* crumpled into the Neretva River.[83] The shelling and final collapse of the bridge was filmed by a number of cameramen,[84] most

---

78 For a detailed account of the destruction of the Old Bridge, see Petrovic, J. 2012. *The Old Bridge of Mostar and Increasing Respect for Cultural Property in Armed Conflict*, particularly 197–264. Leiden: Martinus Nijhoff. The Muslim-Croat War ended with the signing of the Washington Agreement in March 1994. See also testimonies in *Prosecutor v. Jadranko Prlić et al*, ICTY Case No. (IT-04-74).

79 Puljić, B. The bridges, in I. Ribarević-Nikolić and Ž. Jurić (eds) 1992 *Mostar '92 – Urbicid.* Mostar: Hrvatsko vijeće obrane općine Mostar/Javno poduzeće za obnvu izgradnju Mostara. Kaiser, C. *War Damage to the Cultural Heritage in Croatia and Bosnia-Herzegovina*, in COE *Information Report*, Doc. 6756, 2 February 1993, No. 128.

80 Makas, E. 2007. *Representing Competing Identities: Building and Rebuilding in Postwar Mostar*, 212; unpublished Phd thesis, citing Avdić, S. 1993. Prvi rat je bio bolji, *Slobodna Bosna/Ljiljan*, 19 July 1993, 17. For other reports of HVO shelling of the Old Bridge, see for example: Washington Post Foreign Service, 1993 Croat forces evict Muslims, shell famed Mostar Bridge, *The Washington Post*, 15 July, A24, and: Sarajevo radio reports Serb/Croat attacks in Northern and Central Bosnia, Mostar, *Radio Bosnia-Hercegovina*, Sarajevo, 6 July 1993, *BBC Summary of World Broadcasts*, EE/1735/C1, 8 July 1993.

81 See: Bosnian authorities protest over destruction of Mostar Old Bridge, *Radio Bosnia-Hercegovina*, Sarajevo, 9 November 1993, *BBC Summary of World Broadcasts*, EE/1843/C, 11 November 1993. See also: *Letter dated 11 November 1993 from the Permanent Representative of Bosnia and Herzegovina Addressed to the President of the Security Council*, UN Doc. S/26729 which refers to the 9 July letter.

82 See Mayor of Mostar appeals for international help to save Old Bridge, *Radio Bosnia-Hercegovina*, Sarajevo, 8 November 1993, *BBC Summary of World Broadcasts*, EE/1842/C, 10 November 1993, on the shelling of the Old Bridge by a tank and demanding urgent measures to save the bridge.

83 For the Bosnian government's letter of protest to the UN Security Council, the director general of UNESCO, the Council of Europe, the Organisation of the Islamic Conference (OIS) and the Aga Khan Trust for Culture regarding the destruction of the *Stari Most*, see *Letter dated 11 November 1993 from the Permanent Representative of Bosnia and Herzegovina Addressed to the President of the Security Council*, UN Doc. S/26729 which refers to the 9 July letter and attaches a letter: *Letter dated 9 November 1993 from the Presidency of the Republic of Bosnia and Herzegovina Addressed to the President of the Security Council.* See also: *BBC Summary of World Broadcasts*, 11 November 1993, Bosnian authorities protest over destruction of Mostar Old Bridge. For the contrasting reporting of the destruction of the Old Bridge from Sarajevo and Zagreb, see Sarajevo, Zagreb radios report destruction of Mostar's Old Bridge, (a) *Radio Bosnia-Hercegovina*, Sarajevo, 14.00 GMT 9 November 1993, (b) *Croatian Radio*, Zagreb, 14.00 GMT, 9 November 1993, *BBC Summary of World Broadcasts*, EE/1843/C, 11 November 1993.

84 A number of versions of films of the shelling and collapse of the *Stari Most* can be viewed on YouTube, among them *Distruzione del ponte di Mostar*, http://www.youtube.com/watch?v=aiO_UqAV0Ng and *Destruction of the Old Bridge in Mostar*, http://www.youtube.com/watch?v=CM3B-6CFo9k.

Fig. 1.19 (a) The Old Bridge (Stari Most) in Mostar before its destruction in 1993. The bridge, dating from 1566, spans the Neretva River. Its architect was Hayruddin, a pupil of the great Ottoman architect Sinan. © Reha Gunay / Aga Khan Trust for Culture (AKTC)

Fig. 1.19 (b) The Old Bridge (Stari Most) in Mostar as it is struck by Bosnian Croat (HVO) shelling on the afternoon of 8 November 1993. The following morning, under continued shelling, the bridge collapsed into the Neretva River. Still from footage used as prosecution evidence at the ICTY in the case of *Prlić et al.* © ICTY

famously Mostar residents Enes Delalić, Eldin Palata and Nedžad Kasumović. A freelance cameraman also recording the destruction was James Wilson, a Scottish ex-soldier then working for the independent British television company Folio Productions, whose footage was one of the first to become widely accessible to the international media.[85] Images of the bridge tumbling into the river, followed by the rattle of what one witness described as 'celebratory' gunfire from the Croat side[86] was viewed around the world. Though the involvement of the Croatian Army in the attack was suspected at the time, this has always been denied by the Croatian authorities.[87]

Slobodan Praljak, the HVO commander held principally accountable for the destruction of the *Stari Most* was famously said to have remarked that he was prepared 'to destroy hundreds of old bridges for the sake of one of my soldiers' little fingers.'[88] At various times Praljak maintained that according to international law it had not been a war crime to demolish the bridge, asserting on one occasion because it should have been marked with the UNESCO shield (the 1954 Hague Convention emblem) to claim protection as cultural property,[89] and on another because it was a legitimate military target (as the Bosnian Army was using it to move men and supplies across the Neretva).[90] Interviewed in March 2004 awaiting the arrival of an expected war crimes indictment from the ICTY in The Hague, Praljak bullishly predicted that no one would be charged for the destruction of the *Stari Most*, notoriously saying that the Old Bridge had been a military facility, and in a war a military facility could be demolished, regardless of its cultural value.[91]

Shortly afterwards, however, Praljak was indicted (with other HVO leaders) by the ICTY with the destruction of the *Stari Most* among the many war crimes for which they were charged – all integral acts, the Criminal Tribunal declared, in a *joint criminal enterprise* which aimed to permanently remove and ethnically cleanse Bosnian Muslims

85 Personal knowledge; see references to the Folio Productions footage also: *Fourth Information Report on war damage to the Cultural Heritage in Croatia and Bosnia-Herzegovina*, Parliamentary Assembly of the Council of Europe (COE), Strasbourg, Doc. 6999, 19 January 1994, No. 69 and *Final Report of the United Nations Commission of Experts Established Pursuant to Security Council Resolution 780 (1992)*, S/1994/674, 27 May 1994, No. 296. Wilson is also mentioned in Praljak, S. 2006, *Kako je srušen Stari* most (see below).

86 Testimony of Enes Delalić, *Prosecutor v. Jadranko Prlić et al*, ICTY Case No. (IT-04-74), 17 May 2007.

87 COE *Fourth Information Report*, Doc. 6999, 19 January 1994, No. 70. The Croatian government and the Croatian Army at that time was closely involved in the conflict in Bosnia and gave substantial support (including armaments) to the HVO.

88 Djikic (Đikić), I. 2001. ABC of Croatian crimes in Bosnia-Herzegovina, *Bosnia Report* [online], New Series No. 21/22, January–May 2001. Translation of an article originally published in *Feral Tribune*, Split, 24 February 2001.

89 The emblem of the 1954 *Hague Convention for the Protection of Cultural Property in the Event of Armed Conflict* (the *Emblem of the Convention*, sometimes also called 'the Blue Shield' or the UNESCO emblem. See *Prlić et al*, Case No. IT-04-74-T, *Final Trial Brief* of Slobodan Praljak, 31 March 2011, No. 350, 114. Praljak also declared this in an interview with Radio Free Europe, March 1999, cited in Lucić, P. 2000. Shit of the Year 1999, *Feral Tribune*, Split, 3 January 2000; translated online version at Ex-Yu Press, http://www.ex-yupress.com/feral/feralgs99.html.

90 *Prlić et al*, Case No. (IT-04-74), testimony Slobodan Praljak, 8 June 2009, 41283; *Final Trial Brief* of Slobodan Praljak, 31 March 2011, Nos 345–9, 113–14.

91 Šagolj, Z. 2004. Jesam li kriv što nisam Bog?, *Slobodna Dalmacija,* 18 March. Praljak said: 'Inače, optužnice za Stari most uopće neće biti. Zašto? Pa, zato jer je to bio vojni objekt. A vojni objekt u ratu, bez obzira na njegovu kulturološku vrijednost, može se srušiti. Dakle, to ni meni, niti bilo kome neće biti predmet optužnice.'

and other non-Croats from the territories which would form part of a Croat para-state of Herceg-Bosna.[92]

Praljak went on to deny any responsibility whatsoever for the destruction of the Old Bridge, and, like Radovan Karadžić before him in relation to the attack on the *Vijećnica*, accused 'the Muslims' of the destroying the Old Bridge themselves, claiming that the Bosnian Army had set explosives to bring the structure down.[93] Praljak worked strenuously to 'prove' this scenario, writing a book setting out his case, which included an analysis of video footage of the collapse of the bridge by explosives experts who also carried out an experiment reproducing the alleged detonation;[94] one expert was later to expand on their findings in Praljak's defence during his trial at The Hague.[95] However, on 29 May 2013, Praljak and his co-defendants were found guilty of participating in a joint criminal enterprise with the destruction of the *Stari Most* specifically addressed in the ICTY Trial Chamber's judgement and mentioned prominently in the judgement summary released to the media.[96]

Destruction of an Identity

Early in the conflict the systematic and deliberate nature of the destruction of the cultural and religious heritage in Bosnia-Herzegovina (most notably in the areas under the control of, or being attacked by, Bosnian Serb forces), manifested both as part of the process of ethnic cleansing and in the targeting of historic monuments and cultural and religious institutions in major centres like Sarajevo and Mostar, was noted and reported on by international human rights observers and the media.[97] Journalist Roger Boyes clearly analysed the process as the UN/EC sponsored London Conference on Bosnia-Herzegovina took place in August 1992:

> The destruction of Sarajevo's 19th-century town hall and the burning of priceless manuscripts in the Bosnian National Library reveal the hidden heart of darkness in the cruel Balkan war. After centuries of intertwined cultures, one ethnic group is trying to wipe out another: not merely its soldiers or civilians, but its memory. The Serbian prison camps, deplorable as they may be, do not deserve the label "genocidal", but the deliberate shattering of churches, mosques and libraries the living history of a nation is a form

92 *Prlić et al, Initial Indictment*, Case No. (IT-04-74), 2 March 2004. The full indictment was for a long list of alleged crimes against humanity, grave breaches of the Geneva Conventions and violations of the laws or customs of war carried out as a joint criminal enterprise was made against Jadranko Prlić, Bruno Stojić, Slobodan Praljak, Milivoj Petković, Valentin Ćorić and Berislav Pušić.

93 Sense Agency 2007 Destruction of Old Bridge seen from two angles, *Sense Tribunal* [online], 17 May.

94 The book and many other documents are available in English translation on Slobodan Praljak's website www.slobodanpraljak.com.

95 Praljak, S. 2006. *Kako je srušen Stari most* [*How the Old Bridge Was Destroyed*]. Zagreb: Oktavijan. The defence witness for Praljak at his ICTY trial was Dr Slobodan Janković; see Sense Agency 2008 How the Old Bridge was destroyed, *Sense Tribunal* [online], 30 June. See also the testimony of Slobodan Janković at the ICTY, *Prlić et al*, 30 June and 1 July 2008.

96 See *Prlić et al*, Case No. (IT-04-74), *Judgement Summary*, 29 May 2013. See Helen Walasek, *Cultural Heritage, the search for Justice and Human Rights* in this volume for more on this case.

97 See, for instance, Gutman, R. 1992. Unholy war; Serbs target culture, heritage of Bosnia's Muslims, *Newsday*, 2 September 1992; *War Crimes in Bosnia-Hercegovina Volume II*, a Helsinki Watch report, New York, Washington, Los Angeles. London: Human Rights Watch, 1 April 1993, 9 and 14.

> of cultural genocide ... This is another dimension of ethnic cleansing. The premise of the London peace conference is that this sinister practice can be stopped and reversed. People herded out of their villages at gunpoint should be allowed to return as soon as the international community can work out a way of protecting them ... The Serb occupiers, however, are working on a different plane altogether. Why else blow up mosques, destroy libraries and trample on ancient cemeteries? They are moving fast to make their occupation permanent. Soon, when there are no more mosques left in Bosnia, they will even be able to rewrite history and declare that the natural religion of the country is the Orthodox faith. Welcome to Greater Serbia.'[98]

These attacks appeared all the more premeditated as the former Yugoslavia had been considered a model of good practice with regard to the 1954 *Hague Convention for the Protection of Cultural Property in the Event of Armed Conflict* at the time. According to a 1989 UNESCO report on international implementation of the convention,[99] officers of the JNA (a conscript army in which many combatants on all sides of the Bosnian War would have served) were well acquainted with their obligations under the convention and official instructions to the armed forces included appropriate provision for the protection of cultural property.

Furthermore, the report observed, there were plans to draw up an inventory of cultural property in Yugoslavia under the provisions of the convention in the coming year with a view to marking the listed properties with the Hague Convention emblem[100] (as institutions in Sarajevo like the Zemaljski Muzej, the National Museum of Bosnia-Herzegovina, were so marked). Three years later, the occupation of Vukovar in Croatia by JNA forces saw the well-organized removal of museum and art collections from the city to Serbia in November 1991 under the direction of the Serbian Minister of Culture.[101] Thus the belief in many Bosnian heritage professionals of the planned nature of the attacks on cultural property and of maps emanating from Belgrade with co-ordinates identifying the locations of cultural institutions in Sarajevo to be targeted for destruction are not without foundation.[102] Whether the lists used to identify the collections at Vukovar were those drawn up under the terms of the Hague Convention is not known. It is certainly believed by some observers that cultural property displaying the Hague Convention emblem attracted deliberate targeting during the conflicts in Croatia and Bosnia.[103]

---

98 Boyes, R. 1992. This is cultural genocide, *The Times*, 28 August.

99 *Information of the Implementation of the Convention for the Protection of Cultural Property in the Event of Armed Conflict, The Hague 1954, 1989 Reports*, CC/MD/11, December 1989, *Yugoslavia*, 39. Thanks to Jan Hladík of UNESCO for pointing this out to the author.

100 The emblem of the 1954 *Hague Convention for the Protection of Cultural Property in the Event of Armed Conflict* (the *Emblem of the Convention* is sometimes also called 'the Blue Shield,' the 'UNESCO shield' or the 'UNESCO emblem.'

101 Reported on Tanjug news agency, 29 November 1991, cited in Pavić, V. (ed.) 2005 *War Damage to Museums and Galleries in Croatia/Ratne štete na muzejima i muzejskoj građi u Hrvatskoj*, Museums Documentation Center (MDC), Zagreb, 138, available at http://www.mdc.hr/RatneStete/eng/fs-glavni.html.

102 See Bevan, R. 2006. *The Destruction of Memory: Architecture at War*. London: Reaktion Books, quoting Behija Zlatar (38–9), Ferhad Mulabegović and Sabira Husedžinović (40). On their part the Serbian authorities said the collections had been moved in order to protect them; the collections that were stored in Serbian institutions have now been returned to Croatia.

103 See, for instance: Boylan, P.J. 2006. *Implementing the 1954 Hague Convention and its Protocols: Legal and Practical Implications.* Conference: Protecting Cultural Heritage: International

In February 1993 the United Nations Special Rapporteur on Human Rights Tadeusz Mazowiecki reported on the continued policy of ethnic cleansing in areas under Bosnian Serb control and noted: 'Massive violations of human rights and international humanitarian law are not simply features of the war in Bosnia & Herzegovina. They are being used deliberately to achieve ethnically homogenous areas,'[104] going on to say that although the conflict 'is not regarded as a religious one, it has been characterized by the systematic destruction and profanation of mosques, Catholic churches and other places of worship, as well as sites of cultural heritage.'[105] In its Final Report of 27 May 1994 the Commission of Experts appointed by the United Nation's Security Council to gather and assess evidence of grave breaches of the Geneva Conventions of 1949 and other international humanitarian law on the territory of the former Yugoslavia graphically described the acts that made up ethnic cleansing,[106] going on to note the

> widespread destruction of villages by systematically burning them to the ground and blowing up all the houses and structures in a given area. This includes cultural and religious monuments and symbols. The purpose of this destruction is to eradicate cultural, social and religious traces that identify the ethnic and religious groups.[107]

Thus the deliberate and systematic targeting of historic and religious structures was unmistakably identified as an intrinsic part of the process of the ethnic cleansing of populations, as well as being grave breaches of the Geneva Conventions and violations of international humanitarian law in their own right.

As Jan Boeles of the European Community Monitor Mission (ECMM) told journalist Robert Fisk in 1994:

> You have to understand that the cultural identity of a population represents its survival in the future. When the Serbs blow up the mosque of a village and destroy its graveyards and the foundations of the graveyards and mosque and then level them all off with a bulldozer, no one can ever, ever tell this was a Muslim village. This is the murder of a people's cultural identity.[108]

Law after the War in Iraq, Chicago, 3 February 2006, 1; Jogan, S. 2002. The procedure for marking cultural property in Slovenia, in Dutli, M.T. *Protection of Cultural Property in the Event of Armed Conflict; Report on the Meeting of Experts (Geneva, 5–6 October 2000)*, Geneva: International Committee of the Red Cross, Advisory Service on International Humanitarian Law, 100.

104 *Situation of Human Rights in the Territory of the Former Yugoslavia, Report on the Situation of Human Rights in the Territory of the Former Yugoslavia Submitted by Mr. Tadeusz Mazowiecki, Special Rapporteur of the Commission on Human Rights, Pursuant to Commission Resolution 1992/S-1/1 of 14 August 1992*, UN Doc. E/CN.4/1993/50, 10 February 1993, No 6.

105 *Situation of Human Rights in the Territory of the Former Yugoslavia*, UN Doc. E/CN.4/1993/50, No. 106.

106 *Final Report of the United Nations Commission of Experts Established Pursuant to Security Council Resolution 780 (1992)*, S/1994/674, 27 May 1994. No 134: '[M]ass murder, torture, rape and other forms of sexual assault; severe physical injury to civilians; mistreatment of civilian prisoners and prisoners of war; use of civilians as human shields; destruction of personal, public and cultural property; looting, theft and robbery of personal property; forced expropriation of real property; forceful displacement of civilian population; and attacks on hospitals, medical personnel and locations marked with the Red Cross/Red Crescent emblem.'

107 *Final Report of the United Nations Commission of Experts*, S/1994/674, No. 136.

108 Fisk, R. 1994. Waging war on history, *The Independent*, 20 June.

Nihad Čengić, art conservator and staff member of Sarajevo's *Office for the Protection of Cultural Property*, said in 1993: 'How do you destroy a people if you can't kill them all? You destroy all the materials which proved their existence. That's why mosques are being destroyed. That's why the Oriental Institute was destroyed. They're destroying these things to reconstruct history … '[109]

Indeed, it was frequently articulated by the perpetrators themselves that the destruction of cultural and religious property was intended to discourage the return of the expelled population, not only because of their association with community identity, but also to obliterate the material evidence of the expelled communities' deep roots in the locality, effecting a transformation of the cultural landscape from one of ethnic diversity into an apparently historically mono-ethnic domain. It was typical of Bosnia-Herzegovina (most noticeably in small towns), for instance, to find a Catholic church, an Orthodox church and a mosque, often situated within metres of each other. Now these proofs of living side by side were being violently eradicated.

The particular targeting of minarets and their removal from the landscape was noted by Colin Kaiser in a report for the ICTY as ' … a kind of architectural equivalent to the removal of the population, and visible proof that the Muslims had left.'[110] By the end of the war, with one notable exception, not a single intact minaret was left standing on territories occupied by Bosnian Serb forces. The sole functioning mosque and its minaret which survived undamaged on the territory of Republika Srpska during the 1992–1995 war was that in the village of Baljvine near Mirkonjić Grad where Bosnian Serb villagers actively protected their Muslim neighbours and the mosque from destruction.[111]

This transformation of the landscape was at times completed by a change of place name. Most enthusiastic in this respect were the Bosnian Serbs whose nationalist ideologues regarded 'Bosnian' in a place name as a symbol of Turkish and Austro-Hungarian oppression. Thus in territories held by the Bosnian Serbs the 'Bosnian' prefixes were eliminated from Bosanska Krupa, Bosanski Novi, Bosanska Dubica, Bosanski Brod, Bosansko Grahovo, Bosanska Gradiška, Bosanski Petrovac, Bosanski Šamac and Bosanska Kostajnica. Other locales were given a more 'Serbian' name: Foča becoming Srbinje (Serb-town), Divič renamed Sveti Stefan, the Banja Luka suburb of Gornji Šeher changed to Srpske Toplice (Serb Spas) and Donji Vakuf to Srbobran.

Heading the name change project was Radoslav Unković, the then Director of the Institute for Protection of the Cultural, Historical and Natural Heritage of Republika Srpska, who reflected the 'ambitions and thoroughness employed in the task of linguistic cleansing of toponyms'[112] when he spoke to the Belgrade daily *Večernje Novosti* in November 1992. Unković explained that ' … all names "which are associated with evil and where the Serbian tradition was effaced will undergo change." "We shall spare no effort in building

---

109 Schiller, B. 1993. Bosnian artists save heritage treasures, *The Toronto Star*, A10, 15 May.

110 Kaiser, C. 2002. *Report on Destruction of Cultural Property 09-Jul-02, Krajišnik* Case No. IT-00-39, 4.

111 Arnautovic, M. 2010. Bosnia: The village where hate never triumphed, *Institute for War and Peace Reporting* [http://iwpr.net], TRI Iss. 642, 10 April. Two historic minarets did also survive. These were the free-standing minaret of the disappeared seventeenth century Osman-paša Mosque at Kazanci (the mosque and surrounding village were destroyed in 1684) and the minaret of the Mujo Kotezlija Mosque at Kotezi, disused since World War II; see Richard Carlton and Helen Walasek, *Topography of Destruction* in this volume for more on the Kotezi mosque.

112 TOL 1993. Renaming Fashion, *Transitions Online*, 12 April.

Fig. 1.20 **(a)** Proofs of living side by side: the small town of Bosanska Krupa in north-west Bosnia, 2007. The town was occupied by the Bosnian Serbs during the war. The partially-restored Roman Catholic Church of the Blessed Virgin Mary, waiting its stucco coat, is on the left, the Serbian Orthodox Church of the Nativity of the Virgin (1882), right, while the reconstructed minaret of the Čaršijska (Gradska) Mosque can be seen centre. Only the Serbian Orthodox church was not damaged during the war. © Mazbln / Wikipedia. **(b)** The shell of the early twentieth century Roman Catholic Church of the Blessed Virgin Mary in October 2000, and **(c)** the ruins of the Čaršijska (Gradska) Mosque recorded during the winter of 1995–1996, an empty shell, its minaret torn down. Probably of eighteenth century foundation, the mosque was rebuilt a number of times, most significantly in 1908. © Richard Carlton © Kemal Zukić / Centra za Islamsku Arhitekturu, Sarajevo

Fig. 1.21 Not a single intact minaret was left standing on territories occupied by Bosnian Serb forces: the Donja Mosque, Kotorsko, Republika Srpska, badly damaged, its dome and minaret destroyed, in June 2001. © Richard Carlton

the toponymy in accordance with the Serbian tradition," he said, adding that "Sarajevo was to meet its fate in this view."'[113]

András Riedlmayer has commented on how both the perpetrators and victims of ethnic cleansing were well aware of the significance of the destruction and obliteration of a community's cultural and religious monuments.[114] Thus statements were made like that of Simo Drljača, the regional police chief responsible for the notorious Bosnian Serb detention camps near Prijedor: 'With their mosques, you must not just break the minarets,' he said, 'You've got to shake up the foundations because that means they cannot build another. Do that, and they'll want to go. They'll just leave by themselves.'[115] Or that of Branko Grujić, the Serb-installed mayor of the formerly Muslim/Bosniak majority town of Zvornik who told journalists in 1993: 'There were never any mosques in Zvornik.'[116] And during a discussion in September 1993 of how it was only a matter of time before Mostar's Old Bridge would be destroyed, Branimir, a Bosnian Croat militiaman remarked: 'It is not enough to cleanse Mostar of the Muslims – the relics must also be destroyed.'[117]

Finally, testifying at the ICTY trial of Tihomir Blaškić, commander of HVO forces in central Bosnia, Colin Kaiser, the longest-serving observer of the destruction to Bosnia's cultural heritage in the field (as a COE consultant and later UNESCO representative in Bosnia-Herzegovina) remarked on another phenomenon. Kaiser commented that it was clear that those who carried out ethnic cleansing (in this case the Bosnian Croats) were trying to remove the targeted community (Bosniaks/Muslims) and destroy its religious and

113 TOL 1993. Renaming Fashion. See also: AP 1992 Report: Serbs would change name of Sarajevo, *Associated Press* [online], 10 November; Doder, D. 1993. Warring Bosnia factions practice linguistic cleansing of geographic names, *Baltimore Sun* [online], 4 May.

114 Riedlmayer, A. 2002b. From the ashes: The past and future of Bosnia's cultural heritage, in *Islam and Bosnia: Conflict Resolution and Foreign Policy in Multiethnic States,* edited by M. Shatzmiller. Montreal: McGill-Queens University Press, 124.

115 Sudetic, C. 1992. Serbs' gains in Bosnia create chaotic patchwork, *New York Times* [online], 21 August.

116 Williams, C.J. 1993a. Serbs stay their ground on Muslim lands, *Los Angeles Times*, 28 March.

117 Block, R. 1993. Croatian death squad talks tough around the pool table, *The Independent*, 6 September.

cultural traces, both as a message that they were unwanted on the territory and as an attempt to remodel the landscape so that no evidence of the expelled community's former existence there remained as a way of saying 'There are no Muslims here.' But he also noted:

> Now, there's another question which is a very important question. I've thought about it a lot. It's common to regard the destruction of somebody else's cultural heritage as an attack on the other person. It is true. It is obvious. But it is more complicated than that. It is not just physically cleaning the society, removing the others, it also means changing the identity of one's own people ... What you're doing to the Croat from this area, is you're taking the Bosniak Muslim out of him. You are creating a new Croat. You are creating a Croat who has not the memory and the experience of having lived with somebody else.[118]

Identifying the Perpetrators

The findings of the UN Commission of Experts, the International Criminal Tribunal for the former Yugoslavia (ICTY) and a 1995 Central Intelligence Agency (CIA) report identifying the perpetrators of ethnic cleansing campaigns in Bosnia-Herzegovina (and thus of the greatest part of the destruction of religious and cultural property) have established that the bulk of such operations were carried out by Bosnian Serb forces, together with Yugoslav People's Army (JNA) forces and paramilitary units from Serbia, targeting mainly Muslim, but also Bosnian Croat populations.[119] On a considerably lesser scale Bosnian Croat (HVO) forces also carried out policies of ethnic cleansing directed chiefly at Muslim, but also at Bosnian Serb populations. While Bosnian government forces (largely Muslim) did commit what were considered grave breaches of the Geneva Conventions, these assessments found that they had no policy of ethnic cleansing and did not engage in such operations.[120]

These findings accord with estimates of destruction assembled in various sources (see *Documenting the Destruction*) and in the reports on the destruction of cultural and religious property prepared by two experienced and long-term observers, András Riedlmayer and Colin Kaiser, in their capacity as expert witnesses for the ICTY, as do their subsequent testimonies at a number of different ICTY prosecutions[121] and Riedlmayer's testimony at the International Court of Justice (ICJ) in the case of Bosnia-Herzegovina *vs* Serbia and Montenegro on a charge of genocide.[122] During his ICJ testimony in 2006 Riedlmayer

118 *Blaškić* Case No. IT-1995-14-T, testimony, 16 July 1998, 10634. Colin Kaiser spent many years assessing and recording sites of destroyed heritage in Bosnia-Herzegovina during and after the Bosnian War in his work for the Council of Europe and UNESCO.

119 *Final Report of the United Nations Commission of Experts, Annex IV The Policy of Ethnic Cleansing.* S/1994/674/Add.2 (Vol. I), 28 December 1994, Introduction; *Sanitized [ ] Version of Ethnic Cleansing Paper*, dated 5 January 1995. See also *Ethnic Cleansing and Atrocities in Bosnia*, Statement by CIA Deputy Director for Intelligence John Gannon, Joint SSCI SFRC Open Hearing, 9 August 1995.

120 *Final Report of the United Nations Commission of Experts, Annex IV The Policy of Ethnic Cleansing.* S/1994/674/Add.2 (Vol. I), 28 December 1994, *Introduction.*

121 See Riedlmayer, A. 2002a. *Expert Report*, Exhibit P486 Date: 08/07/2003, *Milošević* case and Exhibit P732, Date: 23/05/2005, *Krajišnik* case; and Kaiser, C. 2002. *Report on Destruction of Cultural Property* 09-Jul-02, *Krajišnik* Case No. IT-00-39. See also their various testimonies at http://icr.icty.org/.

122 For Riedlmayer's testimony at the ICJ, see *Application of the Convention on the Prevention and Punishment of the Crime of Genocide (Bosnia and Herzegovina v. Serbia and Montenegro)*, CR 2006/22, International Court of Justice, The Hague, 17 March 2006.

noted that in the 26 municipalities which he had surveyed so far that were Serb-occupied during the war, almost 100 per cent of all Muslim places of worship and more than 75 per cent of all Roman Catholic churches were found to have been either seriously damaged or totally destroyed, of which almost 60 per cent were historic structures.[123] An assessment by Riedlmayer in 2005 of sacral structures belonging to non-Serb communities destroyed in 20 municipalities (18 of them Serb-occupied) found that 85 per cent of the destruction had taken place in 1992 (thus during the period of the most intensive campaigns of ethnic cleansing).[124] Kaiser testified at the ICTY that he held the Bosnian Serbs responsible for the majority of the destruction of cultural and religious property in Bosnia-Herzegovina.[125]

Asked during his testimony at the trial of Radovan Karadžić in 2011 if he had observed any patterns with respect to the cultural destruction in the large number of sites he had researched, András Riedlmayer commented that the pattern of destruction almost entirely covered the municipalities that had come under the control of Bosnian Serb forces which he had surveyed:

> If one mapped the destroyed mosque sites, it was almost possible to draw a territorial map of what areas had been conquered by the Bosnian Serb forces. On the other side of that, there would be damage just beyond the confrontation line and only scattered damage elsewhere, but in the areas that were controlled by Bosnian Serb forces, there were virtually no mosques left intact.[126]

The destruction of religious and cultural property by Bosnian Croat/HVO forces was examined in some detail during the long and wide-ranging case at the ICTY which charged the high-ranking Bosnian Croat political and military leadership (Jadranko Prlić, Bruno Stojić, Slobodan Praljak, Milivoj Petković, Valentin Ćorić and Berislav Pušić: *Prlić et al*) with participating in a *joint criminal enterprise* '… to politically and militarily subjugate, permanently remove and ethnically cleanse Bosnian Muslims and other non-Croats who lived in areas on the territory of the Republic of Bosnia and Herzegovina which were claimed to be part of the Croatian Community (and later Republic) of Herceg-Bosna … '[127]

Herceg-Bosna was to be defined by an exclusively Croat/Catholic identity and looking at selected municipalities in Herzegovina and central Bosnia, the trial revealed clearly how destruction of religious and cultural property was incorporated into the Herceg-Bosna leadership's plans for the ethnic cleansing of the territory and was carried out between January 1993 to April 1994 in conjunction with the widespread deportation, expulsion, imprisonment and persecution of non-Croats solely on ethno-national grounds.[128] The final judgement in *Prlić et al.* on 29 May 2013 showed how the religious and cultural property

---

123 *Application of the Convention on the Prevention and Punishment of the Crime of Genocide (Bosnia and Herzegovina v. Serbia and Montenegro)*, CR 2006/22, 17 March 2006, 17–18.

124 Riedlmayer, A. 2005. *Expert Report: Statistical Addendum: Sacral Sites of the non-Serb Communities in the Surveyed Municipalities Damaged or Destroyed in 1992*, ICTY Exh. P732.B, 22 May 2005.

125 *Karadžić and Mladić*, Case Nos IT-95-18-R61 and IT-95-5-R61, *Rule 61 Hearing*, Colin Kaiser testimony, 2 July 1996, 437.

126 *Karadžić*, Case No. IT-95-5/18-T, Riedlmayer testimony, 9 December 2011, 22551.

127 *Prlić et al, Initial Indictment*, Case No. IT-04-74-T, par. 15. See Walasek, *Cultural Heritage, the Search for Justice and Human Rights* in this volume for more on this case.

128 *Prlić et al, Final Prosecution Trial Brief*, 31 January 2011; Prlić *et al, Jugement, Vol. 3*, Case No. IT-04-74-T, 29 May 2013.

targeted did not constitute military objectives and were variously mined, dynamited or set on fire, frequently after the Muslim population had been expelled from a district.[129] In Mostar alone, the judgement held the HVO directly responsible for demolishing or badly damaging 11 historic mosques and the Old Bridge between May and December 1993.[130]

The *Prlić et al.* prosecution did not investigate the destruction of Serbian Orthodox religious and cultural property, much of which had taken place in 1992, before the HVO's ethnic cleansing campaigns of 1993. However, András Riedlmayer has noted that his assessments found the majority of the attacks on and destruction of Serbian Orthodox churches were carried out by Bosnian Croat forces, especially in Herzegovina and in the Posavina to the north, much of it in the context of fighting with JNA and Bosnian Serb forces during 1992, which also included the intentional destruction of the major monuments of Žitomislić monastery and the Serbian Orthodox cathedral in Mostar.[131]

Thus the early hypothesis of an equivalent and mutual destruction of religious and cultural heritage by all three principal warring parties in the conflict (Bosnian government/'Muslims,' Bosnian Serbs, Bosnian Croats) is not borne out by the evidence and expert assessment. However, as discussed elsewhere in this volume, this assumption continues to have a long afterlife.[132] During their war crimes trials at the ICTY both Slobodan Milošević and Radovan Karadžić attempted to portray the destruction of religious heritage as mutual, reciprocal and carried out equally by all parties – 'tit-for-tat' reprisals for damage previously done to the religious property of the perpetrators of the destruction.[133]

Apart from the evidence, which does not support the premise that most destruction of religious and cultural property could be characterized as reprisal attacks and was mutual and equivalent, it is difficult to find a single statement made by the perpetrators of the destruction or their military and political associates or superiors to support such a contention, but many which support the opposite: that the greatest part of the destruction had been intentional, systematic and an integral part of a programme of the forced removal of particular ethnic groups (see *Destruction of an Identity*, above).

Milan Tupajić, former chief of the crisis staff and wartime president of the Bosnian Serb municipality of Sokolac, articulated this clearly when he testified at the ICTY about the massacre of the Muslim villagers of Novoseoci and the destruction of its mosque by engineers of the Second Romanija Motorized Brigade of the Bosnian Serb Army in September 1992 and stated how within the space of a few days there were no mosques left in Sokolac municipality. Asked why he thought the mosques had been destroyed Tupajić explained: 'There's a belief among the Serbs that if there are no mosques, there are no Muslims and by destroying the mosques, the Muslims will lose a motive to return to their villages.'[134]

As András Riedlmayer pointed out during his testimony at the Milošević trial, while he acknowledged that some tit-for-tat reprisals against religious property did occur, in the vast majority of cases the systematic and uniform nature of the destruction (in this case in Bosnian Serb-held territories), the fact that much of this destruction occurred not in the heat of conflict, but at times when the areas in question were under control of the Bosnian Serb authorities,

129 *Prlić et al, Jugement, Vol. 3.*

130 *Prlić et al, Jugement, Vol. 3*, Nos 1608–11.

131 *Milošević*, ICTY Case No. IT-02-54, Riedlmayer testimony, 8 July 2003, 23865–6.

132 See Richard Carlton and Helen Walasek, *Topography of Destruction* and *Documenting the Destruction.*

133 *Milošević*, ICTY Case No. IT-02-54, 8 July 2003; *Karadžić*, Case No. IT-95-5/18-T, 9 December 2011.

134 *Krajišnik*, Case No. IT-00-39-T, Tupajić, 29 June 2005, 15431.

as well as factors such as the use of explosives and the amount of organization involved, suggested that this was not merely a matter of mob anger but of some sort of directed policy. Lastly, he remarked, if there had been a real tit-for-tat relationship, one could expect the same pattern and degree of destruction on all sides and this was clearly not the case.[135]

Indeed, in 1994 Radovan Karadžić himself had spoken rather differently in his stronghold of Pale than he did nearly two decades later in a courtroom at The Hague. When Michael Williams, UNPROFOR's Director of Information, met with the Bosnian Serb leader in autumn of that year and raised the issue of the deliberate destruction of places of worship and of historical importance in areas controlled by the Serbs, he was taken aback when Karadžić told him that he did not deny that mosques had been deliberately destroyed. The reasoning Karadžić offered was not that mosques had been destroyed in retaliation for the destruction of Serbian Orthodox churches, but because they were symbols of Turkish imperialism. 'He said quite explicitly that it would be a good thing if all aspects of that heritage were erased,' Williams testified at the ICTY.[136]

During his trial Radovan Karadžić attempted to deny that ethnic cleansing had taken place at all and to portray the widespread departure of Muslims from territory held by the Bosnian Serbs as analogous to a natural disaster when it was necessary to evacuate populations, which the ICTY Prosecutor declared, gave rise to many questions, first among them: if that were so, why then had the active destruction of virtually all Muslim religious sites and mosques in those territories occurred?[137] Citing testimony and evidence, the Prosecution noted: 'the destruction of mosques was to eradicate or delete all traces of Muslim presence in the area and, of course, to intimidate the remaining Muslims into leaving the area. The ethnic cleansing was the reason for the destruction of the mosques.'[138]

Various justifications were brought forward during ICTY prosecutions for the destruction of religious property. One was to place the blame for the destruction on 'uncontrollable elements' – in Karadžić's case, he alleged that the destruction had been carried out by Serb paramilitaries or Serb refugees over whom the local Bosnian Serb authorities had no control.[139] Another was that mosques had been used by the Muslims to store weapons, ammunition and explosives and that minarets had been used as machine-gun and sniper nests.[140] This claim was made as the reason behind the destruction of the Aladža Mosque in Foča on 2 August 1992. The ICTY Prosecutor enquired of one witness making this assertion, if that were the case, given that Foča had been in secure control of the Bosnian Serbs since mid-April and there were certainly no Muslim fighters left in the town, why it would have been necessary to blow up the mosque as late as August?[141]

However, as noted above, while the majority of the deliberate destruction of religious and cultural property took place as part of campaigns of ethnic cleansing, some attacks can be characterized as reprisals. Thus, for instance, in late summer and early autumn 1995 as Bosnian Army (ARBiH), HVO and Croatian Army offensives in western Bosnia-Herzegovina regained territory occupied by Bosnian Serb forces since 1992 (and where

135 *Milošević*, ICTY Case No. IT-02-54, Riedlmayer, 8 July 2003, 23848–9, 23882–3.

136 Michael Williams testimony, *Milošević*, ICTY Case No. IT-02-54, 24 June 2003, 22951.

137 *Karadžić*, Case No. IT-95-5/18-T, *Rule 98 bis hearing*, 13 June 2012, 28639.

138 *Karadžić*, *Rule 98 bis hearing*, 13 June 2012, 28639–40.

139 For example, *Karadžić*, 9 December 2011, 22561.

140 See for example, *Milošević*, 8 July 2003; *Karadžić*, 9 December 2011.

141 *Krunarac et al*, ICTY Case No. IT-96-23-T, Velimir Djurovic testimony, 18 July 2000, 5052, 5055–6, 5062–5.

most mosques and many Catholic churches had been destroyed or badly damaged), there were attacks on Orthodox sacral structures (as well as on other Serb properties).

There were certainly assaults by Bosnian Army (ARBiH) forces and Muslims/Bosniaks on Bosnian Serb/Orthodox and Bosnian Croat/Catholic religious and cultural property. Many of them occurring in rural settings, a great proportion of such attacks tended to be at varying degrees of vandalization, desecration and damage to interiors.[142] On this, a Franciscan monk and former guardian of Guča Gora monastery commented at the ICTY 'It is interesting that the Muslim troops, by and large, did not cause any major damage to objects [structures], to the exterior of the objects ... ' and that mostly damage was done to interiors.[143] But there were more serious cases.

A severe attack early in the war was on the Serbian Orthodox church of St George at Donja Sopotnica near Goražde, the earliest part of which dates to the fifteenth century and which between 1519 and 1526 was site of the first printing press in the region. The church was shelled in September 1992 by Bosniak troops and subsequently set alight, destroying the roof and leaving the interior open to the elements. During the final offensives of the war in western Bosnia-Herzegovina described above, the Orthodox church of St Tsar Lazar in Benakovac between Bihać and Sanski Most in north-west Bosnia was burned out by Bosnian Army forces as they advanced in the last month of the war in 1995. In Bosnian government-controlled Maglaj the Orthodox church of St Elijah the Prophet (1909) was desecrated and vandalized, though the town was also subjected to persistent bombardment by the Bosnian Serb Army; by the war's end the church was a roofless shell with a damaged steeple.

In central Bosnia, the Roman Catholic Church of the Transfiguration of Our Lord (1924) in Borovica was mined and burned by Bosnian government forces after they overran the village in November 1993. They encountered no resistance and the Croat inhabitants were forced to flee. The rectory adjacent to the church was also destroyed, along with the parish records and library.[144] During fighting between the HVO and the ARBiH in October 1993, the Catholic church of All Saints in the tiny Croat village of Donja Drežnica between Jablanica and Mostar was desecrated and used by Zijad Kurtović, a military police commander in the Bosnian Army, as a setting for the abuse of male Croat detainees and prisoners of war, often using religious objects as instruments of torture.[145] The case of the Franciscan monastery at Guča Gora and its desecration and vandalization by *mujahedin* fighters will be described in detail below.

It is true to say that where ICTY indictments included charges of destroying religious and cultural property, defendants from all three warring parties were able to produce at least some documentary evidence that orders had been issued by the various political and military authorities prohibiting attacks on the religious structures of other ethnic groups.

142 See Colin Kaiser's observations, *Sixth Information Report on War Damage to the Cultural Heritage in Croatia and Bosnia-Herzegovina*, Parliamentary Assembly of the Council of Europe (COE), Doc. 7133, Strasbourg, 31 August 1994, 15–16. Kaiser also testified at the ICTY that during all his missions he had never seen a building destroyed by explosives in territory held by the Armija (Bosnian Army/ARBiH): *Karadžić and Mladić*, Case Nos IT-95-18-R61 and IT-95-5-R61, *Rule 61 Hearing*, 2 July 1996, 437.

143 Fra Stjepan Neimarević, a Franciscan priest was testifying as a defence witness at the ICTY trial of Bosnian Croat leader Dario Kordić: *Dario Kordić and Mario Čerkez* Case No. (IT-95-14/2) *'Lašva Valley,'* 4 July 2000, 22000.

144 Živković, I. (ed.) 1997 *Raspeta Crkva u Bosni i Hercegovini.*

145 Case No. *X-KRŽ-06/299 – Kurtović Zijad*, Court of Bosnia and Herzegovina, 2007. See Walasek, *Cultural Heritage, the Search for Justice and Human Rights for more on this case.*

Fig. 1.22 The burned-out Serbian Orthodox church of St Tsar Lazar, Benakovac, on the road from Sanski Most to Bihać in July 2002. Graffiti on the exterior reads: 'Kuća bosanskog genocida' = 'Home (or house) of the Bosnian genocide.' The church was burned by Bosnian Army troops who took Benakovac in the final month of the war. © András Riedlmayer

On 12 May 1993, for instance, a week after the two historic mosques of the Ferhadija and Arnaudija had been blown up and demolished in Banja Luka, Radovan Karadžić himself went so far as to issue an official order that 'all places of worship in the town and its surroundings shall be secured and protected from increased terrorist activity.'[146] There was no indication of who the 'terrorists' might be (reports in the Bosnian Serb media at the time attributed the destruction to 'Islamic fundamentalists') and further mosques and other Ottoman/Islamic and Catholic sacral and secular heritage continued to be dynamited and destroyed in the area over the succeeding months and years. The motivation for issuing this order is unclear, given that it was plainly not followed and given Karadžić's statement to Michael Williams the following year, though it may have been prompted by negative reactions to the demolitions from Serbs in Banja Luka and Belgrade.

However, the paucity of official prohibitions on the destruction of religious and cultural property or records of official investigations following any such incidents coming from the Bosnian Serb and Bosnian Croat political and military authorities that were produced as defence evidence at the ICTY in relation to the considerable numbers of such documents emanating from Bosnian governmental authorities and the Bosnian Army is noticeable.

It is clear that neither the Bosnian government nor the Bosnian Army had any policies of encouraging attacks on the sacral structures of other faiths, let alone systematically destroying them. On the contrary, there is ample evidence not only of official policies and orders proscribing such attacks, but of actively providing protection for Christian religious structures and investigating any attacks that did take place.[147] This protection of Serbian

146 *Letter signed by Radovan Karadžić, dated 12 May 1993*, ICTY Exh. D159a; see also *Order from General Momir Talić, Commander 1st Krajina Corps*, 3 August 1992, ICTY Exh. D29-1085A. With regard to the HVO, see *Order from Žarko Šarić, Commander of the 2nd Battalion*, Bila, 21 June 1993, ICTY Exh. D375b.

147 See, for instance, *Prosecutor v. Enver Hadžihasanović and Amir Kubura*, ICTY Case No. (IT-01-47): Testimony of Fikret Ćuskić, former commander 17th Krajina Brigade and Operational Group Bosanska Krajina, based in Travnik, on protection of the religious and cultural heritage by the Bosnian Army and authorities, 22 November 2004, 12080–84; Exh. 00226.E, *Letter to All Chiefs,*

Fig. 1.23 **(a)** The Roman Catholic church of the Transfiguration of Our Lord (1924), at Borovica near Kraljeva Sutjeska, before its destruction. **(b)** The shell of the church after it was burned and mined by Bosnian Army forces when they overran Borovica in November 1993. Both © Biskupska Konferencija Bosne i Hercegovine / ICTY

Security Service Centres, To All Commanders, BH Army Corps, dated 17 June 1993, signed by Bakir Alispahic, 17 June 1993, signed by Interior Minister Bakir Alispahić and ARBiH Commander Rasim Delić after the ARBiH offensive of June 1993 to (among other provisions) 'protect citizens and their property, in particular clergy and places of worship of all religions,' to assist refugee clergy to return to their parishes and to 'Ensure their complete personal safety and the security of places of worship'; Exh. DH186, *ECMM Report dated 19 June 1993*, 4, on churches during the ARBiH offensive of June 1993 near Travnik; Exhs 1135, DH64/E, DH66/E, P811/E. See also the case of the Franciscan monastery at Guča Gora discussed below.

Orthodox and Roman Catholic churches on the part of the Bosnian government authorities was noted by COE consultant and UNESCO representative Colin Kaiser who found it striking.[148] Kaiser noted as well the generally respectful attitude of Bosnia's Muslims to the sacral structures of other faiths.[149]

Thus, Christian sacral structures (both Catholic and Orthodox), by and large, survived well in areas under Bosnian government control, particularly in urban milieux. Apart from damage caused by Bosnian Serb shelling, in central Sarajevo churches and monasteries remained intact, as did churches in Muslim majority towns like Bihać, Zenica, Travnik, Visoko, Tešanj, Gračanica, and Tuzla, where the Social Democratic municipal government tried hard to maintain the city's multiethnic character throughout the conflict. In Sanski Most, Ključ and Bosanska Krupa in north-west Bosnia which had been occupied by the Bosnian Serbs for most of the war, during which time their mosques and Catholic churches had been deliberately destroyed, the local Serbian Orthodox churches remained intact after Bosnian Army troops retook the towns during the final months of the conflict in 1995. This was certainly not the case in territory held by the Bosnian Serbs (which comprised 70 per cent of the territory of Bosnia-Herzegovina) where, according to Riedlmayer, virtually 100 per cent of mosques and 75 per cent of Catholic churches were destroyed or badly damaged.

Museums, Libraries, Archives and Movable Heritage

Attacks on cultural institutions were, of course, not simply attacks on buildings, but also on what the buildings contained. Ivan Lovrenović called the phenomenon 'the hatred of memory' when he described how books, documents and paintings became the target.[150] Museums were sometimes deliberately attacked. A much cited story (based on a news report from September 1992 by BBC journalist Kate Adie) alleges the intentional targeting of the Zemaljski Muzej in Sarajevo by Bosnian Serb artillery.[151] The country's foremost

148 *Prosecutor v. Tihomir Blaškić*, Case No. IT-1995-14-T, Colin Kaiser testimony, 16 July 1998, 10695–700; in his testimony Kaiser described one occasion on a survey mission in central Bosnia when he went to look at the church in Dolac in June 1994: ' … it was somewhat vandalised, not dramatically damaged, and I remember always a soldier, a Bosniak soldier from the Krajine, and he was a refugee, carrying a load of wood. He walked past us and he began sort of shouting at me. He said, "Why aren't you going down to the Lasva valley and looking at the dynamited mosques? Why are you looking always at the churches here?" And then he pointed at the church, and he said, "This is the house of God. We will not destroy it."' See also Kaiser's comments in COE *Sixth Information Report*, Doc. 7133, 31 August 1994, 15–16.

149 *Blaškić* Case No. (IT-95-14), Colin Kaiser testimony, 16 July 1998, 10695–700. See also, for instance, testimonies *Hadžihasanović and Kubura* Case No. (IT-01-47): Munir Karić, 9 November 2004, 11458 and Džemal Merdan, 9 December 2004, 13141.

150 Lovrenović, I. 1994. The Hatred of Memory: In Sarajevo, Burned Books and Murdered Pictures, *The New York Times*, 28 May.

151 This story has unfortunately never been definitively referenced but much repeated. The account has Adie asking a Bosnian Serb artillery commander on the heights above Sarajevo why the Holiday Inn where she and other foreign correspondents were living was continually being shelled. The officer apparently apologized, explaining that they had been aiming at the roof of the national museum opposite the hotel and had missed. Based on a brief news report in *The Art Newspaper* (Bosnia's written history in flames? 1 October 1992), the citings of this story are too numerous to list and appear to have gained some embroidery in the retelling. However, the story is also referred to in Wenzel, M. 1993b. Eyewitness in Sarajevo, our brother and sister curators and art historians: How do ordinary people live in extraordinary and terrible times?, *The Art Newspaper*, No 32, London, November 1993.

Fig. 1.24 **(a)** The Franciscan monastery and classical gymnasium (secondary school) of St Bonaventure and **(b)** the Serbian Orthodox church of St Procopius, both in Visoko, 2013. Both © Richard Carlton

heritage institution, the Zemaljski Muzej (the National Museum of Bosnia-Herzegovina)[152] with its internationally important collections housed in a quadrangle of four Austro-Hungarian period pavilions, stands alongside the modernist Historical Museum on the boulevard notoriously known as 'Sniper Alley' during the conflict. Positioned between the Tito Barracks in one direction and the Parliament Building in the other, the museum was at the epicentre of pitched battles fought at the beginning of the conflict which left its walls punctured by tank shells.

152 The museum is sometimes referred to in sources as the *Land Museum*, the *Territorial Museum* and *Earth Museum* (translated versions of the original Austro-Hungarian designation *Landesmuseum* or its Serbo-Croat variant *Zemaljski Muzej*)

In these dangerous circumstances it was impossible to remove collections or the museum's important library from the buildings. Apart from a few smaller treasures like the museum's most famous artefact, the fourteenth-century Jewish manuscript known as the Sarajevo Haggadah, which were rescued at a critical point at the beginning of the siege of the city and taken to the vaults of the National Bank,[153] collections were stored on site; smaller items were removed to basement depots, while larger exhibits were simply moved to safer inner rooms. Fragile entomological collections in their original display cases were left in place, despite their vulnerable position opposite Bosnian Serb lines – the far greater risk to specimens was moving them at all.

Over the course of the siege of Sarajevo the Zemaljski Muzej was struck by hundreds of shells and small arms fire from Bosnian Serb Army lines directly behind it across the Miljacka River, shattering the glass roofs of the museum's main exhibition halls where the rains and snows of three years caused substantial damage to the fabric of the building. Between April 1992 and March 1993 alone more than 300 mortar shells fell on and around the museum complex. Around 40 of these and five tank shells directly struck the museum buildings.[154] There was another blow in December 1993 when the museum's director, Dr Rizo Sijarić, was killed by a shell blast while walking across a Sarajevo park to a meeting with the head of the state heritage protection institute who had been helping him to obtain plastic sheeting from humanitarian aid agencies to cover the many holes in the museum's walls and windows.[155] Dr Sijarić, who had become director in October 1992, had been working tirelessly to protect the museum and its collections and to raise awareness of its plight internationally.[156] But the museum's dedicated staff (of all ethnicities), depleted from a pre-war total of 95 to 31, continued to work as best they could.[157]

Adjacent to the Zemaljski Muzej was the Historical Museum (the pre-war Museum of the Revolution), its entrance in full view of Bosnian Serb snipers based in multi-story buildings across the Miljacka River, was in an even more perilous position, receiving only one visitor during the entire three and a half years of the war. Yet staff turned up for work daily, even carrying out major repairs to the damaged building wearing the overalls worn by Serb prisoners of war in an attempt to mislead snipers. Historical Museum staff continued to collect actively – now material and artefacts relating to life of Sarajevo's citizens during the siege.[158]

---

153 See more on the Sarajevo Haggadah during and after the war below.

154 Sijarić, R. 1996. Zemaljski Muzej Bosne i Hercegovine u ratu 1992/93 godine. *Glasnik Zemaljskog muzeja Bosne i Hercegovine*, new series, 47, 7–39.

155 Obituary: Dr. Rizo Sijaric, Director of the Zemaljski Muzej, Sarajevo. Killed in Sarajevo, 10 December 1993. *Museum Management and Curatorship*, 13:1, 79–80. Obratil, S. 1996. In memoriam: Dr Rizo Sijarić. *Glasnik Zemaljskog muzeja Bosne i Hercegovine*, new series, 47, 441–7; the article states the date of Dr Sijarić's death was 6 December.

156 See Sijarić, R. 1993. Update on the Zemaljski Muzej, Sarajevo. *Museum Management and Curatorship,* 12(2), June, 195–9.

157 Sijarić, R. 1996. Zemaljski Muzej Bosne i Hercegovine u ratu 1992/93 godine. *Glasnik Zemaljskog muzeja Bosne i Hercegovine*, 30. See also Fahy, M. and Mogul, J. 1995. An interview with Lidija Fekeza: An archaeologist in Sarajevo: Culture under siege, *The Journal of the International Institute* [online], 3:1, Fall 1995.

158 The material collected formed the basis of a highly-regarded permanent exhibition 'Sarajevo under Siege' after the end of the war. An online version titled 'Sarajevo Survival Tools' created in association with the Faculty of Electrical Engineering at Sarajevo University can be found at http://h.etf.unsa.ba/srp/index.htm.

Fig. 1.25 **(a)** The Botanic Garden of the Zemaljski Muzej, Sarajevo, February 1995, facing the Ethnology pavilion, the blown out glass in its windows replaced with plastic sheeting. The medieval tombstones (stećci) displayed in the garden have been protected with sandbags and tarpaulins. The damaged Parliament building rises in the background. © Leif Pareli. **(b)** Interior of the Natural History pavilion of the Zemaljski Muzej, February 1995. Directly facing the confrontation line, the building's glass roof was shattered and unrepairable due to the danger from sniper fire. Staff rigged UNHCR plastic sheeting into huge chutes that drained rain and snow into water tanks. Some of the original Austro-Hungarian period display cases protected with tarpaulins can be seen centre. © Leif Pareli

However, it was the Olympic Museum (by 1992 an avant-garde art centre exhibiting artists like Robert Mapplethorpe) which had the dubious honour of being the first cultural institution in Sarajevo to be intentionally targeted by JNA/VRS forces. The museum's turreted Austro-Hungarian villa was struck by five direct shells at the end of April 1992 and completely burned out while adjoining houses and apartment blocks were left untouched. The museum's director Edin Numenkadić has related the bizarre tale of how as the museum burned, JNA officers – still living nearby in the early days of the conflict – stood firing their revolvers at the burning museum such was their loathing for all it represented.[159] The worst damage to collections was at the Museum of Herzegovina in Mostar. The Regional Museum at Serb-controlled Bileća (not in a conflict zone) was completely burned out in 1993 and its collections totally destroyed – the only such case in Bosnia-Herzegovina.

However, though many museum buildings were badly damaged, particularly major institutions in Sarajevo and Mostar, across the country few were totally destroyed. Overall, museum collections survived the war comparatively well, with relatively rare episodes of looting or destruction or damage to collections.[160] Far more, like the Museum of the City of Sarajevo and the Museum of East Bosnia in Tuzla, suffered from being dispossessed of their premises in the transition from social ownership of property and from collections being stored in inevitably poor environmental conditions. The loss of qualified professional staff due to death and ill-health, displacement, or departure also contributed to museums' problems.

Most regional and local museums held mixed collections of archaeological, historical and ethnographic material. While many were located in areas where the worst ethnic cleansing had taken place, such as Banja Luka, Prijedor, Bijeljina, Zvornik, Foča and Trebinje, museum collections appear not to have been similarly 'cleansed,' even if certain types of artefacts were removed from display. Indeed, curators in Republika Srpska questioned by the author in 1996 for a Council of Europe report made a point of asserting the integrity of their pre-war collections, one declaring 'we have preserved everything – Serbian, Muslim, Croatian – they are a part of our cultural past and we mustn't throw anything away' and 'we keep the cultural heritage of all here; as professional people we are obligated to look after the museum's collections,' it was just that their priorities had changed with regard to what they exhibited.[161] Thus, museum collections hold perhaps the best remaining material evidence of Bosnia's multiethnic/multiconfessional past.

Bosnian heritage professionals have pointed out that cultural institutions across the country had war-preparedness plans in place before the beginning of the conflict. All museums in Sarajevo displayed UNESCO Blue Shield flags which had been sent by the Museum Documentation Center in Zagreb – in theory to confer protection under the 1954

159 Walasek, H. and Wenzel, M. *Report on Museums in Bosnia-Herzegovina during the Present War*, in *Ninth Information Report on War Damage to the Cultural Heritage in Croatia and Bosnia-Herzegovina*, Parliamentary Assembly of the Council of Europe (COE), Strasbourg, Doc. 7464, 19 January 1996, *Museum of the XIV Winter Olympics*. Fortunately, due to Numenkadić's foresight the collections were almost entirely preserved.

160 Walasek, H. and Wenzel, M. COE *Ninth Information Report*, Doc. 7464, 19 January 1996, and Walasek, H. 1997, *Museums in Bosnia-Herzegovina II, A Report on Museums in Republika Srpska with an Update on Museums in Sarajevo*, in *Tenth Information Report on War Damage to the Cultural Heritage in Croatia and Bosnia-Herzegovina*, Parliamentary Assembly of the Council of Europe (COE), Strasbourg, Doc. 7740, 24 January 1997.

161 Walasek, H. COE *Tenth Information Report*, Doc. 7740, 24 January 1997. 4.1 *Damage to and Loss of Collections/Documentation.*

Hague Convention for the Protection of Cultural Property in the Event of Armed Conflict when so marked.[162] But the volatile political situation meant they were discouraged from being seen to be openly preparing for war and received no assistance from the government authorities responsible for culture.[163]

The protection of the relatively small collections of the Olympic Museum were an exception: under the ruse that the museum needed repainting, the museum's director obtained the containers he needed to pack exhibits safely away, thus surviving the direct attacks on the museum building. But the state Institute for the Protection of the Cultural, Historical and Natural Heritage of Bosnia-Herzegovina, after receiving no reply to an official request made in October 1991 for assistance in protecting their documentation, gave up and diverted funds meant for an exhibition to purchase 70 storage containers, preserving this important resource for the country's heritage.[164] At the Zemaljski Muzej, despite never receiving the packing materials and storage containers requested from the Ministry of Culture, basement depots were prepared and some collections moved there. Senior staff have spoken of their dilemma about where to store the precious Sarajevo Haggadah – the obvious place was the vaults of the National Bank, but with the outcome of fighting raging across the centre of Sarajevo between government loyalists and secessionist Serbs uncertain, who could say who would hold the National Bank?[165]

Following the outbreak of war in April 1992 the protection of the cultural heritage in Sarajevo was at first very much the initiative of groups of individuals. Mevlida Serdarević has written on the confusion, disorganization and paralysis that affected many levels of government and institutions in Sarajevo at the time as official structures and hierarchies broke down, not only because of the war situation with shelling, fighting in the city centre, blockades and breakdown of communications, but also because of the departure of experts and officials, department heads, directors and senior staff of institutions, and even government ministers, as they fled the country, were mobilized, or – not infrequently – joined the side of those attacking the city. It was often impossible to gain official authorization to take action (whether due to inefficiency or for more malicious reasons), and as a newly independent state, there was a lack of knowledge, for instance, of how to seek application of international conventions concerning the protection of cultural property such as the 1954 Hague Convention.[166] There were problems simply finding secure accommodation to store the city's important movable heritage, with many potential depots in now inaccessible locations, taken over by refugees or damaged – let alone dealing with the difficulties in finding vehicles (and fuel) to transport such material during open conflict. And overriding all other considerations for officials was the desperate battle for Sarajevo.

Thus, once war broke out it was Sarajevo's heritage experts and other concerned citizens who moved to preserve the city's heritage.[167] On 14 May 1992 the *Commission for the*

---

162 Walasek, H. and Wenzel, M. COE *Ninth Information Report*, Doc. 7464, 19 January 1996, 3. *Museums in War*.

163 See Serdarević, M. 1997. *Pravna zaštita kulturno-historijskog naslijeđa BiH sa posebnim osvrtom na staro urbano jezgro Sarajeva: (nastajanje, očuvanje, destrukcija)*, Sarajevo: Međunarodni centar za mir; Begić, A. 1995 and Walasek, H. 1996.

164 Serdarević, M. 1997, 115.

165 Walasek, H. and Wenzel, M. COE *Ninth Information Report*, Doc. 7464, 19 January 1996, 3. *Museums in War*.

166 Serdarević, M. 1997, 118–20.

167 Among the many active in the rescue of Sarajevo's heritage in the early stages of the war key figures were actor Josip Pejaković, art gallery curator Azra Begić, archaeology professor Enver

Protection of the Heritage of Sarajevo (*Komisija za zaštitu baštine grada Sarajeva*), which later received official accreditation at central state level, was formed. The Commission had a sub-committee for rescuing movable heritage headed by art historian and senior curator Azra Begić which brought together representatives of state and municipal cultural institutions.[168] Teams of volunteers rescued and put into safe storage vast amounts of important cultural material, including the Sarajevo Haggadah and other significant artefacts from the Zemaljski Muzej, the most precious manuscripts from the Gazi Husrev-beg Library, the most valuable works and periodicals from the National Library shortly before it was attacked, Jagomir Studio's notable film archive, and the documentation of the Institute for the Protection of the Cultural, Historical and Natural Heritage of Bosnia-Herzegovina.[169] The period between 14 May and 14 September 1992 was the most active phase of the Commission's work.

A state-level *Office for the Protection of Cultural Property* (Štab za zaštitu dobara culture) headed by the well-known actor Josip Pejaković was later established by the Presidency of Bosnia-Herzegovina which worked with Bosnian Army units to rescue and store movable heritage.[170] At one critical point an UNPROFOR[171] officer arranged a temporary ceasefire with Bosnian Serb forces so that the heritage teams could complete their work.[172] As the siege 'stabilized' the activities of these *ad hoc* organizations tapered off and official bodies resumed a more active role, turning to address the question of protecting Sarajevo's damaged built heritage.[173] By the end of 1993 the *Office for the Protection of Cultural Property* was reported to be looking after ' ... around 7,000 paintings and icons, 1.2 million books and manuscripts, and sundry other items like old coins and pieces of sculpture, spread throughout nine depositories across the city, graded according to safety, and guarded by 70 policemen.'[174] But as the war went on, the question of protecting these movable objects in their usually damp and otherwise unsuitable depots became a problem.[175]

One particularly badly affected regional museum was the Museum of Herzegovina. Located in Bosnian government-held East Mostar, the museum lost all of its buildings due to a combination of war damage and the return of its premises, the Ćejvan Ćehaja Mosque and its adjoining Austro-Hungarian period Mekteb (1899), to the Islamic Community.

Imamović (later a wartime director of the Zemaljski Muzej), lawyer Mevlida Serdarević, conservator Nihad Čengić, art historian Snežana Mutapčić and librarian Edina Vlašić.

168 Serdarević, M. 1993, Postscript, *Museum Management and Curatorship*, 12:2, June 1993, and Serdarević, M. 1997, 124.

169 *Third Information Report on War Damage to the Cultural Heritage in Croatia and Bosnia-Herzegovina*, Parliamentary Assembly of the Council of Europe (COE), *Strasbourg*, Doc. 6904, *20 September 1993, 5 and* 19–20 and Begić, 1995. By 1993 the Commission had evolved into *AIASN Asocijacija za interkulturne aktivnosti i spašavanje nasledje Republike BiH* (*The Association for Intercultural Activities and Rescue of the Heritage of the Republic of Bosnia and Herzegovina*).

170 Begic, A. 1995; Serdarević, M. 1997, 128–30; Karup-Druško, D. 2000. Ram za sliku opljačkog Sarajeva. *Dani*, 23 June, 52. The work of the *Office for the Protection of Cultural Property* is described in: Schiller, B. 1993. Bosnian artists save heritage treasures, *The Toronto Star*, A10, 15 May 1993 and Traynor, I. 1993b. Culture warriors, *The Guardian*, 25 November, 10; unfortunately Traynor describes the *Office for the Protection of Cultural Property* as a 'vigilante' group, though it is clear from the article that it was an official body.

171 UNPROFOR = United Nations Protection Force, the multinational peace-keeping force.

172 Personal interview by author with Josip Pejaković, November 1997.

173 Serdarević, M. 1997. 115–35.

174 Traynor, I. 1993b. Culture warriors.

175 See Bollag, B. 1995. 'We must save what is left', *The Independent*, 20 April.

Fig. 1.26 Mostar. The front of the former Museum of Herzegovina, the Austro-Hungarian period Ćejvan Ćehaja Medresa, its shell damage still apparent in October 1995. © Marian Wenzel / Helen Walasek

Most of its collections were destroyed, with some particularly important objects lost to theft (see below). The museum's archaeological collections were buried under a collapsed building, which itself later required excavation. Like institutions across the country it suffered from the loss of staff, too. By the end of 1995 the museum was entirely run by personnel with no professional experience. Meanwhile, across the Neretva River in West Mostar four highly-qualified experts, three of them pre-war employees of the Museum of Herzegovina (one the former archaeology curator), sat in the empty Museum of Herceg-Bosna, a newly-created museum with (virtually) no collections established by the Bosnian Croat authorities. The sole artefacts it held were items of jewellery stolen by looters from the ethnographic collection of the Museum of Herzegovina during the war and later recovered on the west side (see below).

The devastation of Bosnia's most important libraries and archives and their collections has been extensively described by Riedlmayer.[176] Some major collections are noted elsewhere in this article. It has been estimated that 481,100 linear metres of archival material were destroyed in attacks on historical archives and local registry offices during the war.[177] Riedlmayer has also noted the fate of scores of smaller collections of manuscripts, rare books and archival material, particularly those held by religious communities. Many

176 See Riedlmayer, A. 2002a, 2.3, Damage to libraries and archives, 17.

177 Kožar, A. 1999. *War Destruction of the Archival Materials in Bosnia & Herzegovina*, unpublished paper cited in Riedlmayer, A. 2001. *Convivencia* under Fire: Genocide and Book-burning

were lost when mosques and churches were blown up or burned, some were looted, while others were simply removed from buildings and intentionally destroyed, like the 3,200 Islamic books and manuscripts belonging to the Atik Mosque in Janja, which included the collections of three noted Muslim scholars.[178]

Crimes in Stolac Municipality 1992–1996 documents the collections of manuscripts in Bosnian, Arabic, Turkish and Persian dating from the seventeenth to nineteenth centuries, rare books and archival material lost forever when the Podgradska and Čaršija Mosques and the offices of the Islamic Community in Stolac were destroyed.[179] The Stolac report uniquely documents the catastrophic loss of the movable heritage of one small but culturally significant town during the conflict. Stolac was home to the family seats of prominent Muslim grandees of the Ottoman period, most famously the Begovina of the Rizvanbegović family on the banks of the Bregava River, one of the best-known (and important) historic monuments in Bosnia-Herzegovina. Apart from their libraries and art collections, the historic domestic interiors of these great houses with their intricate wooden panelling were treasure houses of furniture, carpets, embroideries and metalwork of Bosnian manufacture that were looted or completely destroyed.[180]

Movement of Cultural Property

Looting

In common with many other conflicts the looting of movable cultural property took place during the Bosnian War. Even less than with regard to the built heritage has there been any serious attempt to assess losses to Bosnia-Herzegovina's movable heritage during the war, particularly in respect of looting, although the Commission to Preserve National Monuments has begun to research this issue. Some looting was opportunistic, like the theft of ten rifles from the Historical Museum in Sarajevo in the early days of the war – some of whose World War II era weaponry was officially commandeered for the defence of the city.[181] Other looting was more systematic. Nevertheless, movements of cultural goods that on occasion have been termed looting by some and variously characterized as 'rescue,' rightful transfer, or legitimate purchase by others, shows definitions can be far from clear cut.

Of the most straightforward type of opportunistic looting from public collections, perhaps the most serious case was the theft of two small paintings by Ferdinand Hodler, around 40 Italo-Cretan icons, and two Islamic manuscripts from the *Umjetnička Galerija* (National Art Gallery of Bosnia-Herzegovina) in Sarajevo after the normally locked doors of its storage depot were left open; the whereabouts of the stolen works continue to be

in Bosnia, in *The Holocaust and the Book: Destruction and Preservation*, edited by J. Rose. Studies in Print Culture and the History of the Book. Amherst, MA: University of Massachusetts Press, 266–91.

178 Riedlmayer, A. 2007. Crimes of war, crimes of peace: Destruction of libraries during and after the Balkan Wars of the 1990s, *Library Trends*, 56:1(Summer), 116.

179 Mulać, E. 2001. *Crimes in Stolac Municipality 1992–1996*, The Revival of Stolac. Sarajevo: Association [Society] for Renewal of Civil Trust in the Stolac Municipality, reprint with addenda, 66. The Čaršija Mosque is sometimes referred to as the Careva or Čaršijska Mosque.

180 See Amra Hadžimuhamedović elsewhere in this volume for more on the destruction at Stolac.

181 Personal knowledge. See also Bensman, T. 1993. Mechanic's skills useful in Bosnia war, *St Petersburg Times*, 10 January 1993.

unknown. A group of Early Christian reliquaries was stolen from a safe in the Museum of Herzegovina in Mostar, as well as (according to varying accounts) all the weapons, numismatics and jewellery from the ethnographic collection, although some jewellery was later recovered. Another well-documented case was the removal of a collection of Roman jewellery from the Museum of East Bosnia in Tuzla by a curator of Serb nationality the day before the city was attacked by Bosnian Serb forces on 15 May 1992.[182]

There were other public institutions holding large collections which were looted and the fate of which remain unknown.[183] The entire art collection of the Branka Šotra Gallery in Stolac was stolen before its home, the eighteenth-century Šarić House, was destroyed. Other art collections were stolen from museums in Drvar and Jablanica and from the art colony at Počitelj. The collections of the AVNOJ Museum in Jajce and the ZAVNOBiH Museum in Mirkonjić Grad were looted; both museums commemorated crucial meetings of the councils of the anti-Fascist resistance (the 'Partisans' led by Tito) in 1943 which led to the formation of the post-World War II federal socialist state of Yugoslavia and the Republic of Bosnia-Herzegovina respectively.[184] The AVNOJ Museum, a popular tourist destination in the former Yugoslavia, lost 80 per cent of its holdings including historic documents, photographs and other archival material, a large library and many important works of art.[185]

Curator Azra Begić, one of the key actors in rescuing much of Sarajevo's important movable heritage, has written on the destruction and damage to and looting from the many important private collections and artists' studios in Sarajevo.[186] Like buildings, objects could become the focus of ethno-nationalist triumphalism and hatred of the other. Ivan Lovrenović, a prominent Sarajevo intellectual of Bosnian Croat background, has described how Serb gunmen forced residents of the Sarajevo district of Grbavica to watch the burning of 'the Ustasha library of Ivan Lovrenović,' a reference to the World War II Croatian fascists.[187]

Other treatments of movable cultural property were more pragmatic. There have been media reports on the looting of works of art (both well-organized and random) from Bosnia-Herzegovina and Croatia by JNA/Bosnian Serb Army and Serbian paramilitary units and

182 Walasek, H. and Wenzel, M. COE *Ninth Information Report*, Doc. 7464, 19 January 1996, 4.1 *Damage to and Loss of Collections/Documentation*. The removal and transfer to Serbia or Serb-held territory of a range of material items running from works of art to industrial machinery was a phenomenon of the wars in Croatia and Bosnia-Herzegovina.

183 Karup-Druško, D. 2000, 50–53.

184 AVNOJ = *Antifašističko Vijeće Narodnog Oslobođenja Jugoslavije* (Anti-Fascist Council of the People's Liberation of Yugoslavia). ZAVNOBiH = *Zemaljsko antifašističko vijeće narodnog oslobođenja Bosne i Hercegovine* (National Anti-Fascist Council of the People's Liberation of Bosnia and Herzegovina). Information from the Commission to Preserve National Monuments of Bosnia-Herzegovina to the author, 12 July 2011. Thanks to Slobodanka Nikolić for providing this information. See also the website of the Commission http://www.kons.gov.ba. The whereabouts of only four paintings from the AVNOJ Museum are known.

185 See *AVNOJ Centre in Jajce, Complete with its Movable Property*, website of the Commission to Preserve National Monuments, http://kons.gov.ba/main.php?id_struct=50&lang=4&action=view&id=1306; Otvoren obnovljeni muzej 2. zasjedanja AVNOJ-a, *http://www.klix.ba/*, 30 November 2008.

186 Begić, A. 1995.

187 Lovrenović, I. 1994.

the transformation of Belgrade during 1991 and 1992 into a mecca for selling stolen art, often in newly-opened 'antique' shops established by members of the security services.[188]

To their distaste, professional curators and gallery owners were frequently called on to value these all-too-obviously looted works.[189] In Sarajevo the studios of non-Serb artists in Serb-held parts of the city were looted, their paintings later appearing for sale on the black market in Belgrade. Thus,167 works by the painter Afan Ramić were stolen from his studio in Grbavica, as was 20 years of work by Ibrahim Ljubović.[190] Jovan Despotović, former curator of the Museum of Modern Art in Belgrade, recalled how

> About a hundred works of a painter, Afan, from Sarajevo circled the city. They were sold at 50 dinars apiece. Even works of Ibrahim Ljubović appeared. I know that a student, a refugee from Sarajevo, offered her professor a 'deal', and he offered it to a friend of his. The offer included works by Mersad Berber, Bukovac and Mestrovic.[191]

Despotović went on to say how all the political parties in Serbia which operated paramilitary units in the war zone had their own depots of looted works of art. Artworks and rare books looted from the library of the Franciscan Theological Seminary in the Sarajevo suburb of Nedžarići which had been occupied by Serb troops and paramilitaries on 8 June 1992 were later offered for sale to the head of the Franciscan Order in Belgrade by a local art dealer.[192]

The new occupiers had ejected the monks and stripped the seminary of its fixtures and valuables, including the library's large and rich collection of 50,000 volumes which disappeared entirely. After the end of the war and their return to the seminary the Franciscans continued to search for the missing library and artefacts. Finally, in 1999 political contacts located books from the seminary in the Central Library (Matične biblioteke) in Lukavica, East Sarajevo, a suburb of the city now part of Republika Srpska. After negotiations that began with an official meeting in Lukavica on 18 October 1999 between representatives of the Franciscans and the Office of the President of the Federation of Bosnia-Herzegovina and the director of the library in Lukavica and the chief of staff of the Vice-President of Republika Srpska, the first consignment of around 5,000 books was returned to the seminary at Nedžarići on 15 December 1999, with the final number amounting to around 25,000–30,000 volumes.[193] The looted rare books and art works offered for sale in Belgrade have never been traced.

Artist Sead Hasanefendić discovered that his paintings had been removed from his studio by a uniformed unit and taken across the confrontation line to the Serb-held side of the city; so well-organized were the looters that when the works later emerged on Sarajevo's black art market, damage suffered by some paintings during removal had been repaired. When Hasanefendić attempted to expose those behind the theft, he was warned against investigating too deeply.[194]

---

188 Igrić, G. 1995. Rendering of accounts, *AIM Press* [online], Alternativna Informativna Mreža (Alternative Information Network) 9, 8 February 1995.

189 Igrić, G. 1995.

190 Igrić, G. 1995.

191 Igrić, G. 1995. Mersad Berber (1940–2012) was probably Bosnia's best-known living artist. Vlaho (Blaise) Bukovac (1855–1922) and Ivan Meštrović (1883–1962).

192 Riedlmayer, A. 2007.

193 Sudbina knjižnice u zadnjem ratu, www.franjevackateologija.ba, November 2009.

194 Karup-Druško, D. 2000.

Only the quick action of the *Office for the Protection of Cultural Property* in 1992 prevented the looting of the icon collection of the Orthodox church of the Transfiguration in the Sarajevo district of Polafići, then on the frontlines. Alerted to unusual activities at the abandoned church which suggested that the icons were being packed, almost certainly with the intention of stealing them, the Office sent a unit to bring them to the National Theatre, then being used as a secure depot for storing cultural goods.[195]

By late 1993, for staff from the *Office for the Protection of Cultural Property* a ' … typical morning's work these days is for them to show up early at a local police station to inspect, identify and store paintings that the police have brought in from the black markets … local criminals rather than shells are now the problem.'[196] Others have spoken of the convoy leaving Sarajevo searched by the authorities and found to be full of empty frames – the pictures (probably stolen from private collections) having been already removed for sale on the black market, probably as far afield as Zagreb, Belgrade and Budapest.[197] Reports also allege the systematic looting of works of art from Serb houses in Mostar by Bosnian Croat paramilitaries accompanied by an art expert.[198]

By all accounts the arrival of UN troops and staff working for international agencies in the country fuelled the black art market, finding ready purchasers with hard currency who were keen to buy, and there are (unsubstantiated) rumours of artworks, valuable carpets, Bosnian metalwork and other artefacts leaving the country in UNPROFOR convoys and diplomatic bags. Linked to this was what could also be considered a form of 'looting' – the questionable practice of the legitimate purchase (but usually illegal export) of works of art and antiquities from desperate owners at severely undervalued prices by UN peacekeepers, diplomats and staff of international organizations and NGOs. Works by some of Bosnia's most famous artists like Gabriel Jurkić (1886–1974) were acquired by avid collectors from the international community and removed from the country by diplomatic bag, while the antique shops and curio stalls of Sarajevo and Mostar were packed with museum-quality examples of Bosnia's material culture selling at knock-down prices during and in the years after the war.[199]

Serbian Orthodox Church Property

It was noted there was one class of cultural property which, by and large, *was* removed to safety in advance of the war, and this was that belonging to the Serbian Orthodox church. The cultural irredentism which occurred in respect of Serbian Orthodox and other Serb heritage in advance of the conflict in both Bosnia-Herzegovina and Croatia has been a source of debate in both countries. In the case of Croatia the existence of the *Commission for the Protection of Cultural Goods of Serbian Origin on the Territory of War Operations* and the removal of scores of Serbian Orthodox icons and other works of art from churches, museums and other collections in Dalmatia and Slavonia in 1991 by the Serbian Ministry

195 Kovačević, Ž. 2010. Josip Pejaković zaštitio dio blaga BiH, *Radio Slobodna Bosna* [online], 13 December.

196 Traynor, I. 1993b. Culture warriors.

197 Karup-Druško, D. 2000.

198 Fesperman, D. 1995. 'Culturecide' taking heavy toll on art in Balkans war zone, *Baltimore Sun*, 13 April.

199 Karup-Druško, D. 2000.

of Culture on the grounds that they needed restoration is well-documented.[200] That a similar removal from Bosnia to Serbia or to Serb-held territory within Bosnia took place has so far been less easy to verify.[201]

However, *Migrating Icons*, an article by Svetlana Rakić, an art historian closely involved with icon collections in Bosnia-Herzegovina in the years before the war, sheds some light on the issue.[202] She relates how at the end of 1991 the conflict raging in Croatia prompted the synod of the Serbian Orthodox church to suggest that all bishoprics in Bosnia-Herzegovina should evacuate their church treasuries to 'safer' locations – suggesting that the church authorities were well aware of which areas of Bosnia-Herzegovina were soon to become 'Serb' territory.[203] Thus, the collections of the recently-opened (1990) Episcopal Museum in Tuzla were transferred by the Orthodox Bishop of Zvornik-Tuzla, Vasilije Kačavenda[204] in 1992 to Bijeljina where he had relocated his episcopal seat and where they remain. The seat of the Orthodox Diocese of Zahum-Herzegovina was likewise transferred from Mostar to Bosnian Serb-held Trebinje and icons and other movable heritage from Žitomislić Monastery, Mostar's Orthodox Cathedral and Old Orthodox church (all subsequently totally or partly destroyed) were moved to that town.

Richest of all was the collection of 642 icons from the Museum of the Old Orthodox Church in Sarajevo. Rakić is vague on the whereabouts of this collection during the war, implying that part was sent to Belgrade for restoration only after the war was over. However the Institute for the Protection of the Cultural, Historical and Natural Heritage of Bosnia-Herzegovina was clear that removal had taken place prior to the outbreak of war (though the icons in the church itself stayed in Sarajevo).[205] Whatever the truth of the matter, though some icons were restored after the war under the auspices of the United Nations Mission in Bosnia-Herzegovina (UNMIBH) and are displayed in the reopened museum in Sarajevo, the most important are now housed in the museum of post-war establishment (2004) of the Archbishopric of Dabar-Bosnia at Dobrun monastery near Višegrad in rural Republika Srpska. It appears unlikely that the icons will ever return to the Bosnian capital – in 2008 museum staff at Dobrun told visitors that the exhibits testified to the presence of the Serbian

---

200 *Commission for the Protection of Cultural Goods of Serbian Origin on the Territory of War Operations* = *Komisija za zaštitu kulturnih dobara srpskog porekla na teritoriji ratnih dijstava.* An exhibition of a selection of icons from Dalmatia was exhibited in Belgrade in 1992: *Ikonopis dalmatinske krajine sa ratom ugroženih Ravnih Kotara* = Icon painting of the Dalmatian Krajina from the regions imperrialled [sic] by the war [autori izlo.be Nikola Kusovac, Milorad Savić] Beograd: Narodni muzej--Beograd, 1992. See also the report of Hans-Christoph von Imhoff in *Seventh Information Report on War Damage to the Cultural Heritage in Croatia and Bosnia-Herzegovina*, Parliamentary Assembly of the Council of Europe (COE), Doc. 7308, Strasbourg, 15 May 1995 in which he describes the vast quantities of Serbian Orthodox art from Croatia that he sees on a mission to Belgrade.

201 Begić, A. 1995.

202 Rakić, S. 2006. 'Migrating icons': Politics and Serbian cultural heritage in Bosnia and Herzegovina before and after 1992, *Mediterranean Quarterly*, 17:3 (Summer), 43–54.

203 Rakić, S. 2006, 44.

204 Kačavenda was notorious for his hardline Serb nationalist views and actions. His new episcopal residence was constructed on the site of expropriated and demolished houses belonging to ethnically-cleansed Muslims. See Helen Walasek, *Domains of Restoration* in this volume for more on Kačavenda.

205 See COE *Fourth Information Report*, Doc. 6999, 34; see also Begić, A. 1995, who believed the same. Marian Wenzel also reports that the museum collection had been removed to Belgrade in COE *Third Information Report*, Doc. 6904, *20 September 1993, Museum of the Serbian Orthodox Church.*

people there from ancient times, describing the region as an area of the country the Serbs live in 'regardless of its present name.'[206]

Of course, it is also true to say that if these collections had not been removed many would not have survived. Žitomislić Monastery, the Orthodox Cathedral and the Old Orthodox church in Mostar were all destroyed by Bosnian Croat forces, while the Old Orthodox Church in Sarajevo was damaged by Bosnian Serb shelling. However Žitomislić Monastery has been reconstructed and at the time of writing, some of the icons had been returned and there were plans to return more.

Rakić writes how, with the permission of the Serbian Orthodox church authorities, she herself on the eve of war in March 1992 drove through Croat paramilitary checkpoints to Livno in western Herzegovina to attempt to remove to safety in Belgrade the important collection of Italo-Cretan icons belonging to the Orthodox church of the Assumption of the Holy Mother of God there – if the Serb parishioners agreed. They refused, preferring to take their chances. The majority of the church's parishioners were subsequently expelled or fled and on 28 June 1992 the church was set on fire and damaged.[207] However, the icons, over 260 books and other important liturgical objects belonging to the church were rescued by the local Catholic parish priest, Fra Marko Gelo, and taken to the Franciscan Monastery of Gorica.[208] On 9 November 2000 the icons and other artefacts were returned to the church,[209] and in 2004 the church and its movable property were designated as National Monuments under Annex 8 of the Dayton Agreement.[210] From 2006 the icons began to undergo conservation with support from the Federation Ministry of Culture and Sport.[211]

Rakić's article raises important questions about the fate of the movable heritage of communities that have been displaced – should, she asks, their cultural property be 'left behind the lines,' or should they move with the communities that have and are not likely to return, however irrational or unacceptable their reasons? The Serbian Orthodox Church (in common with other religious authorities) regards itself and its parishioners as legal owner of the icon collections, free to do with its cultural and religious property as it likes; those who are trying to restore the interwoven religious and cultural diversity of Bosnia-Herzegovina to a semblance of its pre-war state think otherwise. Rakić describes how intimately the icons and the formation of the collections are bound up with Serb national identity and history, but admits she can offer no solution to the dilemma. On the one hand, removing the icons diminishes the possibility of the mixed pre-war communities of Bosnia-Herzegovina ever being restored: 'by cutting the roots we remove the very chance for new life; we even remove the trace that that life was ever there.' On the other hand, she goes on, 'do those

206 Dugalić, L. 2008. At the table on the bridge: Fifteenth international painters' gathering in Višegrad. *Serbia National Review* [online], No. 9.

207 The church was not destroyed as Rakić writes. *Parohija Lijevanjska*, http://omlivanjska.com/parohija-lijevanjska/parohija-lijevanjska/. See the Commission to Preserve National Monuments entry on the church Assumption of the Holy Mother of God and the icon collection at http://kons.gov.ba/main.php?id_struct=50&lang=4&action=view&id=2485.

208 *Parohija Lijevanjska*; see also: Livanjski franjevci sačuvali ikone srpske pravoslavne crkve, *www.poskok.info* [online], 19 September 2007; *Pravoslavna crkva u Livnu*, http://fotolivno.wordpress.com/2012/03/07/pravoslavna-crkva-u-livnu/.

209 *Parohija Lijevanjska*; Commission website cited above Note 137.

210 On 7 July 2004; see Commission website cited above Note 137.

211 *Parohija Lijevanjska*. The icons are being conserved at the Center for Conservation and Restoration at the Museum of Republika Srpska in Banja Luka.

roots (icons) make sense when detached from the tree (Serbian people), or should they go together with the tree?'[212]

Yet if Rakic had been successful in removing the icons from the church at Livno it seems unlikely that they would have ever been returned. Today the Orthodox Church of the Assumption of the Holy Mother of God has been repaired and the Serb presence in Livno is slowly being been restored. In 2009 the church celebrated its 150th anniversary with a gathering attended not only by members of the Orthodox hierarchy, but by representatives of the Catholic and Muslim clergy, as well as local and Federation officials.[213] Addressing the gathering Bishop Hrizostom of Bihać-Petrovac stressed the important role Serbian merchants had played in Livno's history when establishing the church in 1859 and paid a heartfelt tribute to Fra Marko Gelo and those who had helped save the church's treasures in 1992. However, in common with many of Bosnia's expelled communities, few of Livno's Serb residents have returned. Nevertheless, again like many other such expelled and absent communities, Serbs from Livno region and their descendants have formed an association (*Udruženje Ognjena Marija Livanjska*)[214] which supports the work of the Serbian Orthodox Parish of Livno, including maintaining its churches and cemeteries, upholding links between the Livno region and the Livno Serb diaspora, and preserving the customs, culture and traditions of the Orthodox Serbs of Livno.[215]

Rakic's analysis also leaves out of the equation the Greater Serbia project with its policy of ethnic separation that drove the conflict and the close involvement of the Serbian Orthodox Church in that project, as well as the Republika Srpska authorities' active discouragement of the return of Serb refugees and displaced people to municipalities in the Federation and the restoration of Orthodox churches there. How this ideology of ethnic separation extended to cultural property was revealed by art gallery directors in Republika Srpska interviewed by the author soon after the war's end who expressed interest in arranging exchanges of art with institutions in the Federation, gathering in works by artists of Serb origin in return for works they no longer wished to hold – that is, those by Muslim and Croat artists.[216]

Other Movement of Cultural Property

The Serbian Orthodox church was not the only religious community to move its collections out of Bosnia-Herzegovina. There were a number of such transfers of cultural property from monasteries and churches belonging to the Franciscan Order, most of which held rich and varied collections of cultural heritage – including many artefacts of national importance to the state of Bosnia-Herzegovina. The modern stained glass windows in the Austro-Hungarian period Catholic church of St Michael in Ovčarevo near Travnik were removed on the outbreak of war in 1992 and taken for safekeeping to the Franciscan monastery at Široki Brijeg in Herzegovina near the border with Croatia.[217] The travels of the movable

212 Rakic, S. 2006, 54.

213 Pravoslavni hram je stup opstanka jednog naroda, *Dnevni List* [online], 9 November 2009.

214 See the website of *Udruženje Ognjena Marija Livanjska*, http://omlivanjska.com/. See more on such community associations in Helen Walasek: *Domains of Restoration* in this volume.

215 *News from the Association*, http://omlivanjska.com/english/, October 2012.

216 COE *Tenth Information Report*, Doc. 7740, 24 January 1997, 4.8 *Suggested Exchanges of Art*.

217 See the monument designation for the Church of St Michael, Ovčarevo, on the website of the Commission to Preserve Monuments. The windows were returned to the church in 2000. http://

heritage of Guča Gora Franciscan monastery will be described in detail below. Elsewhere, approximately half the artworks and artefacts (as well as part of the library) from the monastery of St Mark the Evangelist at Plehan were moved to Slavonski Brod in Croatia by the Franciscan community before the outbreak of war and the subsequent destruction of the church and monastery by Bosnian Serb forces on 2 July 1992.[218] The fate of that part of the collection which had *not* been evacuated took another turn. In 2007 the Republika Srpska Minister of Education Anton Kasipović announced the return of 374 works of art to Plehan Monastery which had been stored for more than a decade in the Archive of Republika Srpska in Banja Luka. These artworks had been removed from the monastery in 1992 by the Third Battalion of the 11th Dubica Brigade of the Bosnian Serb Army, eventually making their way to the RS Archive in whose care they remained for the next 15 years.[219] Rescued or looted – depending on one's perspective – the collection had certainly been recognized as cultural property worthy of preservation. The reunited monastery collections were set to be displayed in a new museum being constructed at Plehan as part of the simultaneous reconstruction of three symbolic religious sites announced by the United Nations Development Programme (UNDP) in 2010.[220]

There were other 'migrations' (and returns) of cultural property during and after the war, again on occasion across Bosnia-Herzegovina's borders. Icons and other movable heritage (books, reliquaries and liturgical articles) were removed from the sixteenth-century Orthodox monastery Church of the Presentation of the Virgin in the Bosnian Serb village of Zavala, not far from Dubrovnik and the Adriatic coast most likely by Croatian Army forces some time during the 1992–1995 war. The village and the monastery buildings (though not the church) were attacked and devastated in 1992 and lay abandoned until after the end of the war.[221] At some point the Zavala collection was taken to the Art Gallery of Dubrovnik where it remained until formally returned to the church on 11 September 2009.[222]

When Tešanj underwent shelling by Bosnian Serb artillery during the conflict, Bosnian Army units and the Tešanj authorities transferred the icons of the Orthodox Church of the Cerement of the Virgin to the vaults of the Privredna Bank for their protection; the icons were returned to the church in January 2000. A year later, however, the icons were moved once more, on the instigation of Bishop Vasilije Kačavenda and without the agreement of the heritage protection authority or the local authorities, this time out of the Federation,

www.kons.gov.ba/main.php?id_struct=50&lang=4&action=view&id=2968.

218 The collection consisted of 395 paintings and 35 sculptures (all twentieth century), 2 items of church plate and 5 items of archival material. On 5 July 2006 the collection was designated a National Monument of Bosnia-Herzegovina by the Commission to Preserve National Monuments despite their location in Croatia. In 2005 these collections and the books were damaged by flooding in Slavonski Brod; see Oštećene knjige i slike Franjevačkog samostana Plehan, *Oslobođenje*, No.21.047, 31 August 2005, 24.

219 The collection consisted of 365 paintings and five sculptures. See Luxmoore, J. 2007. Bosnian Serbs make overtures to church, *The Tablet*, 3 March; and Šarić, Ž. 2007. Zbirka vraćena nakon 15 Godina, *Nezavisne Novine*, No. 3067, 22 February, 23.

220 *Reconstruction of Three Religious Monuments – Symbols of Multiculturalism Supported by MDGF Culture and Development Programme*. UNDP Press release, Sarajevo, 20 September 2010. See Helen Walasek, *Domains of Restoration* in this volume for more on this subject.

221 See Tina Wik in this volume for more on the restoration of the church at Zavala and the revival of the monastery and village.

222 Information from the Commission to Preserve National Monuments of Bosnia-Herzegovina to the author, 29 September 2010; thanks to Tarik Jazvin and Aida Bućalović who prepared the data.

to his episcopal palace at Bijeljina in Republika Srpska.[223] Meanwhile in Visoko (in the Federation), the icons from the Orthodox church of St Procopius, one of the largest and most famous collections of icons in Bosnia, were moved by Bosnian Army troops to the basement of the Interior Ministry building in the town for safekeeping and later to the Franciscan Monastery there where storage conditions were more suitable. After the war the icons were transferred to the Serbian Orthodox Church authorities in Sarajevo and eventually returned to St Procopius.[224] There were two controversial movements of particularly significant cultural artefacts that will be discussed in detail below.

The Bones of the King

The first was the transfer of the remains of King Stjepan Tomašević from Jajce to Split (in Croatia) in 1995. Stjepan Tomašević was the last ruler of the medieval Bosnian state before it fell to the Ottomans in 1463; betrayed to the forces of Mehmed II, the king was brought to his royal seat of Jajce and beheaded. A skeleton which has long been believed to be the remains of Stjepan Tomašević was excavated near the town in 1888 and had rested ever since in Jajce's Franciscan Monastery, displayed in a glass sarcophagus. Over the years the king's remains had become a potent emblem both of Croat/Catholic identity and of Bosnian statehood, to which many felt the newly independent Bosnia-Herzegovina was the heir – Bosniak/Muslim nationalists, in particular, felt that Stjepan Tomašević was *their* king. But Stjepan Tomašević had a crucial significance in Serbian history, too: before the debacle at Jajce he had been the final *despot* (ruler) of the slender remnants of an independent Serbia before its final extinction by the conquering Mehmed II in 1459 and Serbia's complete absorption into the Ottoman Empire. Thus all three main parties in the Bosnian conflict felt they had a claim on the king's bones.

After months of bombardment by Bosnian Serb forces the defence of Jajce failed on 29 October 1992, setting off one of the largest population exoduses of the conflict (of both Muslims and Croats). However, before Jajce fell, the HVO, the municipal authorities and the Franciscan hierarchy had made a decision *not* to evacuate the bones of Stjepan Tomašević from the besieged city, feeling that the removal of this powerful symbol would have a disastrous impact on morale.[225] On the Bosnian Serb side, addressing his troops before the final push on Jajce, Colonel (later General) Stanislav Galić declared that among the 'rights' they had to 'free' Jajce was an historic right – as the seat of the medieval Bosnian-Serbian kings.[226]

With the collapse of the city imminent, the withdrawal of the civilian authorities and rumours circulating of a massacre of the monks planned by the invaders, engulfed in the widespread panic of the time, the remaining Franciscans felt they had no choice but to

223 Information from the Commission to Preserve National Monuments of Bosnia-Herzegovina.

224 Except for three which are on display at the Museum of the Old Orthodox Church in Sarajevo. Information from the Commission to Preserve National Monuments of Bosnia-Herzegovina.

225 For an extensive discussion of the fall of Jajce in 1992 and the fate of the bones of King Stjepan Tomašević during that time see: Raguž, J. 2013. Sudbina posmrtnih ostataka kralja Stjepana Tomaševića 1463.–1888.–1992.–1999. (The Fate of Posthumous Remains of King Stjepan Tomašević, 1463 – 1888 – 1992 – 1999) in *Stjepan Tomašević (1461.–1463.) – slom srednjovjekovnoga Bosanskog Kraljevstva : zbornik radova sa Znanstvenog skupa održanog 11. i 12. studenoga 2011. godine u Jajcu*, edited by A. Birin. Zagreb and Sarajevo: Hrvatski institut za povijest u Zagrebu : Katolički bogoslovni fakultet u Sarajevu, 235–82.

226 Quoted in Raguž, J. 2013, 261.

depart Jajce, leaving the bones of the king behind. The occupying Bosnian Serb forces celebrated the liberation of the 'Serbian royal city'[227] and embarked on the destruction of the town's Muslim and Catholic religious structures (the town's Orthodox church had been blown up by Bosnian Croat extremists just two weeks before), the monastery was turned into a prison and its artworks, museum collections and archives looted. Yet before dynamiting the Franciscan church of the Assumption of the Blessed Virgin Mary, the glass sarcophagus holding the bones of King Stjepan Tomašević was removed and taken to the basement of the town library for safe-keeping.[228]

Three years later, Jajce was retaken by Croatian Army/HVO forces on 13 September 1995. With them came Fra Stipo Marčinković, guardian of the Franciscan monastery, who discovered that the king's remains had survived. It was soon announced that on the initiative of the Croatian authorities and the Jajce Franciscan community, the bones of Stjepan Tomašević had been taken out of Bosnia into Croatia[229] to the Museum of Croatian Archaeological Monuments in Split for reasons given variously as security, analysis and conservation.[230] Fra Stipo explained that with the chaotic conditions in the newly-liberated city and the remains clearly in need of attention, it was felt the bones of the king would be safer out of Jajce.[231]

But the news triggered angry protests from official bodies and others in Sarajevo at what they considered the illegal removal of cultural heritage of national importance from Bosnia-Herzegovina with no consultation with the relevant authorities. Enver Imamović, director of the Zemaljski Muzej, compared the removal of the bones of Stjepan Tomašević from Bosnia to the removal of museum collections from Vukovar in Croatia by the JNA in 1991.[232] Over the following years there was bitter public debate between Bosniaks and Bosnian Croats on the subject.[233]

Nevertheless the bones of the king *did* return to Jajce in 1999 on the fourth anniversary of the liberation of the city. But so bound up were the remains with ethno-national identity, that the ideas of nationalist Bosnian Croats for orchestrating a triumphal return clashed with the Franciscan order's desire for the return to be a religious occasion without nationalism, one which also marked their return to a rebuilt and re-sanctified monastery. Thus two separate events marked the return of the bones of King Stjepan Tomašević to Jajce. The first, on 13 September, attended by Croatian and Bosnian Croat military-political figures,

227 Raguž, J. 2013. 261.

228 Raguž, J. 2013. 261.

229 However, in West Mostar in October 1995 (just over a month after Jajce was retaken) the author heard another account of how and when the bones of the king left Jajce from an informant who had recently returned from that city where he had been assessing damage to its cultural heritage. She was told that those caring for the bones of Stjepan Tomašević during the Serb occupation of the city had arranged with a sympathetic Bosnian Serb officer that should Jajce be about to come under attack, he would inform them of the best moment to evacuate the king's remains from the city to Croatia and thus preserve them for posterity. This the officer duly did as the final battle for Jajce was in the offing in August 1995, making arrangements for the vehicle carrying the remains of Stjepan Tomašević to pass safely through the frontlines under cover of darkness to Croat-held territory.,

230 Jurišin, P. 1995. Bosanski kralj u Splitu. *Oslobođenje* [online], No.17011, 18 October, 12.

231 Raguž, J. 2013. 263.

232 Raguž, J. 2013. 267. See earlier in this chapter for more on the removal of museum collections from Vukovar.

233 See for instance, Pandurević, T. 1995. Euforični postupak, *Oslobođenje*, 14 October, 12; and Muhamed Hamidović in S.Š. 1998. Nasilje nad spomenicima: kulturni vandalizam, *Svijet*, 06/12/1998, 13.

saw the sarcophagus paraded through Jajce accompanied by two columns of men wearing 'historic' Croat costumes and was replete with overt ethno-nationalism. The second, held the following day on 14 September, was a solemn religious ceremony presided over by Fra Petar Anđelović, head of the Franciscan order in Bosnia, and attended by both Bosniak and Croat politicians and dignitaries (among them Alija Izetbegović), along with foreign ambassadors and representatives of international organizations.[234]

Jasenovac Memorial Site

The second case grew out of issues of the ownership of collections – legally, ethically and morally – in the successor states of the former Yugoslavia and came to involve the Republika Srpska authorities, the government of Croatia and the United States Holocaust Memorial Museum (USHMM) in Washington DC. The Jasenovac Memorial Site (*Spomen-područje Jasenovac*) commemorated the World War II Croatian Ustaša extermination camp where thousands, principally Serbs (but also Jews, Roma and regime opponents) were brutally murdered between 1941 and 1945.[235] Though relatively few now deny that Jasenovac was the scene of mass murder targeted first and foremost at those of Serb ethnicity, estimates of the numbers who died there is the subject of continued highly charged and politicized dispute and the complex has come to be seen by some as a site of Serb martyrdom.[236]

The memorial site consisted of the extensive grounds of the former camp complex and lay partly in Croatia and partly across the Sava River in Bosnia-Herzegovina – where one of the largest killing fields had been at Donja Gradina, now in the entity of Republika Srpska. Though the main museum buildings were on Croatian territory, many of the museum's artefacts had been excavated from Donja Gradina. With Croatia's declaration of independence in June 1991, the Jasenovac Memorial Site now stood in two different countries.[237] Even before this, however, since the rise of Franjo Tuđman and his election as president of Croatia in 1990, the history of Jasenovac appeared to be in the process of being re-written and the memorial itself under threat. An ardent Croat nationalist, Tuđman's views on the country's WW2 Ustaša regime were not altogether antipathetic and he vigorously disputed Serb estimates of the numbers who had been killed there.[238] The status

234 See for instance: IKA. 1999. U Jajce donešen sarkofag s posmrtnim ostacima posljednjeg Bosanskog kralja Stjepana Tomaševića, *Informativna katolička agencija (IKA)*, 14 September; Zlatović, M. 1999. Sudbina Bosne zrcali se u Jajcu, *Slobodna Dalmacija*, 16 September; Lovrenović, I. 1999. Bosanski Pompeji, *Feral Tribune*, 20 September; Lovrenović, D. 1999. U susret boljoj proslosti, *Dani*, 15 October.

235 The Ustaša (sometimes referred to as Ustaše or Ustasha/e) was a Croatian Fascist movement which during WWII governed the so-called Independent State of Croatia (*Nezavisna Država Hrvatska* or NDH), a Nazi puppet-state established in Axis-occupied Yugoslavia. Pursuing an ideology of racial purity, it was extremely ethno-nationalist and Catholic in character.

236 For varying views of Jasenovac see, for instance, the official website of the renewed Jasenovac Memorial Site in Croatia http://www.jusp-jasenovac.hr and that of the US-based Jasenovac Research Institute http://www.jasenovac.org.

237 That is in Croatia and what remained of Yugoslavia, including Bosnia-Herzegovina which did not become an independent state until April 1992.

238 Though many authorities judge some Serbian estimates of Jasenovac victims to be highly overstated.

of Jasenovac Memorial Site was downgraded to that of a nature park, and the Croatian parliament voted to cut its funding.[239]

The Assistant Director of Jasenovac Memorial Site was Simo Brdar, a Bosnian Serb who lived in Bosnia. Brdar continued to cross the Sava to work at the main site until August 1991. However, he feared for the safety of the artefacts for which he was responsible, and with little faith in the Croatian authorities' commitment to preserving them and the war drawing closer, on his last day at the site he took part of the archives with him. In September 1991 Croatian forces entered the Jasenovac Memorial Site. During their occupation the museum building was vandalized and its exhibits and documentation destroyed, damaged or looted; as the troops withdrew they blew up the bridge over the Sava linking the two parts of the complex. One month later the JNA (Yugoslav People's Army) took control of the main site and Simo Brdar returned to remove what he could of what remained of the collections, taking them to be stored in his home. In February 1992 Yugoslavia submitted a formal protest to the United Nations and UNESCO regarding the Croatian devastation of Jasenovac Memorial Site.[240] The same month a European Community Monitor Mission (ECMM) team visited the museum and memorial complex, confirming the damage and looting of the collections.[241] The whereabouts of the material which had been looted before Brdar's return remains unknown at time of writing.

The Jasenovac artefacts and archival material remained at Simo Brdar's flat throughout the war and for years after, the RS authorities apparently taking little interest in them. Finally, however, a decision was made to house the collection in the Archive of Republika Srpska in Banja Luka in 1999.[242] The following year the United States Holocaust Memorial Museum became aware of the Jasenovac collections in Banja Luka, and considering them to be neglected and at risk, began negotiations for their transfer to the United States. Republika Srpska Prime Minister Milorad Dodik (without the knowledge of the RS Assembly) signed an agreement on 27 October 2000 relinquishing ownership of the collection to USHMM completely – Mr Dodik allegedly believing that it would be permanently preserved and displayed at the US museum. Shortly afterwards American SFOR troops arrived at the RS Archive to remove the artefacts which travelled to the United States via diplomatic pouch.[243] However, USHMM had no intention of keeping the collection, and unbeknownst to Dodik, had already signed a contract with the Croatian Minister of Culture, Antun Vujić, on 26 October 2000, agreeing to return the artefacts to Croatia one year later after cataloguing and

239 Kolovrat, I. 1999. Arhiv i muzejska zbirka Spomen-područja Jasenovac od 1991 us stanu bivšega kustosa Sime Brdara u Bosanskog Dubici!, *Globus*, 10 September 1999, 30–32.

240 Tanjug 1992a. Yugoslav government sends memorandum to UN and others on 'genocide' in Croatia, *Tanjug*, 31 January. *BBC Summary of World Broadcasts*, 3 February. Tanjug 1992b. Yugoslav government memorandum on 'genocide' in Croatia delivered to UNESCO, *Yugoslav News Agency (Tanjug)*, 5 February, BBC Summary of World Broadcasts, 7 February.*Memorandum of the Government of Yugoslavia on the Crime of Genocide in Croatia in 1991 and 1992 and the Vandalising of the Memorial of Jasenovac*, FSRY, Belgrade, 31 January 1992, Presented to the United Nations Commission on Human Rights, Forty-eighth session, 18 February 1992, E/CN.4/1992/71.

241 *Fifth Information Report on War Damage to the Cultural Heritage in Croatia and Bosnia-Herzegovina*, Parliamentary Assembly of the Council of Europe (COE), Doc. 7070, Strasbourg, 12 April 1994, 2.B.6.

242 Anđelić, S. 2000. Arhivsku građu iz Jasenovaca prebacio u Muzej holokausta u Washingtonu!, *Slobodna Bosna*, 28 December, 35–7.

243 *Croatian WWII Concentration Camp Records Made Available for First Time by United States Holocaust Memorial Museum*. United States Holocaust Memorial Museum, Washington DC, Press Release, 13 November 2001.

conservation, on the understanding that they would be eventually sent to a re-established Jasenovac Memorial Site.

The US museum appeared to have minimal acquaintance with the history of Jasenovac and who had been the main victims there (being under the impression they had been chiefly Jews). The USHMM authorities believed Croatia to be the legal owner of the collections (though the memorial site had been an institution of the state of Yugoslavia) and that the collections had been 'stolen' from Croatia by the Serbs.[244] They seemingly had no knowledge of the 1992 Yugoslav Memorandum regarding Croatian devastation of the museum and archive. Much was made by the USHMM of its 'discovering' and 'rescuing' the 'hidden' archive – although its whereabouts were public knowledge and Simo Brdar had been widely interviewed by both Croatian and foreign journalists during the Zagreb-based war crimes trial of Dinko Sakić (a former commander of Jasenovac in WW2) in 1999.[245]

As for Prime Minister Dodik, when news of the agreement reached the RS Assembly in December 2000, an uproar followed and a special commission was created to investigate the circumstances of the transfer of the Jasenovac collections to USHMM, chaired by RS MP and lawyer Miroslav Mikes. The ever-slippery Dodik claimed he had only made the agreement because there was a real danger of SFOR troops storming the RS Archive and removing the Jasenovac material.[246] Dodik stated he had received requests for the material from the Croatian authorities on a number of occasions and that representatives of the OHR and other international organizations had all said the collections had been wrongfully removed from Jasenovac and Donja Gradina.[247] He further implied to the commission that he had been bulldozed into the agreement by US Ambassador Thomas Miller, that it had been Miller's initiative, and when American troops turned up demanding documents he certainly was not going to refuse them.[248] Miller denied Dodik's allegations, saying he had acted only as an intermediary – but USHMM declared the opposite, claiming all requests for them to take the material came from Banja Luka and the US Embassy in Sarajevo.[249]

The RS commission asked that the issue of ownership of the Jasenovac collection be resolved by international arbitration and Mikes travelled to Washington to meet with USHMM officials.[250] After Mikes learned there of the agreement with Croatia, the RS authorities called again for the material to come under international protection and remain in Washington until a decision was made on its ownership, believing UNESCO would play a role in any such arbitration.[251] Mikes travelled to attend the UNESCO General Conference in Paris in October 2001, hoping that an official request from the government of Bosnia-Herzegovina for the collection to be returned would soon be arriving at UNESCO.[252] For

244 Reljić, S. and Gojgić, L. 2001. Skriveni zloin, *NIN*, Belgrade, 15 March. English translation [Hidden crime] at *www.ex-yupress.com/*.

245 Including the Croatian magazine *Globus* and television crews from CNN, ABC News and SKY. See Kolovrat, I. 1999, 31.

246 Reljić, S. and Gojgić, L. 2001. RTRS 2001b. Bosnian Serbs request international arbitration on Jasenovac memorial, *Radio Televizija Republike Srpske (RTRS)*, 23 February, *BBC Monitoring*.

247 RTRS 2001b.

248 Reljić, S. and Gojgić, L. 2001. RTRS 2001b.

249 Reljić, S. and Gojgić, L. 2001. RTRS 2001b.

250 RTRS 2001b; Jasenovac: Muzej holokausta vratice arhivu iz Jasenovca u Hrvatsku, *Novine* [online], Toronto, 23 October 2001.

251 Novine 2001. Jasenovac: Muzej holokausta.

SRNA 2001a. Bosnian Serbs ask Holocaust Museum to keep Jasenovac camp documents. *SRNA News Agency*, Bijeljina, 22 October. *BBC Worldwide Monitoring*, 22 October.

252 Novine 2001. Jasenovac: Muzej holokausta.

such were the RS government's concerns about the imminent transfer to Croatia, that despite its usual disinclination to support actions by Bosnia-Herzegovina as a sovereign state on the international stage, on 23 October 2001 the RS entity Minister of Science and Culture Mitar Novaković had written to the Council of Ministers of Bosnia-Herzegovina in Sarajevo requesting them to intervene on behalf of Republika Srpska in the Jasenovac collections case, stating it was the constitutional and moral duty of the joint Bosnia-Herzegovina authorities to protect the interests of the entities, which were also the interests of Bosnia-Herzegovina.[253] The RS authorities, of course, realized that UNESCO would only deal with such issues at member state level, but this was a rare acknowledgement by Republika Srpska of the central state government and its role representing Bosnia-Herzegovina in the international realm. The chairman of the Council of Ministers, Zlatko Lagumdžija claimed he had not received the letter from Novaković.[254]

Simo Brdar declared that while no one disputed Croatia's right to the Jasenovac material, since most of the collection's exhibits had come from Donja Gradina, the same right could not be denied Republika Srpska and Bosnia-Herzegovina.[255] The director of the RS Archive Dušan Vržina agreed with Brdar, saying that while the agreement was good for the collection in terms of conservation and restoration, it had been bad in terms of Republika Srpska giving up its rights to the material and called for international intervention if USHMM refused to return it. He believed that ownership of the Jasenovac material would be dealt with as a succession issue as property belonging to the former Yugoslavia before its disintegration, and that it would be shared between RS and Croatia.[256]

Succession issues had recently been under discussion under the aegis of the Office of the High Representative (OHR) and an *Agreement on Succession Issues* had been adopted by the successor states of the former Yugoslavia in Vienna on 29 June 2001.[257] Article 3 (2) of Annex A on movable and immovable property dealt with the question of tangible movable State property 'of great importance to the cultural heritage of one of the successor States and which originated from the territory of that State.' Here, ownership of such cultural property would not simply pass to the successor State on whose territory that property was situated on the date on which it proclaimed independence if it was identified as falling into the category defined in Article 3 (2) by the State making a claim to it.[258]

---

253 SRNA 2001. Bosnian Serbs urge central government to get back concentration camp documents, *SRNA News Agency*, Bijeljina, 27 November, *BBC Worldwide Monitoring*, 27 November.

254 SRNA 2001. Bosnian Serbs urge central government to get back concentration camp documents.

255 RTRS 2001b.

256 BH Press News Agency. 2001. Bosnian Serb official criticizes former government for transfer of archives, *BH Press News Agency*, 19 January, *BBC Worldwide Monitoring*, 19 January.

257 *Agreement on Succession Issues*, agreed in Brussels, 10 April 2001, adopted at the Conference on Succession Issues, Vienna, 29 June 2001 by the representatives of Bosnia-Herzegovina, the Republic of Croatia, the Republic of Macedonia, the Republic of Slovenia and the Federal Republic of Yugoslavia.

258 *Agreement on Succession Issues, Annex A. Movable and Immovable Property, Article 3 (2).* The text of the article states: 'Paragraph (1) of this Article does not apply to tangible movable State property of great importance to the cultural heritage of one of the successor States and which originated from the territory of that State, such as: works of art; manuscripts, books and other objects of artistic, historical or archaeological interest to that State; and scientific collections and important collections of books or archives which shall pass to that State. Such property shall be identified by the successor State concerned as soon as possible, but not later than 2 years after the entry into force of this Agreement.' The Agreement entered into force on 2 June 2004.

However, in November 2001 the return of the Jasenovac collection to Croatia was announced by USHMM and a selection of artefacts put on display for the benefit of journalists. By this time the main parties involved seemed to have accepted the return of the artefacts to the Jasenovac Memorial Site. Media coverage toned down the alleged 'theft' of the artefacts from Croatia during the conflict and noted the role of Simo Brdar in protecting them.[259] The Croatian forces damage to and looting of the museum in 1991 was not mentioned. On 5 December 2001 the collections were returned to the Jasenovac Memorial Site.[260]

It is not clear how deeply the United States Holocaust Memorial Museum investigated the background to the movement of the Jasenovac collection during the conflict or the legal issues concerning its ownership and its actions seem a combination of naïve and verging on the unethical. Furthermore, at least part of the collection rescued by Simo Brdar could arguably have come within the terms of the *Agreement on Succession Issues* of 2001.

The Sarajevo Haggadah 1

The potency of the object was demonstrated by the wartime events involving what is arguably Bosnia's most famous cultural artefact: the fourteenth-century Jewish-Spanish Passover manuscript known as the Sarajevo Haggadah. Over the years the manuscript had become a symbol of Bosnia's diverse past and its story during (and after) the 1992–1995 war only added to its mystique. Rescued from the war-swept Zemaljski Muzej at the beginning of the siege of Sarajevo and taken to the vaults of the National Bank, the manuscript was to become a pawn in the propaganda war being waged by Serb lobbyists against the beleaguered Bosnian government.

In late 1994 rumours began to surface in the international media that the Sarajevo Haggadah had been destroyed or was damaged beyond restoration. The authorities in Sarajevo refused to comment for security reasons. Then a provocative article by Chicago Tribune journalist Thom Shanker appeared implying that the Bosnian government had sold the manuscript for arms – and if they had not, the Muslim Minister of Culture was not likely to have much interest in preserving a Jewish treasure.[261] Now the whereabouts of the Sarajevo Haggadah became a worldwide *cause célèbre* and the government was forced to act. On 16 April 1995 at a Passover *seder* in Sarajevo's Ashkenazi synagogue attended by President Izetbegović and a host of national and international dignitaries, the manuscript was brought to the ritual by armoured car and revealed to the world's media.[262]

The Sarajevo Haggadah retained its power as a culturally desirable object which transcended the politics of ethnic separation after the war when in 1998 the Bosnian Serb member of the country's tri-partite Presidency, Momčilo Krajišnik[263] demanded the manuscript be 'shared' between the country's three constituent peoples, suggesting that it

259 See Kovac, V. 2001. After decade abroad, World War II archive returns home to Croatia, *Jewish Telegraphic Agency*, 6 December; and Higham, N. 2001. Croatian holocaust still stirs controversy, *BBC* [online], 29 November.

260 See the Jasenovac Memorial Site website http://www.jusp-jasenovac.hr.

261 Shanker, T. 1995. Sarajevo postcard: Missing pages, *New Republic*, 13 February 1995.

262 See for instance: Cohen, R. 1995b. Bosnia Jews glimpse book and hope, *New York Times*, 16 April; Serotta, E. 1995. From the ravages of war, Sarajevo Haggadah appears, *Jewish Telegraphic Agency* [online], 18 April 1995.

263 Krajišnik is presently a convicted war criminal serving a 20-year sentence for crimes against humanity.

Fig. 1.27 The Sarajevo Haggadah revealed to the world's media at the Passover seder held at Sarajevo's synagogue on 16 April 1995. Holding the manuscript is Bosnian Ministry of Culture official Rešad Gogalija. © Edward Serotta / Time Magazine

be displayed every third year in the Serbs' 'capital city' Banja Luka, rotating with Mostar (as the Croat 'capital') and Sarajevo (presumably, the 'Muslim' capital).[264]

Reactions to the Destruction

When reports of the devastation being wreaked on Bosnia-Herzegovina's cultural patrimony with the crushing assaults on mosques, churches, libraries and museums began to come through there was an early reaction from intergovernmental organizations with a remit for the heritage. Most had already condemned the destruction of cultural property going on in Croatia from the autumn of 1991, particularly the JNA bombardment of Dubrovnik throughout October to December[265] and to where UNESCO had sent an observer mission in November.[266] Thus on 7 May 1992 the Director-General of UNESCO issued a

264 Davis, D. 1998. Serbs demand their 'share' of rare Sarajevo Haggadah, *Jewish Telegraphic Agency* [online], 24 December and Ahmetašević, N. 2002. Kada je Klein stigao I BiH, Hagada je, srećom već bila spašena!, *Slobodna Bosna*, 12 December, 31.

265 See, for instance: *Records of the General Conference of UNESCO*, Twenty-sixth Session, Paris, 15 October to 7 November 1991, 1, Resolutions, *0.10 Appeal by the General Conference Concerning the Natural and Cultural Heritage in Yugoslavia.*

266 *Information on the Implementation of the Convention for the Protection of Cultural Property in the Event of Armed Conflict, The Hague 1954, 1995 Reports*, UNESCO, Doc. CLT-95/WS/13, Paris, December 1995, No. 18, 7. The report includes a description of UNESCO's activities in Croatia.

statement expressing concern about the situation in respect of the destruction of cultural and religious monuments in Bosnia-Herzegovina,[267] while the Committee of Ministers of the Council of Europe strongly condemned 'the destruction of an irreplaceable cultural and architectural heritage.'[268]

Over the three years of the war a succession of declarations, appeals and resolutions either explicitly focused on the cultural destruction in Bosnia or incorporated into more wide-ranging statements on the conflict emanated from intergovernmental organizations which had a concern for heritage protection like UNESCO and the Council of Europe, as well as from those with a broader sphere of activity such as the Organisation of the Islamic Conference (OIC) and the Conference on Security and Co-operation in Europe (CSCE).[269] Most stressed the diverse character of Bosnia-Herzegovina's cultural identity. Pronouncements from international professional bodies committed to preservation of the cultural heritage such as ICOMOS, ICCROM, ICOM and the World Heritage Committee were considerably slower to emerge.[270] Yet set against similar declarations on other war crimes and human rights abuses going on in Bosnia-Herzegovina, those devoted solely to wrongs against the cultural heritage were a fraction of the total.

But the incident that prompted the greatest outpouring of condemnation around the world was the destruction of a single iconic monument: the *Stari Most* (Old Bridge) in Mostar on 9 November 1993 by the HVO.[271] As described elsewhere in this volume, the

267 *Hundred and Thirty-ninth Session of the Executive Board, Item 7.5 The Provisional Agenda: The Situation of the Cultural and Architectural Heritage as well as of Educational and Cultural Institutions in Bosnia-Herzegovina* (139 EX/33), Paris: UNESCO, 18 May 1992. Proposal made by Mr Talat Halman (Turkey) and Mr Ahmed Fathi Sorour (Egypt).

268 *Minutes of the 90th Session of the Committee of Ministers, Final Communique, Appendix 2, 15 Situation on the territory of the former SFR of Yugoslavia*, CM(92)PV1-2-3E, Strasbourg: Council of Europe (COE), 7 May 1992.

269 See for instance: Statements by the Director-General of UNESCO on the situation in Bosnia-Herzegovina, 7 May 1992 and 21 May 1992; *Decision 7.5 The Situation of the Cultural and Architectural Heritage as well as of Educational and Cultural Institutions in Bosnia-Herzegovina* (139 EX/33), *Decisions Adopted by the Executive Board at its 139th Session*, EX139/Decisions, Paris: UNESCO, 18–27 May 1992; *Third European Conference of Ministers Responsible for the Cultural Heritage: Proceedings, Malta, 16–17 January 1992, Resolution No. 4 on Conservation in Situations of Conflict*, Strasbourg: Council of Europe (COE), 1993, 60; *476th Meeting, Decisions Adopted by the Committee of Ministers, Written Question No 343 by MM Baumel and Tummers on the Cultural Heritage and Former Territory of Yugoslavia*, 7d, CM/Del/Dec(92)476, Strasbourg: Council of Europe, 22 May 1992; *Resolutions Adopted by the Fifth Extraordinary Session of the Islamic Conference of Foreign Ministers, Resolution No. 1/5-EX, Situation on Bosnia-Herzegovina*, No. 8, Istanbul: Organisation of the Islamic Conference (OIC), 17–18 June 1992; *Resolution no.19/21-C, Destruction of Mosques and Islamic Monuments in Bosnia-Herzegovina, The Twenty-first Islamic Conference of Foreign Ministers*. Karachi: Organisation of the Islamic Conference (OIC), 25–29 April 1993; *Third Meeting of the CSCE Council, Decisions, 1 (4) Regional Issues, Former Yugoslavia*, Stockholm 14–15 December 1992 (the CSCE is now the Organization on Security and Co-operation in Europe – OSCE).

270 ICOMOS = International Council on Monuments and Sites; ICCROM = International Centre for the Study of the Preservation and Restoration of Cultural Property; ICOM = International Council of Museums. See for instance: *Resolutions of the 1993 General Assembly, No. 12*, Colombo: ICOMOS, 30 July–4 August 1993; *17th Session of the World Heritage Committee*, WHC-93/CONF.002, *Annex V World Heritage Committee Declaration on Bosnia*, Cartagena: World Heritage Committee, 6–11 December 1993; *18th General Assembly, Resolution no. 2: Protection of the Cultural Heritage during Armed Conflict*, Stavanger: ICOM, 7 July 1995.

271 For a selection of the reactions to the destruction of the *Stari Most*, see COE *Fourth Information Report, 3. Reactions to the Destruction of the Old Bridge of Mostar on 9 November 1993*,

HVO were supported and armed by Croatia and the Croatian Army was initially believed to be implicated in the attack. The Parliamentary Assembly of the Council of Europe's Committee on Culture and Education was among the first to react, sending a letter on 10 November directly to the person they felt ultimately responsible for the bridge's demise – Croatian President Franjo Tuđman. The letter condemned the destruction in the strongest terms and drew President Tuđman's attention to Croatian responsibilities in respect of the Old Bridge, calling on him 'to withdraw immediately any fighting units and military material that we have been informed you have on the territory of Bosnia-Herzegovina.'[272] A declaration from UNESCO's Director-General Federico Mayor followed on 11 November that accused the perpetrators of 'trying to eradicate the history of a country and its people' and 'destroying the bridges of mutual understanding built by people of different origins and religious beliefs who had learnt to live together in harmony.'[273]

There was extensive coverage of the assault and the final collapse of the Old Bridge's in the international media, not simply through short news reports but frequently with long feature articles[274] and a string of NGOs issued protests.[275] But some of the most heartfelt protests came from the countries of the former Yugoslavia, most notably Serbia and Croatia.

Regional Voices

The role of the officially controlled media in Serbia and Croatia (and especially in Serbia), the two countries actively fomenting the war in Bosnia-Herzegovina and pursuing an ethno-nationalist/exclusivist agenda, particularly in spreading propaganda, disinformation (especially of purported atrocities against Serbs/Croats) and discrediting opponents, has been extensively analysed.[276] Slobodan Milošević's control and manipulation of the state media in creating the conditions that enabled ethnic cleansing to take place appeared on the

Doc. 6999, 19 January 1994; COE *Fifth Information Report, 3. Further Reactions to the Destruction of the Old Bridge of Mostar*, Doc. 7070, 12 April 1994.

272 COE *Fourth Information Report*, Doc. 6999, 19 January 1994, *3. Further Reactions: A. The Parliamentary Assembly of the Council of Europe; Message to President Franjo Tudjman from Mrs Leni Fischer, MdB, Chairperson of the Committee on Culture and Education*, 10 November 1993.

273 COE *Fourth Information Report*, Doc. 6999, 19 January 1994, *3. Further Reactions: UNESCO*; *Declaration of UNESCO Director-general Federico Mayor*, 11 November 1993.

274 See for instance: Williams, C.J. 1993b. Bosnia's hopes fall with historic bridge in Mostar, *Los Angeles Times*, 12 November; Sudetic, C. 1993b. Mostar's Old Bridge battered to death, *The New York Times*, 10 November; Traynor, I. 1993a. Shells destroy Mostar bridge, *The Guardian*, 10 November, Foreign page, 11; Naudet, J-B. 1993. La destruction du pont de Mostar est un sacrilège irréparable pour les Musulmans de Bosnie, *Le Monde*, 16 November; Matvejevic, P. 1993. La virtud de un Puente, *El Pais*, *Tribuna*, 14 December; Koprüyü sonunda yıktılar, *Milliyet* (Istanbul), 11 November, 15.

275 Among them Bosnia-Herzegovina Heritage Rescue (BHHR), The Society for the Protection of Ancient Buildings (SPAB), and Europa Nostra; see COE *Fourth Information Report*, Doc. 6999, 19 January 1994.

276 Most famously by Mark Thompson in *Forging War: The Media in Serbia, Croatia & Bosnia-Hercegovina*. Luton: University of Luton Press, 1999, but see also: Kurspahic, K. 2003. *Prime Time Crime: Balkan Media in War and Peace*, Washington, DC: United States Institute of Peace Press; Kurspahić was editor-in-chief of the main Sarajevo daily *Oslobodjenje* 1989–1994. On 23 May 2011 the board of Radio Television Serbia (RTS) issued a public apology to Serbia and neighbouring countries for the role played by RTS and Radio Television Belgrade (RTB) in spreading hate speech, propaganda and slander during the 1990s; see Tanjug 2011. State broadcaster 'sorry' for 1990s, *B92* [online], 24 May 2011.

charge sheet against the Serbian leader at the ICTY.[277] Yet despite government domination of most forms of the media (especially television) in these countries, a small outspoken independent media nevertheless existed that was prepared to report on what was going on in Bosnia-Herzegovina, challenging the official discourse on the war and publishing the views of those opposed to it.[278] This included reporting on the widespread and intentional destruction of cultural and religious property in Bosnia and Croatia by JNA/Serb/Bosnian Serb and HVO/Croatian Army forces and authorities and the motivations behind it.

Most notable in Serbia in the print media were the daily *Borba* (to November 1994), the weekly *Vreme* and the periodical *Republika*; in broadcast media, the radio station B92 and the television station Studio B. In Croatia the Split-based weekly *Feral Tribune* was the most vigorous of the independent media outlets. As early as May 1992 *Vreme* was publishing on the destruction of cities in Bosnia and Croatia;[279] in November 1992 the magazine wrote on the attacks on cultural heritage in Bosnia, describing the burning of the National Library in Sarajevo and the Oriental Institute.[280] *Borba* was the first Serbian media outlet to report on the demolition of Bijelina's five mosques in a single night on 13 March 1993, with *Vreme* following up on 22 March with a long article describing the widespread destruction of mosques across Bosnia, as well as attacks on Orthodox and Catholic sacral structures.[281] Journalists from *Feral Tribune* were first to reveal in May 1994 (with accompanying photoreportage) the full extent of the destruction that the HVO had inflicted on Stolac's Ottoman/Islamic heritage the previous summer.[282]

One prominent public figure in Serbia who was prepared to speak out on the destruction of cities in both Croatia and Bosnia was the renowned architect and urban theorist Bogdan Bogdanović. Bogdanović, a former mayor of Belgrade, was a vehement opponent of Serbian nationalism. Outside the region he was probably best-known as the designer (between 1960 and 1981) of more than 20 of the astonishing memorial monuments constructed at crucial locations across Yugoslavia. Most famous is his 'flower' memorial (1966) at Jasenovac Memorial Site (*Spomen-područje Jasenovac*), now in Croatia; in Bosnia-Herzegovina,

---

277 *The Prosecutor v. Slobodan Milošević, Amended Indictment 'Bosnia and Herzegovina'*, ICTY Case No. IT-02-54-T, 22 November 2002, 25.g.

278 All worked under considerable financial pressures and political interference. Analysts have noted that not only did the independent print media have small circulations, the economic situation was so bad that many potential readers could not afford even the price of a newspaper, and thus for most television was their main source of information; however, Eric Gordy has argued that despite small circulations, *Borba*'s readership, for instance, was largely educated, urban and influential, and therefore troublesome to the authorities: Gordy, E.D. 1999. *The Culture of Power in Serbia: Nationalism and the Destruction of Alternatives*. University Park, PA: Pennsylvania State University Press, 83. For more on the opposition media in Serbia and Croatia, see articles by Milan Milošević (on the media wars), Ivan Torov (on Serbia, including the government-inspired legal troubles and takeover of *Borba* from November 1994) and Sven Balas (on Croatia) in J. Udovicki and J. Ridgeway (eds) 2000. *Burn This House: The Making and Unmaking of Yugoslavia*, Durham, NC: Duke University Press. Writings by some of Republika's main contributors can be found in Popov, N. (ed.) 2000. *The Road to War in Serbia: Trauma and Catharsis*. Budapest: Central European University Press.

279 Stefanovic, N. Lj. 1992. The destruction of cities: We built them – we can demolish them, *Vreme News Digest Agency* (Belgrade), No 34, 18 May.

280 Stitkovac, E. 1992. Destruction of heritage: More than hatred, *Vreme News Digest Agency*, No 61, 23 November.

281 Mlakar, M. 1993. Bijeljina dynamiters: The night the minarets fell, *Vreme News Digest Agency*, No 78, 22 March.

282 Malić, G. 1994. Aladin više ne stanuje ovdje: Stolac grad bez prošlosti, *Feral Tribune*, 2 May, 14. See Amra Hadžimuhamedović in this volume on the *Feral Tribune* article.

his Partisan Memorial Cemetery (*Partizansko spomen-groblje*) in Mostar (1965) is considered one of the country's greatest twentieth-century cultural sites and is a designated National Monument.[283]

Bogdanović had long thought about and written on the notion of urbicide. Now it seemed as if his thesis was being put into practice before his eyes across the territory of what had once been Yugoslavia. On 28 April 1992 *Borba* published Bogdanović's article *Ritualno ubijanje grada* (*The Ritual Murder of the City*)[284] where he describes the 'grand masters of destruction' and the 'city despisers and the city destroyers,' of the 'New Serbia,' comparing the attack on Dubrovnik to the act of 'a madman who throws acid in a beautiful woman's face and promises her a beautiful face in return;' Mostar and Sarajevo, he wrote, were waiting their turn.[285] The article was highly influential and went on to be published in many translations around the world, beginning with *El Pais* (Madrid) on 23 May 1992.[286]

Bogdanović commented to journalist Roy Gutman in September 1992 on the destruction of Bosnia's cultural patrimony by JNA, Bosnian Serb forces and Serbian paramilitary units: 'It is terrible. It is unbelievable from the point of view of the history of art and architecture. Beautiful cities have been completely destroyed, and an enormous number of historic buildings. It is a great crime against civilisation, a disgrace for my people and for the army which does that.'[287] Such views led to Bogdanović being vilified in Serbia and he went into exile in 1993.

The official media frequently did not report not at all on the destruction of cultural and religious property – the demolition of the mosques in Bijeljina, for instance, was not mentioned on TV Belgrade, the biggest television station in the former Yugoslavia.[288] Or if such events were reported, it was often in a calculated misrepresentation of what actually happened. Thus, TV Belgrade *did* report the burning of the National Library (*Vijećnica*) in Sarajevo in August 1992 (after being intensively bombarded with incendiary shells from Bosnian Serb Army positions) but asserted that it was 'difficult to establish exactly what caused the fire,' that no signs of shell damage could be seen and 'the flames licked from within' – all pointing to 'another Muslim manipulation like the one in Dubrovnik when car tyres were set on fire;'[289] referring to the Dubrovnik 'hoax' allegedly carried out by Croats

283 See *Partisan Memorial-Cemetery, the Architectural Ensemble*, on the website of the Commission to Preserve National Monuments: http://kons.gov.ba/.

284 *Bogdanović*, B. 1992. Ritualno ubijanje grada, *Borba* (Belgrade), 28 April. The article went on to be published in many translations around the world, beginning in *El Pais* (Madrid) on 23 May 1992.

285 From the English translation of *Ritualno ubijanje grada*: Bogdanovic, B. 1993. *Murder of the city. New York Review of Books*, 40: 10 (27 May 1993).

286 *Bogdanović*, B. 1994. *Grad i Smrt*. Belgrade: Beogradski krug, 13.

287 Gutman, R. 1992. Unholy war.

288 Pesic, M. 1994. *Manipulations on Television Belgrade*, MA thesis, Department of Journalism, City University, London, September 1994, 25.

289 de la Brosse, R. 2003. *Political Propaganda and the Plan to Create 'A State For All Serbs,' Consequences of Using Media for Ultra-nationalist Ends*, expert report for the ICTY, Exhibit D01645, January 2003, 74–75. The TV Belgrade commentary is from the documentary *Images and Words of Hate: Year Three*, ARHITEL/B92 (producers Lazar Lalić and Isidora Sekulić), 1999, cited in de la Brosse, 75; the documentary was one of a series reporting on Serbian state television's warmongering coverage of the Wars of Yugoslav Succession.

during the attack on that city to give the false impression to western journalists that the city was in flames.[290]

However, there were two instances of the destruction of iconic structures that provoked official condemnation in Belgrade and Zagreb. The first was the demolition of the historic Ferhadija and Arnaudija Mosques in Banja Luka by Bosnian Serb forces/authorities on 7 May 1993. The destruction led Dobrica Ćosić, the then president of the rump-Yugoslavia and widely acknowledged as principal architect of the Greater Serbia project, to issue a strong statement condemning the assault on the mosques as an 'act of barbarity' and a 'final warning' to all warring sides in the Bosnian conflict to 'act resolutely, immediately and with all means at their disposal to stop the war and destruction.'[291] It was noted, however, that neither Ćosić, nor any other government official, had similarly denounced the blowing up of five mosques in Bijeljina a few month earlier.[292] The Yugoslav government helpfully forwarded censure of the Banja Luka demolitions that had appeared in the Belgrade media to the Council of Europe, including a report published by the government's long-time critic, *Borba*. An article from Tanjug, the state-run news agency, quoted experts from the Institute for Protection of Monuments of Serbia who commented that the assault on the mosques 'underlined the fact that the destruction of cultural monuments is not only a barbaric act, but also the proof of the shaking of elementary humanism and respect for the past era of historic peoples.'[293]

However, it is clear this unusual interest had little to do with a new concern for Bosnia's Islamic/Ottoman heritage, but more to the overwhelming refusal of the Assembly of Republika Srpska in Banja Luka to endorse the Vance-Owen Peace Plan the day before, a peace plan that their leader Radovan Karadžić had agreed to in principle and which both Dobrica Ćosić and Slobodan Milošević had heavily supported and intensively lobbied for. Milošević, Ćosić and Greek prime minister Constantine Mitsotakis came personally to Banja Luka to add their voices to the calls for acceptance of the plan and sat in the front row listening as it was rejected;[294] Milošević was apparently 'radioactive,'[295] Ćosić 'ashen-faced.'[296] The same day as Ćosić's statement on the Banja Luka mosques, the Yugoslav and

---

290 This alleged ruse is referred to in the official report of the Federal Republic of Yugoslavia (Serbia and Montenegro) in *Information on the Implementation of the Convention for the Protection of Cultural Property in the Event of Armed Conflict, The Hague 1954, 1995 Reports*, UNESCO, Doc. CLT-95/WS/13, Paris, December 1995, 26. See also for instance: Perlez, J. 1997. Serbian media is a one-man show, *The New York Times* [online], 10 August: '"People here still don't believe that Dubrovnik was shelled," said Veran Matic, the founder of B-92, the only independent radio network in Serbia. The Yugoslav army attacked the Croatian port town in 1991. "Belgrade TV showed Dubrovnik with columns of smoke and then said that it was caused by the local people burning tires," he said.' And see also testimonies during the trials of Pavle Strugar and Miodrag Jokić that comprehensively dismiss the claim: *Prosecutor v. Pavle Strugar*, ICTY Case No. IT-01-42-T, and *Prosecutor v. Miodrag Jokić*, ICTY Case No. IT-01-42.

291 Kinzer, S. 1993c. Conflict in the Balkans; 2 Major mosques blown up by Serbs, *The New York Times*, 8 May.

292 Kinzer, S. 1993c. Conflict in the Balkans; 2 Major mosques blown up by Serbs.

293 COE *Fourth Information Report*, Doc. 6999, 19 January 1994, *Reactions in the Republic of Serbia to the Destruction of the Ferhadija and Arnaudija Mosques in Banja Luka*, Nos 83–4.

294 Kinzer, S. 1993a. Conflict in the Balkans; Serbs reject Bosnia pact, defying friends and foes, and insist on referendum, *The New York Times*, 6 May 1993.

295 Cornwell, R, Savill, A. and Brown, C. 1993. Belgrade to cut supplies to Bosnian Serbs: Rejection of peace angers Milosevic – US demands Europe back quick action, *The Independent*, 7 May.

296 Serbs reject peace plan – Parliament ignores Western military threats, sets referendum, *The Palm Beach Post, Palm Beach Post Wire Services*, 6 May 1993.

Serb governments announced the cutting of all aid (except humanitarian aid) to the Bosnian Serbs in the hope of pressuring them to accept the plan.[297]

Inevitably, in common (it seemed) with the rest of the world, the single act of heritage destruction that provoked widespread protest in both Croatia and Serbia was the destruction of the Old Bridge at Mostar on 9 November 1993 after continued shelling from a Bosnian Croat (HVO) tank. To outside observers grief over the destruction of this iconic monument, which had been called one of the most beautiful bridges in the world and whose sublime elegance was held in such high regard across all the former Yugoslavia, appeared to briefly reunite the warring parties in the Bosnian conflict.[298] There were reports of how people of all nationalities burst into tears when they heard the news.[299]

It was perhaps not so unlikely that state-run television in Serbia would screen extensive coverage of the destruction of this Ottoman monument – the Serbs (for once) had not been cast in the role of destroyers.[300] After seeing the footage, a Belgrade student expressed what many clearly felt, saying 'I felt sick when I saw the pictures. No one has the right to destroy our history.'[301] Bogdan Bogdanović said the bridge was 'like a heavenly arch' and lamented 'It is as if one of my closest relatives has died.'[302] An appeal from Serbian professional heritage and museum bodies called for a stop to the destruction of cultural heritage in Bosnia-Herzegovina, citing the collapse of the Old Bridge as the culmination of that destruction; the appeal was addressed to 'the masters of war, and particularly to the institutions around the world that have not been sufficiently active in preserving this cultural and historic heritage.'[303]

In Croatia, official reaction to the destruction of the Old Bridge oscillated uneasily. Though responses generally expressed shock at the disappearance of the celebrated structure, Kemal Kurspahić has written on what he calls 'the-bridge-simply-fell' reporting of the assault on the Old Bridge in the mainstream Croatian media and the characterization of the final collapse in the leading pro-government daily *Vjesnik*, for instance, as being the result of artillery battles between Bosnian ('Muslim') government and HVO forces;[304] other Croatian reports also emphasized the Old Bridge's alleged military importance and that the collapse was simply the outcome of regular heavy fighting in the area of the bridge (rather than because of deliberate targeting).[305] However, it became difficult to maintain this

297 Kinzer, S. 1993b. Conflict in the Balkans; The Serbs; Belgrade vows to cut Bosnia aid, prompting skepticism and hope, *The New York Times*, 7 May.

298 Williams, C.J. 1993b. Bosnia's hopes fall with historic bridge in Mostar, *Los Angeles Times*, 12 November.

299 See for instance: Bosnia loses five-century-Old Bridge in fighting, *National Public Radio (NPR)*, Morning Edition, 12 November 1993, host Bob Edwards interviews Jerrie Hulme, UNHCR Mostar representative; Drakulic, S. 1993. 'How is it possible that the people of Bosnia agonise over the loss of a bridge while they become indifferent to the loss of a human life?,' *The Observer*, 21 November, News, 21; *Prlić et al*, testimony Philip Watkins, 22 May 2007, 18894 and 18900.

300 Williams, C.J. 1993b. Bosnia's hopes fall with historic bridge in Mostar.

301 Williams, C.J. 1993b. Bosnia's hopes fall with historic bridge in Mostar.

302 Sudetic, C. 1993b. Mostar's Old Bridge battered to death.

303 COE *Fifth Information Report*, Doc. 7070, 12 April 1994, *3. Further Reactions following the Destruction of the Old Bridge of Mostar, A. Appeal to the Yugoslav and World Public to Stop the Destruction of Cultural Heritage (Belgrade, 27 November 1993).*

304 Kurspahic, K. 2003. *Prime Time Crime: Balkan Media in War and Peace*, Washington, DC: United States Institute of Peace Press, 132.

305 See: Sarajevo, Zagreb radios report destruction of Mostar's Old Bridge, *Croatian Radio*, Zagreb, 14.00 GMT, 9 November 1993, *BBC Summary of World Broadcasts*, EE/1843/C, 11 November 1993.

position in the face of mounting evidence to the contrary and Kurspahić records the Bosnian writer Miljenko Jergović's observations on the evolution of how the bridge's destruction was reported in Croatia:

> 'Croatian Radio reported at first that Muslims had dynamited the bridge in order to blame Croats. Then, that same evening – thanks to a French cameraman who had taped the moments of the destruction – international TV reported that Croatian forces had done it. Late that night Croatian TV's "Frame on Frame" program showed the footage that forced the propagandists to look for new excuses. General Slobodan Praljak's "explanation" was that the bridge had to be destroyed because Muslims used it to bring ammunition with which to kill Croats. It was more than obvious to the public that there were a lot of lies in all of that.'[306]

Yet the day after the end of the Old Bridge, *Vijesnik* ran an editorial on the destruction unusually critical of the Bosnian Croats, demanding that those responsible be identified, 'especially if they are among us.'[307] While it was no surprise that opposition figures wrote elegies for the Old Bridge in the independent media like that by Predrag Matvejević in *Feral Tribune*,[308] it was more unexpected that American journalist Carol J Williams was able to find a Croatian government official willing to condemn the destruction as 'a tragic loss for everyone, not just the Muslims but for the Croats who live there as well,' though he added that part of the tragedy for the Croatian people was 'everyone will think that we did it.'[309] The Society of Croatian Art Historians issued a protest on the destruction of the Old Bridge, a 'magnificent monument of culture' that 'proved the possibility of co-existence of different faiths and nations of this region. It was a monument of understanding, ending as a victim of intolerance and hatred.'[310]

Whatever Croatian president Franjo Tuđman's real thoughts on the matter,[311] for public consumption he expressed regret over the destruction of the Old Bridge and hopes that

---

306 Kurspahic, K. 2003. *Prime Time Crime: Balkan Media in War and Peace*, 132.

307 Buric, N. 1993. Croatian daily suggests Croats destroyed 16th century bridge, *Associated Press*, 10 November.

308 Matvejević, P. 1993. Most, *Feral Tribune*, 16 November 1993. Croatian writer Slavenka Drakulić's widely-quoted elegy to the Old Bridge was published in a number of versions, including *The Observer* ('How is it possible that the people of Bosnia agonise over the loss of a bridge while they become indifferent to the loss of a human life?,' 21 November 1993, News, 21) and in *The New Republic* (Falling down: A Mostar Bridge elegy, 13 December 1993, 14–15).

309 Williams, C.J. 1993b. Bosnia's hopes fall with historic bridge in Mostar.

310 *Old Bridge – Monument of Peace*, exhibition catalogue, 2004 version, Mostar: Centar za Mir, *Protest of the Croatian Society of Historians of Art Regarding the Destruction of Old Bridge*, 26 November 1993, 12–13. Also referred to in COE *Fourth Information Report*, Doc. 6999, 19 January 1994, 3. *Reactions to the Destruction of the Old Bridge of Mostar on 9 November 1993.*

311 The so-called 'Tuđman Tapes,' secret recordings of hundreds his meetings made and kept by President Tuđman, were revealed by his successor Stjepan (Stipe) Mesić in 2000. One captured a meeting held by Tuđman with Bosnian Croat leaders, among them Mate Boban and Jadranko Prlić, on 10 November 1993, the day after the destruction of the Old Bridge. Tuđman enquires who *had* destroyed the Old Bridge and Boban blames the collapse on heavy rains. From transcripts of the ensuing conversation Tuđman's principal concern appears to be damage limitation in terms of the reactions of the international community and media to the destruction. See: Hedl, D. 2000b. Bosnian trade off, *Institute for War and Peace Reporting* (*IWPR*), 4 September 2000; *Santic, N.* 2005. Thousand pages about division, *Novi List*, 26 February 2005; Vele, F. 2011. Presuda Jadranku Prliću i ostalima potvrdit će da je Hrvatska bila agresor na BiH!, *Dnevni Avaz*, 23 April 2011.

the perpetrators would be brought to justice, while categorically denying any involvement of Croatia or the Croatian Army.[312] Meanwhile, on 24 November 1993 a commission of Croatian experts for reconstruction of the Old Bridge was established at the request of Žarko Domljan, Vice-President of the Croatian Parliament.[313]

The International Response During the War

To heritage professionals the attacks on Croatia's historic buildings – particularly the bombardment of the World Heritage Site of Dubrovnik – had seemed horrifying enough. But the news from Bosnia-Herzegovina of the overwhelming scale and intentional nature of the assaults was even more shocking. Most alarming of all, as the war went on and the destruction continued, the complete ineffectiveness of the international legal instruments for protecting cultural heritage in times of conflict that had been developed in the wake of World War Two was laid bare. The Ferhadija Mosque and the Old Bridge were not destroyed until 1993, a year and more after the war began, yet the international community seemed unable to prevent such acts happening. Far from being a deterrent and protection for historic monuments, when it came to the test the best humanitarian law like the 1954 *Hague Convention* appeared to offer was grounds for indictment in some future war crimes prosecutions – should the international community choose to pursue them.

The nature and extent of the destruction was not in doubt. The reports of the United Nations' Special Rapporteur on Human Rights Tadeusz Mazowiecki in February 1993[314] and the interim (1993) and final reports (1994) of the Commission of Experts appointed by the UN Security Council clearly identified the deliberate and systematic targeting of historic and religious structures as a fundamental aspect of ethnic cleansing, as well as being grave breaches of the Geneva Conventions and violations of international humanitarian law in their own right. The Commission of Experts' Final Report of 27 May 1994 included a section on destruction of cultural property.[315] So overwhelming was the amount of information accumulated on destruction of cultural property that the Commission of Experts did not have the resources to investigate and chose instead to focus on two iconic cases: the attacks on the World Heritage Site of Dubrovnik in Croatia in 1991 and the destruction of the Old Bridge (*Stari Most*) in Bosnia-Herzegovina in 1993. The study was to evolve into Annex IX (*Destruction of Cultural Property Report*) to the Commission's Final Report of 28 December 1994.[316]

The ferocity of the destruction in Bosnia-Herzegovina was to be a wake-up call for organizations and professionals concerned with the protection of cultural heritage. Yet ultimately what form did their action take in confronting and addressing the devastation in

312 COE *Fourth Information Report*, Doc. 6999, 19 January 1994, *3. Reactions to the Destruction of the Old Bridge of Mostar on 9 November 1993.*

313 COE *Fourth Information Report*, Doc. 6999, 19 January 1994, No. 70.

314 *Situation of Human Rights in the Territory of the Former Yugoslavia, Report on the Situation of Human Rights in the Territory of the Former Yugoslavia.* UN Doc. E/CN.4/1993/50, 10 February 1993.

315 *Final Report of the United Nations Commission of Experts*, S/1994/674, 27 May 1994, IV J, *Destruction of Cultural Property*, 66, Nos 285–97.

316 *Final Report of the United Nations Commission of Experts Established Pursuant to Security Council Resolution 780 (1992), Annex XI (Destruction of Cultural Property Report)*, S/1994/674/Add.2 (Vol. V) 28 December 1994

Bosnia? Despite the torrent of condemnation, in the event most major intergovernmental and international professional bodies with a concern for heritage protection were to demonstrate their inadequacy in responding to the greatest destruction of cultural heritage in Europe since World War Two. Even carrying out activities entirely outside Bosnia-Herzegovina, campaigns to help the built heritage, museum collections or support Bosnian heritage professionals were all but non-existent, unlike in similar spheres such, as for example, those concerned with libraries and librarians.[317]

The devastation of the wars in Bosnia and Croatia did spur efforts in policy-making on the protection of cultural property in times of conflict or disaster, as well as an upsurge of conferences and widespread debates of the issue, as discussed elsewhere in this volume. But while these activities conspired to give the impression that action was being taken in responding to the crisis, in reality most heritage bodies were unable or unwilling to offer help where help was needed – in Bosnia itself, despite the continual and desperate requests for assistance coming from Bosnia.[318] As late as mid-June 1993 (more than a year after the war began) Sejdalija Mustafić, the then director of the state *Institute for the Protection of Cultural, Historical and Natural Heritage of the Republic of Bosnia-Herzegovina* (*Zavod za zaštitu kulturno-historijskog i prirodnog naslijeđa BiH*) wrote in a letter faxed to Paris (the first official contact with the Bosnian heritage authorities) in response to an appeal from Council of Europe consultant expert Colin Kaiser for information on damage to the cultural heritage:

> Our Institute regularly informed Unesco and other international organisations dealing with the problem of destroying and ruining cultural heritage in B&H from the beginning of the aggression in April 1992, on the basis of data collected in the last year. However, we have not received any answer up to now.[319]

When even establishing contact with the relevant authorities in Bosnia seemed to pose insuperable difficulties, providing emergency aid in Bosnia-Herzegovina before the fighting stopped (and even after) was to going to prove still more challenging for most intergovernmental, governmental and the large international professional organizations concerned with the heritage, despite the presence of the multinational UN peacekeeping force UNPROFOR and innumerable international humanitarian aid organizations

317 Which was not the case with, for instance, Bosnia's libraries, especially the National Library, which attracted practical support from around the world which will not be dealt with here. See http://www.kakarigi.net/manu/projlist.htm. UNESCO eventually entered the arena its own campaign http://www.unesco.org/webworld/sarajevo/library.htm. The *Bosnian Manuscript Ingathering Project* attempted to reconstruct the holdings of the Oriental Institute through searching out photographs, microfilms and photocopies made by scholars over the years; founded by András Riedlmayer, Amila Buturović and Kemal Bakaršić, its website can be found at http://bmip.info/. The *International Federation of Library Associations* and Institutions (IFLA)/UNESCO *Solidarity with the Librarians in Sarajevo* campaign launched in January 1995 gave personal financial support to librarians.

318 COE *Second Information Report*, Doc. 6869, 29, letter dated 30 May 1993: 'There have been no official, cultural missions to Sarajevo from the outside world, and the condition of the cultural heritage in all senses of the term is exceedingly critical here, and gets more critical every day, despite all our efforts on its behalf. We desperately need this kind of foreign mission because we feel our culture has been abandoned by the world.'; Serdarević, M. 1997, 118–19.

319 *COE Third Information Report*, Doc. 6904, *20 September 1993, Letter to Dr Kaiser from Dr Sejdalija Mustafić, Director of the Institute for the Protection of the Cultural-historical and Natural Heritage of Bosnia and Herzegovina, Sarajevo (June 1993).*

operating across the country, chief among them the United Nations' refugee agency UNHCR[320] and the International Committee of the Red Cross (ICRC), along with the European Community Monitor Mission (ECMM) and state organizations like the UK's ODA (Overseas Development Administration) which carried out projects on all sides of the confrontation lines. There were literally hundreds of NGOs of all sizes providing some form of humanitarian aid (many of them run on a voluntary/activist/solidarity basis) that carried out operations in the country.[321] However, with the single exception of Bosnia-Herzegovina Heritage Rescue (BHHR), there were to be no NGOs working in the field of heritage protection in Bosnia during the war.

While this lack of action was in part due to the low priority given to cultural property protection by the multinational and intergovernmental organizations already operating in Bosnia-Herzegovina, especially in the early phase of the conflict, it undoubtedly had much to do with the mindset of those who ran heritage protection bodies. For instance, at a coordination meeting on the cultural heritage of former Yugoslavia of such bodies convened by the Parliamentary Assembly of the Council of Europe in Paris in June 1993, although a significant level of activity was planned in Croatia (always the easier option for international organizations), with regard to Bosnia there was none. The then director of ICOMOS, Leo van Nispen, revealed (remarkably) that his organization felt they had no role to play in assisting in the former Yugoslavia and that his priority was to raise the profile of cultural heritage protection to that of wildlife conservation.[322]

ICOM took no action in aiding museums in Bosnia-Herzegovina until the dramatic appearance of Dr Enver Imamović, director of the Zemaljski Muzej, at the organization's 18th General Assembly in Stavanger on 7 July 1995 after he had managed to leave Sarajevo and make his way to Norway. Imamović attended the Stavanger meeting as a result of the visit of Leif Pareli a curator from the Norwegian Folk Museum to Sarajevo in February 1995. Typically, Pareli's visit was not an initiative coming from the museum community; he travelled to Sarajevo as part of a Norwegian cultural delegation whose organizers thought it would be a good idea to include a museum professional in the group. Prompted into taking some form of action, ICOM got behind a survey of the situation of museums in Bosnia-Herzegovina in partnership with the Parliamentary Assembly of the Council of Europe.[323]

The Parliamentary Assembly of the Council of Europe and its Information Reports

In the event, the Parliamentary Assembly of the Council of Europe was to be the sole international body to take any form of early action in response to the assaults on Bosnia-Herzegovina's and Croatia's heritage, actively monitoring and documenting the damage

320 UNHCR = The United Nations High Commissioner for Refugees.

321 Estimates vary between 300–530 NGOs operating in Bosnia; see *Introduction*, note 22, for references.

322 COE *Third Information Report*, Doc. 6904, *20 September 1993*, 38.

323 The survey (the first of two) was carried out by Helen Walasek and Marian Wenzel of Bosnia-Herzegovina Heritage Rescue (BHHR) in October 1995. See COE *Ninth Information Report*, Doc. 7464, 19 January 1996, *2. Report on Museums in Bosnia-Herzegovina during the Present War*. ICOM had funded a survey of the situation in Croatian museums in October 1993; see COE *Fifth Information Report*, Doc. 7070, 12 April 1994, 1. *War Damage to the Museums, Galleries and Collections in the Republic of Croatia, Report by the International Council of Museums (ICOM) (February 1994)*.

through its ten *Information Reports*.[324] The reports had come about in 1992 as it became increasingly evident to the parliamentarians of the Council of Europe that the machinery of the 1954 Hague Convention was not functioning in relation to the destruction of cultural property in Bosnia and Croatia.[325] Furthermore, there was little about the destruction in the media and little interest being shown by the international community – even by the COE's own Committee of Ministers and Council for Cultural Co-operation (CDCC).[326] Accurate information was hard to come by, and the Assembly wanted to take the initiative urgently. By compiling independent assessments (based as far as possible on fact-finding missions and investigations) and circulating them widely, the COE Parliamentary Assembly hoped the international community could not say it did not know the true scale of the destruction and would make efforts to stop it happening. The COE parliamentarians aimed, too, at galvanizing the international conservation community into active intervention to protect the cultural heritage that had already been damaged. Unfortunately, their hopes were to be largely unrealized.

Although the bombardment of Dubrovnik had begun in October 1991 and the Assembly Sub-Committee on the Architectural and Artistic Heritage received a mandate to investigate in February 1992, and reports of massive destruction of the heritage in Bosnia coming through by May 1992, it was not until the end of November 1992 that the COE's consultant expert Dr Colin Kaiser was able to make a first fact-finding mission (the first by any international or intergovernmental organization) to assess and report on the destruction of cultural property and to make links with local organizations. Due to the difficulties described below, this first mission was only able to visit the regions of Dubrovnik and Mostar. In February 1993 the COE's first *Information Report on War Damage to the Cultural Heritage in Croatia and Bosnia-Herzegovina* with its shocking assessments finally appeared.[327]

The introduction to the first *Information Report* by the Rapporteur to the Committee on Culture and Education, the veteran French politician Jacques Baumel, makes instructive reading in decoding the lack of action in respect of the built heritage in Bosnia, as do the reports of fellow MEP Leni Fischer[328] and the reports and testimony of the COE's principal consultant expert Colin Kaiser during war crimes prosecutions at the ICTY.

Baumel noted the 'continued intergovernmental reticence' (and even outright opposition) in the both the COE and UNESCO[329] to taking action in what he called 'a cultural catastrophe in the heart of Europe.'[330] There was confusion, too, on what basis help would be offered – as emergency aid or as cultural co-operation? It was odd, he continued, that they remained so ignorant of the extent of the destruction, considering that UNPROFOR,

---

324 The Parliamentary Assembly of the Council of Europe published ten *Information Reports on War Damage to the Cultural Heritage in Croatia and Bosnia-Herzegovina* published between February 1993 and January 1997. See *Documenting the Destruction* in this volume for more on the reports.

325 See Colin Kaiser's testimonies at the ICTY, *Karadžić and Mladić*, Case No. IT-95-18-R61 and Case No. IT-95-5-R61, *Rule 61 Hearing*, 2 July 1996, 426–8 and *Blaškić* Case No. IT-1995-14-T, testimony, 16 July 1998, 10580–83.

326 COE *Information Report*, Doc. 6756, 2 February 1993, 2.

327 COE *Information Report*, Doc. 6756, 2 February 1993.

328 Fischer, L. 1994a. *Report on the Cultural Situation in the Former Yugoslavia*, Doc. 6989, Strasbourg: Parliamentary Assembly of the Council of Europe (COE), 14 January 1994 and 1994b. *Addendum to the Report on the Cultural Situation in the Former Yugoslavia*, Doc. 6989 Addendum, Strasbourg: Parliamentary Assembly of the Council of Europe, 13 April 1994.

329 COE *Information Report*, Doc. 6756, 2 February 1993, 2.

330 COE *Information Report*, Doc. 6756, 2 February 1993, 3.

UNHCR, and the European Community Monitor Mission (ECMM) were active throughout much of the war-torn territory and he remarked, too, on these organizations unwillingness either to share information or help fact-finding missions trying to report on the heritage.[331]

Humanitarian Aid to Heritage

Baumel's introduction raised early on what were to be *leit-motifs* throughout the conflict and afterwards: the need to integrate support for the cultural heritage into other humanitarian aid activities and the importance of giving emergency technical assistance *before* the conflict ended. A famed Resistance leader, well aware of the reality of war, Baumel also urged those who ought to be providing assistance to beware of hiding behind 'false reasons' for not intervening at an early stage, nor to feel ashamed of being concerned for the cultural heritage while men, women and children were suffering, pointing out the social, economic and psychological dimensions of the destruction: after all, the historic buildings attacked were where those same people had lived and worked before being turned into refugees, while the obliteration of cultural and religious monuments was an assault on a community's very identity. COE parliamentarians continued to press for emergency humanitarian aid to the damaged built heritage and protection for movable cultural property.[332]

But notwithstanding Baumel's words, concerns over being seen to be privileging buildings and objects over people remained a live issue throughout the war for international organizations involved with historic buildings and museums which unquestionably contributed to a reticence in acting, even though customary humanitarian law clearly mandated protection of a people's cultural property.[333] Even the committed parliamentarians of the COE were unable to transform their calls for emergency aid for the heritage into concrete actions on the ground.

For despite the publicity given to the destruction of iconic cultural and religious monuments in Bosnia, when it came to actually providing assistance, bodies with a remit for heritage protection felt prey to negative reactions from public opinion, the media and politicians when television screens showed the suffering of Bosnia's people night after night. For instance, the international condemnation that followed the destruction of the Old Bridge in Mostar in November 1993 was seen by some as 'emotional outcries' displaying 'disturbing signs of indifference to human suffering.'[334] Observed one (unnamed) Belgrade-based European diplomat recalling a similar outpouring after the JNA shelling of Dubrovnik in 1991: 'The Mostar bombing has sparked another round of debate over whether the world is more interested in preserving stones or people.'[335]

Yet such concerns were rarely an issue with Bosnians themselves who were usually only too glad for anyone to show an interest in preserving their heritage. As Roger Shrimplin,

331 COE *Information Report*, Doc. 6756, 2 February 1993, 5.

332 See, for instance, *Cultural situation in the Former Yugoslavia*, Recommendation 1239 (1994), Strasbourg: Parliamentary Assembly of the Council of Europe, 14 April 1994.

333 Even much more recently, when there is considerable data to validate the inclusion of restoration of cultural heritage as an important part of the recovery process in post-conflict societies, it seems still necessary to preface discussions of the subject with an apologia for doing so. Stanley-Price, N. 2007. The thread of continuity: Cultural heritage in post-war recovery, in *Cultural Heritage in Post-war Recovery*, ICCROM, papers from the ICCROM Forum held in Rome 4–6 October 2005, Rome.

334 Williams, C.J. 1993b. Bosnia's hopes fall with historic bridge in Mostar, *Los Angeles Times*, 12 November, 11.

335 Williams, C.J. 1993b. Bosnia's hopes fall with historic bridge in Mostar.

Chairman of the East Europe Committee of the Royal Institute of British Architects (RIBA), commented after his mission to Sarajevo in June 1993 with BHHR (see more below):

> … resentment in Bosnia about the rest of the world not taking any interest in what happens to their ruined monuments is really very deep … *Our* view is that *people* suffering is of first priority, never mind the monuments. But that is not *their* view. *They* take global destruction of their monuments very seriously indeed. It is time that their attitude about what is happening to their cultural heritage should be taken seriously by us.[336]

The Problem of Information Gathering and Access

Baumel's words also pointed up another feature of attempts (even at intergovernmental level) to assess the crimes against cultural and religious property going on in Bosnia-Herzegovina and Croatia: how those concerned with making authoritative official assessments remained so ignorant of the extent of the destruction, despite the presence of so many international organizations on the ground. The absence of coordination of the collection of data on such attacks throughout the war is striking. Yet information regarding the state of cultural and religious property *was* being collected at local level and reported on by representatives of many of the external actors in Bosnia-Herzegovina, including (among others) ECMM, UNHCR, ICRC and UNPROFOR, with reports travelling through administrative systems and up hierarchies where decisions were made with what to do with such information. British UNPROFOR battalions, for instance, produced daily reports (military information summaries, or *milinfosums*), as did ECMM monitors.[337]

Operating behind confrontation lines, officials of humanitarian organizations were often the only independent international witnesses to the destruction of cultural property or its aftermath. With their ability to move around difficult to access areas, staff of these organizations were frequently the first sources journalists and observers turned to for reliable information on violence against cultural and religious property. Reporting on the Serb-led assault on Bosnia's Muslim heritage in September 1992, Roy Gutman went to UNHCR in Geneva for comments and its spokesman Ron Redmond noted that so far no international agency had been able to study the extent of the damage. Yet Redmond himself had not long been in Sanski Most in north-west Bosnia, forcibly occupied by JNA and Serb forces at the end of April 1992, where he had observed a dynamited mosque: 'Someone had gone to quite some length to reduce it to a pile of stone. But I can't say how widespread the practice was,' he added.[338]

But the link between the destruction and ethnic cleansing soon became clear to these humanitarian aid representatives. On a field trip in early 1993 Francis O'Donnell, chief of the United Nations Volunteers (UNV) Humanitarian Relief Unit, passed through Serb-held Bosanska Gradiška, noting in his mission report it was ' … a pretty town, with tree-lined streets and squares, cafés, bistros, and all the allure of a prosperous cosmopolitan

336 COE *Third Information Report*, Doc. 6904, *20 September 1993*, 11.

337 The UNPROFOR and ECMM reports were noted in COE *Information Report*, Doc. 6756, 2 February 1993, No. 109. Considerable use of these reports has been made during war crimes prosecutions at the ICTY regarding the state of cultural property, as for instance, in the case of Guča Gora Franciscan monastery (discussed in detail below).

338 Gutman, R. 1992. Unholy war.

western European town. However, the mosque was now a heap of rubble, and the Muslim population all but eliminated.'[339]

UNHCR and ICRC officials were based in Bosnian Serb-controlled Banja Luka, scene of some of the most notorious episodes of ethnic cleansing and heritage destruction of the war, throughout the period of the demolition of all the city's 15 mosques and the majority of its Islamic and Ottoman heritage between April and December 1993. When the last three mosques in Banja Luka were burned down in September 1993, it was an ICRC official in the city who confirmed their destruction, telling a reporter: 'We saw one burning last night and the other one which was destroyed we saw this morning on our way to the office.'[340]

After a final round of demolitions on 15 December 1993 that included such important monuments as the *turbe* (mausolea) of Ferhad-paša and his *bajraktars* (standard-bearers) in the cemetery at the site of the already-destroyed Ferhadija Mosque, local UNHCR representatives (led by its head of office, Louis Gentile) were arrested and detained by Bosnian Serb police as they attempted to investigate and prevent the bulldozers of the municipal utility company from removing the structures' remains.[341] The destruction of these monuments brought Nicholas Morris, UNHCR Special Envoy to the former Yugoslavia, to Banja Luka where he challenged its notorious mayor Predrag Radić over the devastation with little result.[342] After his visit Morris stated it was apparent the Banja Luka authorities wanted not only to remove all minorities from the area, but to erase any trace that they had ever lived here.[343]

Thus it is clear that considerable amounts of information in relation to crimes against cultural and religious property were being gathered by many different international bodies in Bosnia-Herzegovina – the NGO Human Rights Watch, for instance, in its 1994 report on Banja Luka cites a report filed by locally-based international field personnel in December 1993 with detailed statistics of the destruction of Roman Catholic churches and ecclesiastical buildings in the municipality.[344] But it appears that such information was rarely being passed on or shared, even with intergovernmental organizations (like the COE) known to be actively collecting such data.[345] Whether this was due to the perceived low status of attacks against cultural property (in comparison with those against persons) among officials

339 O'Donnell, F. 1993. *UNV/HRU Report on Mission to Croatia and Bosnia-Herzegovina,* 30 April 1993, cited in Otim, B.B. 1996. Caught in the Crossfire: Dilemmas of Human Rights Protection in Former Yugoslavia, in *Volunteers Against Conflict: United Nations Volunteers* edited by United Nations Volunteers (UNV), Tokyo: United Nations University Press.

340 Judah, T. 1993. Banja Luka loses last mosque, *The Times*, 10 September.

341 Gentile, L. 1994. In Banja Luka, terror seems uncannily normal, *The New York Times*, 14 January.

342 Gušić, B. 1995. Život i smrt u okupiranoj banjaluci, *Oslobođenje*, 15–23 March, English translations 'Testimony of Bedrudin Gusic,' parts 1 and 2 available online at https://groups.google.com. Morris was UNHCR Special Envoy to the former Yugoslavia over 1993–1994.

343 *Banja Luka. Ethnic Cleansing Paradigm or Counterpoint to a Radical Future*, Briefing of the Commission on Security and Cooperation in Europe, Washington D.C.: CSCE, 1996, 10

344 Human Rights Watch 1994. *War Crimes in Bosnia-Hercegovina: UN Cease-Fire Won't Help Banja Luka*, D608, 1 June 1994.

345 Though UNPROFOR did pass on information about attacked cultural and religious property to ECMM monitors for investigation; see COE *Sixth Information Report*, Doc. 7133, 31 August 1994, 27. Information on attacks to cultural property was also being passed on the United Nations Commission of Experts gathering information on war crimes committed in the countries of the former Yugoslavia; see *Final Report of the United Nations Commission of Experts, Annex XI (Destruction of Cultural Property Report)*, S/1994/674/Add.2 (Vol. V), 28 December 1994, I.A.

at the higher levels of these bodies, or for other reasons, would bear investigation. At the early stages of the conflict COE consultant Colin Kaiser noted the daily reports produced by UNPROFOR and ECMM, his inability to access the information relating to cultural property in those reports, and the importance of the data that ECMM in particular was collecting.[346]

In his testimony at the ICTY in 1998 Kaiser outlined the problems those trying to assess the damage to the heritage in Bosnia-Herzegovina had faced.[347] To observers working in heritage preservation it was very soon apparent that the war was not simply about military objectives but also about identity, and cultural and religious monuments were being targeted as symbols of people's identity. But, he recalled, 'It was very clear to us that the international community, at the beginning of the Bosnian war, wasn't very interested in these things.'[348] There was a flurry of interest when the *Vijećnica* (the National Library) in Sarajevo was set on fire as 'an archetypal crime against cultural heritage.' Yet, such crimes were being carried out across Bosnia-Herzegovina.

Kaiser found it extremely difficult to organize a fact-finding mission as none of the intergovernmental organizations already operating in Bosnia (naming particularly UNPROFOR, UNHCR and ECMM) believed their mandate included the cultural heritage and none were willing to assist. So, he said ' … it took awhile to get into the field.'[349] Yet the ECMM teams, which first arrived in the former Yugoslavia to monitor implementation of the Brioni Accord (including human rights violations) some months before UN forces arrived,[350] but later carried out a range of humanitarian activities, did have a mandate to monitor the state of cultural heritage as early as autumn 1991 and did sporadically carry out such monitoring, largely in response to requests from the various religious authorities and from its own Humanitarian units.[351] The ICRC, too, had unmistakable responsibilities towards the protection of cultural and religious property under the terms of the Geneva Conventions.[352]

The problem of gaining access to Bosnia was mentioned again and again. In a January 1994 report COE MEP Leni Fischer commented on the obstacles to making on the spot observations, among them ' … the red-tape and lack of resources of the UNCHR and UNPROFOR that control access … ' and notes 'It is not for want of trying that a mission of the Committee on Culture and Education has not yet been able to set foot in Sarajevo.'[353] The imposition of a kind of cultural isolation on Bosnia by the United Nations – at least in the early part of the war – was mentioned by observers.[354]

346 COE *Information Report*, Doc. 6756, 2 February 1993, No. 109.

347 *Blaškić* Case No. IT-1995-14-T, 16 July 1998, 10580–83.

348 *Blaškić* Case, 10580.

349 *Blaškić* Case, 10581.

350 The ECMM was established under the terms of the Brioni Accord of 7 July 1991.

351 COE *Fifth Information Report*, Doc. 7070, 12 April 1994, 31–3. See also, for instance, the testimony of former ECMM monitor Hendrik Morsink, *Prosecutor v. Dario Kordić and Mario Čerkez 'Lašva Valley,'* ICTY Case No. (IT-95-14/2), 12 October 1999, 8210–11, regarding investigations of alleged attacks religious property in response to information received.

352 The ICRC's responsibility came from the Additional Protocols I (Article 53) and II (Article 16) to the Geneva Conventions of 1977. See the Introduction for more on this issue.

353 Fischer, L. 1994a. *Report on the Cultural Situation in the Former Yugoslavia*, COE Doc. 6989, 14 January 1994, *II Explanatory Memorandum*, Nos. 8–9.

354 Shrimplin, R. 1993. *Note on a Visit to Sarajevo to Review Damage to the Cultural Heritage and to Assess Prospects for a Repair/Restoration Programme June 1993*, COE *Third Information Report*, Doc. 6904, *20 September 1993, 33.*

At first Kaiser was reduced to assessing unverifiable information from a variety of sources with all the problems and inaccuracies described elsewhere in this volume.[355] On the importance of fact-finding missions and gathering information for his COE reports he recalled:

> The point of writing information reports was to hopefully build up some sort of reaction on the part of government members' states of the Council of Europe, not so much public opinion. If you didn't have good information, you couldn't hope that your reports were going to have much of an impact.[356]

But in 1994 Kaiser began to find an 'echo' from the ECMM coming, he said, from the monitors in the field themselves.[357] Thus with this strong interest from monitors on the ground and continued pressure from Kaiser, and with the more pressing claims of their political/military role now diminished, the ECMM did begin to systematically monitor and collect information on the state of the cultural heritage in Bosnia-Herzegovina (and Croatia) as part of its humanitarian monitoring activities, making cultural heritage monitoring a Standard Operating Procedure (SOP) from March 1994,[358] when it also began to provide operational assistance to the COE consultant. Local religious, cultural and political authorities were also encouraged use the ECMM for gathering information on damage to religious and cultural property.

The ECMM began publishing and disseminating its *Cultural Heritage Reports* from December 1994; the Chief of its Humanitarian Section noted that it was 'the only international body in the former Yugoslavia to attempt to monitor the cultural heritage, which' he continued, 'is an enormous task, considering the extent of the damage.'[359]

By summer of 1994 protection and restoration work in disastrously damaged Mostar *did* become possible after the Washington Agreement and the advent of the European Union Administration in Mostar (EUAM). The EUAM had a small budget for restoring historic buildings and funded the restoration of 14 structures, divided equally between the east and west sides of the city; one of the buildings restored was the Roznamedži Ibrahim-efendija mosque.[360] With funding from the Italian government, essential works consolidating the domes and minarets of the Karađoz-beg and Koski Mehmed-paša mosques began under the direction of Carlo Blasi of University of Florence, who later elaborated a plan for the protection and restoration of Mostar's historic core, training young local architects and urban planners through the process.[361]

In Sarajevo the Office of the Special Coordinator William Eagleton was tasked by the United Nations from April 1994 to mobilize and coordinate international support for the

---

355 COE *Second Information Report*, COE, Doc. 6869, 17 June 1993, 4. See also Richard Carlton and Helen Walasek, *Documenting the Destruction* in this volume.

356 *Blaškić* Case, 10582.

357 *Blaškić* Case, 10582.

358 COE *Sixth Information Report*, Doc. 7133, 31 August 1994, 2.

359 *Eighth Information Report on War Damage to the Cultural Heritage in Croatia and Bosnia-Herzegovina*, Parliamentary Assembly of the Council of Europe (COE), Doc. 7341, Strasbourg, 28 June 1995, 2.

360 Yarwood, J. 1999. *Rebuilding Mostar: Urban Reconstruction in a War Zone*, Liverpool: Liverpool University Press, 67. COE *Eighth Information Report*, Doc. 7341, 28 June 1995, report of Colin Kaiser, *4. The situation in Mostar*.

361 COE *Eighth Information Report*, Doc. 7341, 28 June 1995, report of Colin Kaiser.

restoration of essential public services in Sarajevo under the action plan *Restoring Life to Sarajevo* financed by funds from donor states and a UN Trust Fund. Help to cultural property appeared under the heading 'transitional needs' (following 'urgent needs') as 'provisional repairs to historic buildings in danger of collapse.'[362] The large City of Sarajevo Project of George Soros' Open Society Fund (OSF), which implemented its activities through its association with the legendary disaster relief specialist Fred Cuny and his organization Intertect, then working in Bosnia under the auspices of the International Rescue Committee (IRC), included emergency repairs to cultural buildings.[363] By mid-January 1995 repairs had been carried out on shell damage to the roof of the Umjetnička Galerija and attempts were being made to undertake repairs to one of the badly damaged roofs of the Zemaljski Muzej; there, however, due to the danger of Bosnian Serb sniper fire from which UNPROFOR could offer no protection, work was not carried out.[364] OSF also widely circulated its publication *Cultural Institutions and Monuments in Sarajevo* of March 1995, which as well as giving damage reports, listed the operational requirements of each institution.[365]

Meanwhile, the need for emergency intervention, if not full-blown restoration, was highlighted when the damaged minaret of the exquisitely frescoed Tabačica Mosque (c.1600) collapsed during high winds in August 1994.[366] In March 1995 Colin Kaiser commented gloomily that 'considering the media coverage that Mostar had in the past, the international attention to monuments is much less than could be expected, and were the EUAM not present it could be wondered if there would be any protection activity at all.'[367] But the prevailing mindset among international organizations was to wait until the war was over and it was frequently voiced that there was little point throwing money and effort into protecting or restoring buildings which might be destroyed again.[368]

Thus during the conflict, there was little will to support or encourage actions to preserve what remained of Bosnia's cultural heritage on the part of the political authorities of the countries contributing troops to UNPROFOR, nor in the other international bodies present in the country. In the United Kingdom, for instance, the efforts of Patrick Cormack MP (President of BHHR) to ensure the inclusion of a cultural heritage dimension to the Overseas Development Administration's (ODA) reconstruction and aid projects being carried out across Bosnia-Herzegovina during the war were completely without success.[369]

362 *Sixth Information Report on War Damage to the Cultural Heritage in Croatia and Bosnia-Herzegovina*, Parliamentary Assembly of the Council of Europe (COE), Doc. 7133, Strasbourg, 31 August 1994.

363 For more on OSF's association with Cuny, see Neier, A. 2011. Helping the Balkans Survive a Decade of War, *www.opensocietyfoundations.org*, 18 October 2011.

364 *Open Society City of Sarajevo Project Update*, 13 January 1995; in author's files. See also: Bollag, B. 1995. 'We must save what is left', *The Independent* [online], 20 April 1995.

365 Čengić, A. and Duraković, F. (eds) 1995. *Cultural Institutions and Monuments in Sarajevo*. Budapest: Open Society Institute, Open Society Project of Sarajevo, March 1995.

366 COE *Seventh Information Report*, Doc. 7308, 15 May 1995, 42 and COE *Eighth Information Report*, Doc. 7341, 28 June 1995, 17.

367 COE *Seventh Information Report*, Doc. 7308, 15 May 1995, 43.

368 Personal knowledge. See also COE *Eighth Information Report*, Doc. 7341, 28 June 1995, report from Marian Wenzel of Bosnia-Herzegovina Heritage Rescue (BHHR),

369 Personal knowledge. Cormack's efforts included arranging a meeting between BHHR representatives and others involved in heritage protection with the then Minister for Overseas Development, Lynda Chalker, on 5 July 1994 to discuss this issue. However, ODA officials in both Bosnia-Herzegovina and Croatia were very helpful to BHHR staff in providing use of office facilities and other assistance.

Fig. 1.28 The ruins of the exquisitely frescoed Tabačica Mosque, Mostar, c.1995–1996. © Kemal Zukić / Centra za Islamsku Arhitekturu, Sarajevo

Support and funding for preserving historic monuments and heritage organizations, like museums, if considered at all, was always a secondary priority in the face of what were deemed to be more pressing humanitarian concerns. However, there is little doubt of the deep sense of abandonment felt by those in Bosnia concerned with trying to protect their country's heritage by this evident lack of interest from the international community in helping them to preserve what remained of their cultural patrimony – especially in view of the circumstances in which it was destroyed.[370]

UNESCO's Response[371]

UNESCO's prominent role in the post-war rebuilding of Mostar's Old Bridge has obscured its lack of real action with respect to the cultural heritage for a considerable part of the war in Bosnia-Herzegovina. The inability of UNESCO (expected by many to assume the lead in mobilizing and coordinating international activities) to take action at an early stage appeared especially remiss in the face of the reports by Human Rights Rapporteur Tadeusz Mazowiecki and the Commission of Experts working within the United Nations framework and when other UN agencies (like UNHCR) had been carrying out operations across the country from the start of the conflict. UNESCO, too, had specific responsibilities for the

370 See, for instance: Bollag, B. 1995. 'We must save what is left', *The Independent*, 20 April.

371 For UNESCO's own account of its activities in respect of cultural property protection during the wars in Bosnia and Croatia, see *Information on the Implementation of the Convention for the Protection of Cultural Property in the Event of Armed Conflict, The Hague 1954, 1995 Reports*, UNESCO, Doc. CLT-95/WS/13, Paris, December 1995.

application of the Hague Convention for the Protection of Cultural Property in the Event of Armed Conflict of 1954.[372]

Decision 7.5 adopted by UNESCO's Executive Board at its 139th session in May 1992 invited the Director-General 'as soon as the situation permits, to send a mission to Bosnia-Herzegovina to determine the damage to educational, historical, archaeological and cultural property in the region and to explore the feasibility of sending emergency assistance to Bosnia-Herzegovina,' requesting him to report back at the 140th session of the board in October 1992.[373] On 16 June 1992 the UNESCO Director-General received a letter from the United Nations' Secretary-General informing him that he had requested the Commander of UNPROFOR to provide UNESCO with the 'assistance it needed for the rapid dispatch of the mission to Bosnia-Herzegovina,' though it was indicated that the situation might not permit such a mission in the immediate future.[374]

By October 1992 the mission had still not taken place and the Executive Board repeated its request that the Director-General implement its decision, asking as well that he 'examine how best the diverse cultural, religious and ethnic heritage of the population of Bosnia-Herzegovina can be safeguarded' and 'the possibility of setting up in coordination with other organizations of the United Nations family a Bosnia-Herzegovina emergency programme to address the immediate issues and long-term rehabilitation needs in the fields of culture, education, science and communication.'[375] With no UNESCO mission to Bosnia yet taking place, the board reiterated its request that 139 EX/Decision 7.5 be implemented at its 141st session in May 1993.[376]

While the COE's first *Information Report* in February 1993 noted that the United Nations seemed to have put a moratorium on UNESCO missions to Bosnia,[377] the will of UNESCO heritage officials to actually travel to Sarajevo could be called into question. In June 1993 Mounir Bouchenaki, Director of UNESCO's Physical Heritage Division, claimed that UNESCO 'lacked means for effective intervention on the spot' and 'had failed to gain access to Mostar or elsewhere in BiH.'[378]

Hoping for implementation of the 1954 Hague Convention for the Protection of Cultural Property in the Event of Armed Conflict, information on Bosnia's cultural heritage had been sent to UNESCO from Sarajevo as early as 6 April 1992.[379] Following the fall of Jajce to the Bosnian Serb Army at the end of October, on 20 December 1992 the city's Franciscan

---

372 *Convention for the Protection of Cultural Property in the Event of Armed Conflict*, The Hague, 14 May 1954. Apart from technical matters, UNESCO's responsibilities lay under Article 22 *Conciliation Procedure*, Article 23 *Assistance of UNESCO*, Article 19 *Conflicts Not of an International Character*.

373 *Decisions Adopted by the Executive Board at its 139th session*, UNESCO, 139 EX/ Decisions, Decision 7.5, (Paris, 18–27 May 1992), Paris, 16 June 1992.

374 *Report by the Director-General on the Implementation of 139 Ex/Decision 7.5*, 140th Session of the Executive Board, Item 8.4 of the provisional agenda, 140 EX/25, Paris, 5 October 1992.

375 *Decisions Adopted by the Executive Board at its 140th session*, UNESCO, 140 EX/ Decisions, 8.4, (Paris, 12–30 October 1992), Paris, 18 November 1992.

376 *Decisions Adopted by the Executive Board at its 141st session*, UNESCO, 141 EX/ Decisions, 9.3, (Paris, 10–28 May 1993), Paris, 17 June 1993.

377 COE *Information Report*, Doc. 6756, 2 February 1993, No. 108.

378 COE *Third Information Report*, Doc. 6904, *20 September 1993*, 37.

379 Serdarević, M. 1997, 118. According to Serdarević the first communication with UNESCO was a letter sent to Lyndel Prott, a legal advisor at UNESCO on 6 April 1992 (apparently by fax) detailing activities to protect Sarajevo's heritage. Other communications describing the losses to Bosnia-Herzegovina's cultural heritage and appeals for help in protecting it followed.

community made a plea to UNESCO hoping to save the monuments of culture and history there and begged the organization to mediate with the Serbian authorities to prevent their demolition. But even as their appeal was being written, the monks later learned, Jajce's Franciscan church had already been destroyed.[380] A stream of other communications from the Bosnian authorities and heritage organizations describing the losses to Bosnia-Herzegovina's cultural heritage and appeals for help in protecting it to UNESCO and other bodies followed with little response.[381]

Bosnia-Herzegovina became a member state of UNESCO on 2 June 1993.[382] Already noted above is the letter sent by the Bosnian authorities on 9 July 1993 to the UN Security Council, UNESCO and the Council of Europe requesting protection for the Old Bridge in Mostar and the presence of a UNESCO team which received no reply. Three days later on 12 July 1993 the state of Bosnia-Herzegovina acceded to all existing UNESCO conventions on protection of cultural property, including the 1954 Hague Convention.[383] When the Old Bridge was destroyed four months later, the Bosnian government sent another letter to the same international bodies reminding them of their lack of action.[384] In the court of public opinion, however, it was UNESCO that took the brunt of the blame for the failure of the international community to protect Bosnia-Herzegovina's most famous historic monument when a few days later a cartoon by Božo Stefanović appeared in the main Sarajevo daily, *Oslobođenje (Oslobodjenje)*, showing UNESCO collapsing along with the Old Bridge.[385]

The unwillingness of UNESCO heritage functionaries to even attempt to send a mission to Sarajevo was regarded with dismay in Bosnia, contributing to an increasingly negative view of the organization which came to be perceived as having very little interest in preserving the country's cultural heritage and even to be actively hostile towards it.[386] Mevlida Serdarević, legal advisor of the state heritage protection institute and one of those most active in trying to protect Bosnia's heritage and alert the international community to its destruction, later wrote of the constant official requests from Bosnia-Herzegovina sent to the UNESCO Director-General to send a mission to Sarajevo and each time receiving the

380 Apel fra Pere Jurišića UNESCO-u od 20.XII.1992, *Bosna Srebrena*, 3 (1993), 22. Cited in Raguž, J. 2013. 261–2.

381 Serdarević, M. 1997, 156.

382 http://www.unesco.org/new/en/unesco/worldwide/europe-and-north-america/bosnia-and-herzegovina/

383 Convention for the Protection of Cultural Property in the Event of Armed Conflict (1954); Convention on Prohibiting and Preventing the Illicit Import, Export and Transfer of Cultural Property (1970); Convention on the Protection of World Cultural and Natural Heritage (1972).

384 See: Bosnian authorities protest over destruction of Mostar Old Bridge, *Radio Bosnia-Hercegovina*, Sarajevo, 21.00 GMT, 9 November 1993, *BBC Summary of World Broadcasts*, EE/1843/C, 11 November 1993: 'Dear Sirs, on 9th July this year the government of Bosnia-Hercegovina sent a request to the UN Security Council, UNESCO and the Council of Europe demanding that they prevent the destruction of the old bridge in Mostar and urgently send a UNESCO team to ensure its protection. The old bridge in Mostar has been destroyed. Today, 9th November 1993, the Croatian Army and the Croatian Defence Council [HVO] destroyed the most important monument of Bosnian culture.' See also: Dr. Silajdžić poziva UN i UNESCO: spasite Stari Most, *Oslobođenje,* 10 July 1993, 8, cited in Makas, E. 2007. *Representing Competing Identities: Building and Rebuilding in Postwar Mostar*, 209.

385 Božo Stefanović cartoon, *Oslobođenje,* November 15, 1993, 2, cited and reproduced in Makas, E. 2007. *Representing Competing Identities*, 209–10.

386 Serdarević, M. 1997, 157. The author heard these views on many occasions from senior heritage professionals during the war.

reply that it was too dangerous.[387] What, Serdarević enquired, would have happened if the officials of UNHCR and other organizations had behaved in the same way?[388]

Thus, as UNESCO made its excuses, it was left to the tiny UK organization BHHR to make the first international mission to Sarajevo to assess the damage to cultural heritage in mid-June 1993 and to hold meetings with the relevant Bosnian authorities and institutions, sending its director Marian Wenzel and Roger Shrimplin, Chairman of the East Europe Committee of the Royal Institute of British Architects (RIBA).[389] At the meetings Bosnian government ministers and the director of the state heritage preservation institute, Sejdalija Mustafić, again pleaded for a mission by a senior UNESCO delegation with experts able to advise on protecting and stabilizing monuments and hoped for a permanent UNESCO presence in Sarajevo, a plea seconded by Wenzel and Shrimplin. Planning minister Mustafa Dizdarević highlighted what they hoped from a UNESCO mission: help in obtaining reliable information on losses to the heritage in occupied areas, help in saving what was left, and help for Bosnian experts to expand their expertise along with technical aid.[390] Local heritage professionals emphasized it was important to recognize that the destruction was not a *result* of the war, but rather the *objective* of the war and argued for a humanitarian aid for monuments, for instance, in rebuilding historic housing complexes.

But UNESCO was not to send its first mission to Bosnia-Herzegovina to report on the state of the cultural and architectural heritage in Sarajevo and Mostar until 9 June 1994, a year after the visit by BHHR and more than two years after the war began and Executive Board Decision 139 Ex/7.5.[391] It was a joint mission with the Council of Europe and followed a visit to Sarajevo by UNESCO Director-General Federico Mayor on 3 May. A UNESCO office was opened in Sarajevo in autumn 1994 and thereafter its activities in the field of cultural heritage began to improve, with, for instance, support put in place for emergency repairs to the badly damaged Zemaljski Muzej via UNESCO and the Open Society Fund to be implemented by Intertect.[392] In April 1995 a UNESCO branch opened in Mostar headed by the COE's consultant expert Colin Kaiser which was to be the focus of UNESCO's

387 Serdarević, M. 1997, 157.

388 Serdarević, M. 1997, 157. Serdarević wrote: 'Šta bi bilo da su na isti način postupali odgovorni UNHCR-a i drugih sličnih institucija?'

389 COE *Third Information Report*, Doc. 6904, *20 September 1993,* 11. Wenzel and Shrimplin were in Sarajevo 12–18 June 1993; Wenzel also visited Mostar on 7–8 June 1993. Wenzel and Shrimplin's reports are included in the COE report.

390 COE *Third Information Report*, Doc. 6904, *20 September 1993, 11.*

391 *The Situation of the Cultural and Architectural Heritage and of Educational Institutions in Bosnia and Herzegovina. 145th Session of the General Conference. Provisional Agenda*, UNESCO, Doc. 145 EX/36, Paris, 7 October 1994, 8.2, No. 5; *Information as an Instrument for Protection against War Damages to the Cultural Heritage, Report from a Seminar*, June 1994, ICOMOS Sweden et al, Stockholm: Svenska Unescoradets skriftserie, 4/1994, 25. UNESCO's Division of the General Information Programme also sent an expert mission to Sarajevo 14–18 June 1994 to discuss the problems of the National Library: Doc. 145 EX/36, 7 October 1994, No. 6; see also *Revival of the National and University Library Bosnia and Herzegovina in Sarajevo, Activities of the Programme*, http://www.unesco.org/webworld/sarajevo/srjevact.htm.

392 *The Situation of the Cultural and Architectural Heritage and of Educational Institutions in Bosnia and Herzegovina. 146th Session of the General Conference: Provisional Agenda*, UNESCO, Doc. 146 EX/46, Paris, 3 May 1995, 9.3. The UNESCO's Division of the General Information Programme also sent George MacKenzie, Deputy Secretary-General of the International Council on Archives (ICA) to assess archives in Sarajevo and Mostar in early February 1995 in accordance with 145/EX/Decision 8.2: Doc. 146 EX/46, 3 May 1995, No. 13; see also: *Archives in Bosnia and Herzegovina, UNISIST Newsletter*, General Information Programme, UNESCO, 23:1, 1995, 17.

cultural heritage activities: coordinating local and international activities, preparing Mostar's nomination for the World Heritage List, providing technical assistance and preparing an action plan for restoration, as well as acting a base for cultural heritage monitoring.'[393]

The Organisation of the Islamic Conference (OIC)

At its Fifth Extraordinary Session of the Islamic Conference of Foreign Ministers in Istanbul on 17–18 June 1992 (called on the initiative of Turkey specifically to discuss the situation in Bosnia-Herzegovina) the Organisation of the Islamic Conference (OIC), like other major intergovernmental bodies, issued a strong condemnation of the 'wanton destruction' of the country's 'cities and historical, religious and cultural heritage.'[394] The OIC had an already established research programme on the history and culture of Bosnia-Herzegovina at its Istanbul-based cultural arm IRCICA (Research Center for Islamic History, Art and Culture) which did much to raise awareness of the extent of the destruction of the country's cultural heritage through its publications and exhibitions.[395]

The OIC's Twenty-First Islamic Conference of Foreign Ministers in Karachi held over 25–29 April 1993 issued Resolution No. 19/21-C on *Destruction of Mosques and Islamic Monuments in Bosnia and Herzegovina* that directed IRCICA to 'take the necessary steps for the assessment of the damage made to the cultural heritage of Bosnia and Herzegovina and to create a fund to facilitate the restoration of these monuments in cooperation with the International Commission for the Preservation of Islamic Cultural Heritage (ICPICH) and UNESCO.'[396] The declaration also called on member states to 'help the restoration of mosques, Islamic monuments, libraries and cultural heritage in Bosnia-Herzegovina.'[397]

IRCICA's major contribution was *Mostar 2004*, a project conceived and driven by Mostar architect Amir Pašić.[398] The series of regular workshops held between 1994 and 2004 aimed to formulate a restoration plan and methodology for the reconstruction of Mostar. The initial workshops took place in Istanbul and did not take place in Mostar itself until 1997. Here again, despite the OIC resolution of April 1993 calling for action by IRCICA, there was to be no work carried out in Bosnia until after the end of the war and relatively little in terms of concrete support for restoration of monuments afterwards.

Though the efforts of IRCICA were welcome, they should be taken in the context of the substantial financial resources of many OIC state members, some of which (like Saudi Arabia) were deeply engaged in providing extensive humanitarian aid in Bosnia-Herzegovina. Few heeded the call to help with the restoration of Ottoman mosques and other Islamic monuments.

---

393 *The Situation of the Cultural and Architectural Heritage and of Educational Institutions in Bosnia and Herzegovina. 145th Session of the General Conference. Provisional Agenda*, UNESCO, Doc. 145 EX/36, Paris, 7 October 1994, 8.12. COE *Eighth Information Report*, Doc. 7341, 28 June 1995, 17.

394 *Final Communique of the Fifth Extraordinary Session of the Islamic Conference of Foreign Ministers*, Istanbul, Turkey, 17–18 June 1992, Ex-ICFM/5-92/FC (Final), 12.

395 IRCICA = *Research Center for Islamic History, Art and Culture*. http://www.ircica.org/. See Richard Carlton and Helen Walasek, *Documenting the Destruction* in this volume for more on IRCICA's publications.

396 *Resolution No. 19/21-C, Destruction of Mosques and Islamic Monuments in Bosnia & Herzegovina* of the Twenty-First Islamic Conference of Foreign Ministers. Karachi, Pakistan, 25–29 April 1993, No. 3.

397 *Resolution No. 19/21-C*, No. 4.

398 The rebuilt Old Bridge was officially opened on 23 July 2004.

Fig. 1.29 The Historical Museum of Bosnia-Herzegovina / Historijski Muzej BiH, Sarajevo. Marian Wenzel, Director of BHHR (right), Soniboj Staničić, Deputy Director (centre) and Stanislav Tomašević, Legal Officer (left) examine war damage to the museum during a mission for the Parliamentary Assembly of the Council of Europe, October 1995. © Helen Walasek

Non-governmental Organizations

As early as November 1992 UNESCO's Director-General Federico Mayor appeared keen to pass responsibility for emergency aid to the cultural heritage in the former Yugoslavia on to non-governmental organizations who were able to act quickly where (apparently) UNESCO could not, envisioning the creation of a cultural heritage equivalent of Médecins sans Frontières.[399] In the event, despite innumerable NGOs of every type, size and degree of formality and informality operating in Bosnia-Herzegovina throughout the war there was to be only one in the field of cultural heritage. The UK-based Bosnia-Herzegovina Heritage Rescue (BHHR) was to be the first and only international group working in heritage protection accredited by the UN's refugee agency UNHCR as a humanitarian aid organization during the Bosnian War.

Founded in September 1992 by art historian Marian Wenzel after desperate pleas from colleagues in Sarajevo, BHHR provided the earliest independent lists of damage to cultural property later that year and sent the first cultural mission to Sarajevo made by any organization in June 1993. Wenzel was one of the few international scholars knowledgeable

399 *Address by Mr Federico Mayor, at the First International Symposium on Human Rights La Laguna (Tenerife)*, UNESCO, Doc. DG/92/44, 1 November 1992.

about Bosnia's medieval culture and heritage and had carried out fieldwork in the country from the 1960s. Operating on limited finances, BHHR's important achievements were (apart from awareness-raising conferences held in London)[400] its efforts to assist the Zemaljski Muzej (which appointed the NGO its official representative outside Bosnia), bringing the museum much-needed funds, technical advice (most especially from Robert Child, Head of Conservation at the National Museums and Galleries of Wales), supplies and equipment and its role as a mediator between local heritage institutions, such as the Historical Museum with whose staff it worked closely, and the confusing array of international agencies operating inside and outside the country.[401]

Demonstrating that it was possible to assist heritage institutions while the war was ongoing, at the request of the Zemaljski Muzej, in March 1995 BHHR brought to Sarajevo half the funds collected by the Swiss National Museum during a public appeal on its behalf which the museum used for constructing temporary roofing for the Natural Sciences and Archaeology buildings, and upgrading the security of its basement storage depots. Among its activities after the war BHHR carried out a survey of mined graveyards in Croatia (1996), assessing how they were prioritized in mine clearance programmes and the humanitarian implications for return of refugees and displaced persons; held the first training days for museum professionals from across Bosnia-Herzegovina since the beginning of the war; and found funding for the professional training of a textile conservator for the Zemaljski Muzej in the UK.

While the French organization Patrimoine sans Frontières (PSF)[402] was founded in 1992 with support from the French Ministry of Culture in response to the heritage crisis in the former Yugoslavia, it did not carry out work inside Bosnia-Herzegovina, though it did facilitate the visit of a group of Sarajevan architects to France in 1994. In Sweden, Cultural Heritage without Borders (CHwB), also reacting to the destruction in Bosnia-Herzegovina, was not founded until 1995 and carried out no work in country until after end of the war.

The International Response After Dayton

After the signing of the Dayton Peace Agreement in December 1995 much was to change regarding the involvement of international heritage organizations in the country. One of 11 annexes to the peace accord, *Annex 8*, sought to address the destruction of cultural heritage during the conflict by the creation of a *Commission to Preserve National Monuments* to be run under the aegis of UNESCO for its first five years.[403] There was, too, a growing (but still slight) interest in cultural heritage issues among the aid agencies representing national governments. Yet, once again, there was to be just one international organization actively

400 On 25 November 1993 at the Courtauld Institute, London and on 15 June 1995 at the Society of Antiquaries, London.

401 Information on the activities of BHHR can be found in COE *Information Reports* 1–6 and 8–10.

402 For more information on the activities of Patrimoine sans Frontières see http://www.patrimsf.org/projet/index.php.

403 See below and Valery Perry in this volume for more on Annex 8 and the Commission to Preserve National Monuments. *Annex 8: Commission to Preserve National Monuments*. The full text of Annex 8 is available at http://www.ohr.int/dpa/default.asp?content_id=376.

implementing the restoration of historic monuments across Bosnia-Herzegovina according to conservation principles – the Swedish NGO Cultural Heritage without Borders.[404]

Inspired by the work of BHHR, CHwB's founding objective was to promote the incorporation of cultural heritage restoration into humanitarian emergency aid and with its exceptional connections to the Swedish government and official heritage bodies it was instrumental in convincing the Swedish International Development Cooperation Agency (Sida) to include cultural heritage restoration in its post-conflict aid programme in Bosnia-Herzegovina.[405] CHwB collaborated with local institutes for the protection of monuments to restore structures of all types, among them the Zemaljski Muzej, the Kuršumlija Mosque in Maglaj and the Orthodox Church of the Presentation of the Virgin in Zavala. Social impacts were a key consideration when identifying buildings to restore and its most ambitious project in this respect was at the medieval town of Jajce where work on a range of buildings in the historic Old Town integrated the return of Muslim refugees (by restoring homes in partnership with a traditional aid organization), the rehabilitation of their cultural environment (restoring two historic mosques) while helping the revival of one of the town's pre-war economic bases – tourism (restoration of parts of the ancient town walls and a seventeenth-century Ottoman house for use as a tourist office).

Following their successful nomination to include the small Ottoman town of Počitelj on the World Monuments Watch 100 Most Endangered Sites list for 1996,[406] the University of York's Post-war Reconstruction and Development Unit (PRDU) prepared an integrated plan for its reconstruction and revitalization.[407] Restoration of the cultural heritage was one element in a strategy that sought to encourage the return of the original Muslim residents while taking into consideration the needs of Bosnian Croats displaced from central Bosnia who had been housed there, repair infrastructure and ensure the economic sustainability of the town and its hinterland. PRDU hoped the plan would be a model for future interventions in the country. PRDU director Sultan Barakat later chaired the experts group appointed to oversee realization of the project beginning in 2001, which was implemented through the Federation's Ministry of Physical Planning and the Environment on the basis of the PRDU study.[408]

Elsewhere, apart from the World Bank-driven projects in Mostar, a mixed bag of individual restoration projects of war-damaged buildings were financed by a variety of international sources for a variety of reasons (for an extensive treatment of post-war restoration, see *Domains of Restoration* elsewhere in this volume). Many projects, particularly Austro-Hungarian period institutional buildings, were part of general reconstruction efforts and were not necessarily seen as heritage restoration. Despite being the focus of international attention for years, the *Vijećnica* (the National Library) could only attract the attention of the Austrian government for the first phase of its reconstruction which was to be bogged down for many years.[409] The restoration of the *Umjetnička Galerija* was entirely due to an

404 See CHwB's website for more information on its activities at http://www.chwb.org.

405 See Tina Wik in this volume for a detailed account of CHwB's work in Bosnia-Herzegovina.

406 See the World Monuments Fund for more on the *100 Most Endangered Sites* list at http://www.wmf.org/watch.

407 Barakat, S. and Wilson, C. 1997. *The Revitalisation of the Historic Settlement of Počitelj*, York: PRDU, University of York.

408 Ward, G. (ed.) 2001. Reconstruction of Pocitelj to go ahead. *Revival*, Iss. 15, August 2001. York: Post-war Reconstruction and Development Unit (PRDU), University of York, Institute of Advance Architectural Studies.

409 See Helen Walasek, *Domains of Restoration*, in this volume.

accident of history; here, funding came from the Swiss Embassy because the gallery holds one of the most important collections of works by the Swiss artist Ferdinand Hodler.[410]

While such types of projects were not unwelcome, they illustrate the randomness of international approaches to heritage restoration. Sultan Barakat, PRDU's founder-director, has commented that the international aid community did not know how to situate heritage in the broader scheme of reconstruction in Bosnia. Further, there was polarization between aid agencies and the heritage community, who somehow thought they operated on a more elevated level – an attitude prevalent among both international and local heritage professionals.[411]

None of the internationally-financed restorations took place in Republika Srpska[412] where help was badly needed by communities of returnees.[413] And nowhere was the model proposed by PRDU (and partially explored by CHwB) used as a basis for intervention in small to medium-sized communities where such interventions could have served to stimulate refugee return and economic recovery in the surrounding area. Here, too, where destruction was actually the greatest, there was ample evidence that the practices of traditional emergency aid organizations aimed at rebuilding houses was contributing to the further degradation of complexes of historic vernacular buildings which could easily have been restored at similar cost.[414] In Počitelj, for example Ottoman houses were repaired by the United Methodist relief agency UMCOR using inappropriate materials with no consideration of their historic nature, unnecessarily devaluing the ensemble as a whole,[415] only one example of a phenomenon noted elsewhere by Hadžimuhamedović and Wik (in this volume) and by others.[416] But it was the main urban centres which continued to receive the bulk of international attention.

Contributing to the restoration of religious buildings was often felt to be problematic by many national governments and international organizations, exposing them (they felt) to accusations of taking sides with one particular religious or 'national' group. The governments of Islamic states and Islamic foundations had no such qualms about involvement with funding religious structures. But this type of support went almost entirely towards building new mosques or rebuilding completely destroyed former historic mosques in forms alien to

410 The collection was formed by Jeanne Charles-Cerani, a former model of Hodler's, who later married the Yugoslav diplomat Mehmed Čišić. On Čišić's retirement the couple lived in Mostar and the collection purchased for Bosnia-Herzegovina in the 1960s.

411 Author interview with Sultan Barakat, 21 December 2011.

412 Of war-damaged structures, that is. In its early phase CHwB undertook some restoration work in Republika Srpska of buildings not damaged by the war, but this was patently not in its mission and ceased with a change of project manager.

413 See Helen Walasek, *Domains of Restoration* in this volume.

414 See Tina Wik and Amra Hadžimuhamedović in this volume on this phenomenon. Also noted by Calame, J. and Sechler, K. 2004. Is preservation missing the point? Cultural heritage in the service of social development, *Future Anterior: Journal of Historic Preservation History, Theory and Criticism, 1*(1), Spring 2004, 57–63.

415 See the Commission to Preserve National Monuments website on Počitelj. http://kons.gov.ba/main.php?id_struct=50&lang=4&action=view&id=780.

416 See Ward, G. (ed.) 1997. *Analysis of Cultural Heritage Damaged by War in Bosnia-Herzegovina*, Synopsis of PRDU Research and Training Project, Bosnia-Herzegovina January-March 1994, Post-war Reconstruction & Development Unit (PRDU), University of York, Institute of Advance Architectural Studies, *Revival*, Iss. 8, August 1997 and Calame, J. and Sechler, K. 2004. on this phenomenon.

the Bosnian tradition.[417] One notorious intervention that resulted from funding from Saudi sources for 'restoration' was the whitewashing of the decorated interior of the monumental Gazi Husrev-beg Mosque (1531) in Sarajevo which provoked a wide public outcry from local residents.[418] Relatively little support from Islamic sources went towards the restoration of historic Ottoman or traditional Bosnian style mosques, with the exception of a handful, mainly in Mostar, where, for instance, the Saudi High Commission for Relief to Bosnia-Herzegovina contributed to the restoration of the Tabačica Mosque[419] and the Nesuh-aga Vučijaković Mosque, to which the Kingdom of Jordan also gave funding.[420]

But it was Mostar's Old Bridge and the city's historic core which were to perfectly fulfil the requirements of the international community in its quest for symbolic value and they became the recipient of a vast outpouring and attention from donors – swallowing by far the greater part of the funds spent in Bosnia-Herzegovina for restoration of the cultural heritage. Yet, there had been any number of attempts to begin the restoration of the Old Bridge that had come to nothing, largely due to perceptions that it was a solely a 'Muslim' project. It was only when the World Bank stepped into the picture in 1998, with UNESCO in tow, that international donors fell in behind. Such was the enthusiasm for participating in the restoration of Mostar's Old Town to the virtual exclusion of all else, that one 'restoration' project carried out in the framework of IRCICA's *Mostar 2004* programme and funded by the ruler of the United Arab Emirates was the complete rebuilding of the Nezir-aga Mosque – a structure which had been ceased being used as a mosque in 1932 and completely demolished in 1950 after extensive damage during World War Two – possibly not the most urgent candidate for heritage reconstruction in the city.[421]

But if participating in the restoration of buildings proved difficult for donors and organizations, holding workshops and training courses did not. These required no long-term commitment and few of the bureaucratic and operational problems that a restoration could incur. Nonetheless, training and professional development were highly needed in a country where a large proportion of senior professionals had fled the country, retired or died, while the younger generation had suffered from years of inadequate and patchy teaching or were simply unable to practice their profession. Coming from the state-controlled heritage hierarchies of socialist Yugoslavia, none had the contractual or proposal-writing skills needed to negotiate the torturous search for donors in the post-war environment.

417 Zubčević, A. 2006. Islamic sites in Bosnia: 10 years after the war, *Islamica Magazine* [online], Iss. 15, 88.

418 The interior has since been redecorated by conservator/restorer Nihad Čengić. For more on the mosque see *Gazi Husrev Bey Complex* on archnet at http://archnet.org/library/sites/one-site.jsp?site_id=9244. See also Naegele, J. 2000. Saudi aid workers bulldoze Balkan monuments, *Radio Free Europe*, Newsline, 8 August; for pictures of the mosque interior after it had been whitewashed see Sells, M.A. 2001. Erasing culture: Wahhabism, Buddhism, Balkan mosques, 2 April, http://www.haverford.edu/relg/sells/reports/WahhabismBuddhasBegova.htm.

419 See *US$11 Million for Educational and Cultural Projects in Bosnia and Herzegovina*, UNESCO Press Release, 8 July 1997. http://www.unesco.org/bpi/eng/unescopress/97-109e.htm; Goodey, B. and Desimpelaere, W. 2003.

420 See designation of the Nesuh-aga Vučijaković Mosque as a National Monument on the website of the Commission to Preserve National Monuments: *Nesuh-age Vučijakovića džamija, graditeljska cjelina*; http://www.kons.gov.ba/main.php?id_struct=6&lang=1&action=view&id=2427

421 See the IRCICA website, *Inauguration of Neziraga Mosque*, http://www3.ircica.org/metinler.php?kat_id=338&grup=EVENTS.

When the Commission to Preserve National Monuments was re-established as a Bosnian state institution on 22 December 2001 following the end of UNESCO's mandate (see below), the governments of Turkey and Sweden supported the costs of international members of the Commission from their respective countries.

One well-developed research and training project funded by the UK's Overseas Development Administration (ODA) was implemented by PRDU (with other UK bodies) in 1997.[422] Intended to focus on the social/economic reconstruction and development of rural settlements, it was anchored in the heritage – the vernacular heritage rather than 'big' monuments.[423] Although an intensive three-week survey of three small towns (Vranduk, Maglaj, Travnik) and one micro-region (Bjelašnica) was carried out as what was meant to be the first stage of a multi-disciplinary rehabilitation project, it never progressed further (the fate of many such initiatives), nor were the results of the survey ever published.[424]

A major initiative was launched by the Council of Europe (COE) following the *Appeal in Aid of the Cultural Heritage of Bosnia and Herzegovina and Croatia* made at the Fourth European Conference of Ministers Responsible for the Cultural Heritage at Helsinki over 30–31 May 1996.[425] The COE's *Specific Action Plan for Bosnia-Herzegovina* which its Cultural Heritage Division's Technical Cooperation and Consultancy Programme for the Integrated Conservation of the Cultural Heritage had been drawing up in consultation with the Bosnian authorities since the country's accession to the European Cultural Convention in December 1994 envisaged a broad range of activities directed at re-establishing and developing heritage preservation structures and professional networks to provide for the future management of the built heritage within the framework of current European conventions and standards.[426] The Action Plan's preliminary phase began promisingly enough with plans for a survey of the architectural heritage across the country starting in 1997.

However, the COE soon ran up against the difficult realities of post-conflict Bosnia-Herzegovina when the participation of the *Institute for the Protection of the Cultural, Historical and Natural Heritage of Bosnia-Herzegovina* (*Zavod za zaštitu kulturno-historijskog i prirodnog naslijeđa BiH*), the republic-wide heritage body before and throughout the war, was abruptly terminated after the Republika Srpska authorities refused to recognize its jurisdiction over the whole country, even though it had participated in the survey's planning stages as the competent state institution.[427] Nevertheless COE external consultant experts proposed to carry out the surveys over the summers of 1997 and 1998 working with two teams of qualified local professionals from the Federation and Republika

422 The PRDU's partner organizations were ICOMOS UK and the Society for the Protection of Ancient Buildings (SPAB).

423 Author interview with Sultan Barakat, director of PRDU, 21 December 2011.

424 *Analysis of Cultural Heritage Damaged by War in Bosnia-Herzegovina*, Synopsis of PRDU Research and Training Project, Bosnia-Herzegovina January-March 1994, *Revival*, Iss. 8, August 1997. York: PRDU, University of York, Institute of Advance Architectural Studies.

425 IVth European Conference of Ministers Responsible for the Cultural Heritage, Helsinki, 30–31 May 1996 – Report by the Secretary General, CM(96)97, 9 July 1996, Appendix III: *Appeal in Aid of the Cultural Heritage of Bosnia and Herzegovina and Croatia*, Appendix 1: *Specific Action Plan for Bosnia and Herzegovina*, 29–32.

426 *Specific Action Plan for Bosnia-Herzegovina, Preliminary Phase, Final Report*, Technical Cooperation and Consultancy Programme related to the integrated conservation of the cultural heritage, Strasbourg: Council of Europe (COE), Cultural Heritage Division, March 1999.

427 Author interview with Muhamed Hamidović (director of the institute) and Philipp Altenburg COE consultant-expert Sarajevo October–November 1997.

Srpska. Then the 1997 field trips were curtailed in Republika Srpska when political tensions made travel outside Banja Luka impossible and the team was unable to reach sites in eastern Bosnia. Despite an International Police Task Force (IPTF) escort, problems were also cited in collecting information at Bosnian Croat-controlled Stolac where monument records note that individual monuments were often unable to be visited due to the tensions in the area.[428] At Čajniče (Republika Srpska) in August 1998 the team recorded that they were only able to take pictures of the remains of the Mir-Muhamed Mosque from the street due to the hostility of the people in the neighbourhood.[429]

Later, there were accusations of a lack of transparency in access to the material collected, and from the errors apparent in the final report, suspicions that the external consultants had been misled on some occasions by local officials in Republika Srpska as to the pre-war existence of mosques. While not intended to be complete inventory, there were conspicuous omissions of major monuments or concentrations of monuments.[430] Finally, there is an evident lack of knowledge of Bosnia's history and its impact on the country's heritage (as well as an absence of any appreciation of its distinctive qualities) on the part of the external consultants, who seemed oblivious as well to the terms of Annex 8 of the Dayton Agreement and the existence of the then UNESCO-led Commission to Preserve National Monuments.

While the COE programme was always aimed at building capacity rather than actually supporting or carrying out restoration projects, its dynamic title raised many false hopes in Bosnia. Despite the publication of the survey in 1999 there was little evidence of any action afterwards and local heritage professionals took to calling it the 'Specific Inaction Plan.'[431] The plan was later superseded by the Regional Programme on Cultural and Natural Heritage in South East Europe (RPSEE).[432]

The Sarajevo Haggadah 2

In 2001–2002 the Sarajevo Haggadah became the poster boy for the culmination of UN Special Representative Jacques Paul Klein's efforts to revive Bosnia's historic cultural and religious pluralism. He had already made an attempt with the built heritage in 2000, announcing the simultaneous reconstruction of four religious sites across Bosnia, to be implemented through the country's Interreligious Council and with Prince Hassan of Jordan, Bianca Jagger and George Soros apparently lined up for the organizing committee.[433]

---

428 Author interview with Philipp Altenburg COE consultant-expert, Sarajevo October–November 1997. Also noted in individual monument records for Stolac in the *Preliminary Heritage Inventory*.

429 Monument record V.O47 in the *Preliminary Heritage Inventory*. The mosque has been misidentified as the Sinan-beg Boljanić Mosque.

430 Personal knowledge of author.

431 See *Report by the Secretary General on the IVth European Conference of Ministers Responsible for the Cultural Heritage*. COE, Doc. Committee of Ministers CM(96)97, Helsinki, 30–31 May 1996.

432 The programme was launched in 2003. See http://www.coe.int/t/dg4/cultureheritage/cooperation/see/default_en.asp.

433 *Simultaneous Reconstruction of Four Religious Sites in BiH*, United Nations Mission in Bosnia-Herzegovina (UNMIBH) press release, Sarajevo, 23 May 2000. The structures to be restored or, rebuilt were the Orthodox Cathedral and Jewish Cemetery in Sarajevo, and the Aladža Mosque in Foča and the Catholic Church of St George the Martyr (Sveti Juraj Mučenik) in Derventa, both totally destroyed.

In late November 2002 came the re-opening of the Museum of the Old Orthodox Church in Sarajevo which the UN Mission in Bosnia-Herzegovina (UNMIBH) had supported, hopefully re-affirming the Bosnian Serb presence in the centre of the capital. The restoration of around 25 of the icons on display had been financed via the UN Trust Fund, though the bulk of its collection remains far from Sarajevo in Republika Srpska (see above).

Four days later on 2 December 2002 Ambassador Klein's pet project came to fruition – a project which he was to regard as the most important thing he did during his time in Bosnia: to put the Sarajevo Haggadah on permanent public display in the Zemaljski Muzej.[434] To Klein, the little manuscript, hidden in a bank vault for ten years and rarely on view before that, perfectly encapsulated the message he wanted to convey to Bosnians, not only of the country's historic cultural and ethnic diversity, but its own survival against the odds because of the efforts of Bosnians of different faiths working together.

So keen was the UN Representative for the US $120,000 project to take place he bypassed the Zemaljski Muzej administration (whom he regarded as too negative) entirely in the negotiations, choosing instead to deal with the Jewish Community of Bosnia-Herzegovina and (under pressure) the Ministry of Culture for the Federation, neither of which had any legal responsibility for the museum, in theory a state institution. UNESCO had put together a team of international experts who travelled to Sarajevo to assess the manuscript's condition, one of them a conservator who later carried out minor repairs to its binding.[435] At the Zemaljski Muzej a small room at the museum where the Sarajevo Haggadah would form the centrepiece of a display of manuscripts from Bosnia's other main faiths – Islam, Eastern Orthodoxy and Roman Catholicism – was elaborately fitted out with climate-control and a sophisticated electronic security system. A glittering opening ceremony was held for the new exhibition which attracted worldwide media coverage.[436]

But by the end of the month the UN Mission in Bosnia was over, Ambassador Klein had departed, the Sarajevo Haggadah was back in the bank vaults and the exhibition closed to the public, a victim of what has been characterized as a classic 'hit-and-run' of humanitarian aid,[437] where international agencies fund projects which hard-pressed local institutions have no hope of sustaining. The museum, one of the state-level cultural institutions that had existed in a budgetary limbo since the signing of the Dayton Peace Agreement, was in no position to maintain the security system which regularly malfunctioned. The Ministry of Culture for the Federation – which did provide some financial support to the museum – was quick to wash its hands of responsibility for maintenance costs, leaving the Zemaljski Muzej with no option but to remove the Sarajevo Haggadah from display and close the room to the public.

---

434 Klein, J.P. 2002. *The Journey of the Sarajevo Haggadah, Inauguration of the Vault Room for the Sarajevo Haggadah and Other Cultural Artefacts, Sarajevo National Museum, 2 December 2002 Address given on 2 December 2002.*

435 The expert commission were Professor Bezalel Narkiss, an expert on medieval Judeo-Spanish manuscripts from the Hebrew University of Jerusalem, who had seen the original manuscript while carrying out doctoral research in the 1960s; Jean-Marie Arnoult, France's Inspector General of Libraries (who had been on many cultural missions for UNESCO); and Andrea Pataki, a German conservator.

436 See for instance: Sarajevo Haggadah goes on display, JTA (Jewish Telegraphic Agency) [online], 3 December 2002.

437 Gienger, V.G. 2003. Out of Sight, *Balkan Reconstruction Report, Transitions Online*, 20 February.

Heritage, Justice and Human Rights

The quest for justice for victims of the Bosnian War was to become a testing ground of international humanitarian law with regard to the protection and preservation of cultural property, the right to a people's enjoyment of their cultural heritage and concepts of cultural heritage and identity. Crucial in this (and treated in detail elsewhere in this volume) were the ground-breaking legal precedents of the International Criminal Tribunal for the former Yugoslavia (ICTY) and the International Court of Justice (ICJ), alongside the decisions of the Human Rights Chamber for Bosnia & Herzegovina (HRC).

But perhaps the greatest acknowledgement of how crucial a role destruction of cultural and religious property had played during the conflict was the inclusion of an annex in the peace agreement that ended the war which attempted to address that destruction. *Annex 8* of the Dayton Peace Agreement[438] provided for the formation of a *Commission to Preserve National Monuments* whose mandate was to receive petitions to designate property of 'cultural, historical, religious or ethnic importance' as National Monuments. The annex (discussed in detail elsewhere in this volume) further committed entity authorities to protect cultural property designated as National Monuments, as well as those on the interim or provisional list awaiting a decision on designation.[439]

The initial five-year term of the Commission under the management of UNESCO was not regarded a great success. Its most significant achievement was probably the preparation and formal adoption of a *Provisional List of National Monuments* at its final meeting on 14 June 2000.[440] Though much criticized for its omissions and duplications, the entry of a cultural property on the Provisional List gave it the same legal protection as if it were a National Monument until a final decision (on which there was no time-limit) was made on its status.[441] However, since the re-establishment of the Commission to Preserve National Monuments as a Bosnian state institution at the end of 2001 it has been at the forefront of heritage protection in the country.[442]

Bosnia-Herzegovina's Heritage after the War: The First Decade

Nearly 15 years after the signing of the Dayton Peace Agreement in a report to the EC/COE Regional Programme on Cultural and Natural Heritage in South East Europe[443] the Commission to Preserve National Monuments noted some of the threats to heritage

438 Formally the *General Framework Agreement for Peace*, or *GFAP*. For the full text and Annexes see http://www.ohr.int/dpa/default.asp?content_id=380.

439 Valery Perry: *Cultural Heritage Protection in post-Conflict Bosnia-Herzegovina: Annex 8 of the Dayton Peace Agreement.*

440 For the complete Provisional List adopted at the meeting (in Bosnian only): *Privremena lista nacionalnih spomenika Bosne i Hercegovine, Kompletna* at the website of the Commission to Preserve National Monuments, http://www.kons.gov.ba/main.php?id_struct=165&lang=1.

441 See: *About the Provisional List*, http://www.kons.gov.ba/main.php?id_struct=108&lang=4

442 The Commission's website can be found at http://kons.gov.ba/. By December 2013 the Commission had designated 743 National Monuments, including buildings, groups of buildings, collections and individual objects.

443 Mulalić Handan, M. 2010. *Report on Assessment of the Architectural and Archaeological Heritage, Bosnia and Herzegovina.* Sarajevo: Commission to Preserve National Monuments of Bosnia and Herzegovina, Regional Programme on Cultural and Natural Heritage in South East Europe, IRPP/SAAH, *Weaknesses in Heritage Management.*

protection in Bosnia-Herzegovina, most of which derived from the war and the changed political structures that came with the peace treaty and the transition from socialism: the absence of adequate documentation (much of which had been destroyed during the war) making it difficult to carry out expert restoration and rehabilitation of monuments; the absence of state-level legislation governing heritage protection and the lack of political will to pass such legislation; the absence or inadequacy of other legislation governing matters that had a direct or indirect impact on heritage protection such as the restitution of property and town and country planning; difficulties in implementing decisions of the Commission designating properties as National Monuments, the prevention of illegal construction, inexpert renovation and lack of maintenance, or other forms of damage; the disparity between the very large number of endangered properties and the funds available for the protection and restoration of the cultural heritage; the reluctance of the religious communities as owners of cultural property to co-operate with heritage protection institutions and their failure to act according to the professional advice they receive. To which one might add the ethno-national political pressures applied to heritage protection institutions and the politicization of restoration. Set against these were the problems of the heritage protection institutions themselves which suffered from inadequate numbers of qualified staff, outmoded organizational structures, poor equipment, and funding shortfalls.

All these threats applied even more intensely to the protection and preservation of Bosnia-Herzegovina's surviving heritage in the turbulent post-conflict decade. The destruction of heritage did not stop with the signing of the Dayton Agreement, despite the huge international presence across the country which directly controlled implementation of the peace treaty for many years after the end of the conflict. Most at risk in districts where populations had been ethnically cleansed and where there was no inclination by the local authorities to carry out preservation measures were the fragile but still substantial remains of mosques and Ottoman structures such as, for instance, the shell of the Mir-Muhamed Mosque in Čajniče and the central domed structure of the Mehmed-paša Kukavica Mosque in Foča,[444] and the many historic housing complexes in Stolac and Gornji Šeher.

There were also more actively destructive actions. In 1998 a multi-story building was constructed on the site of the Zamlaz Mosque in Zvornik (RS).[445] The eighteenth-century Opijač House in Stolac (Federation), part of a complex of listed Ottoman buildings known as the Ada residential ensemble and on the Provisional List of National Monuments, was torn down in 2002 by its new Croat owners who had purchased it from Serb residents of Stolac forced to leave in 1992.[446]

444 The case of the Mehmed-paša Kukavica Mosque will be discussed in detail in *Domains of Restoration*. No work of any kind had been carried out to preserve the remains of the Mir-Muhamed Mosque (also known as the Mehmed or Mustaj-beg Mosque) as late as June 2008 when it was visited by an assessment team from the Commission to Preserve National Monuments prior to its designation as a National Monument; the mosque was reconstructed, however and formally reopened in July 2012.

445 *Islamic Community vs Republika Srpska (Zvornik Mosques), Decision on the Admissibility and Merits (Delivered on 9 November 2000)*, Human Rights Chamber (HRC), Case no. CH/98/1062, VI 2, 6.

446 See *Monuments at Risk in Bosnia and Herzegovina, 2. The Ada Complex in Stolac* on the Commission website http://kons.gov.ba/main.php?id_struct=161&lang=4 and Mulać, E. 2002. Destruction of the Opijač House in the Ada Residential Ensemble in Stolac, *The Revival of Stolac*, Iss. No. 17, Sarajevo: Association [Society] for the Renewal of Civil Trust in the Stolac Municipality, March 2002. *Udruženje za obnovu građanskog povjerenja općini Stolac; Udruženje* is sometimes translated as 'Association.'

Fig. 1.30 **(a)** The shell of the elegant Franciscan church of Saints Peter and Paul (Sv. Petra i Pavla), Mostar (1866) after bombardment by JNA and Bosnian Serb forces on the night of 9–10 May 1992. Although badly damaged, its bell tower survived and plans had begun for its reconstruction. © Ćiril Ćiro Raič / Mostar City Council / ICTY. **(b)** The enormous 107.2 metre high bell tower of the new church of Saints Peter and Paul looms over Mostar in 2001, dwarfing every other structure. The 1866 church was torn down and a new church constructed, beginning with a huge 1,000 seat crypt, in the autumn of 1995. © András Riedlmayer

Religious communities were on occasion responsible for destroying the restorable remains of their own historic structures. Even as fundraising was underway to reconstruct the elegant classical Franciscan church of Ss Peter and Paul in Mostar which had survived JNA bombardment in 1992, in 1997 its surviving façade, walls and bell tower were demolished by the Franciscan authorities and replaced with a concrete church whose gargantuan bell tower towered over the city in a calculated gesture of cultural domination. In Ustikolina the Islamic Community removed the substantial fragments of the destroyed Turhan Emin-beg Mosque, the oldest mosque in Bosnia-Herzegovina, building a brand new mosque in its place. The mosque fragments (mainly of the minaret) were pulverized and used in a local road building project by French SFOR troops.[447] The illegally-constructed new mosque at Ustikolina, and in particular its minaret, became a highly politicized struggle between the Islamic Community and the Federation authorities in official attempts to have it demolished.[448]

Concerns were beginning to be expressed, too, about the quality of some of the restoration work carried out on the remains of historic structures and in rebuilding those which had been completely destroyed and whether they conformed to international conventions and standards for conservation in both the documentation and construction phases. Local actors (politicians, religious leaders and returnees) and international donors alike were keen for such projects to finish quickly once a commitment to carry them out

447 Fočo, M. 2005. National monuments endangered for lack of measures to save them from further dilapidation. *Baština*, The Annual of the Commission to Preserve National Monuments. Sarajevo: The Commission to Preserve National Monuments of Bosnia and Herzegovina, I/2005, 29–45.

448 For a full account see Aksamija, A. 2010. Contested identities: Identity politics and contemporary mosques in Bosnia and Herzegovina, in *Mosques in Europe: Why a Solution has Become a Problem*, edited by S. Allievi in cooperation with Ethnobarometer, Network of European Foundations Initiative on Religion and Democracy in Europe. London: Alliance Publishing Trust, 325–32.

Fig. 1.31 **(a)** Ruins of the small Serbian Orthodox Church of the Transfiguration of Christ at Klepci, near Čapljina, in 2003. The church was destroyed in 1992. **(b)** The reconstructed church in 2007. Both © Commission to Preserve National Monuments of Bosnia & Herzegovina

had been made. Gilles Péqueux, the French engineer leading the reconstruction of the Old Bridge in Mostar until early 2003, has spoken of the pressures from the international community to complete the project quickly and how its focus appeared to be entirely on the opening of the completed bridge.[449]

The efforts of the reconstituted Bosnian-led Commission to Preserve National Monuments since 2002 to prevent the reconstruction of completely destroyed provisional or designated National Monuments as structures wholly unlike the original, often in styles alien to Bosnian/Herzegovinian and local traditions, or to restore partially or totally destroyed ones with inappropriate materials or additions was (and is) ongoing,[450] as are the efforts of the Commission to assist in such reconstructions.

For instance, initial plans drawn up for rebuilding the Orthodox Church of the Transfiguration of Christ in Klepci near Čapljina in the Federation by the Transportation Institute (Saobraćajni institut) in Belgrade in 2003 at the request of the Serbian Orthodox Church were rejected by the Commission as failing to comply with the principles of rehabilitation outlined in the Law on the Implementation of Decisions of the Commission in respect of alterations to the dimensions of the building, changes to the materials used (which were of different materials from those used for the original building), and the use

449 Pequeux, G. 2003. Rencontre avec Gilles Péqueux, ingénieur responsable de la reconstruction du pont de Mostar, *Courrier des Balkans* [online], 1 March; Rossini, A. 2003. Ricostruire il ponte di Mostar seguendone lo 'stato dello spirito', *Osservatorio Balcani e Caucaso* [online], 11 April.

450 This, of course, is often the case even in stable societies where there is strong heritage protection legislation and institutions.

of modern structures and building techniques.[451] The church had been totally destroyed in 1992 and designated a National Monument in 2003. The small limestone single nave barrel-vaulted structure with a semi-circular apse and arched bell holder (of the 'preslica' type) above the entrance façade dated originally from the sixteenth century and was typical of many Serbian Orthodox churches throughout Herzegovina. The church was subsequently successfully reconstructed as a project of the Institute for Protection of the Cultural, Historical and Natural Heritage of Republika Srpska over 2005–2007.

The Commission intervened in the case of the partially-destroyed Ali-beg Kapetanović Mosque (1858) at Vitina near Ljubuški where in 2008, while awaiting a decision on funding from the German Embassy and without consulting the heritage authorities, the Medžlis of the Islamic Community of Ljubuški had begun unsuitable repair work using a donation from the Embassy of the Islamic Republic of Iran. After the Commission intervened, work was stopped and a new project drawn up and carried out in association with the Commission. The Ali-beg Kapetanović Mosque was finally restored with funding from the Federal Republic of Germany, the Federation Ministry of Physical Planning (Ministarstvo prostornog uređenja), the Commission to Preserve National Monuments and the Medžlis of the Islamic Community of Ljubuški.[452]

At a meeting of the Commission with leaders of Bosnia-Herzegovina's religious communities on 8 December 2003 the Commission stressed that it would endeavour to implement the legal requirement that every property damaged or destroyed during the war and designated a National Monument was to be reconstructed or renovated in the same form, on the same site, with the same dimensions, of the same materials, and using the same techniques as the original, wherever possible. It also indicated it would rigorously block the addition of new symbols to National Monuments that could be manipulated for political ends. At the same time, the Commission recognized that the majority of National Monuments were owned by the religious communities and respected the function of such buildings and the needs of their users. They acknowledged, as well, the demands to quickly renovate destroyed religious buildings as part of the process of return of refugees and displaced persons.[453]

For where ethnically cleansed populations began to return to the towns and villages from which they had been expelled, the restoration and rebuilding of destroyed religious structures was often a crucial factor in marking the re-establishment of the community. As shall be discussed elsewhere in this volume, where communities returned to municipalities controlled by the ethno-national group responsible for the destruction, their initial battle was frequently to obtain permission to rebuild from the local authorities in the first place.[454]

451 See the designation of the church as a National Monument by the Commission on 9 October 2003: *Church of the Transfiguration of Christ in Klepci, School Building and Burial Ground, the Site and Remains of the Architectural Ensemble*, at http://www.kons.gov.ba.

452 See: *Restoration of the Ali-bey Kapetanović Mosque in Vitina, Ljubuški*, 11 November 2008, at the website of the the Commission, http://www.kons.gov.ba/main.php?mod=vijesti&extra=1075460509&action=view&id_vijesti=631&lang=4. See also: Kapetanović, E., Šarančić Logo, A., Numanagić, H. and Vesković, E. 2008. The Ali Bey Kapetanović Mosque in Vitina: Restoration of the national monument/Džamija Ali-bega Kapetanovića u Vitini: Restauracija nacionalnog spomenika, *Baština (Heritage)*, The Annual of the Commission to Preserve National Monuments, IV/2008, 335–64.

453 *Collaboration: 12/8/2003 – Meeting with Leaders of the Religious Communities of Bosnia and Herzegovina*, http://www.kons.gov.ba/.

454 See Helen Walasek, *Domains of Restoration*.

Thus most of such reconstruction at this early post-war period proceeded without support and advice from heritage protection institutions which were in a state of disarray and dysfunction, and might be unable (or possibly unwilling) to help.

The division of Bosnia-Herzegovina into two entities and the devolving of most political powers to the entities under the terms of the Dayton Peace Agreement and the ethno-national forces that actively worked against the existence of any institutions that represented the country as a state had serious implications for the heritage preservation structures of the country. There was no state-level legislation on protection and preservation of the cultural heritage, and at time of writing this was still the case, though there have been many efforts at drafting such legislation, including those of the Commission to Preserve National Monuments working in association with the Council of Europe.

The *Institute for the Protection of the Cultural, Historical and Natural Heritage of Bosnia-Herzegovina* had been the republic-wide preservation body for all of Bosnia-Herzegovina before, during the war and in the immediate post-war years.[455] However, the refusal of Republika Srpska to recognize its state-wide remit led not only to the body being starved of funding, but to the termination of the institute's representation of Bosnia-Herzegovina at international level and its eventual downgrading to a 'heritage centre.' In 2003 the Institute was subsumed into the entity-level *Institute for Protection of Monuments* of the Federation (a post-war creation) under the aegis of the Federation Ministry of Culture and Sport.[456]

In Banja Luka, the entity-level *Republic Institute for Protection of the Cultural, Historical and Natural Heritage*, an organization within the RS Ministry of Education and Culture, was a former regional institute whose authority was extended after the war to all of the newly-created Republika Srpska.[457] And the institute came to claim an even more extensive jurisdiction. Despite Republika Srpska's refusal to recognize the authority of the former state institute in Sarajevo, the RS institute assumed a *de facto* responsibility for Serbian Orthodox monuments across all of Bosnia-Herzegovina, highlighting again the critical nature of cultural heritage in the ethno-nationalist identity politics that underpinned the conflict.

Thus the *Commission to Preserve National Monuments of Bosnia-Herzegovina*, widely regarded as a Dayton success story, is now the sole state-wide body concerned with cultural heritage protection and preservation in Bosnia-Herzegovina.[458] Its position and authority is fiercely contested by the entity heritage preservation institutes and it is under constant threat by manoeuvring from ethno-nationalist politicians, particularly those from Republika Srpska and Milorad Dodik's SNSD party which works tirelessly to devolve powers to the entities from the central state and its institutions.[459] SNSD party member Sredoje Nović has

455 *Institute for the Protection of the Cultural, Historical and Natural Heritage of Bosnia-Herzegovina = Zavod za zaštitu kulturno-historijskog i prirodnog naslijeđa BiH.*

456 *Institute for Protection of Monuments = Zavod za zaštitu spomenika.* See Mulalić Handan, M. 2010. *Report on Assessment of the Architectural and Archaeological Heritage, Bosnia and Herzegovina.* Sarajevo: Commission to Preserve National Monuments of Bosnia and Herzegovina. Regional Programme on Cultural and Natural Heritage in South East Europe, IRPP/SAAH, 17 and 27. There are also a number of cantonal institutes and agencies for protection of heritage in the Federation most notably in Sarajevo, Mostar, Tuzla, Bihać and Jajce.

457 *Republic Institute for Protection of the Cultural, Historical and Natural Heritage = Republički zavod za zaštitu kulturno-istorijskog i prirodnog nasljeđa.*

458 *Commission to Preserve National Monuments = Komisiji za očuvanje nacionalnih spomenika.*

459 SNSD = *Savez nezavisnih socijaldemokrata* [Alliance of Independent Social Democrats].

held the central state Ministry of Civil Affairs (MoCA) for some years and continues to do so in 2013; Nović has sought to undermine the position of the Commission in a number of ways, such as for instance through the drafting of a rival state-level law on cultural heritage that would strip the Commission of its authority and transfer all heritage protection to the entities.[460] In 2013 the Commission to Preserve National Monuments began to come under attack in the media from the Federation Minister for Culture and Sports, Salmir Kaplan.[461] Such strategies do not bode well for the continued survival of Bosnia-Herzegovina's much-assaulted cultural heritage.

The Case of Guča Gora Franciscan Monastery

One case which brought into play many of the issues described above was that of Guča Gora Franciscan monastery near Travnik in central Bosnia. Guča Gora monastery did not receive the intense media coverage of Sarajevo's National Library (*Vijećnica*) or the Old Bridge (*Stari Most*) in Mostar, in part because of its rural location and relative distance from the centres where international journalists congregated. Nevertheless, Guča Gora monastery was to become emblematic of the attacks on cultural/religious property during the conflict owing to the convergence of a number of factors. One was the central place Franciscan monasteries held in Bosnia-Herzegovina as living symbols, not only of the Catholic/Croat tradition, but of Bosnia's wider identity and the continuity of its historic cultural and religious diversity. With an unbroken presence of seven centuries in Bosnia, reaching back to the medieval period when Bosnia had been an independent kingdom and through the centuries of Ottoman rule, and famed as centres of learning and education that contained, as well, some of the country's greatest cultural treasures, the influence of the Franciscan monasteries was felt across all communities.[462] Not only that, the head of the Franciscan Order in Bosnia-Herzegovina, Fra Petar Anđelović,[463] openly supported Bosnia's independence as a multiethnic state, as did many Franciscan priests (particularly in central Bosnia) leading some to be branded 'traitors' to the cause of Croat ethno-nationalism.[464]

Yet the alleged 'destruction' of Guča Gora monastery and its church was attributed by nationalist Croats to Bosnian Army forces – predictably labelled as 'Muslim' forces. As the Bosnian government had throughout the conflict asserted that, far from attacking and destroying Catholic or Orthodox sacral structures, they actively protected them (unlike the Bosnian Serbs and Bosnian Croats) the case of Guča Gora became a prized exemplar for

460 Personal knowledge. See also *Wikileaks*: Bosnia – Two New Serb Challenges To Dayton, Cable reference id: #08SARAJEVO287, 13 February 2008. http://www.cablegatesearch.net/cable.php?id=08SARAJEVO287. This issue is discussed in more detail by Valery Perry elsewhere in this volume.

461 See for instance: *Komisija za očuvanje nacionalnih spomenika demantirala Kaplana*, Dnevni Avaz [online], www.avaz.ba, 11 June 2013.

462 See for instance, the testimony of Fra Franjo Krizanac, former guardian of Guča Gora monastery, *Prosecutor v. Enver Hadžihasanović and Amir Kubura*, ICTY Case No. (IT-01-47), 17 December 2003.

463 Fra Petar Anđelović (Andjelović) was provincial (head) of the Franciscan Province of Bosna Srebrena.

464 See for instance, *HVO Commander 3rd OG Report 16 June 1993*, *Hadžihasanović and Kubura*, Exhibit DH1192e.

those opposing them to hold up to the international community. It became, too, a case able to be utilized in the argument that held all sides in the conflict behaved equally badly in respect of war crimes and monument destruction.

The events at Guča Gora Monastery took place in the context of the armed conflict between the once allied Bosnian Croat forces (HVO) and the Bosnian Army (ARBiH) in spring 1993 as the HVO attempted to force implementation of the Vance-Owen Peace Plan to divide Bosnia into ethnically-dominated cantons,[465] a strategy which lead to the notorious ethnic cleansing of the Bosniak/Muslim communities from the Lašva Valley by the HVO, not far from Guča Gora.

The events in the Bosnian Croat village of Guča Gora and its Franciscan monastery were examined over the course of several prosecutions at the International Criminal Tribunal for the former Yugoslavia (ICTY) in relation to a number of dimensions of the conflict, making it at time of writing one of the most extensively described and analysed cases of the destruction of cultural and religious property in Bosnia-Herzegovina examined at the tribunal or elsewhere.[466]

A crucial reason behind much of the intensive scrutiny of the case of Guča Gora was the ICTY's investigation of the contentious role played by *mujahedin* groups during the war in Bosnia-Herzegovina, specifically to try to establish whether the doctrine of 'command responsibility' applied in a number of prosecutions – that is, whether these groups had been under the command and control of the Bosnian Army at the times when alleged war crimes were committed by the *mujahedin*. The *mujahedin* were militant foreign Muslim volunteer fighters (mainly from Arab and other Islamic countries) who came to Bosnia in defence of its beleaguered Muslim population and who were to become a thorn in the side of the Bosnian government.[467]

As discussed earlier, allegations of damage to cultural and religious property coming from the parties in the conflict were often regarded with suspicion by informed observers as to their accuracy or possible exaggeration stemming from their use in the propaganda war over heritage destruction, particularly of religious sites. Inevitably these allegations were linked with accounts of other war crimes, atrocities and ethnic cleansing. But such charges could have other agendas, too. Thus, another issue examined during the ICTY prosecutions were allegations that the HVO propagated exaggerated and false claims about war atrocities, including the destruction of Catholic religious sites (among them Guča Gora monastery), to instill fear in Croat populations living in Muslim-controlled areas as a strategy for encouraging them to leave their homes in central Bosnia and move to Croat-controlled districts elsewhere (mainly Herzegovina) in pursuit of HVO attempts to consolidate the territory of the secessionist so-called Croatian Republic of Herceg-Bosna.[468]

A question that frequently arose during the ICTY trials was the comparison of the treatment of religious sites by the Bosnian Army and authorities and Bosnian Croat forces

465 See the Introduction for more on the so-called 'Muslim–Croat War.'

466 See ICTY Cases: *Prosecutor v. Tihomir Blaškić* Case No. (IT-95-14) *'Lašva Valley,' Dario Kordić and Mario Čerkez* Case No. (IT-95-14/2) *'Lašva Valley,' Jadranko Prlić et al.* Case No. (IT-04-74), *Enver Hadžihasanović and Amir Kubura* Case No. (IT-01-47), *Rasim Delić* Case No. (IT-04-83).

467 The number of *mujahedin* in Bosnia has been estimated in the low thousands. These Islamist groups also attempted to propagate their hardline conservative interpretation of Islam among the local Muslims with little success, though a small number did join these groups. The *mujahedin* were required to leave Bosnia-Herzegovina under the terms of the Dayton Peace Agreement.

468 See the testimony of Hendrik Morsink, *Blaškić*, 2 July 1998.

and authorities. Testimonies and documents produced as exhibits revealed generally respectful attitudes and policies in the Bosnian Army and authorities towards the cultural and religious property of Christian communities (Bosnian Croat/Catholic, in the case of Guča Gora), their willingness to investigate alleged attacks on such property, and their attempts to maintain the possibilities of continued co-existence.[469] This was certainly in marked contrast to the responses of HVO commanders to reports of the deliberate destruction of Muslim religious structures by HVO forces which were characterized by one witness (a member of the European Community Monitor Mission (ECMM) at the time) as disinterest, denial, or placing the blame on 'uncontrolled elements' (who in fact were regular HVO forces), with no interest in investigating such charges.[470]

Comparisons were made, too, between the degree of damage to Catholic and Muslim religious sites by the opposing forces. As a number of witnesses testified, the two were not comparable. If Catholic structures in ARBiH-controlled territory were damaged, they said, it was usually in a minor way, often because of collateral damage during military operations, though vandalization of church interiors did occur; whereas there were many examples of mosques and minarets which had been deliberately and completely destroyed in HVO-controlled territory.[471]

Extensive testimonies made during the course of the ICTY prosecutions by many of the different actors involved, from former ECMM monitors, UNPROFOR peacekeepers, Franciscan priests, Bosnian Army and HVO soldiers and officers, to a one-time member of a *mujahedin* group, alongside contemporaneous documents produced as exhibits and news reports, allow a Rashomon-like reconstruction of the fortunes of one highly symbolic heritage structure over the course of the conflict.

What Happened at Guča Gora?

The narrative begins in early June 1993 during fighting between the Bosnian Army (ARBiH) and the HVO in the Travnik region of central Bosnia when disturbing claims were made by the Bosnian Croat (HVO) authorities of the destruction of religious property by ARBiH forces alongside atrocities committed against Croat civilians and Catholic clerics. There were accusations, too, of ethnic cleansing by ARBiH troops as thousands of Croats fled the region. Nationalist Bosnian Croat leader Mate Boban published a dramatic open letter to Bosnian president Alija Izetbegović and international mediators claiming that the Croat people in central Bosnia were on the verge of total extinction.[472]

---

469 See, for instance, *Hadžihasanović and Kubura*: Exhibits DH234e, P893e, DH1159e, DH1223e, P427e, DH64e, P811e. See also many testimonies such as, for instance: *Hadžihasanović and Kubura*, testimonies Džemal Merdan, 15 December 2004; Hendrik Morsink, 25 May 2004; Remzija Šiljak, 22 October 2004; and *Blaškić*, testimony Hendrik Morsink, 2 July 1998.

470 Hendrik Morsink, *Blaškić*, testimony 2 July 1998, describing the reaction of Mario Čerkez to allegations of the deliberate targeting of the mosque at Kruščica by HVO forces.

471 Hendrik Morsink, *Blaškić*, testimony 2 July 1998; Colin Kaiser, *Blaškić*, testimony 16 July 1998; Fra Stjepan Neimarević, *Kordić and Čerkez*, testimony 4 July 2000.

472 Beelman, M.S. 1993. Muslims win fierce battle for key Bosnian town, *Seattle Times*, 8 June. *ECMM Special Report on Travnik dated 8 June 1993* by Jean-Pierre Thébault, Philip Watkins and Hendrik Morsink, *Hadžihasanović and Kubura*, Exhibit DH185. See also Mate Boban's letter to UNPROFOR Bosnia-Herzegovina Commander General Philippe Morillion of 22 June 1993, *Hadžihasanović and Kubura*, Exhibit 1912e, which made similar claims.

Most emotive and most repeated were the atrocities alleged to have taken place at Guča Gora (and all of which were false), including the supposed destruction of the Franciscan Monastery (which was described as being in flames), the execution of the priests and nuns there by Bosnian Army forces, and even the killing and setting on fire of Croat captives by local 'extremists' and *mujahedin*.[473] United Nations spokesman James Myles was among those claiming 'strong evidence of atrocities' and that British troops had seen Bosnian government forces machine-gunning fleeing Croat civilians; statements which were later shown to be untrue.[474]

These claims, while made in the context of an ARBiH offensive, came barely eight weeks after the massacre of 116 Muslim civilians by HVO troops at Ahmići, near Travnik and not far from Guča Gora, and the total deliberate destruction of the village, including its two mosques – the culmination of the ethnic cleansing of the Lašva Valley and one of the most infamous episodes of the war, one which continued to reverberate across the region and internationally.[475]

The allegations about what had supposedly taken place at Guča Gora and the wider area were almost immediately established as almost entire untrue by monitors from the European Community Monitor Mission (ECMM),[476] and UN officials were forced to deny the allegations made by its spokesman James Myles.[477] A special report on the situation in the Travnik region prepared by the ECMM regional centre at Zenica headed by French diplomat Jean-Pierre Thébault declared such allegations against the Bosnian Army as intentionally exaggerated – 'when not completely false.'[478] But once made in the media, the charges had a continuing life and impact in the wider world.[479] So much so that ECMM staff became alarmed that their superiors and others in the international community chose to accept as true what the monitors had characterized as sensationalized, grossly exaggerated and unsubstantiated news reports that indiscriminately repeated allegations emanating from the efficient HVO propaganda machine,[480] rather than believing its own staff's reports of their contemporaneous assessments on the ground such as those made by the British and Dutch monitors Philip Watkins and Hendrik Morsink at Guča Gora on 8 June 1993, the day of the alleged atrocities and destruction.[481]

---

473 See *ECMM Report dated 19 June 1993* by Jean-Pierre Thébault, Philip Watkins and Torbjorn Junhov, *Hadžihasanović and Kubura*, Exhibit DH186; Muslim offensive drives Croats out of 32 villages, 250 killed, *Croatian Radio*, Zagreb, 8 June 1993. *BBC Summary of World Broadcasts*, EE/1711/C1, 10 June 1993.

474 Frost, B. 1993. As thousands flee, the dead must lie where they fall in the bloody fields of Travnik, *The Times*, 9 June, 1.

475 The Ahmići massacre took place on 16 April 1993.

476 *ECMM Special Report on Travnik dated 8 June 1993* and *ECMM Report dated 19 June 1993*.

477 Sudetic, C. 1993a. Croats in Bosnia Walk Out of Talks with Government, *New York Times* [online], 10 June 1993.

478 *ECMM Special Report on Travnik dated 8 June 1993*, covering letter by Jean-Pierre Thébault.

479 See testimony Colin Kaiser, *Blaškić*,16 July 1998. See also, for instance, U.S. Department of State, *Dispatch*, 26 July 1993, 4:30, reporting on war crimes in the former Yugoslavia repeating the allegations made by UN spokesman, James Myles, on 8 June 1993, that were subsequently denied as untrue by the United Nations.

480 *ECMM Report dated 19 June 1993*. See also the testimony Hendrik Morsink, *Blaškić*, 2 July 1998. For the beliefs of the Bosnian authorities see, for instance, *Hadžihasanović and Kubura*, Exhibit DH234e, DH1223e.

481 *ECMM Special Report on Travnik dated 8 June 1993*.

So disturbed were the monitors that they wrote a follow-up report to the events at Guča Gora and in the wider Travnik area on 19 June, systematically refuting the HVO claims with their concrete eyewitness observations and including a section on allegations of atrocities and damage/desecration of churches.[482] The ECMM monitors further believed these false HVO claims of atrocities and destruction had a clear purpose (a belief held by others in the international community): that they were made with the intention of inciting a mass exodus of the Croat population, the first of HVO-controlled moves of whole communities, ultimately towards Croat-held territory in Herzegovina where a Bosnian Croat para-state of Herceg-Bosna could be consolidated and from where the Muslim and Serb populations had largely been expelled.[483] Fra Franjo Krizanac, guardian of Guča Gora monastery and the priest at nearby Brajkovići at the time, later agreed in testimony at the ICTY that false and misleading propaganda had been a factor in instilling fear in the Croat population.[484] Fra Krizanac himself and his fellow priest and two nuns had been incorrectly reported as killed and the church at Brajkovići destroyed in Croat media stories.

So what was the sequence of events at Guča Gora Franciscan monastery in early June 1993? It was widely reported on 8 June 1993, via information emanating from Croatian and Serb sources and repeated by HVO leaders, that the monastery was totally destroyed and in flames and that the local Croat villagers, along with priests and nuns at the monastery, had been massacred by Bosnian Army ('Muslim') forces.[485] After ECMM monitors Philip Watkins and Hendrik Morsink heard this claim from an HVO liaison officer called Darko Gelić in Vitez on 8 June, they drove directly to Travnik to confront Bosnian Army officers with the allegations.[486] The monitors met with Colonel Mehmed Alagić, commander of the ARBiH Third Corps Operational Group *Bosanska Krajina*, who categorically denied the rumours were true and suggested they visit the monastery immediately together, bringing with them two representatives from Travnik's Croat community to confirm the monastery's condition.

By chance, Major Roy Hunter, 'A' Company commander with the British UNPROFOR battalion based at Stari Bila, not far from Guča Gora, was in Travnik that afternoon on a religious affairs liaison visit to the local Catholic priests, accompanied by the battalion's chaplain and education officer.[487] Now Hunter was directed to ARBiH headquarters to collect the deputation and transport them to Guča Gora in his armoured vehicles. Adding to the party Fra Mirko Ivkić, one of the Catholic priests, Salko Beba, head of the Travnik municipal commission for prisoner of war exchanges and a representative from UNHCR, the group set off.[488]

---

482 *ECMM Report dated 19 June 1993.*

483 *ECMM Special Report on Travnik dated 8 June 1993* and *ECMM Report dated 19 June 1993.* See also the testimony Hendrik Morsink, *Blaškić*, 2 July 1998. For the beliefs of the Bosnian authorities see, for instance, *Hadžihasanović and Kubura*, Exhibit DH234e, DH1223e.

484 Testimony Franjo Krizanac, *Hadžihasanović and Kubura*, 17 December 2003, 1139.

485 See for instance, Sudetic, C. 1993a. Croatian forces face Muslim army, *New York Times*, 9 June. Sky News, *Martin Brunt Reporting from Central Bosnia*, 8–9 June 1993, television news footage.

486 Testimony Hendrik Morsink, *Blaškić*, 2 July 1998.

487 Thanks to Roy Hunter for clarifying many points regarding British Army involvement and the events at Guča Gora on 8–9 June 1993. Email communications with Roy Hunter 9–16 July 2013. The British UNPROFOR troops (BritBat) were the 1st Battalion of The Prince of Wales' Own Regiment of Yorkshire (1PWO).

488 See *Hadžihasanović and Kubura, ECMM Special Report on Travnik dated 8 June 1993.* See also testimonies of Hendrik Morsink, *Blaškić*, 2 July 1998 and *Hadžihasanović and Kubura*, 25 May 2004; Tomislav Rajić, *Hadžihasanović and Kubura*, 16 and 17 February 2004.

At his war crimes trial at the ICTY Colonel Tihomir Blaškić, Commander of HVO forces in central Bosnia, testified that on 8 June 1993 he had sent a request to Lieutenant-Colonel Alastair Duncan, commander of the British UNPROFOR unit at Stari Bila, to send armoured vehicles to Guča Gora to protect the monastery and its priests and Croat civilians taking refuge there and that Duncan had responded by sending Warrior armoured vehicles and tanks to the village.[489] Duncan's later testimony at the ICTY about the events makes no mention of a request from Blaškić. However, his statements that his troops on the ground had detected there was 'going to be a problem' in Guča Gora and that the Croat population 'could well be massacred,'[490] and the emotive tone of the battalion's usually succinct military information summary (*milinfosum*) for the day which spoke of 'atrocities,' 'many civilian casualties including old ladies' and 'fears for the lives of the trapped civilians'[491] suggest the influence of HVO propaganda, particularly in light of what actually did take place at Guča Gora.

Vaughan Kent-Payne, then a major in the British battalion commanding 'C' Company,' testified that the sound of firing from the area of Guča Gora could be heard at their base, and after what had happened at Ahmići, their policy was to investigate all such incidents. On the evening of 7 June Major Hunter had tasked a small patrol of armoured vehicles to gather information on the fighting in the Bila Valley between the rapidly-advancing Bosnian Army and the HVO. Led by Lieutenant Jason Medley, the patrol set off the next morning but got no further than Guča Gora. Arriving in the village about 13.00 they were fired upon – apparently by *mujahedin* fighters.[492] But it was clear their help was needed here.

Gathered near the large twin-towered stone church beside the monastery which dominated Guča Gora, the British troops found the remaining one third of the population of the village, mainly women, older men and children. The others had left in a column before eight that morning to escape the fighting, among them a group of HVO troops and two priests, one of them the monastery's guardian, or abbot, Fra Stjepan Neimarević,[493] leaving behind only two nuns at the monastery.[494]

In the midst of those assembled near the church were a dozen or so local HVO soldiers, who had been unaware of what had happened earlier in the day and that they had been abandoned by their colleagues. The HVO soldiers quickly disarmed and changed out of their uniforms and into civilian clothes, apparently at the suggestion of the British patrol

489 Testimony of Tihomir Blaškić, *Blaškić* 18 March 1999. Blaškić was tried at the ICTY (*Prosecutor v. Tihomir Blaškić* Case No. (IT-95-14) *'Lašva Valley'*) and found guilty on counts of persecution and crimes against humanity, among them the destruction and wilful damage done to institutions dedicated to religion or education and sentenced to nine years imprisonment. See also Walasek, H., *Cultural Heritage, the Search for Justice and Human Rights* in this volume.

490 See testimony of Alastair Duncan, *Delić*, 30 Aug 2007. Duncan later accepted that evidence contradicted his initial view that Croats could have been massacred; see *Delić*, 31 August 2007, 2030, referred to in *Delić*, *Defence Closing Brief*, 30 May 2008, No. 197.

491 PWO Milinfosum, 8 June 1993, *Delić*, Exhibit 00292. The assessment of atrocities and 'many' casualties was based on the (initial) discovery of one dead male civilian who had been shot in the head.

492 See *Hadžihasanović and Kubura*: Testimonies of Vaughan Kent-Payne, 22 March 2004, and Tomislav Rajić, 16 February 2004.

493 The title of the priest in charge of a Franciscan monastery in Bosnia was 'guardian' (*gvardijan*).

494 Testimonies: Tomislav Rajić, *Hadžihasanović and Kubura*, 16 and 17 February 2004; Stjepan Neimarević, *Kordić and Čerkez*, 4 July 2000.

Fig. 1.32 UNPROFOR BritBat armoured vehicles outside the Franciscan monastery at Guča Gora after the arrival of the ECMM deputation and Colonel Alagić on 8 June 1993 escorted by British troops. The photograph was used as prosecution evidence at the ICTY case against Dario Kordić and Mario Čerkez. © Hendrick Morsink / ICTY

commander, enabling them to come under UN protection as non-combatants.[495] Among them was Guča Gora resident Tomislav Rajić who later testified about the events at the ICTY. The villagers were moved into the church for their safety, while the British troops positioned their Warrior armoured vehicles around the monastery as a defensive shield.

At around 16.00 the ECMM-led deputation from Travnik with Colonel Alagić arrived at Guča Gora to find the monastery and village intact and undamaged, the UNPROFOR armoured vehicles stationed in front of the monastery, and nearly 200 Croat villagers sheltering in the church, as well as the two nuns and the disarmed HVO soldiers.[496] The flames which had been seen from a distance by members of the international press corps and reported as coming from Guča Gora, were, in fact, the nearby Muslim village of Bandol which had been attacked by HVO forces and set afire.[497] There had certainly been fighting in Guča Gora during which, it was discovered the following day, seven people had been

495 Testimony of Tomislav Rajić, *Hadžihasanović and Kubura*, 16 and 17 February 2004; PWO Milinfosum, 8 June 1993, *Delić*, Exhibit 00292. Personal communication Roy Hunter.

496 See *Hadžihasanović and Kubura*: Testimony of Hendrik Morsink, 25 May 2004; Tomislac Rajić, 16 February 2004 and *ECMM Reports 8 June 1993* and *19 June 1993*. See also PWO Milinfosum, 8 June 1993, *Delić*, Exhibit 00292. Some accounts vary between 181 and 187 people being in the church; see testimony Alastair Duncan, *Delić*, 30 August 2007.

497 Testimony of Hendrik Morsink, *Blaškić*, 2 July 1998; *Delić*, OG Bosanska Krajina Security News, 8 June 1993, Exhibit 304.

killed (five HVO soldiers discovered in trenches near the village and two male civilians).[498] But there had been no destruction of the monastery (or the village), no massacre, and no executions.[499]

Now a decision had to be taken whether the villagers should remain in the area (now completely under the control of the Bosnian Army) or be evacuated. Colonel Duncan ordered the UNPROFOR armoured vehicles, with Major Hunter in command, to remain in Guča Gora overnight as a protective measure,[500] while the ECMM and UNHCR representatives planned to return the following day to discuss a possible evacuation.

On the morning of 9 June Colonel Duncan himself appeared in Guča Gora, having ordered Major Kent-Payne to come with as many armoured vehicles as he could muster once an evacuation had been decided on and which could be used to transport the Croat villagers.[501] UNHCR representative Margaret Green arrived, as did the ECMM monitors Philip Watkins, Hendrik Morsink and Torbjorn Junhov. Also in the village that day to discuss the proposed evacuation were Bosnian Army Third Corps Deputy Commander Colonel Džemal Merdan, there at UNPROFOR's request to ensure safe passage for the civilians and who had arrived in a British armoured personnel carrier, as well as another Bosnian Army officer, Third Corps 306th Brigade commander Esed Sipić.[502] And sensing an interesting news story (particularly with UK troops involved), a group of British journalists had travelled to Guča Gora to record the day's events, with a television crew in tow.[503]

Discussions began with the villagers to determine whether they wished to stay or go. UNHCR wanted to be clear that any mass departure would be of their own free will – it did not want to find itself appearing to be a participant in ethnic cleansing. Commander Sipić argued that the Croat population had no need to leave their homes and would be protected by the Bosnian Army and assurances were given for their safety.[504] But the villagers were clearly terrified of remaining, despite these assurances.[505] Among their anxieties was the presence of *mujahedin* fighters in the area.[506] The main *mujahedin* base

498 Testimony of Tomislav Rajić, *Hadžihasanović and Kubura*, 17 February 2004. Personal communication Roy Hunter.

499 Many houses in Guča Gora were, however, looted and a significant number burned over the following months when the area was in the control of the ARBiH.

500 PWO Milinfosum, 8 June 1993, *Delić*, Exhibit 00292.

501 See testimonies of Vaughan Kent-Payne, *Hadžihasanović and Kubura*, 22 March 2004; Alastair Duncan, *Delić*, 30 August 2007; PWO Milinfosum 9 June 1993, *Delić*, Exhibit 00293, personal communication Roy Hunter. Kent-Payne says sixteen armoured vehicles went to Guča Gora; other news accounts say nineteen and twenty-four.

502 See *Hadžihasanović and Kubura*: Testimonies of Hendrik Morsink, 25 May 2004 and Džemal Merdan, 9 December 2004.

503 Among them were British journalist Maggie O'Kane and BBC correspondent Allan Little; personal communication from Roy Hunter. For print media accounts of the evacuation see Stephen, C. 1993. Croat refugees flee Muslim vengeance, *IPS-Inter Press Service*, 10 June; O'Kane, M. 1993. Eyewitness: Peace eludes the people of St Francis, *The Guardian*, 10 June, 24.

504 Testimony of Hendrik Morsink, Hadžihasanović and Kubura, 25 May 2004. See also Reuters, Bosnia-Herzegovina/Croatia: British U.N. troops evacuate 181 Croatians from monastery near Travnik/Refugees from fighting in the town cross the border into Croatia, 9 June 1993, television news footage, Reuters Ref. 604260160; Stephen, C. 1993.

505 O'Kane, M. 1993.

506 See testimony Stjepan Neimarević, *Kordić and Čerkez*, 4 July 2000.

was at Poljanice, not far from Guča Gora, and fears of their unpredictable and violent behaviour were not unfounded.[507]

Around midday heavy machine gun fire rang out from Bosnian Army positions, striking the monastery and two of the Warriors responded with their chain guns.[508] To the sounds of sporadic gunfire, battalion chaplain Carson Nicholson read a funeral service for the seven killed during the fighting the day before as they were buried in the grounds of the church, their graves marked with simple wooden crosses. UNHCR official Margaret Green warned local residents that if they left they would probably not come back.[509] She gave them ten minutes to decide whether to leave under UN protection and join the thousands of the homeless displaced, or stay under the protection of the Bosnian Army.[510]

After an emotional debate, the villagers decided finally to leave Guča Gora and their homes behind. The disarmed HVO soldiers were also permitted to leave, including Tomislav Rajić, who was well-known to Colonel Merdan from their earlier work contacts in Travnik. Rajić and Merdan spoke to each other, Rajić asking Merdan to ensure the civilians reached Vitez safely. An as an HVO soldier, he offered to surrender, but Merdan indicated he had no wish to hold Rajić as a prisoner of war.[511]

As the television crew filmed, the British troops began moving the villagers out of the church and into the waiting armoured vehicles.[512] With them came what few religious artefacts from the monastery and church could be squeezed into the cramped Warriors. Among the items that travelled with the community into displacement was a nearly life-size statue of St Francis which had stood in the apse of the church, and the leather-bound Bible from the altar.[513] The nuns themselves carried away the priest's vestments and the births and deaths register.[514] Major Hunter climbed into the belfry to reclaim the UN flag he had hung there the day before, and the convoy finally departed, leaving the monastery and Guča Gora empty of its inhabitants.[515]

---

507 For an example of the violent behaviour of the *mujahedin* see a description of events surrounding the execution of four Croat men from the village of Miletići (not far from Guča Gora) on 24 April 1993 by the *mujahedin*, see *Hadžihasanović and Kubura*, Judgement 15 March 2006, 295–307.

508 Personal communication, Roy Hunter. See also Stephen, C. 1993 and ITN *News at Ten*, 9 June 1993, ITN Ref: BSP090693016 0.

509 Stephen, C. 1993.

510 O'Kane, M. 1993.

511 See *Hadžihasanović and Kubura*: Testimony of Tomislav Rajić, 16 February 2004; and Džemal Merdan, 9 December 2004.

512 Testimony of Hendrik Morsink, *Hadžihasanović and Kubura*, 25 May 2004; *Blaškić*, 2 July 1998; and *Kordić and Čerkez*, 12 Oct 1999.

513 See *Hadžihasanović and Kubura*: Testimonies Vaughan Kent-Payne, 22 March 2004; Tomislav Rajić 17 February 2004. O'Kane, M. 1993. Personal communications from Roy Hunter. Some of the artefacts can be seen in footage of the evacuation screened on ITN *News at Ten*, 9 June 1993, ITN Ref: BSP090693016 0. A contemporaneous photograph of the statue and other religious object taken during 8–9 June can be seen on different pages at http://www.arrse.co.uk/bosnia-former-yugoslavia-op-grapple-op-resolute/197526-20-years-ago-morning%3B-ahmici-bosnia-1993-a.html. Other items included a smaller statue of the Virgin Mary and some chalices. All were taken to the priest at the Catholic church of St George (Sv. Juraj) in Vitez for safe-keeping and later returned to the monastery.

514 According to O'Kane, M. 1993. Stephen, C. 1993 reports two life-size statues of the Virgin Mary were brought from Guča Gora, but this is not correct.

515 See television news footage: Bosnia: British rescue civilians, News at Ten, *ITN*, 9 June 1993, Terry Lloyd reporting, ITN BSP090693016 0; Bosnia-Herzegovina/Croatia: British U.N. troops evacuate 181 Croatians from monastery near Travnik, *Reuters*, 9 June 1993. See also *Hadžihasanović and Kubura*: *ECMM Reports* dated *8 June 1993* and *19 June 1993* and testimonies of Vaughan Kent-Payne,

The Bosnian authorities quickly went on the media offensive to refute the HVO claims of Guča Gora's destruction and accounts of supposed atrocities there,[516] bringing a group of journalists and cameramen to the intact monastery on 10 June[517] where they encountered a group of foreign *mujahedin* engaged in looting goods from Croat houses.[518] The *mujahedin* soon moved into the monastery and embarked on its desecration and vandalization, attacking and defacing religious pictures in the church, throwing red paint at the large contemporary fresco by Zlatko Keser on the apse behind the altar, painting graffiti in Arabic on walls, damaging statues, destroying the organ and confessional booth, and overturning and smashing pews – smearing some with excrement.[519] A bronze statue of St Francis which stood in the monastery courtyard was attacked and shot at and even the simple wooden crosses marking the graves of those buried in the churchyard on 9 June were pulled up.[520]

Even more alarming, the *mujahedin* declared their intention to destroy the church, in retaliation, they told Džemal Merdan (who had been alerted to their plans as he passed through Guča Gora) for the many mosques destroyed by the HVO[521] and in doing so, as one-time *mujahid* Ali Ahmad Ali Hamad has related, to send out a message to the Catholic Church and the Vatican.[522] Merdan talked to the *mujahedin* for several hours to try to dissuade them from their plans, but on his return to headquarters in Zenica he immediately informed Third Corps commander Enver Hadžihasanović of the situation and they concluded that military police should be sent to provide security for the monastery.[523]

Warned of the threat to the monastery, an angry Mehmed Alagić travelled to Guča Gora to confront the *mujahedin* and found them setting explosives in the church walls.[524] A violent argument followed, with Alagić ordering the *mujahedin* to stop planting explosives, threatening to order his entire corps to attack them if they did blow up the church. The *mujahedin* were undeterred, refusing to take orders from him and instead threatened Alagić himself.[525] Fortunately, the *mujahedin* leader, who had not been present during this episode, took a more strategic view on his return, deciding it was more important to preserve the

---

Džemal Merdan and Tomislav Rajić. See Associated Press 1993. Croats besieged in Bosnia monastery get lift out from British, *Los Angeles Times* [online], 10 June 1993. Personal communication Roy Hunter.

516 See *Hadžihasanović and Kubura*: *Daily Operations Report, 3rd* Corps, 11 June 1993, Exhibit P196e; *Report on Implementation 3rd Corps Order 18 June 1993*, Exhibit DH1269e.

517 See testimony Halim Husić, *Delić*, 11 and 12 March 2008.

518 See testimonies Halim Husić, *Delić*, 11 and 12 March 2008; Izet Mahir, *Hadžihasanović and Kubura*, 4 March 2005.

519 See particularly testimonies of Stjepan Neimarević, *Kordić and Čerkez*, 4 July 2000 and *Hadžihasanović and Kubura*, Vaughan Kent-Payne, 22 March 2000. See also *Hadžihasanović and Kubura*: Testimonies of Jasenko Eminović, 20 April 2004; Izet Mahir, 4 March 2005; Samir Sefer, 19 November 2004. See also *Hadžihasanović and Kubura, Report from Croatian Defence Council, Frankopan Brigade to Croatian Defence Council Central Bosnia Operative Zone (re Damage to Guca Gora Church as of 14/07/1993)*, dated 27-Jul-93, Exhibit P735e.

520 Testimonies of Stjepan Neimarević, *Kordić and Čerkez*, 4 July 2000 and *Hadžihasanović and Kubura*, Vaughan Kent-Payne, 23 March 2000.

521 Testimony Džemal Merdan, *Hadžihasanović and Kubura*, 9 December 2004.

522 See *Excerpt from Book by Ali Hamad, Delić*, Exhibit 00332E. Former *mujahedin* fighter Ali Ahmad Ali Hamad was also known as Ali Ahmad Ali Hamad, Ali Hamad Ubeid and Abu Ubaidah al-Bahraini.

523 Testimony Džemal Merdan, *Hadžihasanović and Kubura*, 9 December 2004.

524 See *Delić*: Testimony Ali Ahmad Ali Hamad, 07 September 2007; *Excerpt from Book by Ali Hamad*, Exhibit 00332E.

525 Testimony Ali Ahmad Ali Hamad, *Delić*, 07 September 2007

group's relations with the Bosnian Army and local people which would be severely tested if they went ahead with their plans to blow up the church and could perhaps result in their expulsion from Bosnia.[526]

The ARBiH Third Corps command moved to secure the monastery sending a Military Police unit on 16 June to eject the *mujahedin* (which was not easily done) and remain there to protect the site.[527] Major Kent-Payne also visited the monastery on 16 June where he found a group of 20 to 30 middle-aged and elderly Muslim women clearing up the damage to its interiors; an attempt had already been made to paint over the graffiti.[528] Kent-Payne was quick to describe these clean-up efforts as a 'cover-up' of probable breaches of the Geneva Conventions. However, as is clear from different testimonies at the ICTY, in the perceptions of the Bosnian Muslims the clean-up of the monastery was certainly not an effort to cover up a war crime, but a gesture of good will. The painting over of the graffiti had been carried out in the presence of a Catholic priest and the Muslim soldiers and officials believed it was a way of showing their regret for the desecration, as well as consideration for the feelings of the clergy who had to witness it.[529]

The monastery continued to be guarded by the Bosnian Army, primarily by the 306th Brigade which relocated its brigade command there to better provide security for the monastery,[530] until its formal return to the Franciscans and the monastery's new guardian, Fra Franjo Krizanac, in September 1994.[531]

The Aftermath

The belief that the monastery had been totally destroyed persisted in the Bosnian Croat community – and much further afield. Yet a succession of visitors came to inspect the monastery. About a month after he had left, Fra Stjepan Neimarević was brought back by UNPROFOR to inspect the monastery.[532] Both Kent-Payne and Merdan visited the monastery later in the summer accompanying Catholic priests who wanted to confirm that the monastery had not been completely demolished so they could reassure their displaced parishioners.[533]

Because of his knowledge of the true state of the monastery Kent-Payne was expressly asked to take the priest from Vitez along with the British regimental chaplain to Guča Gora at the end of June 1993.[534] As he testified: 'rumours were rife and the Croats believed that the Bosnian Muslims had completely destroyed the monastery and they'd burned it down,

526 See *Delić*: Testimony Ali Ahmad Ali Hamad, 09 September 2007; *Excerpt from Book by Ali Hamad*, Exhibit 00332E.

527 See *Hadžihasanović and Kubura*: Testimony Izet Mahir, 4 March 2005; Trial Judgement 15 March 2006

528 See *Hadžihasanović and Kubura*: Testimonies Vaughan Kent-Payne, 22 and 23 March 2004; Franjo Krizanac, 17 December 2003; Tomislav Rajić, 17 February 2004; Jasenko Eminović, 20 April 2004; Munir Karić, 9 November 2004; Samir Sefer, 19 November 2004;. Izet Mahir, 4 March 2005.

529 See *Hadžihasanović and Kubura*: Testimonies Jasenko Eminović, 20 April 2004 and Samir Sefer, 19 November 2004.

530 Testimonies *Hadžihasanović and Kubura*: Munir Karić, 9 November 2004; Fahir Čamdžić, 16 November 2004.

531 See testimony Franjo Krizanac, *Hadžihasanović and Kubura*, 17 November 2003.

532 Stjepan Neimarević, *Kordić and Čerkez*, 4 July 2000

533 See *Hadžihasanović and Kubura*: Testimonies Vaughan Kent-Payne, 22 March 2004 and Džemal Merdan, 9 December 2004.

534 *Hadžihasanović and Kubura*: Testimony Vaughan Kent-Payne, 22 and 23 March 2004.

and the object of taking the priest there was to show him the damage that had been done but so he could go back to reassure his people that actually the building had not been razed to the ground, as many people believed.'[535]

Merdan accompanied a delegation led by Canadian ECMM monitor Remi Landry bringing Father Stjepan Radić from Zenica to inspect the condition of the monastery on 3 August 1993. Like many others, the Franciscan priest believed the monastery had been completely demolished.[536] At the monastery the group saw Bosnian Army soldiers at work repairing the damage to the interiors. Father Stjepan was apparently pleased at the presence of the military police giving protection to the monastery, telling the ECMM monitors how cooperative Colonel Merdan was in protecting religious property.[537]

Yet the lack of knowledge about what had happened to Guča Gora monastery extended as far as professionals based in western Europe working for international bodies involved in assessing damage to the cultural heritage in the former Yugoslavia. Colin Kaiser, at the time a Council of Europe consultant reporting on damage to the cultural heritage in Bosnia-Herzegovina and Croatia, testified that they had received information of Guča Gora monastery's total destruction. Nevertheless, Kaiser emphasized, despite the presence of international organizations and scores of UN convoys moving about the area, those in heritage protection 'didn't really know what was going on.'[538]

Thus, it was not until Kaiser finally managed to carry out a mission to central Bosnia in May–June 1994 under the aegis of the ECMM was he able to confirm the survival of Guča Gora monastery for himself.[539] Yet it was ECMM monitors who had one year earlier filed a report on the day of the alleged destruction of the monastery, refuting these claims and describing the true state of affairs, a report that was widely distributed within the ECMM (including to its Zagreb HQ) and externally as far as Geneva[540] and which Colin Kaiser (who had many contacts within the ECMM system) had apparently never seen.

How information such as the desecration and vandalization of the monastery by the *mujahedin* was passed up the hierarchy of the international organizations operating in Bosnia was of interest to ICTY judge Albert Swart who asked Major Vaughan Kent-Payne whether he had officially reported his findings and how they would have been disseminated.[541] Kent-Payne replied that while he would certainly have written a report, but whether it was included in the daily military information summary (*milinfosum*) was a matter of judgement for the military information officer. Milinfosums, however, were distributed to Brigade headquarters in Split and to the UN military headquarters in Kiseljak at the very least.

Meanwhile, the HVO were well aware of the true chain of events at Guča Gora, even making a formal (and accurate) report on the desecration of the monastery and state of churches in Brajkovići, Dolac and Ovčarevo as of 14 July 1993.[542] However they continued

535 *Hadžihasanović and Kubura*: Testimony Vaughan Kent-Payne, 22 March 2004, 4817.

536 *Hadžihasanović and Kubura*: Testimony Džemal Merdan, 9 December 2004, 13128. See also testimony Džemal Merdan 14 December 2004 on the visit.

537 *Hadžihasanović and Kubura*: *Daily Report from European Community Monitoring Mission Team VI dated 3/Aug/93B/C/S*, Exhibit P164; testimony Džemal Merdan, 9 December 2004, 13127.

538 See testimony Colin Kaiser, *Blaškić*, 16 July 1998.

539 COE *Sixth Information Report*, Doc. 7133, 31 August 1994.

540 *ECMM Special Report on Travnik dated 8 June 1993*. See addressees for report which included ECMM in Zagreb and LO Geneva.

541 *Hadžihasanović and Kubura*: Testimony Vaughan Kent-Payne, 23 March 2004.

542 *Report from Croatian Defence Council Frankopan Brigade to Croatian Defence Council Central Bosnia Operative Zone (re damage to Guca Gora Church as of 14/07/1993)*, dated 27-Jul-

to exploit Guča Gora in their propaganda battles, now with an additional factor, using the protective presence of the 306th Brigade to declare (falsely) that the ARBiH Third Corps had established its headquarters in the monastery – implying a breach of the 1954 Hague Convention for the Protection of Cultural Property in the Event of Armed Conflict.[543] This was despite that (in clear breach of the Hague Convention) the Frankopan Brigade of the HVO actually *had* been headquartered in the monastery until the end of May 1993, though by June only a communications cell consisting of two local men remained in place.[544]

Major Kent-Payne was to remonstrate with Colonel Merdan about the desecration of the monastery by the *mujahedin*, believing that such acts compromised the Bosnian government's position regarding the destruction of religious sites and their accusations against the Serbs and Croats in this respect.[545] Merdan pointed out that the desecration had not been carried out by ARBiH troops, who, far from damaging religious buildings, had orders to take measures to protect them, and that the acts at Guča Gora monastery had been carried out by the *mujahedin*. While Kent-Payne (like many in the international community) had formed the impression that the *mujahedin* were under the command and control of the ARBiH, Merdan knew this was not the case, but believed they had to be brought under control and such breaches of international humanitarian law prevented.

The Movable Heritage of Guča Gora Monastery

The narrative of Guča Gora monastery's important movable cultural property during the conflict took a somewhat different path. Like most Franciscan monasteries in Bosnia it held large collections of paintings and other art objects, archival material and early books. The monastery had suffered terrible losses to its heritage in February 1945, including the archive of the bishop and noted scholar Fra Marijan Šunjić (1798–1860),[546] when it and the church were set on fire by the Partisans. Now, during the current conflict, the monastery was within range of Bosnian Serb artillery entrenched on the heights of Mt Vlašić. Fra Franjo Krizanac (guardian of the monastery from July 1994) testified that in autumn 1992, under the impact of shelling from Mt Vlašić and with the catastrophic destruction of the monastery during World War II in mind, a group of Franciscan clergy had met at Guča Gora Monastery to discuss protection of the significant movable cultural heritage held there.[547] At the meeting a decision was taken to transfer the most precious items in the monastery collections to a safer place.

93, *Hadžihasanović and Kubura*, Exhibit P 735e. The information in the report on Guča Gora may have been based on information received from Fra Stjepan Neimarević who was brought to inspect the monastery by UNPROFOR approximately one month after his departure on 8 June; see note 64. The report noted the church in Brajkovići damaged by shelling, a chapel at Bukovica burnt down, the church at Dolac damaged on the outside and the church at Ovčarevo undamaged.

543 See testimony Tihomir Blaškić, *Blaškić*, 7 April 1999. Blaškić was informed of this by HVO military intelligence in December 1993.

544 Testimony Dragan Radić, *Hadžihasanović and Kubura*, 26 February 2004.

545 See *Hadžihasanović and Kubura*: Testimonies Vaughan Kent-Payne, 22 March 2004 and Džemal Merdan, 13 December 2004.

546 Fra Marijan Šunjić was a noted Catholic bishop and Apostolic Vicar and activist on behalf of the Catholic community in Bosnia during the Ottoman period, as well as being an important linguist (in Oriental languages), scholar and writer.

547 See *Hadžihasanović and Kubura*: Testimonies Franjo Krizanac, 17 December 2003, and Tomislav Rajić, 17 February 2004; testimony Franjo Krizanac, *Kordić and Čerkez*, 25 September 2000.

However, the guardian of Guča Gora in 1992 was Fra Stjepan Neimarević. In his ICTY testimony Neimarević did not mention the autumn 1992 meeting, nor a decision taken at that time to relocate the monastery's movable heritage.[548] He testified only that he had received instructions (apparently in early December 1992) from his provincial (Fra Petar Anđelović in Sarajevo)[549] via Fra Blašević, the then director of Caritas in Split,[550] that the art objects and other valuables belonging to the monastery should be moved to preserve them.[551]

No matter when the decision to relocate the important artefacts was initially made, immediately after receiving the instructions from his provincial, on 8 December 1992 Fra Neimarević made an urgent request to the HVO authorities in Travnik seeking official written approval for the removal of the artefacts that could be shown at checkpoints and while crossing the border unto Croatia, as well as HVO assistance in ensuring there were no problems with the transfer.[552] The necessary permit was issued by the President of the Travnik Municipality HVO the same day and two days later an armed military escort was approved to accompany the evacuation of the monastery's treasures to Imotski in Croatia which took place soon afterwards.[553] The cultural property relocated included paintings and other works of art, archival material and books, among them works in Bosančica,[554] all of which, according to Krizanac, was stored in the parish office in Vrgorac near Imotski until their return to the monastery in 1998.[555]

Since Fra Neimarević and Fra Krizanac were appearing as defence witnesses in the trial of Bosnian Croat nationalist politician Dario Kordić for war crimes committed during the ethnic cleansing of the Muslim population of the Lašva Valley,[556] ICTY prosecutor Susan Somers pressed both priests on the timing of the transfer of the monastery's important artefacts to Croatia. Why had it been carried out in December 1992? This was well before open conflict between the HVO and ARBiH, but at a time, however, when relations were increasingly strained. Had the Franciscans at Guča Gora prior knowledge of impending events and perhaps, she implied, sympathy with the policies of the illegal Bosnian Croat para-state of Herceg-Bosna (in which Kordić was a leading figure) and its wish for eventual

---

548 See testimony of Stjepan Neimarević, *Kordić and Čerkez*, 4 July 2000.

549 His provincial (superior) was Fra Petar Anđelović head of the Franciscan Province of Bosna Srebrena.

550 Caritas is the Catholic charity that provided humanitarian aid across Bosnia-Herzegovina during and after the 1992–1995 war. According to his CV, Fra Božidar Blažević was head of Caritasa Vrhbosanske Nadbiskupije za BiH in Split during 1991 and head of the charitable society *Kruh sv. Ante* in Split 1991–1994; http://www.kroatien-links.net/vidosi/%8EIVOTOPIS_FR_%20BO%8EO_NOVI.doc.

551 Testimony Stjepan Neimarević, *Kordić and Čerkez*, 4 July 2000.

552 *Kordić and Čerkez*, Guca Gora Monastery valuables transport request, Exhibit 306-4a.

553 See *Hadžihasanović and Kubura*: Exhibit DH 344e *Approval* dated 8/12/1992; and Exhibit DH 1696e, *Approval for Soldiers Franjo Neimarevic and Stipe Lujanovic from Guca Gora to be an Armed Escort during the Transport of Monastery Value from Guca Gora to Imotski* dated 10/12/1992. Testimony Stjepan Neimarević, *Kordić and Čerkez*, 4 July 2000.

554 See *Hadžihasanović and Kubura*: Testimonies Franjo Krizanac, 17 December 2003, and Tomislav Rajić, 17 February 2004; *Kordić and Čerkez*, testimonies Franjo Krizanac, 25 September 2000, and Stjepan Neimarević, 4 July 2000. Bosančica is a now extinct form of medieval Bosnian Cyrillic script.

555 *Hadžihasanović and Kubura*: Testimony Tomislav Rajić, 17 February 2004.

556 *Dario Kordić and Mario Čerkez* Case No. (IT-95-14/2) *'Lašva Valley.'* Dario Kordić was sentenced to 25 years imprisonment and Mario Čerkez sentenced to 6 years imprisonment on 17 December 2004.

union with Croatia?[557] But both denied this had been the reason behind the removal of the monastery treasures. It was, they said, the threat from the Bosnian Serb heavy weapons on Mt Vlašić that had prompted the move, and, as Fra Krizanac pointed out, the Partisan units which had destroyed the monastery in World War Two were predominately Serb.[558]

In that case, the ICTY prosecutor asked, if the threat from Bosnian Serb forces had been felt to be so great, why were the items not moved earlier in 1992? However, the instruction to relocate the monastery treasures had come from the Sarajevo-based Franciscan provincial Fra Petra Anđelović who (as mentioned above) was not a supporter of Herceg-Bosna policies. The timing of the artefacts transfer to Croatia was indeed during a period of intense shelling from Mt Vlasic when (in breach of a ceasefire) a Bosnian Serb offensive had begun in the area and there were fears that Travnik would fall.[559] This could well have given urgency to what appeared an increasingly precarious situation for the monastery and that there was no time to lose in removing the collections from harm's way, especially given the overwhelming and deliberate destruction Bosnian Serb forces had inflicted on the Franciscan monastery at Plehan in June 1992.

Nevertheless, the greater part of the library of eight to nine thousand books and other archival material remained in the monastery, along with other less valuable objects. When on 9 June 1993 the British UNPROFOR troops began to evacuate the remaining civilian Croat population from Guča Gora, Major Vaughan Kent-Payne was asked if some of the monastery's and church's religious artefacts could be taken with them. Here the testimonies of Kent-Payne and Tomislav Rajić (a local man) and Fra Stjepan Neimarević, the guardian/abbot of the monastery at the time differ.

Kent-Payne relates in great detail how the 'abbot in charge of the monastery'[560] had asked their battalion chaplain Carson Nicholson if some of the religious artefacts could be brought with them to the church in Vitez for safekeeping, describing how the 'abbot' had explained the monastery's antiquity and its importance as a religious site to Bosnian Croats, even taking the British officers to the library and showing them the thousands of books there. According to Kent-Payne, the 'abbot' and Captain Nicholson finally locked the door of the monastery and were last into the armoured vehicles.[561]

Yet both Tomislav Rajić and Fra Stjepan (the monastery's actual guardian/abbot at the time) testified that the priests, one of whom was Fra Stjepan himself, had left Guča Gora early on the morning of 8 June and only two nuns remained in the monastery.[562] It was the nuns who had decided which artefacts should be taken with the community.[563] Major Hunter also confirmed the departure of the priests and that the two nuns were the senior Catholic religious figures left.[564] However Rajić testified that Ivo Marković, the Catholic priest from

557 See *Kordić and Čerkez*: Testimonies Franjo Krizanac, 25 September 2000, and Stjepan Neimarević, 4 July 2000; Trial Judgement, 26 February 2001, para. 491 on the intention of Herceg-Bosna to secede from Bosnia-Herzegovina and seek unification with Croatia.

558 *Kordić and Čerkez*: Testimony Franjo Krizanac, 25 September 2000,

559 See Burns, J.F. 1992. Cease-fire or no, Serbian shells fall, *New York Times*, 24 November; Fisk, R. 1992. Travnik trusts in Allah as Serbs advance, *The Independent*, 8 December.

560 The title of the priest in charge of a Franciscan monastery in Bosnia was 'guardian' (*gvardijan*).

561 Testimony Vaughan Kent-Payne, *Hadžihasanović and Kubura*, 22 March 2004.

562 Testimony Tomislav Rajić, *Hadžihasanović and Kubura*, 16 February 2004, 2819; testimony Stjepan Neimarević, *Kordić and Čerkez*, 4 July 2000, 22001, 22046–7.

563 Testimony Tomislav Rajić, *Hadžihasanović and Kubura*, 16 and 17 February 2004

564 Personal communications, Roy Hunter.

Dolac (which had come under attack by Bosnian Army forces), was present and a news report mentions a Franciscan monk among those evacuated.[565]

The library remained at the monastery, more or less intact, throughout its brief occupation by the *mujahedin* and the longer occupation by the ARBiH forces installed for its security. When Munir Karić commander of the ARBiH 306th Brigade moved into the monastery he found the library and archival material, including documents from the Vatican and the local church, in a state of disarray.[566] Karić had no idea of the library's original contents or whether items were missing, but had what was there put into some kind of order and the books were used as a reference library.

The monastery was due to be returned to the care of the Franciscans in the autumn of 1994 and the new guardian Fra Franjo Krizanac had been making frequent visits from July onwards, negotiating with Mehmed Alagić, now commander of the Third Corps. Then unaccountably, shortly before the handover the library was packed up by the 306th Brigade and taken to its new headquarters in a school[567] and later transferred to the Archives of Central Bosnia.[568] No explanation was given for this peremptory removal.

When Fra Krizanac arrived on 11 September 1994 to officially take charge of the monastery, the library (which had been there days earlier) was nowhere to be seen. Fra Krizanac made repeated requests for the return of the books, at first to General Alagić and later to others following his subsequent investigations and discovery of their whereabouts. But it was only after exerting constant pressure for the books' recovery that the last part of the library (now somewhat depleted) was finally returned to the monastery by the Head of Travnik Municipality, Besim Halilović – according to Krizanac in 1998, but according to news reports, in March 2000.[569] ICTY Prosecutor Mathias Neuner and Judge Jean-Claude Antonetti believed that the monastery library could be considered part of the cultural heritage of the Croat people of Bosnia and Karić was asked what his thoughts were on the books which had apparently gone missing; he was not, however, asked why the books had been removed from the monastery in the first place.

The Hadžihasanović and Kubura Case

The issue raised by Major Kent-Payne of command responsibility for the desecration that had taken place at Guča Gora monastery was to rebound on senior Bosnian Army commanders. In 2001 ARBiH Generals Enver Hadžihasanović and Mehmed Alagić were indicted by the ICTY on (among a number of other counts) charges of violations of the laws or customs of war (destruction or wilful damage done to institutions dedicated to religion) under Article 3(d) of the 1993 ICTY Statute for their alleged omissions as commanders in relation to the 'destruction or wilful damage of Bosnian Croat institutions dedicated to religion,' allegedly by ARBiH forces under their command and control and specifically citing the cases of Guča Gora monastery and the Catholic church in Travnik (which had

565 Testimony Tomislav Rajić, *Hadžihasanović and Kubura*, 16 February 2004; Stephen, C. 1993. Croat refugees flee Muslim vengeance, *IPS-Inter Press Service*, 10 June 1993.

566 See Munir Karić, *Hadžihasanović and Kubura*, testimony 10 November 2004.

567 *Hadžihasanović and Kubura*: Testimonies Munir Karić, 10 November 2004 and Franjo Krizanac, 17 December 2003

568 *Hadžihasanović and Kubura*: Testimony Munir Karić, 10 November 2004.

569 OHR *BiH TV News Summary*, 13 March 2000 on the return of the books by Halilović. See *Hadžihasanović and Kubura*: Testimonies of Franjo Krizanac, 17 December 2003 and Munir Karić, 10 November 2004; Krizanac testified that the books were finally returned by Halilović in 1998.

also been desecrated by the *mujahedin*). A co-defendant was Amir Kubura, who at the time of the charges under consideration commanded the ARBiH Third Corps Seventh Muslim Mountain Brigade. This was the first case to be tried at the ICTY of crimes alleged to have been committed solely by the indictees' subordinates.

But the prosecution failed to prove their case and, though Mehmed Alagić died before his case was brought to trial, Enver Hadžihasanović was found not guilty 'of failing to take the necessary and reasonable measures to prevent or punish the destruction of or wilful damage to institutions dedicated to religion in Guča Gora and Travnik in June 1993.'[570] The judgement further found that the damage to the monastery of Guča Gora in June 1993 (which constituted desecration) was the work of foreign *mujahedin* who were not subordinated to the Third Corps at that time, and thus not under the control and command of Hadžihasanović.[571] An anomaly in the indictment noted by the judges was that though the desecration of the monastery at Guča Gora was alleged by the prosecution to have been carried out by ARBiH units, among them the Seventh Muslim Mountain Brigade, yet Kubura had not been indicted on this charge; it was established during the trial that the Seventh Muslim Brigade

Fig. 1.33 Inside the church of the Franciscan monastery of Guča Gora in 2012. The red paint defacing the fresco in the apse remains, the statue of St Francis evacuated in 1993 is back in place. © Peter Lippman

570 *Hadžihasanović and Kubura*, Judgement 15 March 2006, Part IX (Disposition), Count 7.
571 *Hadžihasanović and Kubura*, Judgement 15 March 2006, 2007–2008.

had been nowhere near Guča Gora on the dates in question.[572] Hadžihasanović and Kubura were found not guilty of the majority of other counts against them.

The Muslim village of Bandol, the flames from which had been mistaken by the international media as coming from Guča Gora on 8 June 1993, was totally destroyed by HVO forces. Bandol was reported by a British UNPROFOR patrol led by Major Kent-Payne on 13 June as 'a scene of total devastation with every house burned and all livestock slaughtered' and a dead body was found that appeared to have been tortured.[573] The mosque was described by Kent-Payne as completely destroyed beyond recognition, its remaining walls left no more than one and a half metres high.[574]

After the events of 8–9 June 1993 many of the abandoned Croat houses were looted and some were burned. The Franciscans returned to the monastery in September 1994. The *mujahedin* maintained a base in Guča Gora until around 1997. The first of the Croat residents who had left in June 1993 began to trickle back to Guča Gora in 1998, though, as elsewhere across Bosnia-Herzegovina, most were not able to return for a variety of reasons. Fortunately many had not moved far and on Catholic feast days they returned to Guča Gora where the monastery had resumed its role as the spiritual and cultural centre of the area. The statue of St Francis which had been evacuated with the villagers was restored to its place in the apse of the church. As it was not possible to remove the red paint which had been thrown by the *mujahedin* on the fresco behind the statue, it was decided to leave it as a reminder of the events of 1993.

572 *Hadžihasanović and Kubura*, Judgement 15 March 2006, 592–4.

573 *Hadžihasanović and Kubura, PWO Milinfosum No. 45 dated 13 June 1993*, Exhibit DH 71. See also testimony of Fikret Čamdžić, 16 November 2004, who visited Bandol on 14 June 1993.

574 Testimony Vaughan Kent-Payne, *Hadžihasanović and Kubura*, 23 March 2004.

Chapter 2
Documenting the Destruction

Richard Carlton and Helen Walasek

Given the scale and significance of the cultural destruction in Bosnia-Herzegovina during the 1992–1995 war, the worldwide media attention and international condemnation that followed and its inclusion in indictments and subsequent war crimes trials at the International Criminal Tribunal for the former Yugoslavia (ICTY), the surveys or scholarly works on the subject are remarkably few. Nearly two decades after the end of the war there has yet to be an independent statistical appraisal of classes of monuments destroyed and damaged or an authoritative evaluation of the regional and global significance of the loss, a state of affairs noted by the Commission to Preserve National Monuments of Bosnia-Herzegovina in 2010.[1]

Thus, while at first sight the heritage destruction in Bosnia appears to have been extensively documented, given the amount of publications and catalogues produced, yet precise figures of destruction or damage are elusive, in part because the terms 'damaged' and 'destroyed' are interpreted very differently by various authorities, who may also disagree on the scale and cause of damage for a given monument.[2] None of the currently available sources are fully accurate or internally consistent in their definitions of damage and destruction, nor are they comprehensive in terms of the range of monument classes they considered.

Lists compiled during the war of losses in areas that had been ethnically cleansed and where the most extensive destruction had taken place (and thus inaccessible to those with most reason to collect such information) relied almost entirely on the accounts of surviving expellees, many of whom were dispersed across the world. Furthermore, wartime damage lists, particularly those produced by (or on behalf of) the religious communities, were to become a politicized tool for claim and counter claim by the warring parties and competing interest groups regarding the pattern of destruction. False claims made by Bosnian Croat/ Croat nationalists of the supposed destruction of the Franciscan monastery at Guča Gora by Bosnian Army forces and how they were used to mobilize international support for a separate Croat para-state of Herceg-Bosna, for instance, are discussed elsewhere in this volume.[3] Such practices led these wartime lists, with their often unverified information,

---

1 Mulalić Handan, M. (ed.) 2010. *Report on Assessment of the Architectural and Archaeological Heritage Bosnia and Herzegovina*, Integrated Rehabilitation Project Plan/Survey of the architectural and archaeological heritage (IRPP/SAAH), Regional Programme on Cultural and Natural Heritage in South East Europe, Sarajevo: Commission to Preserve National Monuments. For more on the Commission to Preserve National Monuments see Valery Perry in this volume.

2 Noted by Richard Carlton and Helen Walasek, *Topography of Destruction* in this volume and by expert witnesses on cultural heritage András Riedlmayer (2002a) and Colin Kaiser (2002) for the International Criminal Tribunal for the former Yugoslavia (ICTY).

3 See Helen Walasek, *Destruction of the Cultural Heritage*. Hrvatsko Vijeće Obrane (HVO) = Croatian Defence Council.

to be regarded with unease by heritage experts[4] and resulted in an early assumption of a pattern of mutual destruction by scholars like Chapman in 1994.[5] Inaccurate data from some of these surveys, still used as sources, continue to have currency in academic articles. After the end of the war, the different religious communities depended heavily on information supplied to them by representatives at a local level which could be at times conspicuously inaccurate, even when given by informants on the ground.[6]

This chapter will provide a review of the principal forms of documentation of the wartime destruction produced by local, regional and international bodies during the war and in the immediate post-conflict years, as well as offering statistical assessments of the damage currently available.

Wartime Reports: Local Sources

> The assault against Bosnia's Islamic heritage has been a largely unreported facet of the 'ethnic cleansing' campaign, for it occurred in areas now occupied by the Serb conquerers. But accounts by refugees and data collected by the Bosnian government, Bosnian scholars, and by Muslim leaders point to an irreplaceable loss.[7] (Roy Gutman, *Unholy war*, September 1992)

That the assault on Bosnia-Herzegovina's heritage (particularly its Islamic heritage) was an objective (and not simply a byproduct) of the war and a fundamental part of the process of ethnic cleansing was clear from the start of the conflict and many different bodies and groups in Bosnia began to collect data of the mounting losses of cultural and religious property and attempt to analyse the pattern of destruction, despite the considerable difficulties they faced. Among those gathering information were the Bosnian government, the different religious communities (Muslim, Roman Catholic and Serbian Orthodox), heritage preservation institutes, architects' professional associations and civil society organizations. It is notable that the first of only three requests made by the Bosnian authorities in June 1993 for assistance from the international community with regard to the cultural heritage was for reliable information on losses in areas now unreachable to them.[8]

Possibly the earliest published damage reports from local official sources were those compiled by the *State Commission for Gathering Facts on War Crimes in the Republic of Bosnia-Herzegovina* which produced five *Bulletins* on the destruction of religious property between October 1992 and August 1993 with lists of damaged and destroyed objects,

4 See, for instance, the concerns of Colin Kaiser in *Fourth Information Report on War Damage to the Cultural Heritage in Croatia and Bosnia-Herzegovina*, Doc. 6999, Strasbourg: Parliamentary Assembly of the Council of Europe (COE), 19 January 1994 on the damage lists of Serbian Orthodox sacral structures in Bosnia-Herzegovina and Croatia, paras 10–12, 14–19.

5 Chapman, J. 1994. Destruction of a common heritage: The archaeology of war in Croatia, Bosnia and Hercegovina, *Antiquity* 68, 120–26.

6 As noted by Carlton and Walasek, *Topography of Destruction* in this volume: see also Riedlmayer, A. 2002a, 23 and following and Kaiser, C. 2002, 3.

7 Gutman, R. 1992. Unholy war; Serbs target culture, heritage of Bosnia's Muslims, *Newsday*, 2 September.

8 *Third Information Report on War Damage to the Cultural Heritage in Croatia and Bosnia-Herzegovina*, Parliamentary Assembly of the Council of Europe (COE), *Strasbourg*, Doc. 6904, *20 September 1993, 17.*

listing Islamic, Catholic and Orthodox structures and including the loss of archives and libraries, but noted the inevitable difficulties in obtaining reliable data in many cases, with a substantial number of entries coming from hearsay reports (most often from ethnically cleansed populations).[9]

The *Institute for the Protection of the Cultural, Historical and Natural Heritage of Bosnia-Herzegovina*, the country's state heritage preservation body, compiled and circulated a number of lists during the war. The first of these lists to reach the COE, then seeking accurate reports on the damage to the heritage, was received by fax in Paris on 16 June 1993; the COE's consultant expert Colin Kaiser, while acknowledging its importance and the seemingly catastrophic damage being done to Bosnia's Islamic heritage, nevertheless found it hard to evaluate, given the difficulties in corroborating some of the information.[10] The Institute's later *Information Document* on the destruction and damage to the cultural-historical heritage of September 1993 was included in the fourth of the Council of Europe's *Information Reports* in January 1994.[11] This long report from the Sarajevo-based Institute listed not only the damage done to Islamic, Catholic, Orthodox and Jewish sacral monuments, but to other historic structures (including urban complexes), archaeological sites, as well as to movable heritage in museums, libraries archives and institutes. The Institute went on to produce the most complete assessment of the damage done to the cultural heritage shortly before the end of the war and which will be described in more detail below.

At the same time, the *Islamic, Roman Catholic* and *Serbian Orthodox Religious Communities* also began to produce lists of damage and destruction to their property. As mentioned above, these relied substantially on information provided by local informants and hearsay reports and their complete reliability, as well as the credibility of the reports, frequently tended to be held in some doubt.[12] Among these were the lists produced by the *Riyasat* [Rijaset] *of the Islamic Community in Bosnia-Herzegovina* on 12 September and 10 November 1992 and 4 February 1993 (on Islamic religious property);[13] the letter of the *Archbishopric of Bosnia*[14] in March 1993 (on Roman Catholic religious property);[15] and the *Report on War Devastation to Orthodox Churches in Croatia and Bosnia-Herzegovina* produced by the *Institute of Protection of Cultural Monuments of the Republic of Serbia* (Yugoslavia) with the *Museum of the Serbian Orthodox Church*, Belgrade, of January 1993 (on Serbian Orthodox religious property).[16]

9 *Bulletins 1–5*, Sarajevo: State Commission for Gathering Facts on War Crimes in the Republic of Bosnia and Herzegovina, 1992–1993.

10 Kaiser, C. Update on war damage to the cultural heritage in Croatia and Bosnia-Herzegovina, COE *Third Information Report*, Nos 29–37.

11 COE *Fourth Information Report, 2. Information Document submitted by the Institute for the Protection of the Cultural-historical and Natural Heritage of the Republic of Bosnia and Herzegovina (Sarajevo, September 1993).*

12 On how information was gathered by the religious communities, personal communications Helen Walasek with Slobodan Mileusnić, author of *Spiritual Genocide*, in Belgrade (via Marko Attila Hoare) and conversation with Fra Marko Karamatić in Sarajevo. See Colin Kaiser in his reports in the COE *Second, Third and Fourth Information Reports* for his comments on various lists from religious communities.

13 *Second Information Report on War Damage to the Cultural Heritage in Croatia and Bosnia-Herzegovina*. Doc. 6869, Strasbourg: Parliamentary Assembly of the Council of Europe (COE), 17 June 1993, No. 33.

14 Archdiocese of Vrhbosna/Vrhbosanska nadbiskupija.

15 COE *Third Information Report*, No. 33.

16 COE *Fourth Information Report*, No. 6.

Some publications were regarded as biased and inaccurate in many respects. One perceived as produced in order to counter-balance accounts of ethnic cleansing by Bosnian Croat forces (HVO) was the Croatian Community of Herzeg-Bosnia's booklet *Ethnic Cleansing of Croats in Bosnia and Herzegovina, 1991–1993* of August 1993 which included details of damage and destruction to Catholic religious property.[17] A similar volume was *Iskorenjivanje Srba u Bosni i Hercegovini, 1992–1993* [*The Eradication of Serbs in Bosnia and Herzegovina, 1992–1993*], by Jovanović, Bundalo and Govedarica, which included a section on '*Destruction of Orthodox Religious Structures*.'[18]

One of the earliest, best illustrated and most inclusive in terms of types of monuments and structures listed among the early local surveys of destroyed architecture was *Mostar '92 – Urbicid*,[19] a catalogue which recorded the heritage and urban structures targeted by Bosnian Serb/ JNA (Yugoslav People's Army) forces in Mostar during the early months of the war and accompanied an exhibition that opened in Zagreb, Croatia in December 1992 and subsequently travelled to other European cities.[20] The exhibition and catalogue grew from an original concept of the *Association of Architects of Mostar* (DAR) and was edited by staff of Mostar's *Institute for the Protection of Cultural Monuments*.[21] The publication is now also remarkable for being published by (among others) the HVO (*Hrvatsko Vijeće Obrane*),[22] the Bosnian Croat nationalist paramilitary defence force which, from early 1993 when fighting broke out in the so-called Muslim/Croat War until the signing of the Washington Agreement of March 1994, was responsible for the destruction of scores of historic monuments (among them the *Stari Most*, or Old Bridge) in Mostar, Stolac and elsewhere.[23] The 1992 catalogue, however, included the city's Ottoman and Islamic heritage and even the remains of the totally destroyed Serbian Orthodox Cathedral (*Saborna Crkva*), the destruction of which has been attributed (until time of writing) to the HVO. Referring in its foreword by the DAR to the 'urban genocide' of the city,[24] *Mostar '92 – Urbicid* is also notable for being probably the earliest use of the term 'urbicide' in relation to the destruction of cities and urban culture during the Wars of Yugoslav Succession; this usage of the term came to be increasingly appropriated by urban theorists in discussions of cities and conflict.[25]

---

17 Ivanović, V. and Vlahušić, A. and Croatian Community of Herzeg-Bosnia 1993 *The Ethnic Cleansing of Croats in Bosnia and Herzegovina, 1991–1993*, Mostar: Office of President of the Croatian Community of Herzeg-Bosnia, August 1993. See COE *Fourth Information Report*, Doc 6999, 19 January 1994, for Colin Kaiser's assessment of the information contained in the booklet.

18 Jovanović, D., Bundalo, G. and Govedarica, M. 1994. *Iskorenjivanje Srba u Bosni i Hercegovini, 1992–1993* [*The Eradication of Serbs in Bosnia and Herzegovina, 1992–1993*], *Destruction of Orthodox Religious Structures*. Belgrade: Rad, 1994, 205–20.

19 Ribarević-Nikolić, I. and Jurić, Ž. 1992. *Mostar '92 – Urbicid*. Mostar: Hrvatsko vijeće obrane općine Mostar and Javno poduzeće za obnvu izgradnju Mostara.

20 COE *Second Information Report*, 26.

21 DAR = Društvo Arhitekata Mostara. Institute for the Protection of Cultural Monuments = *Zavod za Zaštitu Spomenika Kulture*.

22 Hrvatsko Vijeće Obrane (HVO) = Croatian Defence Council.

23 After the Vance-Owen Plan to divide Bosnia into ethnically pure cantons was put forward, encouraging nationalist Bosnian Croats forces to turn on their Bosnian government allies. See *Introduction*.

24 Ribarević-Nikolić, I. and Jurić, Ž. 1992. *Mostar '92 – Urbicid*, 6.

25 Most notably by the British academic Martin Coward. For an early use see Coward, M. 2001. *Community as Heterogeneous Ensemble: Mostar and Multiculturalism*, ISA Annual Convention, Chicago, 21–24 February 2001.

The *International Centre for Peace* (*Međunarodni centar za mir*), a Sarajevo-based civil society group, produced two long reports on war crimes committed in Bosnia-Herzegovina in 1992 which were circulated internationally and included information on destruction of the cultural and historical heritage, in particular that of Sarajevo.[26] In October 1993 the *Association of Architects of Bosnia-Herzegovina* (*DAS–SABIH*) mounted the exhibition: *Warchitecture – Urbicide Sarajevo*[27] with an accompanying catalogue that surveyed destroyed architecture in Sarajevo. In the process the Sarajevo architects coined a new term: 'warchitecture,' that (like 'urbicide') came to be much employed in the following years.[28] The catalogue was later reprinted in various versions as the exhibition travelled to (among other countries) France, Germany, Austria, Sweden and the United States over the following years. In March 1995 the *Open Society Fund* published *Cultural Institutions and Monuments in Sarajevo* as part of its City of Sarajevo Project which, along with listing the needs of institutions, also included damage reports.[29]

However, the most complete assessment of all types of cultural and religious heritage was *A Report on the Devastation of Cultural, Historical and Natural Heritage of the Republic of Bosnia and Herzegovina*, published by the *Institute for the Protection of Cultural, Historical and Natural Heritage of Bosnia-Herzegovina* in 1995 (with funding from UNESCO), which attempted under difficult conditions to document destruction and damage to the cultural heritage up to 5 September 1995.[30] With the war then still in progress, the report drew on a few site visits to easily accessed localities in Sarajevo and elsewhere in the Federation (one of the two entities into which Bosnia-Herzegovina is divided, the other being Serb-dominated Republika Srpska), but largely relied on the documentation of the *State Commission for Gathering Facts on War Crimes* (see above), reports from religious communities and other similar sources.

However, both the introduction and summary of the report make no claims for the accuracy and completeness of data due to the difficulties already described of acquiring and verifying information during the war. Indeed, long after the conflict ended it remained difficult and potentially dangerous (for Muslims in particular) to attempt to gather information on destroyed Islamic monuments in Republika Srpska and some parts of the Federation. Although the lists of monuments in the report provide a good framework for

---

26 Serdarević, M. 1997. *Pravna zaštita kulturno-historijskog naslijeđa BiH sa posebnim osvrtom na staro urbano jezgro Sarajeva: (nastajanje, očuvanje, destrukcija)*, Sarajevo: Međunarodni centar za mir, 123–4.

27 Čurić, B. et al. 1994. *Warchitecture – Urbicide Sarajevo*, exhibition catalogue, originally published by the Association of Architects of Bosnia-Herzegovina (DAS–SABIH), Sarajevo, October 1993, reprint Munich, with the assistance of Baumeister, May 1994. DAS–SABIH = *Društvo arhitekata Sarajevo : Savez arhitekata BiH.*

28 Most notably by the US academic Andrew Herscher. See, for example: 2008 Warchitectural Theory, *Journal of Architectural Education*, 61:3, February 2008, 35–43.

29 Čengić, A. and Duraković, F. (eds) 1995. *Cultural Institutions and Monuments in Sarajevo.* Budapest: Open Society Institute, Open Society Project of Sarajevo, March 1995.

30 Pusilo, N. et al. 1995. *A Report on the Devastation of Cultural, Historical and Natural Heritage of the Republic of Bosnia and Herzegovina (From April 5, 1992 until September 5, 1995)*, Naše Starine, Vanredni broj XXI, Sarajevo: Institute for the Protection of Cultural, Historical and Natural Heritage of the Republic of Bosnia-Herzegovina [Zavod za zaštitu kulturno-historijskog i prirodnog naslijedja BiH]. A revised Bosnian language only edition was published in 1997: *Izvjestaj o devastaciji kulturno-historijskog i prirodnog naslijedja Bosne i Hercegovine (1992–1995)*, Sarajevo: Institute for the Protection of Cultural, Historical and Natural Heritage of Bosnia-Herzegovina [Zavod za zastitu kulturno-historijskog i prirodnog naslijedja BiH], Centar za naslijedje BiH.

further investigation, they are by no means complete and there are numerous duplications of monuments under slightly different names, omissions of monuments and inaccuracies as to the level of damage.

Wartime Reports: International Sources

Outside Bosnia, the earliest damage lists compiled from a variety of Bosnian sources were circulated in 1992 by the UK-based NGO *Bosnia-Herzegovina Heritage Rescue* (*BHHR*).[31] This was followed by the list reproduced in the April 1993 newsletter of *IRCICA*, the Research Center for Islamic History, Art and Culture of the Organisation of the Islamic Conference (OIC).[32] IRCICA included a further updated list in a book on Islamic architecture in Bosnia-Herzegovina in 1994.[33] Once again, these early lists suffer from the difficulties of collecting verified information during the war and cannot be regarded as accurate and authoritative sources.

But what of institutions representing the international community (such as UNHCR, ICRC, ECMM and UNESCO) which had a certain amount of freedom of movement across Bosnia even during the war and complete freedom of movement afterwards and whose representatives were based for many months in contentious locations across the country, largely inaccessible to Bosnians of the 'wrong' ethno-national group?[34]

Despite not having a presence in the enormous multinational effort in Bosnia-Herzegovina during the war, the most wide-ranging sources of information were the ten widely distributed *Information Reports on War Damage to the Cultural Heritage in Croatia and Bosnia-Herzegovina* produced from February 1993 to January 1997 for the *Committee of Culture and Education of the Parliamentary Assembly of the Council of Europe* (COE). While these were largely reports of specific missions made by COE consultant experts, principally Colin Kaiser,[35] they also included a variety of reports and letters sent by governmental and non-governmental bodies in Bosnia-Herzegovina and elsewhere and later the reports of the European Community Monitor Mission (ECMM). They did not include photographic documentation and were geographically limited to areas consultants could access as the war was in progress. The reports and the difficulties faced by the principal consultant expert in gathering information for them are described elsewhere in this volume.[36]

Under pressure from COE consultant Colin Kaiser, the ECMM did begin to systematically monitor and report on damage to religious and cultural property. The ECMM focus was mainly on sacral structures, partly because these were the most attacked and partly because

31 The BHHR report was later published in the Parliamentary Assembly of the Council of Europe's *[First] Information Report on War Damage to the Cultural Heritage in Croatia and Bosnia-Herzegovina*, Doc. 6756, Strasbourg: Parliamentary Assembly of the Council of Europe (COE), 2 February 1993.

32 Ihsanoglu, E. (ed.) 1993. Bosnia-Hercegovina: History, culture, heritage. *Research Center for Islamic History, Art and Culture Newsletter* 31, April 1993, Istanbul: IRCICA.

33 Pašić, A. 1994. *Islamic Architecture in Bosnia and Herzegovina*. Istanbul: IRCICA.

34 For more on this issue see Walasek, *Destruction of the Cultural Heritage* elsewhere in this volume.

35 Kaiser went on to head the UNESCO antenna in Mostar (1995) and later became the organization's representative for Bosnia-Herzegovina, based in Sarajevo.

36 See Walasek, *Destruction of the Cultural Heritage*.

of the difficulties for the monitors in identifying civil heritage.[37] The *European Community Monitor Mission*'s five *Cultural Heritage Reports on the Situation in Croatia and Bosnia and Herzegovina* produced from 31 December 1994 to 17 December 1996 included some photographic records of damage.[38] However, as it was carried out by minimally-trained ECMM staff without backgrounds in cultural heritage, the reports were far from comprehensive and monitoring of the cultural heritage (certainly not an ECMM priority) was hampered by the limitations on the operational capabilities of the ECMM and the areas made accessible to them at any time by the warring parties; ECMM had only infrequent access to Republika Srpska, for instance. The final report notes the difficulties even in the post-war period of accessing cultural sites.[39] The ECMM Reports are now unobtainable and their database consisting of more than a thousand entries, including data from other sources, appears to no longer exist.[40]

However, given the conditions under which they were produced, neither the COE nor the ECMM reports were intended to be attempts to make a professional, systematic and comprehensive evaluation of damage.

Post-war Reports: Local and Regional Sources

After the end of the war a number of assessments of varying usefulness were produced by government departments or heritage institutes. At entity-level, the *Ministry of Education, Science, Culture and Sports of the Federation of Bosnia-Herzegovina* compiled a report in 1999 on *Cultural heritage of special value destroyed in the war 1992–1995* on the condition of objects of the cultural and historical heritage in the Federation which consisted simply of lists of classes of monuments in the different cantons of the entity and their status in 1995 and 1999.[41] No similar report was produced by the Republika Srpska entity Ministry of Education and Culture. The *Institute for the Protection of the Cultural-historical and Natural Heritage of the Canton of Sarajevo* produced *Katalog obnove kulturno-historijskog i prirodnog naslijedja Kantona Sarajevo, 1996–2000*, a report on war damaged cultural heritage in the canton listing monuments by classes and dates with their damage category in 1996 and their condition in 2000.[42]

In the immediate post-war period a group of local or regional publications produced by, or on behalf of, the three main religious communities of Bosnia-Herzegovina appeared: *Raspeta Crvka u Bosni i Hercegovini/The Crucified Church in Bosnia-Herzegovina*

37 COE *Eighth Information Report on War Damage to the Cultural Heritage in Croatia and Bosnia-Herzegovina*, Doc. 7341, 28 June 1995, *Cultural Heritage Report by the European Community Monitoring Mission (Humanitarian Section)*, No. 5.

38 See Walasek, *Destruction of the Cultural Heritage* in this volume for the background to the ECMM Reports.

39 The ECMM *Cultural Heritage Reports Nos 2* and *3* were also published in the eighth and tenth COE *Information Reports*.

40 ECMM *Cultural Heritage Report No. 5*, 3, on size of database. Personal communication, Helen Walasek with ECMM HQ Sarajevo in 2000. The staff at that date had no knowledge of the existence of such a database despite making some efforts to enquire within the organization.

41 Husedžinović, S. 1998. *Cultural Heritage of Special Value Destroyed in the War 1992–1995*, Sarajevo: Federation of Bosnia and Herzegovina, Ministry of Education, Science, Culture and Sports.

42 Čelić-Čemerlić, V. (ed.) 2000. *Katalog obnove kulturno-historijskog i prirodnog naslijedja Kantona Sarajevo, 1996–2000*, Sarajevo: Kantonalni zavod za zastitu kulturno-historijskog i prirodnog naslijedja Sarajevo.

edited by Živković (documenting Catholic structures and produced in Croatia),[43] *Duhovni Genocid/Spiritual Genocide* by Mileusnić (documenting Orthodox structures in both Croatia and Bosnia-Herzegovina and produced in Serbia)[44] and *Prilozi Izučavanju Genocida nad Bosnjačima 1992–1995/Contributions to the Study of Genocide against the Bosniaks 1992–1995* by Omerdić[45] and *Slike Zločina: Rušenje Islamskih Vjerskih Objekata u Bosni i Hercegovini/Pictures of Crime: The Destruction of Islamic Religious Buildings in Bosnia-Herzegovina* by Zukić[46] (both documenting destruction to Islamic property, the first as one part of a book documenting all war crimes against Bosnian Muslims/Bosniaks,[47] the second covering mainly the Federation and both produced in Sarajevo).

All these publications were similarly based on compilations of often uncorroborated and occasionally erroneous reports gathered from a variety of informants during the war, usually local imams and priests or other members of the religious communities. However, taken overall, the volumes are not significantly inaccurate.

In *Raspeta Crkva* the reports on Catholic churches and sacral structures appear to be accurately recorded and are usually illustrated with before and after photographs, making it easy to identify structures and remains in their settings. There are a number of statistical tables.

The damage to the Orthodox heritage has, in the main, also been reasonably accurately recorded in *Spiritual Genocide*, though not well accompanied with photographic evidence to back up reports. Early editions had included sites known to have been destroyed in World War II, which aroused wide criticism at the time, but these anomalies were largely edited out of the last edition in 1997. However, even that edition included inaccurate entries such as, for instance, that on the monastery at Zavala in Herzegovina which states that the sixteenth-century monastery church of the Presentation of the Virgin had been demolished by the Croatian Army, which is incorrect. Though the other monastery buildings were badly damaged or destroyed (as the entry also notes), the church itself was not attacked.[48]

Islamic sites in the Federation were largely well-documented, but while Omerdić listed municipalities across Bosnia-Herzegovina using information collected shortly after the end of the war, the records suffered from lack of access to large areas of Republika Srpska (RS) where enormous devastation had been inflicted on the Islamic/Ottoman heritage. Omerdić's volume has little photographic evidence, though it includes many statistical tables and graphs and usefully includes the total numbers of structures by type at the start of the war. Zukić's publication covers solely the Federation (with the exception of Banja

43 Živković, I. (ed.) 1997. *Raspeta Crkva u Bosni i Hercegovini: Uništavanje Katoličkih Sakralnih Objekata u Bosni i Hercegovini (1991–1996)*, Banja Luka, Mostar, and Sarajevo: Biskupska Konferencija Bosne i Hercegovine, Hrvatska Matica Iseljenika Bosne i Hercegovine, and Zagreb: Hrvatski Informativni Centar.

44 Mileusnić, S. 1997. *Duhovni Genocid / Spiritual Genocide 1991–1995*. Belgrade: Museum of the Serbian Orthodox Church.

45 Omerdić, M. 1999. *Prilozi Izučavanju Genocida nad Bosnjačima 1992–1995*. Sarajevo: El-Kalem.

46 Zukić, K. 1999. *Slike Zločina: Rušenje Islamskih Vjerskih Objekata u Bosni i Hercegovini* (*The Evidence of Crime: The Destruction of Islamic Buildings in Bosnia & Hercegovina*), Sarajevo: Centar za islamsku arhitekturu.

47 The name many Bosnian Muslims choose to identify themselves with.

48 See Tina Wik in this volume for more on the monastery at Zavala and the conservation of the church.

Luka Muftijstvo in RS) and is well illustrated with photographs taken by the author on field trips over 1995–1996.[49]

Since these publications of the immediate post-war period, no further updated global lists and assessments have been published by the religious communities, although local officials are now in a position to provide accurate information. In an attempt to collect more precise figures of heritage destruction, in 2010 the Ministry of Civil Affairs of Bosnia-Herzegovina suggested that the Interreligious Council of Bosnia-Herzegovina or the religious communities submit up-to-date information on their properties.[50]

Other local publications documenting the destruction featured a range of the best-known monuments or single monuments.[51] *Crimes in Stolac Municipality (1992–1994)* and its later versions, published by the *Society [Association] for Renewal of Civil Trust in the Stolac Municipality*, was uniquely exemplary in its treatment of Stolac, integrating the destruction of cultural heritage in the context of other war crimes committed in the municipality and providing a detailed overview of a wide range of monument classes and portable items damaged, destroyed or lost.[52] Others, like Tucaković's *Aladža Džamija – Ubijeni Monument*, recorded Islamic monuments destroyed in particular communities (in this case Foča) as well as documenting other war crimes against Muslims.[53] Another notable publication was Ravlić's *Banjalučka Ferhadija: Ljepotica Koju su Ubili* on the Ferhadija Mosque; Ravlić, a Bosnian Serb journalist and historian from Banja Luka, was forced into exile because of his protests against the destruction of the Islamic heritage there.[54]

Post-war Reports: International Sources

After the war ended the *Cultural Heritage Division* of the Council of Europe (COE) under their *Technical Co-operation and Consultancy Programme* carried out a *Preliminary Heritage Inventory* of Bosnia-Herzegovina during 1997–1998 as the Preliminary Phase of its *Specific Action Plan for Bosnia-Herzegovina* (see above).[55] But the objectives of the project meant that the survey was intended neither to be comprehensive (recording only 315 religious and 211 secular monuments), nor to focus on war damage (or who committed it), but to form a basis for proper future management of the architectural heritage. There were

49 Many of the photographs from this publication can be seen online at www.rijaset.ba.

50 Mulalić Handan, M. (ed.) 2010 *Report on Assessment of the Architectural and Archaeological heritage*, 6 (Note 7).

51 Such as Rizvanbegović, F. 1999. *Begovina – Stolac*. Sarajevo: Federation of Bosnia and Herzegovina, Ministry of Education, Science, Culture and Sports; Čamdžić, E. Ef. 2000. *Sa* Džamijama i Povratak. Banja Luka: Banja Luka Muftistvo.

52 Mulać, E. (ed.) 1996. *Crimes in Stolac Municipality (1992–1994)*, The Revival of Stolac, Iss. No. 1, Sarajevo: Association for Renewal of Civil Trust in the Stolac Municipality. Reprinted with additional material in 2001 (*Crimes in Stolac Municipality (1992–1996)*). *Society* for Renewal of Civil Trust in the Stolac Municipality = *Udruženje za obnovu građanskog povjerenja općini Stolac; Udruženje* is sometimes translated as 'Association.'

53 Tucaković, Š. 1998. *Aladža Džamija – Ubijeni Monument*. Sarajevo: Institut za istraživanje zločina protiv čovječnosti i međunarodnog prava.

54 Ravlić, A.A. 1996. *Banjalucka Ferhadija: Ljepotica Koju su Ubili/ Banja Luka's Ferhadija Mosque: A Murdered Beauty*. Rijeka: AARis.

55 *Specific Action Plan for Bosnia-Herzegovina, Preliminary Phase, Final Report*, Technical Cooperation and Consultancy Programme related to the integrated conservation of the cultural heritage, Strasbourg: Council of Europe (COE), Cultural Heritage Division, March 1999.

problems gathering information in certain locations (described elsewhere in this volume).[56] Even so, there were glaring omissions for which no explanation was made in the report, such as the inclusion of only five monuments in Sarajevo of which four are nineteenth century, and the inclusion of only the 'historic core' of Mostar, as well as numerous errors in identifying sites. Furthermore, inaccuracies regarding the pre-war existence of mosques at a number of sites in Republika Srpska led to suspicions that the external consultants had occasionally been deceived by local authorities. There were also misattributions such as identifying the surviving simple rectangular shell of the Mir-Muhamed Mosque in Čajniče as the remains of the Sinan-beg Boljanić Mosque (1570) which had been of a complex cube construction as can be seen in pre-war photographs; it had been dynamited in 1992 and the site levelled.

Two expert reports based on a combination of site visits and assessments of reliable already collected data were prepared in 2002 for the Office of the Prosecutor (OTP) of the *International Criminal Tribunal for the former Yugoslavia* (ICTY) on destruction of cultural property: Colin Kaiser's *Report on Destruction of Cultural Property* and András Riedlmayer's *Destruction of Cultural Heritage in Bosnia-Herzegovina, 1992–1996: A Post-war Survey of Selected Municipalities*.[57] However, although Riedlmayer continued to assess a number of further municipalities,[58] both experts surveyed only those municipalities selected by the ICTY as relevant to their prosecutions against Serb/Bosnian Serb defendants and focused almost entirely on Islamic and Catholic sacral structures.

Statistics of Destruction

As noted above, precise figures of destruction or damage are elusive. Bearing in mind the limitations described above, at November 1995 the (incomplete) figures compiled by the *Institute for the Protection of Cultural, Historical and Natural Heritage of Bosnia-Herzegovina* recorded 2,771 cultural properties were damaged or destroyed during the war, 713 were totally destroyed and 554 were set on fire and are unusable.[59] Analysing this documentation, the Commission to Preserve National Monuments in its *Report on assessment of the architectural and archaeological heritage Bosnia and Herzegovina*,[60] found that out of 60 valuable urban nuclei, 49 were destroyed or very badly damaged; these agglomerations

56 See Walasek, *Destruction of the Cultural Heritage*.

57 Kaiser, C. 2002. *Report on the Damaging and Destruction of Islamic and Roman Catholic Sacral Buildings in the Municipalities of Bosanski Novi, Donji Vakuf, Kljuc, Kotor Varos, Prijedor and Sanski Most in 1992 (April to December) and Preliminary Remarks on Damaging and Destruction of Islamic and Roman Catholic Sacral Buildings in the Municipalities of Bijeljina, Brcko, Bratunac, Cajnice, Doboj, Foca, Ilijas, Rogatica, Visegrad and Zvornik*, (*Report on Destruction of Cultural Property* 09-Jul-02, Exhibit P906, *Krajišnik* Case No. IT-00-39); and Riedlmayer, A. 2002. *Destruction of Cultural Heritage in Bosnia-Herzegovina, 1992–1996: A Postwar Survey of Selected Municipalities*, (*Milošević* Case No. IT-02-54, Exhibit P486, Date: 08/07/2003, and *Krajišnik* Case No. IT-00-39, Exhibit P732, Date: 23/05/2005). Riedlmayer's records for each monument with photographs are also available at the ICTY Court Records website http://icr.icty.org/. Riedlmayer initially assessed 19 municipalities (392 sites assessed). He was later requested by ICTY to survey a further seven, bringing the total of sites assessed to 452.

58 See for instance: Destroying all traces of mosques and. Churches, *Sense Tribunal* [online], 2 June 2010. Personal communication Helen Walasek with Riedlmayer, 22 May 2012.

59 Mulalić Handan, M. (ed.) 2010, 5.

60 Mulalić Handan, M. (ed.) 2010, 5–7.

included many individual structures assessed as heritage of international importance such as the Old Bridge (*Stari Most*) at Mostar. All 9 valuable urban-rural ensembles were seriously damaged. The Commission further noted that out of the 58 the most valuable mosques and *tekije*, 22 were totally demolished, including 13 Category One buildings. Of the 40 most valuable churches and Orthodox and Catholic monasteries (fifteenth – nineteenth centuries), 5 were destroyed and 4 damaged. Many cemeteries and funerary monuments suffered major damage, and a number listed as Category One were totally destroyed. From the Austro-Hungarian period 16 Category One buildings were damaged and 4 destroyed. In October 2010 the Commission to Preserve National Monuments noted that of 582 National Monuments so far designated by the Commission,[61] 246 suffered varying levels of war damage; of those 92 were totally destroyed or very badly damaged, 143 damaged, 5 ensembles with at least one building in the ensemble destroyed and 6 historic urban area and 2 larger architectural ensembles heavily destroyed or damaged (adding up to hundreds of buildings destroyed or damaged).[62]

Figures collated from Husedžinović's 1999 report for the *Ministry of Education, Science, Culture and Sports of the Federation* on heritage destruction/damage in the Federation found 371 structures *uništeno* (destroyed) or *teško oštećeno* (severely damaged), including 31 'other sites,' urban agglomerations, and so on, of which 253 were either Islamic (157), Catholic (57), Orthodox (36) or Jewish (3) sacral structures, the majority (220) being mosques, churches or synagogues. It found 171 *oštećeno* (damaged) monuments, including 29 'other sites,' urban agglomerations, and so on, of which 103 were Islamic (63), Catholic (26) or Orthodox (14) sacral structures, 84 of which were mosques, churches or synagogues.[63] The report also reports figures for *lakše oštećeni* (slightly damaged) and *manje oštećeni* (hardly damaged) monuments, but as there are no definitions for any of the different levels of damage it is difficult to distinguish the difference between the two. As noted above, no similar assessments with statistics of heritage destruction were produced by the Republika Srpska entity Ministry of Education and Culture.

For figures published on behalf of the different religious authorities with regard to damaged/destroyed sacral structures and discussed above, it should be noted that only a proportion of these would be considered historic monuments and that definitions of levels of damage and destruction are not defined, not necessarily consistent and listings are not always accurate. Only Omerdić gives the total number of sacral structures by type that existed at the start of the war. The figures given below do not include destruction and damage to other buildings owned by the religious communities, such as parish houses or *mektebs*.

Reporting on attacks on Islamic sacral structures Omerdić uses only two damage levels: *uništeno* (destroyed) and *oštećeno* (damaged).[64] According to Omerdić in the region of 832 mosques (*džamija*) and the type of small mosque known as a *mesdžid* (usually without a minaret and frequently without a minber, or pulpit) out of a pre-war total of 1,701 were 'destroyed' and 348 'severely damaged,' totalling 1,180 mosques destroyed or damaged. Of the attacks on *džamija* Omerdić attributed 85 per cent of destruction and damage to

61 Further National Monuments are designated at each session of the Commission. See Valery Perry in this volume for more on the working of the Commission. See *Sessions of the Commission* on its website at http://www.kons.gov.ba/main.php?id_struct=66&lang=4.

62 Information provided by the Commission to Preserve National Monuments in October 2010.

63 Husedžinović, S. 1998.

64 Omerdić, M. 1999, 469, 476.

Serbs (58 per cent + 27 per cent) and 15 per cent to Croats (9 per cent + 6 per cent); for the type *mesdžid* he attributes 75 per cent of destruction and damage to Serbs (67 per cent + 8 per cent) and 25 per cent to Croats (17 per cent + 8 per cent). Among the other classes of Islamic sacral monuments, 4 dervish tekkes (*tekija)* were 'destroyed,' 5 'severely damaged,' while 37 mausolea (*turbe*) were 'destroyed' and seven 'severely damaged.' Drawing on additional sources for destruction to Islamic structures, Riedlmayer amends these figures slightly to a total of 1,186 mosques, 15 *tekije* and 90 *turbeta* destroyed or damaged.[65]

Recording attacks on Roman Catholic sacral structures Živković uses three levels of damage: *potpuno uništeno* (completely destroyed), *teško oštećeno* (severely damaged) and *oštećeno* (damaged).[66] Although no figures are given for the total number of Catholic structures by type at the start of the war, according to Živković, 188 parish and other churches and chapels were 'completely destroyed', 162 'severely damaged,' and 230 'damaged.' Of these figures 86 per cent of those 'completely destroyed' were attributed to Serb forces and 14 per cent to Muslim forces, of the 'severely damaged' 69 per cent were attributed to Serb forces and 31 per cent to Muslim forces, while in the 'damaged' category 60 per cent was attributed to Serb forces and 40 per cent to Muslim forces. Eight monasteries and eight graveyards were described as 'completely destroyed' and 14 monasteries and 53 graveyards were 'severely damaged.'

With regard to attacks on Serbian Orthodox (Serb) parish and other churches, chapels and monasteries, Mileusnić lists two levels of damage: *porušene* (destroyed) and *oštećeno* (damaged). According to Mileusnić 121 churches in eparchies/dioceses in Bosnia-Herzegovina (including monasteries) were 'destroyed' and 169 'damaged.'[67] The total figures are not broken down into percentages of which forces perpetrated the damage or destruction, nor are there figures for the total number of Orthodox structures by type at the start of the war. The reports for the eparchy of Zahum-Herzegovina include a number of parishes in Croatia which are a part of this diocese and which have been excluded from this total.

However, despite the lack of an independent statistical appraisal of classes of monuments destroyed and damaged in Bosnia-Herzegovina during the war or an authoritative evaluation of the regional and global significance of the loss, and despite the limitations described above of what documentation exists, it is more than apparent that the scale and extent of the destruction to the country's cultural heritage was undeniably without parallel since the Second World War.

65 Riedlmayer, A. 2002. From the Ashes: The Past and Future of Bosnia's Cultural Heritage, in: *Islam and Bosnia: Conflict Resolution and Foreign Policy in Multi-Ethnic States,* edited by M. Shatzmiller. Montreal: McGill-Queens University Press, 99.

66 Živković, I. (ed.) 1997, 357.

67 Mileusnić, S. 1997, 235–6.

Chapter 3
Topography of Destruction: Post-conflict Fieldwork Assessing the Cultural Heritage in Bosnia-Herzegovina

Richard Carlton and Helen Walasek

In October 2000 and June 2001 the authors made two fieldtrips to Bosnia-Herzegovina[1] to carry out an independent assessment of the uneven, occasionally inaccurate and sometimes contradictory information available from various published and unpublished sources then available (and by 2012 still not yet superseded)[2] on the damage and destruction caused to the country's cultural and religious heritage during and immediately after the 1992–1995 Bosnian War. In conjunction with a close appraisal of the record described in Chapter 2 above,[3] the fieldtrips also allowed a critical examination of the validity of the highly charged and politicized wartime claims and counter claims of monument destruction made by competing interest groups.

Thus, the intention was not to complete a comprehensive survey, but rather in the short time available, to visit and photograph sites (many in isolated rural locations) for which there still appeared to be no verified data. The authors hoped as well to gain a visual and anecdotal impression of the impact of the destruction of Bosnia-Herzegovina's cultural heritage outside the major urban centres of Sarajevo and Mostar which had been exhaustively recorded and publicized. These trips, which proved to be the first attempt to carry out a field survey of the breadth of destruction of the Islamic/Ottoman heritage across the entity of Republika Srpska, were also to vividly illustrate the problematic nature of recording that destruction even five to six years after the end of the war.

The photographic and other site records collected on the two fieldtrips were made available to the International Criminal Tribunal for the former Yugoslavia (ICTY). This material was later used as documentation in the testimony of the ICTY expert witness András Riedlmayer to whom the authors provided advice, useful contacts, and information in advance of his 2002 mission to survey the destroyed cultural and religious heritage in selected Bosnian municipalities for the Tribunal which was to form the basis of evidence in subsequent war crimes prosecutions.

1 The authors are grateful to the Bosnian Institute (London) for its support for their fieldtrips in 2000 and 2001.

2 See Mulalić Handan, M. 2010. *Report on Assessment of the Architectural and Archaeological Heritage Bosnia and Herzegovina*. Integrated Rehabilitation Project Plan/Survey of the architectural and archaeological heritage (IRPP/SAAH), Regional Programme on Cultural and Natural Heritage in South East Europe, Sarajevo: Commission to Preserve National Monuments, 6.

3 See Richard Carlton and Helen Walasek, *Documenting the Destruction* in this volume.

Theories of Destruction

The ideology and policies underpinning both the destruction of buildings and removal of human populations in Bosnia during the wartime campaigns of ethnic cleansing have been extensively detailed.[4] One of the first scholarly (and influential) examinations of monument destruction during the conflict by Chapman in 1994, based on information derived up to late 1993[5] (by which time most damage had been done but very little independently verified) presented a theory of cultural destruction which has been widely cited and continues to have a long afterlife in academia. Chapman argued that the destruction of historic monuments formed part of the deliberate destruction of historical identities forged by association with place and landscape, reinforced by the longevity of visible monuments within that landscape. In a conflict such as the Bosnian War, between identity groups which tended to distinguish themselves mainly (or ostensibly) on the basis of religion, the main physical symbols of such ethno-religious identity, sacral monuments, are likely to be targeted most aggressively. Chapman predicted a wave of indiscriminate destruction in which each group targeted each others sacral symbols of identity: 'Mosques for Serbs and Croats, Orthodox churches for Muslims and Croats, Catholic monasteries for Serbs and Muslims – each monumental symbol fatally attracts the cultural warriors.'[6]

Until the end of the conflict and freedom of movement restored, however, it was difficult to validate whether the pattern of mutual and equivalent destruction by all three main warring parties assumed by Chapman and others had indeed occurred. The authors' fieldtrips (and subsequent visits), in addition to verifying the nature and scale of monumental destruction of all types, have also allowed scrutiny of assertions made during the war by Bosnian government authorities that Christian sacral monuments were protected by Bosnian Army (ARBiH) forces.

Fieldwork Methodology

Before their first fieldtrip in 2000 the authors had consulted the various available reports on the destroyed and damaged cultural and religious heritage in Bosnia-Herzegovina to plan a strategy for sites to visit in both of the two political entities into which the country is now divided: the Federation of Bosnia-Herzegovina (composed of ten cantons, either predominately Bosniak (Muslim) or Bosnian Croat and Republika Srpska (overwhelmingly populated by Bosnian Serbs).[7] However, because of the disparate, incomplete, inaccurate

4 For instance Cigar, N. 1995. *Genocide in Bosnia: The Policy of 'Ethnic Cleansing,'* College Station: Texas A&M Press and Sells, M. 1997. Religion, History, and Genocide in Bosnia-Herzegovina, in *Religion and Justice in the War over Bosnia*, edited by G.S. Davis, New York: Routledge, 23–43.

5 Chapman, J. 1994. Destruction of a common heritage: The archaeology of war in Croatia, Bosnia and Hercegovina, *Antiquity* 68, 120–26. Mostly sourced to Ihsanoglu, E. (ed.) 1993. Bosnia-Hercegovina: History, culture, heritage, *Research Center for Islamic History, Art and Culture Newsletter*, 31 (April 1993), Istanbul: IRCICA.

6 Chapman, J. 1994, 122.

7 See Carlton and Walasek, *Documenting the Destruction* in this volume. Primarily the Parliamentary Assembly of the Council of Europe's ten *Information Reports on War Damage to the Cultural Heritage of Croatia and Bosnia-Herzegovina*,1993–1997; *Specific Action Plan for Bosnia-Herzegovina*, Council of Europe (COE), 1999; Pusilo, N. et al. 1995. *A Report on the Devastation of Cultural, Historical and Natural Heritage of the Republic of Bosnia and Herzegovina, From April 5, 1992 until September 5, 1995*; Mileusnić, S. 1997. *Spiritual Genocide 1991–1995* on Serbian

and uncorroborated nature of many of the sources (compiled at various stages of the conflict), visits to organizations and bodies in Bosnia which had assembled data to see if further information had been collected or material updated since the publication of their reports (particularly with any photographic documentation of destruction or damage to monuments) were essential. This was to be followed by field visits to as many unverified and un-photographed sites as possible, covering what appeared to be the most neglected areas in the record over a wide geographical spread.

A visit to the state heritage preservation body, the *Institute for the Protection of the Cultural, Historical and Natural Heritage of Bosnia-Herzegovina* (*Zavod za zaštitu kulturno-historijskog i prirodnog naslijeđa BiH*) in Sarajevo yielded largely 'negative' information. However, the visit pointed up some of the many difficulties faced by heritage bodies operating in the conflicted political environment of post-war Bosnia. With the devolving of most political powers to the entities and continued pressure from ethno-nationalist politicians who worked actively against the existence of any organizations representing Bosnia-Herzegovina as a nation, pre-war state institutions like the institute were experiencing severe problems. Institutionally and financially weakened by the loss of its premises and equipment (though fortunately not its documentation) during the siege of Sarajevo,[8] the institute suffered as well from the refusal of the Republika Srpska (RS) authorities to recognize the institute's country-wide remit after the war, thus starving it of funding. Because of this refusal the institute was barred from working with international bodies like the Council of Europe (COE) with which it had formerly co-operated as a representative state institution, and, among other sanctions, was forbidden to participate in the 1997–1998 cultural heritage survey carried out by the COE's Technical Cooperation and Consultancy Programme, despite its active involvement in the project's preliminary stages.[9]

Ferhad Mulabegović, the institute's director, explained that the organization did not have the capacity to update the information published in its 1995 'White Book': *A Report on the Devastation of Cultural, Historical and Natural Heritage of the Republic of Bosnia and Herzegovina.*[10] This report was the most useful listing then available of Bosnia-Herzegovina's important cultural property, but had a number of shortcomings (described elsewhere in this volume) including inaccuracies as to the extent of damage as well as the duplicate listing of monuments. However, Mr Mulabegović updated the authors on monuments he had visited personally, as well as on the shortcomings of the Commission to Preserve National Monuments created under Annex 8 of the Dayton Peace Agreement which was then about to complete its first five-year term under the aegis of UNESCO.[11]

The *Centar za islamsku arhitekturu* (*Center for Islamic Architecture*) in Sarajevo, linked with the Islamic Community in Bosnia-Herzegovina and then directed by Kemal Zukić, had actively visited and recorded sites of destroyed Islamic structures across the Federation. However, at that time it could offer no information on damage to Islamic monuments and sacral structures in Republika Srpska except for Banja Luka. A limited amount of documentation (not all with photographs) covering the Federation had been collected in a

---

Orthodox religious sites; Živković, I. 1997. *Raspeta Crkva u Bosni i Hercegovini: Uništavanje Katoličkih Sakralnih Objekata u Bosni i Hercegovini (1991–1996)* on Catholic sacral structures.

8 See COE *Third Information Report*, Doc. 6904, 20 September 1993, 20, regarding the rescue of the Institute's documentation.

9 Personal conversation, Helen Walasek with Muhamed Hamidović, then director of the Institute, 4 November 1997, Sarajevo.

10 Pusilo, N. et al. 1995.

11 See Valery Perry in this volume for more on the Commission to Preserve National Monuments.

publication by Mr Zukić which he provided.[12] It was clear that similar visits in Republika Srpska, though much-needed, had not taken place due to the (probably well-founded) fear of travelling about that entity for Muslims, particularly on such a contentious issue. Staff at the centre expressed their alarm at the authors' intention to visit Islamic sites in RS and strongly urged them not to go. However, they did supply what was to prove one of the most useful set of addresses they were to acquire: contact details for local *medžlis* (Islamic Community officials) offices around the country and especially in Republika Srpska, all of whom proved exceptionally helpful when contacted and who frequently proved to be the only Muslim returnees in many districts at the time.

The *Vakufska Direkcija* (*Vakuf Directorate*), which manages the vakuf endowments and property of the Islamic Community, and its then director Mehmed Hodžić was supportive of the fieldwork objectives and provided a number of photographs of mosques before their destruction. Mr Hodžić also acquired for the authors one of the few bound copies of the recently produced documentation of the *Banja Luka Muftijstvo* on destroyed mosques in the city and its suburbs.[13] A visit was also made to *The Institute for the Research of Crimes against Humanity and International Law* (*Institut za istraživanje zločina protiv čovječnosti i međunarodnog prava*) at University of Sarajevo, where they received from its director Prof. Smail Čekić, Tucaković's *Aladža Džamija – Ubijeni Monument*,[14] published by the institute, which contained a valuable survey of destroyed Islamic structures in Foča and its surroundings. On later occasions the authors received information from Dr Sabira Husedžinović, advisor on heritage to the *Federation Minister of Education, Science, Culture and Sports* who provided her 1998 report on war-damaged cultural heritage in the Federation[15] and Fra Marko Karamatić of the *Franciscan Theological Seminary* (*Franjevačka teologija*), Sarajevo, who updated them on documentation of the destruction to Catholic sacral sites.

Outside Sarajevo, at *Muzej Pounja*, the regional museum in Bihać, curator Enisa Jusić allowed the authors to photograph pictures and record documentation on panels from an exhibition held in 1997 on destroyed and damaged mosques in Una-Sana Canton. The authors also visited the *Mostar Muftijstvo* for lists of mosques outside Mostar that had been attacked, as well as the *Institute for Preservation of Cultural-Historical and Natural Heritage of Mostar*.

In Banja Luka, the *Republic Institute for Protection of the Cultural, Historical and Natural Heritage* (*Republički zavod za zaštitu kulturno-istorijskog i prirodnog nasljeđa*), the former regional institute whose jurisdiction was extended after the war to the whole of the newly-created entity of Republika Srpska, was unable (or unprepared) to provide documentation on the condition of Serbian Orthodox monuments across Bosnia-Herzegovina when asked, although the authors had been informed that they did hold such material.[16] Nor

12 Zukić, K. 1999. *Slike Zločina: Rušenje Islamskih Vjerskih Objekata u Bosni i Hercegovini (The Evidence of Crime: The Destruction of Islamic Buildings in Bosnia & Hercegovina)*. Sarajevo: Centar za islamsku arhitekturu.

13 Čamdžić, E. Ef. 2000. *Sa Džamijama i Povratak*, Banja Luka: Banja Luka Muftistvo.

14 Tucaković, Š. 1998. *Aladža Džamija – Ubijeni Monument*. Sarajevo: Institut za istraživanje zločina protiv čovječnosti i međunarodnog prava.

15 Husedžinović, S. 1998. *Cultural Heritage of Special Value Destroyed in the War 1992–1995*, Sarajevo: Federation of Bosnia and Herzegovina, Ministry of Education, Science, Culture and Sports.

16 Personal communication to Helen Walasek from Tina Wik, June 2000, then Project Manager based in Sarajevo for Cultural Heritage without Borders (CHwB).

would staff members provide any information on destroyed cultural property in Republika Srpska (which would have been substantially Muslim/Ottoman or Croat/Catholic).

They did comment, however, that the impact of the conflict upon archaeological sites in the entity had been relatively slight (although some on the Romanija plateau had been mined) and declared that sites of buildings which had been listed monuments before the war (such as destroyed mosques like the Ferhadija) retained that protected status and that nothing could be built on them (although this did in fact occur in a number of locations in RS, see below). A visit was also paid to the director of the *Archive of Republika Srpska* (*Arhiv Republike Srpske*) which the authors had been reliably informed held many pictures of Banja Luka's mosques before their destruction.[17] However, they were told this was not the case and were directed back to the Republika Srpska heritage protection institute.

The additional information provided on these visits to these organizations and official bodies was added to and compared with published sources in order to develop a picture of the state of knowledge across the entire country. It became clear that important locations such as Sarajevo, Mostar, Jajce, Počitelj and Stolac had already been well-documented. Overall, damage to Catholic churches and sacral structures had been well-recorded (and photographed),[18] as had damage to the Orthodox heritage (though not so well accompanied by photographic evidence).[19]

However, while Islamic sites in the Federation were also largely well-documented (if not all photographed), the standard of documentation for Islamic sites in Republika Srpska where the destruction was extensive was extremely patchy or non-existent. Thus, in the event, it was to prove that the majority of the sites most needing visual verification were Islamic/Ottoman sacral and secular monuments across large areas of Republika Srpska, still at that time problematic or even dangerous for Muslims to visit, and it was to these areas that the authors turned their focus. It was also clear that the destruction and damage to Bosnia-Herzegovina's varied vernacular architecture and rural ambient ensembles had barely been recorded in either entity and special attention was directed to these types of structures. Visits were also made *en route* to sites in the Federation, principally to the important urban ensembles of Mostar, Stolac, Jajce, Bihać and Prusac.

A wide geographical area was covered visiting sites in Republika Srpska, ranging from the far north-west along the Sava (Bosanski Novi, Bosanska Kostajnica, Modriča), other sites in the north-west (Mrkonjić Grad, Banja Luka, Kotor Varoš, Prijedor, Kozarac), the north (Derventa, Kotorsko, Doboj), the north-east going south (Bijeljina, Janja, Zvornik, Divič, Srebrenica, Žepa, Rogatica, Višegrad, Rudo, Foča, Knežina), to the far south within reach of the Adriatic (Nevesinje, Plana, Bileća, Kotezi, Trebinje and its region). A full list of locations visited during the fieldtrips can be found at the end of the chapter.

Site Identification

The nature and scale of the systematic attack on the Islamic heritage throughout much of Bosnia-Herzegovina led to concerns in advance of fieldwork about the problems of

17 Personal communication to Helen Walasek from András Riedlmayer (7 September 2000) who indicated that the then director of the archive had told him in 1999 that the RS archive held many photographs of all the demolished mosques and Islamic monuments in Banja Luka and urged Riedlmayer to visit and make copies of them.

18 Primarily in Živković, I. 1997. *Raspeta Crkva*.

19 Primarily in Mileusnić, S. 1997. *Spiritual Genocide*.

identifying sites of destroyed monuments, particularly in Republika Srpska where reports suggested that elements of the population remained hostile to any kind of interest in the Islamic heritage. A variety of maps, including large-scale British War Office and Austro-Hungarian military maps, as well as modern road maps provided a reasonable basis for fieldwork, but did not display towns in detail nor include all sacral monuments in rural locations. Pre-war large-scale town maps were extremely useful for the few locations covered, as well as photographs in scholarly monographs and other publications. Particularly valuable were Mujezinović's works on Islamic epigraphy in Bosnia-Herzegovina,[20] Hasandedić's volumes on Islamic heritage in Herzegovina,[21] Ayverdi on Ottoman architecture in Yugoslavia,[22] Bećirbegović on mosques with wooden minarets in Bosnia-Herzegovina,[23] articles from the journal *Naše starine*,[24] and tourist publications with photographs of monuments in their pre-war condition which were helpful in locating sites by their contexts (as at Knežina from a photograph in the indispensable *Treasures of Yugoslavia*[25] and at Banja Luka, after the unexpected acquisition of a pre-war guidebook[26] which was then still freely available on sale at the city tourist office). Aerial photographic sources were explored as a back-up to map evidence, but the most promising source, the *Geodetski Institut* in Sarajevo, held photographs for only parts of the country and could not supply these quickly.[27]

A problem that frequently arose in identification was the use of multiple names for mosques, so that published sources occasionally listed different names for the same building, or informants frequently did not know the 'official' name for a mosque used by heritage authorities, but knew only the local name. It was sometimes the case that mosques had three or more alternative names and this was often the reason for duplicate entries in lists of destroyed and damaged structures (such as the Handanija Mosque in Prusac which was named variously in different sources as the Handanagina, Handan-Bey, Hajdar Ćehaja or Čaršija Mosque). Another riddle was the use of the designation *Stara džamija* (Old Mosque) to buildings which were patently only a decade or so old. The writers were to discover this was to differentiate the structure from a second *Nova džamija* (New Mosque) built nearby: the 'Old Mosque,' though itself a recent structure, was always on the site of the community's original mosque, usually of ancient foundation. At Oborci in central Bosnia an old man explained: pointing at the 'Old Mosque,' a newly-built structure in the upper village, he told the authors that there had been four 'Old Mosques' on that site in his time: the first in traditional Bosnian style with a pitched wooden roof and minaret, the second dating from 1956 had a tiled roof and wooden minaret, the third constructed around 1985–1986 had been damaged in the 1992–1995 war and its concrete minaret lay

20 Mujezinović, M. 1974–1982. *Islamska Epigrafika u Bosne i Hercegovine, Vols 1–3*, Sarajevo: Biblioteka Kulturno Nasjleđe, Veselin Masleša.

21 Such as Hasandedić, H. 1990. *Muslimanska Baština u Istočnoj Hercegovini*. Sarajevo: El-Kalem.

22 Ayverdi, E.H. 1981. *Avrupa'da Osmanli Mimâri Eserleri, Yugoslavya*. Vol. II Part 3, Istanbul: Istanbul Fetih Cemiyeti.

23 Bećirbegović, M. 1990. *Džamije sa Drvenom Munarom u Bosni i Hercegovini*. Sarajevo: Biblioteka Kulturno Nasljeđe, Veselin Masleša.

24 The annual of the *Institute for the Protection of the Cultural, Historical and Natural Heritage of Bosnia-Herzegovina*.

25 Tomašević, N. et al. (eds) 1980. *Treasures of Yugoslavia*. Belgrade: Yugoslaviapublic.

26 Horozović, I. and Vukmanović, M. 1984. *Banjaluka*. Zagreb: Turistcomerc.

27 Furthermore, it was suggested by staff at the Geodetski Institut that aerial photographs held in Pale, now in RS, had been altered to remove mosques and other Islamic sacral buildings. There was no opportunity to verify this.

Fig. 3.1 The fourth 'Old Mosque' at Oborci in October 2000, built since the end of the 1992–1995 war, with the concrete minaret of the third 'Old Mosque' destroyed during the conflict still lying on the ground beside it. © Richard Carlton

on the ground at the side of the fourth, of domed construction, built since the end of the conflict. The 'New Mosque' was seen under construction in the lower village.

With regard to the accuracy of reports of damage and destruction not accompanied by photographic documentation, the writers learned for themselves how information supplied to religious communities by representatives at a local level could on occasion be wildly inaccurate for no apparent reason, even when given by informants on the ground. Told by a local informant that the Donja Mosque at Kotorsko was totally destroyed, one of the authors later saw the mosque for himself – badly damaged, but substantially intact, save for its roof and its minaret.

In the event, mosque sites were located by a combination of map evidence, identification of surroundings from pre-war photographs, deduction, questioning local people and other informants and using archaeological investigation techniques. In Republika Srpska local Muslim contacts, where not too fearful to point out mosque sites, were particularly helpful, especially representatives of the Islamic Community's *medžlis* offices. Some of these offices had produced their own documentation of destroyed sacral buildings in their area, occasionally with accompanying before and after photographs. Notable was Osman ef. Mulahuseinović, the imam from Modriča, who guided one of the authors to the sites of all the destroyed and damaged mosques in his district, standing to be photographed in front of each holding a picture of the structure before its destruction. However, Muslim informants could be difficult to locate and generally it was found that the most reliable informants were elderly non-Muslim residents who were frequently willing to discreetly point out site locations and could often give an insight into methods of destruction and other matters of interest, such as local feeling on refugee returns.

In north-west Bosnia a network of personal contacts among returned refugees and displaced people was crucial for the authors in locating many sites in Republika Srpska, especially in Banja Luka, Prijedor and nearby Kotor Varoš. Other helpful informants were Bosnian employees of the Office of the High Representative (OHR) and Organization for Security and Co-operation in Europe (OSCE) who, if they could not provide information themselves, often knew who could. The OSCE's Srebrenica office contacted Olivera Stanić, head of the NGO Želja in Skelani, who later accompanied the authors in identifying mosque sites along the Drina River where it borders Serbia, including locating (after enlisting the help of an elderly Serb resident working in his garden) the crumbling remains of the

Fig. 3.2 Imam Osman ef. Mulahuseinović from Modriča at the site of the completely demolished mosque at Modrička Lug in June 2001 holding a photograph of the structure before its destruction. What remains of the mosque can be seen behind him. © Richard Carlton

sixteenth-century *Stara džamija* at Peći, described in Bećirbegović,[28] hidden deep in dense young woodland grown up since the mosque's destruction in May 1992. Other humanitarian NGOs and human rights organizations and, on one occasion, local Social Democratic Party officials (in Trebinje) also provided information, assistance and advice.

Often however, in the absence of informants, archaeological techniques of field detection had to be employed to identify sites from their surviving remains and contexts. In many cases an archaeological awareness was the only way to detect, or suggest possible locations for monuments that until recently formed an integral part of long-established communities. The majority of the most completely destroyed sites where mosques had been razed and the rubble cleared away were in urban centres where they provided open spaces for car parking and miscellaneous activities such as markets, wood cutting and storage. Most common of all on these centrally-located completely razed sites (and frequently a good identifier) was the presence of large communal garbage containers and/or accumulations of strewn rubbish. Other former mosque sites were found already built upon, such as at Divič and the site of the Dašnica Mosque at Bijeljina.

Visible remains of mosques, or of the monumental ensembles they formed part of, most commonly included cemeteries (*harem/haremi* when part of the mosque enclosure, *mezaristan/mezaristani* when separate burial grounds), which generally, although by no means universally, survived better than the mosques themselves and often included mausolea (*turbe/turbeta*), as at Knežina where the domed sixteenth-century Sultan Selim II (Selimija) Mosque had been razed to the ground. Mausolea or tombstones (*nišan/nišani*) could be not always be used to locate mosques by association, however, since burials are

28 Bećirbegović, M. 1990, 115.

Fig. 3.3 The mihrab of the sixteenth-century Stara Džamija (Old Mosque) at Peći, not far from Srebrenica, in October 2000, whose crumbling remains were hidden in dense young woodland grown up since the mosque's destruction in May 1992. © Richard Carlton

Fig. 3.4 The fate of the majority of completely demolished mosque sites in town centres: the rubble removed to provide open spaces for parking lots, kiosks, storage, and as rubbish dumps. The site of the Mehmed Čelebija Mosque in Kozluk, Republika Srpska, in October 2000. © Richard Carlton

Fig. 3.5 **(a)** The Selimija (Sultan Selim II) Mosque in Knežina before conservation in 1969 when the shingled roof over the dome was removed and replaced with lead. © Institute for the Protection of the Cultural, Historical and Natural Heritage of Bosnia-Herzegovina

Fig. 3.5 **(b)** Looking for the site of the Selimija Mosque in Knežina, Republika Srpska, in 2001. The sixteenth-century structure had been one of the more important domed mosques in Bosnia-Herzegovina. Although the mosque had been completely demolished and its remains removed, some gravestones and a turbe still stand nearby. © Helen Walasek

often placed individually or in family groups close to settlements or along roadsides – indeed, many such groups survive in Republika Srpska in similar contexts to small clusters of the Bosnian medieval tombstones known as *stećci* (singular *stećak*). Nonetheless, in some cases gravestones provided the first clue to the locations of destroyed mosques – for example, one mosque site at Rogatica, used as a car park, was associated with a few surviving *nišani*, as were some of the similarly razed sites in Foča.

Other remains of mosques seen included the lower courses of walling, foundations or foundation platforms. At Trebinje the only visible trace of the eighteenth-century Careva Mosque (also known as the Sultan Ahmed, Atik, or Stara Mosque), quietly pointed out to the authors by an elderly Serb resident, was a line of stone foundations in an open space being used as a car park. At Bosanska Kostajnica, where a local woman confirmed the site, the foundations of the Azizija Mosque (1862) were visible only in conditions of low light as slight undulations in grassland. The Azizija was one of a number of mosques of the same name built in northern Bosnia with state funding and named after the Sultan Abdul Aziz at a period when an influx of Muslim refugees from Serbia settled in the region.

However, despite the presence of the many empty spaces where a mosque had formerly stood in 1992, given the overwhelming impression presented by reports (even as late as 2000) that all mosques and Islamic sacral buildings in Republika Srpska had been razed to the ground, the authors found this to be far from the case, since substantial numbers of structures survived, albeit badly damaged and beyond use. On the other hand, although the level of survival of Islamic structures may have been higher than expected from sources published hitherto, the overall impression was of damage ranging in intensity from severe to complete demolition and razing of the structure.

As described above, it was frequently impossible to identify sites of mosques and other significant monuments being searched for without local assistance, so completely had traces of structures disappeared. Some were in isolated rural areas where there was no one to question and lack of time prevented further exploration. Photographs of sites frequently needed to be taken discreetly, or under the pretext of photographing something completely different, such as the conveniently positioned low-level farm buildings directly behind the substantial upstanding remains of the Careva Mosque in Bileća. In urban settlements it was usually difficult to explore and photograph beyond a certain period of time without attracting unwelcome attention and raised suspicions, such as at Zvornik where one of the authors' activities attracted police interest. At Rudo, half a mile from the Serbian border and with night drawing in, the authors had identified the mesdžid in Staro Rudo and the Kara Mustafa-paša Mosque on the outskirts of town, but could not locate the site or remains of the mosque in nearby Sokolovići, birthplace of the sixteenth-century Ottoman Grand Vizier Mehmed-paša Sokolović, patron of many building projects in Bosnia (most famously the bridge on the Drina at Višegrad). Closely watched by groups of young men gathered in the evening gloom on Rudo's main street, the search was abandoned.

A final fieldwork hazard was the potential presence of landmines, about which the authors sought advice from the *Mine Action Centre* in Sarajevo, which provided maps showing the locations of the greatest concentration of minefields – normally along the former confrontation lines. While the authors were held up on the road to Prusac in central Bosnia by mine-clearing activities, in practice most sites visited for the purposes of buildings survey were in areas of Republika Srpska over which control had been swiftly established at the beginning of the conflict by Bosnian Serb forces with the help of Yugoslav People's Army (JNA) and Serb paramilitary forces and therefore contained no minefields. However, before venturing off track in country districts local people were always asked about the danger of landmines.

Fig. 3.6 (a) Using archaeological field detection techniques to identify sites: the foundations of the Azizija Mosque (1862) at Bosanska Kostajnica in October 2000 were visible only when low light revealed slight undulations in grassland. © Richard Carlton

Fig. 3.6 (b) The Azizija Mosque at Bosanska Kostajnica before its destruction. The mosque was rebuilt and officially opened in July 2008. © Gazi Husrev-beg Library, Sarajevo

The Nature and Scale of Architectural Losses

The fieldtrips confirmed the huge scale of destruction caused to a wide range of monuments, but in particular Muslim or Ottoman sacral and secular monuments. The magnitude of destruction can hardly be overstated. The authors found great tracts of territory utterly devastated and depopulated, particularly in eastern Bosnia and the north, where much of the Posavina region, previously a majority Catholic/Croat and Muslim/Bosniak region, was abandoned save for some larger villages and towns. Here the ruins of thousands of smallholdings within sight of the Doboj to Bosanski Brod highway were rapidly becoming swallowed by afforestation, and mile after mile of ruinous buildings were common sights in eastern Bosnia and parts of the Bosnian Krajina in the west. The transition from the bustling towns and villages of the Federation, where many displaced survivors of ethnic cleansing were then living, to the all but empty countryside and quiet settlements of Republika Srpska was often striking. More than a decade later, in 2011, while much of the Krajina has been repopulated and rebuilt, and some return has occurred to larger settlements such as Kotorsko in the Posavina, many smaller former Bosnian Croat and Muslim settlements in that northern region remain abandoned, or visited only as weekend or summer homes by former residents now settled elsewhere in Bosnia, in Croatia and further afield.

Among the range of Islamic sacral monuments found to have been targeted, most prominent were the following classes of buildings: *dzamija* (mosque), *mesdžid* (mosque without a minber and normally without a minaret), *tekija* (tekke or dervish meeting house), *mekteb* and *medresa* (religious schools), *gusulhana* (mortuary house for ritual bathing, usually part of a mosque complex), *abdesthana* (place for ritual washing next to a mosque, often associated with cisterns), *šadrvan* (fountain with a basin for ritual washing in the precincts of a mosque), *turbe* (mausoleum) and *nišan* (gravestone). In addition, structures associated with mosques, particularly imam's houses, or with other distinctively Muslim functions, were also targeted. Some of the types of structure listed above would not, of course, be considered historic monuments of the highest order in their own right, but were important as constituent parts of historic ensembles.

Among non-sacral monuments, targets for intentional destruction were primarily those buildings with Muslim or Ottoman associations: *sahat kula* (clock tower); *česma* (public water fountain, often associated with cisterns, or *čatrnje*); *hamam* (public bath), *musafirhana* (inn) and *konak* (lodge); and grand Ottoman houses, of timber-framed or stone construction and varied design, depending on their age and locality. Where traditional Ottoman-period (or Ottoman-style) houses survived, those left empty after their owners had been expelled, departed under threat or killed, were seen to be rapidly decaying. In this class was a riverside group in Ilidža *mahala* at Gornji Šeher, a suburb of Banja Luka, among them the important eighteenth-century Šeranić House. Ilidža *mahala*, an area of thermal springs (*ilidža* = spa or hot springs), is known for its still surviving collection of Ottoman period covered baths of a type known as *hauz*, either as freestanding buildings or incorporated into residential properties; the group and other historic structures associated with the springs are now designated as a National Monument under Annex 8 of the Dayton Peace Agreement.[29] This collection of houses appeared to have survived destruction because they were of use to the city authorities and its remaining (mainly Serb) residents, when

29 See the website of the Commission to Preserve National Monuments: *Spa in Ilidža mahala in Gornji Šeher, the architectural ensemble* at http://www.kons.gov.ba/main.php?id_struct=50&lang=4&action=view&id=2526

Fig. 3.7 Abandoned and decaying, but still surviving. The group of Ottoman houses in Ilidža *mahala* at Gornji Šeher, Banja Luka, in October 2000. Many have internal or freestanding bathhouses that made use of the area's hot springs. At far right is the eighteenth-century Šeranić House. © Richard Carlton

Banja Luka was subjected to an influx of Serb refugees, particularly from the Croatian Krajina after the Croatian Army's Operation Storm in August 1995.[30] The Šeranić House had such temporary occupants and its ground floor was turned into a cowshed; the house was abandoned again after the end of the war in 1995.[31] Indeed, it appeared to be a more general pattern that where non-sacral classes of Ottoman/Muslim monuments served a clear function their survival was more likely to be ensured. Two Ottoman buildings which survived in Republika Srpska because of their functions as a nightclub and a restaurant respectively were the Mehmed-paša Kukavica Han in Foča and the Ottoman-style house, Dom Lovaca,[32] in Prijedor.

In eastern Herzegovina several examples were observed of a particular, though largely unrecorded, class of strong-house, usually with walls over one metre thick, vaulted ground floor rooms, defensible doorways and small windows only at first floor level. These survive in locations such as Ošaniči and Kotezi and do not appear to have been deliberately targeted, even where apparently associated with religious buildings or former Ottoman landowners, as at Kotezi,[33] although there is some evidence that they were targeted when used as places

30 Information from a local resident.

31 Emergency structural repairs have been carried out on the Šeranić House by the Commission to Preserve National Monuments in 2008 funded by the US Ambassadors Fund for Cultural Preservation. See Ćorović, A. 2008. Šeranića kuća u Gornjem Šehera kod Banka Luka / The Šeranić House in Gornji Šeher near Banja Luka. *Baština* [*Heritage*], The Annual of the Commission to Preserve National Monuments, Sarajevo, IV/2008, 199–218. See also Mulalić Handan, M. et al. 2005. *Preliminary Technical Assessment of the Architectural and Archaeological Heritage in South East Europe: Historic Building, Seranic's House in Banja Luka*. European Commission/Council of Europe Joint Programme on the Integrated Rehabilitation Project Plan /Survey on the Architectural and Archaeological Heritage (IRPP/SAAH), Regional Programme on Cultural and Natural Heritage in South East Europe 2003–2006, Sarajevo: Commission to Preserve National Monuments.

32 The building is now a café bar called Life Club.

33 Hasandedić records that a religious school, two wells, three towers, several manor houses and other residential buildings formed the sacral and vernacular ensemble at Kotezi during the Ottoman

of refuge during wartime artillery attacks, as at Ošaniči. Other, essentially private and vernacular structures which survived unless damaged collaterally or through neglect include watermills (*mlinovi* or *mlinice*) and numerous kinds of farm buildings, particularly those away from main centres of population, such as čardak (a freestanding agrarian building, sometimes used as a dwelling) and shepherd's hut, which tend not to be charged with ethno-religious identity issues. Bridges (*most/mostovi*) provide particularly good examples of vernacular buildings and structures preserved for pragmatic reasons. The Mehmed-paša Sokolović Bridge (1578) over the river Drina at Višegrad made famous by Ivo Andric's *The Bridge on the Drina*[34] and scene of notorious massacres of the town's Muslim residents during the 1992–1995 war, appeared to be preserved as the most convenient means of crossing the river. The equally famous bridge in Mostar linked only Muslim-held territories either side of the Neretva and was destroyed by Croatian Army/HVO[35] forces in 1993 (as were all of Mostar's bridges). Other notable Ottoman bridges the authors found surviving in present non-Muslim majority areas are the Arslanagić Bridge near Trebinje (c.1573/4), and the sixteenth-century Bridge on the Žepa (both moved to their present positions in the twentieth century). The Inat Bridge in Stolac, believed to be of medieval origin, also survived along with at least two others of Ottoman date.

The clock towers (*sahat kule*) of Bosnia-Herzegovina occupy a particularly vulnerable position in terms of perceived ethno-religious affiliation since, although vernacular and introduced to Ottoman Bosnia as an essentially Western, Christian concept, all were built during the Ottoman period and most form part of distinctively Ottoman urban ensembles. Among those clock towers extant up to the outbreak of war in 1992, the most ancient, at Banja Luka (before 1587) was completely destroyed. Yet the authors found others survived in Republika Srpska at Trebinje, Nevesinje and at Foča, despite the severe damage to the town's surrounding Ottoman core. In the Federation clock towers damaged by shelling were seen at Prusac and Gradačac.

It was apparent that the architecture of the Bosnian Muslim/Ottoman sacral and secular heritage had, indeed, been devastated in all areas controlled for any period during the war by Bosnian Serb or Bosnian Croat forces. Furthermore, monuments within territory held by the Bosnian government were subjected to bombardment, as at Sarajevo, Mostar, Gradačac, Tešanj and Prusac, where the rich complex of sixteenth – seventeenth century sacral and other monuments described by Čepalo,[36] including mosques, a dervish lodge, several religious schools, inns, tower houses, a library, public baths and clock tower, as well as cemeteries and a medieval castle, was badly damaged. Here, imam Sakib ef. Began took the authors to view the major monuments of which most mosques had already been restored, including the complex of the seventeenth-century Hasan Kafija Mosque with its *medresa/ tekija* (dervish lodge), *turbe* and *mekteb*. However, Efendi Began asked for assistance in rebuilding the still unrestored and heavily damaged Handanija Mosque (1617) whose roof was completely destroyed, its façade punctured with huge shell impacts. This important small stone-built mosque has several unique features, including its porch with its arcade of pointed arches and interior wooden dome. Moreover, the Handanija was also starting point of the yearly pilgrimage to *Ajvatovica*, the largest Muslim pilgrimage in Europe;

period. He also reports that none of these has survived intact, although at least two, substantially unmodified strong-houses (presumably his 'manor house' type) were present in 2010. Hasandedić, H. 1990. *Muslimanska Baština u Istočnoj Hercegovinu*. Sarajevo: El-Kalem, 285.

34 Andrić, I. 1945. *Na Drini Ćuprija*. Beograd: Prosveta.

35 Hrvatsko Vijeće Obrane = Croatian Defence Council (the Bosnian Croat defence force).

36 Čepalo, H. 1990. *Prusac – Istorijski Spomenici*. Prusac: Mala Biblioteka.

banned during the Communist era, it had been revived in 1990 and was now one of the most important events in the Bosnian Muslim year.[37]

The heaviest losses within Republika Srpska and nationalist-dominated Bosnian Croat areas of the Federation appeared in older urban settlements, especially those based on defended medieval settlements which later developed as Ottoman administrative and market centres. In many of these Muslims traditionally formed a high proportion of the population, and even where the balance of population changed due to state-sponsored influxes of non-Muslim groups in the Austro-Hungarian, Kingdom of Yugoslavia and (most notably) post-World War II Communist periods, the Ottoman character of the urban nucleus often remained. The survival of this distinctive Ottoman (and therefore, by association 'Muslim' or 'Turkish') character led inevitably in the 1992–1995 war to the gutting of the ancient cores of many towns and villages, leaving incongruous empty spaces inside apparently modern settlements, disorientating for the visitor accustomed to identifying urban centres by a standard range of expected features.

The most completely destroyed in this way were Foča (in Republika Srpska) and Stolac (in a hardline Croat-controlled municipality of the Federation in Herzegovina). Stolac was, and to an extent remains, a highly nucleated complex of four mosques (sixteenth – eighteenth century), a *mekteb*, several important discrete Ottoman residential complexes, including those of the Mahmutćehajić family and the Begovina of the Rizvanbegovićs (one of the most significant such complexes in Bosnia-Herzegovina), important lodges (*konak/konaci*), a market, public baths, mills and bridges. All of the above and many other monuments in the vicinity, including the impressive medieval and later castle at its centre, were stone-built and collectively recognized as a site of international significance. This outstanding and harmonious complex of Ottoman urban architecture was, with the exception of its mills and bridges, destroyed to a sequence of crumbling ruins, with many monuments completely razed to the ground, by HVO forces in summer 1993 and was accompanied by the expulsion of Stolac's Muslim residents.[38]

Other important Ottoman urban complexes settlements attacked by Bosnian Croats/HVO in Herzegovina in this way included major sites such as Počitelj and Mostar,[39] both of which suffered severe and well-documented damage, but were not substantially razed as at Stolac. However, the authors also found smaller rural ambient ensembles with historic value badly damaged at Ošanići, Mostaći and Blagaj. Elsewhere, in all towns that fell for any period under Bosnian Serb control, virtually all Muslim sacral buildings had been destroyed or survived as heavily damaged, burned out shells. But since most such buildings were dispersed within mainly vernacular complexes of modern character, wholesale destruction on the scale of Stolac had generally not occurred, leaving instead a patchwork of open areas where mosques and other sacral buildings formerly stood. The exceptions to this were Foča (see below) and Prijedor, where apart from the levelling of the mosques, the entire *Stari Grad* or Old Town, a Muslim neighbourhood on the banks of the Sana River, had been razed to the ground, leaving a solitary Ottoman-style building, *Dom Lovaca*, described

37 Walasek alerted Tina Wik of Cultural Heritage without Borders (CHwB) about the state of the Handanija Mosque and it was eventually restored by CHwB. See Wik in this volume on the mosque's restoration.

38 See Mulać, E. (ed.) 2001. *Crimes in Stolac Municipality (1992–1996)*, Reprint with addenda, Sarajevo: Association [Society] for the Renewal of Civil Trust in the Stolac Municipality, 36–73. See Amra Hadžimuhamedović on Stolac in this volume. Stolac is currently on the tentative World Heritage Site list.

39 Mostar was, of course, attacked in the early stages of the war by Yugoslav People's Army (JNA) and Bosnian Serb forces and suffered major damage to its cultural heritage.

above. Here, however, in 2000 the authors saw Muslim returnees rebuilding their houses, which was certainly not the case in Foča at that time.

Foča, in north-eastern Herzegovina, in the pre-war period a town retaining a strong Ottoman character, must be mentioned as a slight anomaly in the overall pattern of destruction. Foča is one of the longest established Ottoman towns in Bosnia-Herzegovina and until 1992 contained the greatest ensemble of surviving sacral and vernacular Ottoman buildings of any town east of Sarajevo. Despite heavy losses during World War II, the Muslim population of Foča and its hinterland had maintained itself at a high level, forming a bridge between the Bosnian capital and Muslim Slav populations in neighbouring Montenegro and Serbia. Early in the 1992–1995 war, the intent of the local authorities to 'Serbianize' the town by removing its Ottoman character and Muslim population was clearly acknowledged when its name was changed from Foča to Srbinje, or 'Serb-town.' It was not clear that this move found support within the wider population, however. Indeed, when the authors asked local residents which name they should use, support for the name change appeared not to be completely whole-hearted. Among the comments were: 'I've used Foča all my life and I'm not going to stop now,' 'There are two names for the town – Foča and Srbinje – you can use either, it doesn't matter' and 'I call it Foča – call it what you want.'

Around the urban core of the town, centred upon a market place near the confluence of the Drina and Ćehotina rivers, spread residential districts or *mahalas* of Ottoman houses, many of them stone-built, interspersed with mosques. While open plots marked the sites of nine of the 11 mosques standing there in April 1992, in 2000 the shells of two survived. Most important were the remains of the domed stone-built Mehmed-paša Kukavica Mosque (1751) standing at the heart of the preserved Ottoman core. Despite clear marks of attempts to blow it up and the total destruction of its minaret and adjoining *medresa*, the Kukavica Mosque's central structure and dome remained intact at that time, although a local woman living across the road from the ruin commented: 'I wish they'd tear it down.'

The second mosque in Foča whose shell survived was the Atik Ali-paša (or Musluk) Mosque (1564), set among ruined houses in the extensive *mahala* of Donje Polje, which though burned out and roofless and its minaret dynamited, retained characteristic architectural features such as its *mihrab* decorated with *muqarnas*. All other mosques were completely razed, most notably the Careva Mosque (1501)[40] and the renowned Aladža or Painted Mosque (1550–1551), some of whose remains were found in 2001 and 2004 on the banks of the Drina during the excavation of suspected mass grave sites of massacred Muslim residents.[41]

Although considerable destruction occurred in the town with some residential districts extremely badly damaged, so overwhelming was the Ottoman character of pre-war Foča that in 2000 many fine Ottoman houses remained, many populated by Bosnian Serb incomers. Around the Mehmed-paša Kukavica Mosque, however, the ancient urban core had been devastatingly gutted, leaving derelict empty spaces interspersed with the few old shop-lined streets, along with two surviving monuments associated with the mosque and also endowed by the great Bosnian governor Mehmed-paša Kukavica: the large *han* or caravanserai (before 1758), one of the few surviving in Bosnia-Herzegovina, and clock tower (before 1761). The remains of the Nakšibendi or Bajezid-baba *tekija* were also found surviving in the main Muslim cemetery.

40 Also known as the Imperial Mosque or the Sultan Bajezid Valija II Mosque.

41 See Helen Walasek *Destruction of the Cultural Heritage in Bosnia-Herzegovina* and *Domains of Restoration* in this volume for more on the Aladža Mosque and its destruction and reconstruction.

Fig. 3.8 **(a)** Walking through the ruined and empty *mahala* of Donje Polje in Foča towards the Musluk Mosque in October 2000. © Richard Carlton

Fig. 3.8 **(b)** The ruins of the Musluk Mosque, Foča, in October 2000, a burnt out roofless shell with its minaret destroyed, but still retaining characteristic features of Ottoman architecture. © Richard Carlton

Fig. 3.9 The Mehmed-paša Kukavica *sahat kula* in Foča (before 1761) in October 2000, one of the surviving Ottoman clock towers in Republika Srpska. © Richard Carlton

Fig. 3.10 The ruined *tekija* in the Muslim cemetery at Foča, October 2000. © Richard Carlton

More typical of the pattern of destruction found consistently throughout Republika Srpska was that exemplified in Banja Luka, where all traces of city centre mosques were entirely removed and a slightly less drastic approach displayed to suburban or rural mosques. Thus mosques like the city centre Ferhadija were entirely razed to foundation level, while some in the wider vicinity, such as the late sixteenth century/early seventeenth-century Behram Efendija mosque in the suburb of Novoselija, survived as recognizable structures, albeit heavily damaged and with their minarets torn down.[42] The mosque, founded by the Sufi Behram Efendi, was a traditional Bosnian style mosque with a wooden minaret which had retained its authenticity and original features, including an underground room (*kuiju*) for solitary worship, until its destruction on 26 May 1992.[43]

Banja Luka, the administrative centre of Republika Srpska and the second largest city in Bosnia-Herzegovina, lost its 15 mosques (12 of which were listed national monuments and included the sixteenth-century Ferhadija and Arnaudija Mosques), along with important mausolea, fountains, clock tower and other associated monuments between April and December 1993.[44] Witnesses have related how the programme of destruction involved dynamiting by military specialists of one or more mosques every few weeks and the later removal of the remains. One recently returned Muslim resident living across the Vrbas river from the Turbe of Halil-paša (after 1590) told the authors how he had witnessed its destruction in December 1993, commenting how it had been reported afterwards on Banja Luka television that local Muslims had blown up the monument themselves.

However, regional variations were apparent in the degree of destruction of Ottoman/Muslim sacral architecture carried out. In north-east Bosnia the most fastidious approach was applied: not a trace survived of a mosque or any other building of discernible Ottoman character in the towns of Bijeljina, Brčko, Janja, Višegrad and Zvornik. In Bijeljina the site of the Dašnica Mosque had been partially built on and in 1999 construction began of a bank building encroaching on the site of the Atik Mosque, the town's central mosque (sixteenth century, extended eighteenth–nineteenth centuries). At Divič on the River Drina south of Zvornik the authors found the newly-built Serbian Orthodox church of Sveti Stefan occupying the site of the destroyed Bosnian style mosque (of sixteenth century origin, rebuilt c.1936) in a blatant assertion of territorial control and dominance. Divič, which had an almost entirely Muslim pre-war population, had been renamed Sveti Stefan and a Serb population moved in. In 2009, with about 400 Muslim residents returned to the village, the

42 This pattern was apparently also found by Colin Kaiser in his expert report on the destruction of mosques and Catholic sacral structures in sixteen municipalities carried out for the Office of the Prosecutor (OTP) of the International Criminal Tribunal for the former Yugoslavia (ICTY), though he later discounts it as a phenomenon: Kaiser, C. 2002. *Report on Destruction of Cultural Property* 09-Jul-02, Exhibit P906, *Krajišnik* Case No. IT-00-39. Kaiser was the UNESCO Representative in Bosnia-Herzegovina and the principal consultant expert reporting on damage to the cultural heritage in Croatia and Bosnia-Herzegovina for the Parliamentary Assembly of the Council of Europe's ten *Information Reports*).

43 See the designation of the Behram Efendija Mosque as a National Monument (in Bosnian): *Behram-begova ili Behram-efendijina džamija, graditeljska cjelina* http://www.kons.gov.ba/main.php?id_struct=6&lang=1&action=view&id=1835.

44 The number is 16 mosques when the mosque at Vrbanja is included. See Helen Walasek *Destruction of the Cultural Heritage* in this volume for more on the destruction of Banja Luka's Islamic/Ottoman heritage. See also *The Islamic Community in Bosnia and Herzegovina v. The Republika Srpska (Banja Luka Mosques), Decision on the Admissibility and Merits, (Delivered on 11 June 1999)*, Human Rights Chamber for Bosnia and Herzegovina (HRC), Case No. CH/96/29.

Fig. 3.11 **(a)** The Behram Efendija Mosque in Novoselija before its destruction in 1992. A modest, but very typically Bosnian mosque type with a wooden minaret whose form had remain unchanged since its construction in the late sixteenth/early-seventeenth century. © Institute for the Protection of the Cultural, Historical and Natural Heritage of Bosnia-Herzegovina

Fig. 3.11 **(b)** The ruins of the Behram Efendija Mosque in the Banja Luka suburb of Novoselija, October 2000. © Richard Carlton

Fig. 3.12 The recently-built Serbian Orthodox church of Sveti Stefan at Divič on the River Drina, south of Zvornik, in 2001. It was constructed on the site of the mosque (of sixteenth-century origin, rebuilt c.1936) destroyed in 1992. © Richard Carlton

Islamic Community controversially paid nearly €200,000 to have the church removed and rebuilt one kilometre away in the interests of good neighbourly relations.[45]

In other areas under Bosnian Serb control it was apparently considered necessary only to deal with key, particularly urban core, monuments so meticulously, the rest merely being destroyed beyond repair or left as piles of rubble. Thus, in Trebinje not a trace of either of the town centre mosques survived, but building platforms and traces of upstanding walls within piles of rubble were visible at the outlying villages of Donje Ćićevo and Pridvorci (1804). Similarly, at Modriča only the town centre Čaršijska mosque was totally lacking upstanding remains, while some of the suburban mosques, such as the Jakiš and Gornji Tarevci mosques, survived to roof level.

Apart from Modriča in northern Bosnia, upstanding mosque shells in Republika Srpska were noted on the outskirts of Prijedor and in the smaller urban settlements of Gornji Šeher, Vrbanja (both suburban Banja Luka) and Kotorsko (near Doboj), while rural/village mosques survived in similar condition at Kotor (Kotor Varoš). It was in this region that the systematic nature of efforts to destroy all Islamic structures in Bosnian Serb-held territory was brought home by seeing the ruins of demolished mosques in

45 See Riedlmayer, A. 2002a. *Destruction of Cultural Heritage in Bosnia-Herzegovina, 1992–1996: A Post-war Survey of Selected Municipalities*, (*Milošević* Case, IT-02-54, Exhibit P486, Date: 08/07/2003, and *Krajišnik* Case No. IT-00-39, Exhibit P732, Date: 23/05/2005), International Criminal Tribunal for the former Yugoslavia (ICTY), 14. See also Kovac, E. 2009 Controversial church relocated, *Southeast European Times* [Online], 8 June.

Fig. 3.13 The shell and collapsed minaret of the Čaršijska mosque at Kotor Varoš, near Prijedor, in October 2000. © Richard Carlton

isolated villages high in the hills near Kotor Varoš. Upstanding remains of urban core mosques were observed only at Doboj, where part of the sixteenth-century Sultan Selim (Selimija) Mosque beneath the medieval castle survived along with destruction rubble, and a second unfinished mosque survived in a residential area where dynamiting would have caused severe collateral damage to neighbouring properties. In south-eastern Bosnia and eastern Herzegovina mosque shells surviving in urban or urban periphery contexts included the Kara Mustafa-paša Mosque at Rudo, the Mehmed-paša Kukavica and Atik Ali-paša (Musluk) Mosques at Foča and the Careva (Gradska/Obradovića) Mosque (1895) at Bileća. The Čaršija (Sinan-kadi effendi/ Čučkova) Mosque (1880) in Nevesinje, disused as a mosque since 1930 and without its minaret, also survived.

Few traces of surviving rural mosques were observed in the districts around Foča or elsewhere in south-east Bosnia and eastern Herzegovina, although a search was carried out for, among others, remains of the *mesdžid* known to have existed at Gornje Garnčarevo and the seventeenth-century Avdić Mosque at Plana. Before the 1992–1995 war the Avdić Mosque had been one of a small surviving group of unique mosques in south-eastern Herzegovina with square church-like 'campanile minarets,' believed by most authorities to have been influenced by the church structures of the nearby Dalmatian coast and possibly constructed by Roman Catholic master builders from the area who were regularly called upon by Ottoman officials to carry out work on Islamic religious structures in Herzegovina.[46]

46 Kiel, M. 2011. The campanile minarets of southern Herzegovina, in *Centres and Peripheries in Ottoman Architecture: Rediscovering a Balkan Heritage*, edited by M. Hartmuth. Sarajevo: Cultural

Fig. 3.14 Kosa Bajalo, sole resident of the village of Kotezi in October 2000, shows the authors the ancient Mujo Kotezlija Mosque. © Richard Carlton

The exceptional rural survival in this region of the upstanding remains of the Old Mosque at Peći, on the west bank of the Drina near Srebrenica, has been described above. A further survival was the Mujo Kotezlija Mosque (probably seventeenth century) in Kotezi at the western end of Popovo Polje, by tradition the oldest mosque in southern Herzegovina and reported destroyed early in the 1992–1995 war.[47] The sole remaining resident of the village, Kosa Bajalo, told the writers that before the recent war Kotezi had been a mixed settlement of Serbs, Croats and Muslims, but that the mosque had been disused since damage in World War II, and had not been attacked in the recent conflict – probably, the authors thought, because its ivy-clad exterior obscured its defining characteristic, a unique low, cone-shaped stone minaret, now one of the last two Ottoman period minarets surviving in Republika Srpska.[48]

Muslim cemeteries and their associated grave monuments, including standing stones (*nišani*) and various forms of mausolea (*turbe*) generally appeared to have survived much better than any other kind of Muslim sacral sites or monuments. Notable monumental survivals seen, albeit in damaged form, included *turbe* at Knežina and Gornji Šeher. Large Muslim cemeteries were seen in Republika Srpska at Knežina, Foča, Višegrad, Rudo, Staro Rudo, Bosanska Dubica, Bosanska Kostajnica, Bosanski Novi, Prijedor, Nova Kasaba,

Heritage without Borders (CHwB), 60–79.

47 The mosque's destruction was reported in *Bulletin No. 1* of the State Commission for Gathering Facts on War Crimes in the Republic of Bosnia and Herzegovina, Sarajevo, October 1992, No. 194.

48 The other being the free-standing minaret of the seventeenth century Osman-paša Mosque at Kazanci; the mosque and the surrounding village were destroyed by *hajduks* from Montenegro in 1684.

Fig. 3.15 Although frequently vandalisd, many Muslim cemeteries survived throughout Republika Srpska, like this one at Doboj, visited in June 2001. © Richard Carlton

Kozluk, Kula Grad (Zvornik), Doboj, Nevesinje, Plana, Trebinje, Gornje Garnčarevo, Gornja Lastva, Žepa and Modriča, providing evidence that many survive.

However, the great majority of cemetery sites visited in 2000 and 2001 displayed evidence of attack in the form of damaged gravestones and internal buildings. In some cases damage was extensive. Other important grave-markers were observed at Kotorsko, Kotezi and Kruševci (near Sokolac) where the mosque had been destroyed and whose destruction was cited in ICTY war crimes indictments for Momčilo Krajišnik, Biljana Plavšić and Radovan Karadžić,[49] but whose site the authors could not identify, despite assistance from Bosnian Serb residents, newcomers displaced from Ilijaš and Vareš. At Kula Grad (near Zvornik) the destroyed mausoleum of Šej Hasan Kaimija (1691) by 2001 had been rebuilt by returned Muslim families.

The usual position of cemeteries on the periphery of settlements may have protected them in the same way that traces of suburban and rural mosques generally survived better than in urban core contexts. However, several informants cited a superstitious unwillingness to disturb the dead as the main reason for the survival of these sites, which were certainly not protected out of respect for the Islamic cultural heritage. A large number of smaller cemetery sites, occasionally comprising a single standing stone, are dispersed through the countryside, often in roadside positions and sometimes next to mosque sites. In many localities they provided the only visible evidence for the historic Muslim presence in what is now Republika Srpska.[50]

49 See *Prosecutor v. Momčilo Krajišnik and Biljana Plavšić, Amended Consolidated Indictment*, ICTY Case No. IT-00-39 & 40-PT, 4 March 2002 and *Prosecutor v. Radovan Karadžić, Prosecution's Second Amended Indictment*, ICTY Case No. IT-95-5/18/PT, 18 February 2009.

50 A single roadside gravestone standing outside Sokolac on the road towards Rogatica was the only Islamic or Ottoman period monument seen in that locality. The pattern of damage to Muslim cemeteries observed by the authors was also noted by Colin Kaiser in *Report on Destruction of Cultural Property* 09-Jul-02, Exhibit P906, *Krajišnik*, ICTY Case No. IT-00-39.

Fig. 3.16 Returnees rebuilding the turbe of Šej Hasan Kaimija at Kula Grad near Zvornik, Republika Srpska, June 2001. © Richard Carlton

Finally, a range of upstanding pre-Ottoman structures, notably castles and towers, but also including monumental late medieval grave-markers (*stećci*) standing singly or in cemeteries and prehistoric earthworks were also noted during the course of survey. In general, it appeared that while such monuments and sites were not deliberately attacked during the 1992–1995 conflict, since most are outside main population centres and are considered 'neutral' in terms of ethno-religious affiliation, inevitably some suffered collateral damage. Exceptions include the numerous medieval and Ottoman military strongholds occupied and defended as artillery positions during the conflict, such as the restored medieval fortress at Ostrožac and smaller castles at Podzvizd and Pećigrad in the Bosnian Krajina region of north-west Bosnia,[51] and at Prusac in central Bosnia where trenching works were still clearly visible in 2000. Recent survey work has begun to assess the nature and extent of damage caused to such monuments in that area.[52] Other castle sites have suffered from the inappropriate addition of structures as ethno-national markers, such as at Kula Grad above Zvornik, where Muslim returnees are confronted by an Orthodox shrine placed there since 1992, and prominent crosses erected at Stolac and Počitelj.

51 For the military use of Ostrožac and other castles in this area during the 1992–1995 conflict see O'Shea B. 1998. *Crisis at Bihac: Bosnia's Bloody Battlefield*. Phoenix Mill, Stroud: Sutton Publishing.

52 Carlton, R. and Rushworth, A. 2009. The Krajina Project: Exploring the Ottoman–Hapsburg borderland, in *The Frontiers of the Ottoman World*, edited by A.C.S. Peacock. London: British Academy.

Fig. 3.17 Sometimes the only visible evidence for the historic Muslim presence in what is now Republika Srpska. A turbaned Muslim gravestone (*nišan*) in a farmyard at Kruševci. © Richard Carlton

The Post-war Situation

The fieldtrip highlighted all the symptomatic problems facing Bosnia's damaged cultural heritage in the post-war period in a climate where cultural and religious heritage issues had become highly politicized and almost every project meant more than simply restoring a building to its pre-war condition. Thus in Mostar the authors saw the Franciscan church of Ss Peter and Paul, whose elegant classical façade, walls and bell tower of 1866 survived bombardment by JNA and Bosnian Serb Forces in May 1992,[53] and which was the subject of fundraising for a restoration project until its substantial and restorable remains were abruptly torn down by the Franciscan authorities in 1997. Its ill-designed concrete replacement sports a grotesquely over-sized bell tower whose primary *raison d'être* appeared to be a show of cultural domination over Mostar's non-Catholic/Croat population, particularly over the rebuilt minarets that had begun to rise again across the city.

Symptomatic, too, were the number of mosques the authors saw being rebuilt or newly-built in the Federation and largely funded by Saudi Arabian, other Middle Eastern, Indonesian, Malaysian or other Islamic sources, a trend also observed in Kosovo.[54] Many destroyed mosques were being rebuilt to Middle-Eastern designs (almost always domed and often with twin minarets) alien to Bosnian mosque-building traditions.[55] These included formerly listed historic buildings like the Austro-Hungarian period Baš (Čaršijska) Mosque in Donji Vakuf, whose rebuilding was funded by Britain's Yusuf Islam, and the Varoška Mosque at Busovača. A rough rule of thumb appeared to be that when rebuilding projects were funded by public subscription, mosques tended to be rebuilt in traditional Bosnian style, with a single minaret and hipped roof.

One such was the mosque at Kozaruša near Prijedor, the first to be rebuilt in Republika Srpska. On 30 October 2000, as the authors were inspecting shattered mosques in suburban Banja Luka, a few miles away high-level guests gathered for a ceremony

53 The church's destruction is still attributed by popular rumour to phosphorus bombs dropped by helicopter, a claim dismissed by informed sources: see Kaiser, C. 1993. *War damage to the cultural heritage in Croatia and Bosnia-Herzegovina* in COE *Information Report on the destruction by war of the cultural heritage of Croatia and Bosnia-Herzegovina*, Doc. 6756, Strasbourg: Parliamentary Assembly of the Council of Europe, 2 February 1993, 45.

54 Ford, P. 2001. Mosques face new danger, *Bosnia Report* (New Series) 23/24/25 (June – October 2001), London: The Bosnian Institute, 46.

55 Only about 30 mosques in pre-war Bosnia were domed.

marking the reconstruction of Kozaruša Mosque, five and a half years after the end of the war.[56] While among those attending were the heads of Bosnia's religious communities (including a Metropolitan of the Serbian Orthodox church), UN and SFOR officials, foreign ambassadors, representatives of many of the international organizations present in Bosnia and members of the international and local press corps, not a single official of Republika Srpska was present.[57]

Within the medieval castle perimeters of Pećigrad in the extreme north-west Federation, the authors visited the historic mosque (c. 1577), which had been shelled in the war and lost its roof. Since the end of the war it had been completely renewed with an inappropriate porch added to the front of the building and the original beautifully worked doors featured in Bećirbegović[58] simply thrown away – supposedly because they were too damaged to retain. More sympathetic examples of mosque renewal were found in Tešanj and Gradačac, where wooden minarets, once common in Bosnia and particularly prevalent in the Bosanska Krajina region, had been reconstructed. More recently, the trend to renovate and reconstruct mosques and other important structures in a way that respects local traditions of workmanship and results in buildings faithful to the originals has continued, notably in eastern Herzegovina. Among those sites visited by the authors and which are now rebuilt are the Gornji Grančarevo mesdžid, east of Trebinje (opened July 2010); the Skočigrm mesdžid (opened August 2010) in the same area; the Čučkova Mosque in Nevesinje; Mostaći mosque; the Avdić Mosque at Plana (nearing completion but stalled due to lack of funds in late 2012); Pridvorci mosque (opened July 2011); the Careva Mosque, Bileća (September 2011); and others in the process of rebuilding at Nevesinje and Gacko. The reconstruction of the Careva Mosque at Trebinje began in November 2011. Significantly, by far the greater share of funding for the above projects has come from the accumulation of small donations made by private citizens dispersed around the world but with family connections to the buildings they support.

Appraisal and Conclusions

By assessing the available literature and published figures from various sources and reviewing the results of their own fieldwork, the authors have attempted to shed light on the scale and character of the damage caused to the built heritage of Bosnia-Herzegovina as a result of the 1992–1995 war and its aftermath. This review suggests that the true scale of the damage is not fully quantifiable, since accurate records of damage were not produced for even the most well-known groups of historic monuments, notably mosques. Needless to say, many other classes of buildings, particularly vernacular buildings such as houses, farm buildings and mills, and ensembles of such structures, particularly in rural contexts, lack even the most basic assessments of damage.

Witness accounts of destruction given to the authors, as well as observations made in the field, have reinforced the view of systematic destruction of cultural and religious property as part of the deliberate destruction of historical identities in a landscape of competing identities. In particular, buildings associated with Muslim populations were

56 The mosque at Kozaruša had been officially reopened for the faithful on 26 August 2000.

57 Tadic, T. 2000. A report from Kozarusa: Of mosques and politics, *AIM (Alternativna Informativna Mreža) Banja Luka* [online], 31 October.

58 Bećirbegović, M. 1990. 70; plates 131, 191 and 192.

destroyed in very large numbers, with sacral buildings particularly vulnerable. Indeed – and notwithstanding some regional variations apparent in the degree of destruction of Ottoman/ Muslim sacral architecture – virtually the entire architecture of the Bosnian Muslim sacral heritage was comprehensively attacked and decimated in all areas controlled for any period during the war by Bosnian Serb forces, and to a slightly lesser extent in areas held by Bosnian Croat forces, amounting to well over 75 per cent of the total land area of the country. Furthermore, many such buildings within territory held by the Bosnian Army, notably Sarajevo and Mostar, were also subject to selective bombardment from remote positions.

However Christian sacral monuments – Orthodox (Serb) and Catholic (Croat) churches – have generally survived well in areas under Bosnian government (largely Muslim) control, particularly in urban contexts.[59] All city centre churches and monasteries in Sarajevo survive intact, for example, as do the town centre churches of both denominations in predominantly Muslim towns such as Bihać, Gračanica, Velika Kladuša, Tešanj, Visoko, Zenica and Tuzla. Where non-Orthodox sacral structures survived unharmed in Republika Srpska, these were almost entirely Catholic, such as the church of Church of the Immaculate Heart of Mary (1904) in Bijeljina and the Cathedral of St Bonaventura (1974) in Banja Luka.[60]

These findings correspond with United Nations reports and testimony at ICTY prosecutions which found no policy of ethnic cleansing and the associated destruction of monuments in areas under Bosnian Army control.[61] The reasons for the tendency of the multiethnic, but Muslim-dominated Bosnian Army and the civilian population under its control not to target Christian monuments are complex, but Mahmutćehajić has suggested that to attack Christian monuments would have been to attack the sense of Bosnia as a cultural and state entity, 'its unity in diversity integral to the survival of the Bosniac nation.'[62] Whatever the reasons, the wave of indiscriminate destruction in which each group targeted each other's sacral symbols of identity anticipated by commentators like Chapman at the early stages of war and still assumed by many (perhaps most) after its end, is not borne out by observations on the ground.

Finally, the present study has drawn attention to the damage that continues to be done to the cultural heritage of post-war Bosnia-Herzegovina through unnecessary or targeted demolition and hasty or inappropriate reconstruction projects, much of which results from ineffective planning controls within a state of divided and ineffective government. However, positive signs were noted in the trend for recent cultural heritage redevelopment projects to respect local traditions of workmanship while aiming faithfully to renovate or reconstruct buildings according to their original appearance. This trend is currently particularly apparent in eastern Herzegovina within communities aiming to rebuild significant buildings through public subscription. The philosophy and economics involved in such local, community-based projects naturally also leads to the preservation of craft and building skills upon which architectural traditions depend for their future and so deserve support through training and expertise (if not funding) from local and international heritage preservation authorities.

59 See Helen Walasek *Destruction of the Cultural Heritage* in this volume for more on this issue.

60 The sole unattacked functioning mosque on the territory of Republika Srpska was that at Baljvine which was protected by Bosnian Serb residents of the village. See Arnautovic, M. 2010. Bosnia: The village where hate never triumphed, Institute for War and Peace Reporting [online], 10 April, TRI Iss. 642.

61 See Walasek, *Destruction of the Cultural Heritage*, in this volume on the subject.

62 Mahmutćehajić, R. 2000. *Bosnia the Good: Tolerance and Tradition*. Budapest: Central European University Press, 33.

Locations Visited on Fieldtrips in 2000 and 2001

Republika Srpska

Šipovo, Mrkonjić Grad, Banja Luka, Gornji Šeher, Kotor Varoš, Kotor, Ravne, Prijedor, Kozarac, Kozaruša, Bosanski Novi, Bosanska Kostajnica, Derventa, Modriča, Kotorsko, Doboj, Bijeljina, Janja, Kozluk, Gornja Kalesija, Zvornik, Kula Grad, Divič, Nova Kasaba, Srebrenica, Dobrak, Skelani, Peći, Kruševci, Knežina, Žepa, Rogatica, Višegrad, Sokolovići, Rudo, Foča, Nevesinje, Plana, Bileća, Kotezi, Mostaći, Trebinje, Pridvorci, Donje Čičevo, Lastva, Gornje Garnčarevo, Skočigrm.

Federation

Sanski Most, Bosanska Krupa, Bihać, Pećigrad, Cazin, Ostrožac, Mala Kladuša, Prusac, Oborci, Donji Vakuf, Jajce, Kiseljak, Tešanj, Maglaj, Dubnici, Trnovo, Sarajevo, Goražde, Ustikolina, Mostar, Stolac.

Chapter 4
Cultural Heritage Protection in Post-conflict Bosnia-Herzegovina: Annex 8 of the Dayton Peace Agreement†

Valery Perry

Introduction

The Dayton Peace Agreement (DPA)[1] that ended the three and a half year war in Bosnia-Herzegovina (BH) in 1995 included many provisions that made the document more ambitious than a simple peace treaty. In addition to ending the fighting among the warring parties and codifying territorial and governance issues, it served as a broad blueprint for the potential post-war reconstruction of an historically diverse, multicultural society. The Agreement outlines general norms and declarations concerning the state of Bosnia-Herzegovina, and includes a substantial set of 11 annexes that commit the country's leaders (the formerly warring parties and subsequent post-war political authorities) and its neighbours Croatia and Serbia to a wide range of activities, including military stabilization, refugee return, compliance with European and international treaties and conventions, the organization of elections and human rights protection.

Regardless of the realistic potential for full and lasting implementation of these ambitious peace-building and state-building activities among post-war political leaders, the authors of Dayton sought compromise among the Bosniaks (Bosnian Muslims), Bosnian Croats (Bosnian Catholics), and Bosnian Serbs (Bosnian Orthodox Christians)[2] by integrating power-sharing mechanisms into the Agreement to ensure the equality of these three 'constituent peoples' in the newly independent state. The protection of human rights was paramount in ensuring that the people of Bosnia-Herzegovina would not only recover from the trauma of war in order to return and rebuild, but also to keep a focus on the long-term goal of integration into Euro-Atlantic institutions, and most importantly, into the European Union. It can also be argued that this broad approach to the peace deal was partly driven by an interest among leaders in the United States and Europe, as well as among some of the BH parties, to attempt to reverse the impact of the wartime ethnic cleansing.

---

† Much of the research for this paper was conducted through the European Centre for Minority Issue (ECMI) programme in Bosnia-Herzegovina November 2001–January 2003. Several workshop reports detailing ECMI's Annex 8 project are available on the ECMI website, at http://www.ecmi.de. An earlier version of this article was published in *The European Yearbook of Minority Issues*, Vol. 2 (2002/2003), Leiden and Boston: Martinus Nijhoff Publishers.

1 The Dayton Peace Agreement, formally known as the *General Framework Agreement for Peace in Bosnia-Herzegovina* (GFAP), entered into force on 14 December 1995. See http://www.ohr.int/dpa for full text.

2 Bosnian Croat and Bosnian Serb are not legal terms, but will be used in this chapter to ensure a distinction from Croats/Croatians in Croatia, or Serbs/Serbians in Serbia.

Two of the Agreement's Annexes were central to this attempted reversal. Annex 7 (Agreement on Refugees and Displaced Persons) guaranteed the right of people to reclaim and return to their pre-war homes, and has been a primary focus of the international community since the beginning of its involvement in Bosnia-Herzegovina. Article 1 of Annex 7 directly links return to lasting peace: 'The early return of refugees and displaced persons is an important objective of the settlement of the conflict in Bosnia and Herzegovina.'[3] This has been a critical component of the peace accord, as approximately half of the country's population of just under four million had been displaced from their homes during the war, either as refugees or internally displaced persons. Support for return included the facilitation of post-war *minority* return – the return of a person to a community where they would now be in a minority.[4] Annex 8 (Agreement on Commission to Preserve National Monuments) guaranteed the right to reconstruct, rehabilitate and protect national monuments that had been damaged or destroyed during the war. Together, these two Annexes provide for a unique right of return, not only of people, but potentially of the culture, history and identity that existed in the region before the damaging results of war.

International and domestic parties had varying interests in a policy rooted in a commitment to the return of people and place. The international community had been shamed by its inability to stave off a war on the European continent, and the collective guilt that accompanied the world's impotence during the three and a half year siege of Sarajevo, a televised European refugee crisis, and the Srebrenica massacre had an impact on the development of peace talks and the scope of plans for peace enforcement and implementation. Support for a policy of return sent a signal that the gains and spoils of aggression would not be simply accepted as defining and legitimizing the new status quo.

Domestic imperatives drove this issue as well. During the war in Bosnia-Herzegovina some of the warring parties (or sub-groups within the warring parties, particularly paramilitary groups) attempted to erase the heritage of other groups through brutal programmes of ethnic cleansing.[5] Not only were people purged from communities, but the architectural heritage that had represented and reflected the multiethnicity of the region over centuries was heavily damaged or destroyed. While all sides suffered attacks on their cultural symbols, the Bosniaks in particular have had a strong interest in this issue, as the impact of ethnic cleansing deeply affected not only the community's population, but also its particular Islamic and Ottoman cultural heritage. Data collected and analyzed by the Research and Documentation Center reveals that the vast majority of civilian wartime deaths were among Bosniaks.[6] In his 2011 testimony at the trial of Radovan Karadžić at the International Criminal Tribunal for the former Yugoslavia (ICTY), cultural heritage expert András Riedlmayer testified that 95 per cent of the mosques he visited in 23 municipalities

3 *General Framework Agreement for Peace*, Annex 7, Article II(5).

4 See for example Toal, G. and Dahlman, C. 2011. *Bosnia Remade: Ethnic Cleansing and its Reversal*. Oxford: Oxford University Press; Phuong, C. 2000. 'Freely to return': Reversing ethnic cleansing in Bosnia-Herzegovina, *Journal of Refugee Studies*, Vol. 13, No. 2, 165–80.

5 See Toal, G. and Dahlman, C. 2011; Burg, S.L. and Shoup, P.S. 2000. *The War in Bosnia-Herzegovina: Ethnic Conflict and International Intervention*. London: M.E. Sharpe; Bose, S. 2002. *Bosnia After Dayton: Nationalist Partition and International Intervention*. London: Hurst & Company.

6 See the Research and Documentation website at http://www.idc.org.ba/; for an analysis of the data see also Hoare, M.A. 2008. What do the figures for the Bosnian war dead tell us?, *Greater Surbiton* [online], 4 January 2008, available at http://greatersurbiton.wordpress.com/2008/01/04/what-do-the-figures-for-the-bosnian-war-dead-tell-us/.

under Bosnian Serb control during the conflict were severely damaged or destroyed.[7] As a result of the social and cultural impact of the war, Bosniak politicians have been the loudest in their support for return, as well as in support of the reconstruction of destroyed sites. Croat and Serb politicians have not driven the return process to the same extent, but have sought to rebuild damaged or destroyed sites, and to assert their right to 'their place' within and throughout the country – although tied up as the reconstruction of cultural and religious monuments was with return, Serb authorities resisted the rebuilding of Serb/Orthodox cultural property outside Republika Srpska in the early post-war years.[8]

Agreements to preserve 'national monuments' or to generally address the issue of cultural heritage are not normally a key element in peace treaties or the standardized 'state-building and democratization kit' that has become part and parcel of post-conflict and development approaches.[9] While some might argue that an issue such as the protection of national monuments is trivial in a country still beset with political problems, unemployment, corruption, human rights abuses and discrimination, the role that protection of national monuments and cultural heritage can play in multiethnic societies should not be underestimated. In a 1993 report detailing the destruction of heritage in Bosnia-Herzegovina and Croatia in the early months of the war, the Council of Europe (COE) affirmed the role of culture in *any* community:

> There is no reason to be ashamed of being concerned for the cultural heritage when men, women and children are suffering in war. When historic villages and residential districts – and we are talking about regions where most people are living in historic buildings – are destroyed or damaged, these people become refugees, reduced to the degrading experience of refugees, nourishing hatred and preparing the wars of tomorrow.[10]

Therefore, Annex 8, while an often overlooked part of the Dayton Peace Agreement, can play an important role in long-term reconstruction and reconciliation prospects in Bosnia-Herzegovina. Due to the nature of this issue, successful implementation of Annex 8 integrates a broad spectrum of inter-related themes including human rights protection, minority rights, education, rule of law, civil society engagement and reconciliation. The Annex 8 implementation experience provides a useful and notable case study of how one focused and specific aspect of Dayton implementation has evolved and matured, successfully making the transition from international community direction to local ownership of an established process driven by the rule of law. It also stands as an example of a state-strengthening endeavour that is increasingly at risk by disintegrative forces.

This article provides a review of the evolution of the Annex 8 legislation and implementation process, including early implementation efforts, the development of a Commission to Preserve National Monuments, and implementation progress to 2012. An understanding of the Annex 8 process is important as the right to cultural identity and the

---

7 Karadzic: Destruction of mosques and Catholic churches, *BIRN's Justice Report* [online], 9 December 2011, available at http://www.bim.ba/en/299/10/33999/.

8 See Walasek, H. *Domains of Restoration*, in this volume.

9 In his book, *Aiding Democracy Abroad: The Learning Curve*, Thomas Carothers reviews many of the standard items on this checklist: judicial reform, parliamentary strengthening, civil society assistance, media work, political party development, civic education and electoral programmes. Washington DC: Carnegie Endowment for International Peace, 1999.

10 *The destruction by war of the cultural heritage in Croatia and Bosnia-Herzegovina presented by the Committee on Culture and Education. Information Report*, Doc. 6756, Strasbourg: Parliamentary Assembly of the Council of Europe, 2 February, 4.

protection of cultural heritage are challenges faced in many post-conflict countries. Bosnia-Herzegovina's experience in this innovative initiative can therefore benefit not only its own citizens, but can offer lessons learned to people and cultures suffering similar challenges across the globe.

Annex 8 of the Dayton Peace Agreement

Dayton's Mandate

It is quite stunning that the Dayton Peace Agreement even included Annex 8, among such crucial issues as military stabilization (Annex 1a), refugee return (Annex 7) the establishment of an international civilian administrator (Annex 10) and establishment of an international police task force (Annex 11). Yet, the concern to address the devastation of Bosnia-Herzegovina's cultural heritage during the war that led ultimately to Annex 8 was under discussion even before the talks at Dayton began. The very brief list of Agreed Basic Principles approved in Geneva on 8 September 1995 by the foreign ministers of Bosnia-Herzegovina, Croatia and Yugoslavia as a basis for further negotiation, included 'the appointment of a Commission to Preserve National Monuments.'[11] Thus the question apparently stood high on the negotiating agenda.

The discussions in Geneva were quite heated – in general, as well as on the topic of national monuments. In his review of the negotiations, Chollet notes that in an informal discussion on the proposed Commission to Preserve National Monuments after the conclusion of negotiations, BH Foreign Minister Muhamed Sacirbey remarked 'that the commission was, in some ways, a "substitute for national institutions,"' suggesting that 'since mosques, churches, bridges and other national monuments were spread around Bosnia, the commission would help preserve them all for posterity and possibly help build some trust along the way.' Yugoslav Foreign Minister Milan Milutinović, however, 'immediately linked the commission to ethnicity and territory: before one could decide which monuments to preserve, one had to determine to whom they belonged, because only their ethnic owners would really care about preserving it.'[12] The decision to include Annex 8 was political, and driven by the demands of the negotiating parties, including concerns about not only destroyed heritage sites, but of cemeteries.[13] UNESCO, the logical international partner for implementation of such an Annex, was named as playing a key role in the operation of the Commission for its first five years. Yet UNESCO was neither involved in nor consulted on the drafting of the Annex, and in fact only learned of Annex 8 and its own responsibilities from the US Observer to UNESCO *after* the peace treaty had been signed.[14]

---

11 Ministerial Conference on Bosnia-Herzegovina, *Agreed Basic Principles*, 3.4, Geneva, 8 September 1995, Annex 1.

12 Chollet, D. and Freeman, B. 2005. *The Secret History of Dayton: U.S. Diplomacy and the Bosnia Peace Process 1995*, National Security Archive Electronic Briefing Book No. 171 [online], 21 November 2005, 97. The author would like to thank to Helen Walasek for pointing out the Commission's first appearance in the Agreed Basic Principles and its discussion in Chollet.

13 Verdery analyses the role of cemeteries and the dead in political mobilization in Yugoslavia and other parts of Eastern and Central Europe in Verdery, K. 2000. *The Political Lives of Dead Bodies*, New York: Columbia University Press.

14 Telephone interview 30 September 2011, Helen Walasek with Mounir Bouchenaki, former Director of the Division of Cultural Heritage, UNESCO, 1992–2000.

Annex 8 of the Dayton Peace Agreement is formally entitled *Agreement on Commission to Preserve National Monuments*, and consists of 11 short articles. It provides a very basic framework, including the following tasks and mandates:

- Establishment of an independent Commission according to specific composition criteria (Articles 1 and 2);
- Funding of the Commission and its staff, to be determined and shared jointly by the entities (Article 3);
- A mandate to receive and decide on petitions for the designation of property as a National Monument (Article 4);
- Cooperation among the parties, cantons, and municipalities with the work of the Commission (Article 8);
- Transfer of the Commission from the parties to the Government of Bosnia-Herzegovina five years after the signing of the DPA (Article 9).

Article 6 concerns 'Eligibility' and is the basis for consideration of a site as a National Monument. The full text of this article declares:

> The following shall be eligible for designation as National Monuments: movable or immovable property of *great importance to a group of people with common cultural, historic, religious or ethnic heritage*, such as monuments of architecture, art or history; archaeological sites; groups of buildings; as well as cemeteries.[15]

This definition, with its reference to 'a group of people,' makes it clear that monuments are defined by a community that holds them in value, and is slightly different than common terms and definitions in cultural preservation circles.[16] While heritage professionals typically use the word 'national monument' to refer to a monument of importance to the state and its citizens as a whole, in Bosnia-Herzegovina the word 'nation' and its derivatives connote *ethnicity*, and refer to the three official 'nations' of Bosnia-Herzegovina – the Bosniaks, Bosnian Croats and Bosnian Serbs – and not to the entire citizenry of the state of Bosnia-Herzegovina. While national monuments in Bosnia-Herzegovina might in fact be appreciated by a diverse cross-section of its citizens, politically (and practically) the term 'nation' continues to be based on the exclusive concept of ethnicity rather than citizenship. These nuances demonstrate that although a 'soft' aspect of the Dayton Agreement, the political implications and consideration of Annex 8 are potentially significant in a divided post-war state.

In addition to the potentially controversial issue of designating sites of importance to one specific national group, the Annex 8 Commission may also provide for the preservation of monuments that are more universal in their significance, and which reflect a community broader than a single ethnic group. Crucially, Article V.3 bestowed legal protection on monuments that were the subject of a petition submitted to the Commission. Like many of the Dayton Peace Agreement's annexes, Annex 8 is short on both length and details, instead laying out broad parameters that can be implemented according to various interpretations, and which leave room for local development of the process. More detailed legislation

15 *General Framework Agreement for Peace*, Annex 8, Article 6 (emphasis added).

16 The author would like to thank to Helen Walasek for originally pointing out these distinctions and the relevance of such nuances in the context of Bosnia-Herzegovina.

and Commission rules and procedures were left for designated experts to develop and parliamentary bodies to adopt. However, this lack of detail combined with the inattention of the international community towards Annex 8 implementation in the years following the end of the war were to store up problems for the future.

The First Five Years: December 1995–December 2000

Article 2 of Annex 8 states that the initial Commission to Preserve National Monuments[17] was to include two members appointed by the Federation of Bosnia-Herzegovina (FBiH), representing the Bosniaks and Croats, and one member appointed by Republika Srpska (RS), representing the Serbs. Two additional members were to be appointed by the UNESCO Director-General; both were non-Bosnians, with one being a UNESCO staff member. The BH Commissioners served for three years, while those appointed by UNESCO served for a five-year term. The UNESCO Director-General was also to designate one UNESCO-appointed member to serve as chair. After the first five years, when responsibility would have shifted from international to domestic authorities, the members of the Commission would be appointed by the Presidency of Bosnia-Herzegovina. UNESCO is the only international organization noted in the Annex, and its mandate and responsibilities are limited on paper to the selection of two Commission representatives and designation of the chair. However, the Commission was effectively run under the aegis of UNESCO during this period.

While not highly visible, Annex 8 implementation began with the establishment of a Commission which held its first meeting in Sarajevo on 31 March 1996 in the presence of the UNESCO Director-General. UNESCO funded the work of the Commission for this initial five-year period.[18] The Commission met approximately every three to four months,[19] and in contrast to the heavy-handed role of the international community in other matters of Dayton implementation (for instance, in military stabilization, election support or policing), the Annex 8 process was initially driven by the domestic representatives, rather than by representatives of the international community. Several individuals interviewed noted confidentially that a significant difficulty during this time was that the representatives were not permanently appointed and did not attend the meetings consistently, making it difficult to move forward, assign responsibility and accountability, and take concrete actions. This was noted to be a particular problem among the often-changing representatives from Republika Srpska, and it was suggested that this might have been a subtle way to obstruct consistency and progress.

The key accomplishment of the Commission during this period was the development of a *Provisional List of National Monuments*. However, while an achievement in itself, the

---

17 The official designation of the Commission is the Commission to Preserve National Monuments. However the author shall use both this title and the term 'Annex 8 Commission' interchangeably throughout this chapter.

18 The domestic Commissioners initially appointed under the aegis of UNESCO were Fahrudin Rizvanbegović and Miroslav Palameta for the FBiH and Simo Brdar for RS. The two international Commissioners appointed by UNESCO were Azedine Beschaouch (a UNESCO staff member) and Léon Pressouyre. Léon Pressouyre was the Chairman. RS subsequently replaced Simo Brdar with a succession of Commissioners: Jelena Božića, Nedeljko Rašula, Momir Jungić and Radenko Ilić. For more information on the Commission at this stage, please see http://kons.gov.ba.

19 For a list of meetings held 1996–2000 and very brief resumés of their proceedings see Goodey, B. and Desimpelaere, W. 2003. *External Evaluation of UNESCO's Action in the Preservation of Cultural Heritage Damaged by Conflict: Bosnia-Herzegovina 1996–2003*, UNESCO Doc. IOS/EVS/PI/24, Appendix II.

terms of Annex 8 did not specifically require that such a comprehensive list be developed. Instead, it mandated that the Commission should proceed by accepting petitions for potential designated National Monuments. Then, based on these petitions and subsequent selection decisions by the Commission, a list could in time be compiled. Some observers and participants have indicated that the Commission members decided *not* to initiate a petition process out of concern that the Commission lacked the financial and human resources to effectively process the large number of petitions it expected to be submitted. Making allowances for these circumstances and limitations, international observers at the time agreed with this decision. While accepting and reviewing petitions may have been preferable, it was not viewed as appropriate or logistically possible under the given constraints.[20]

Therefore, rather than working through a petition process, the Commission members developed a Provisional List of potential National Monuments, building on existing pre-war documentation (including the 1986 inventory of the country's protected historic monuments and sites)[21] and consultations with experts. The process of developing the list continued through the year 2000, including a series of 17 joint meetings with representatives of the various preservation institutes (at the entity and canton levels) in Bosnia-Herzegovina. Development of the Provisional List was a controversial and delicate exercise. It was also very political, as political concerns and imperatives related to ethnic cleansing and return of refugees led to the inclusion of some sites, and the non-inclusion of others – decisions based not solely on the objective heritage value of sites, but on the symbolic importance of the noted sites in the post-war environment. As there was a clear reluctance by the Republika Srpska authorities at that time to participate in any efforts that could legitimize the existence of the state of Bosnia-Herzegovina, and as the Bosnian Croats were also somewhat obstructionist of state-building activities that could jeopardize their position as the numerically smallest of the three constituent peoples, their enthusiasm in this exercise was lower than among Bosniaks who were perceived to have the most to gain from state-level efforts. At one stage the Commission was even 'unwilling to publicize the contents of the list, due to the well-founded fear that ultra-nationalist groups or individuals might attempt to destroy monuments and sites that reflect a different group's heritage.'[22] However, in the absence of a petition process, the list development constituted a small but important first step in simply defining cultural heritage sites throughout Bosnia-Herzegovina, as all sites included on the list would receive provisional status and protection as National Monuments.[23]

20 For a basic point of reference, claims or petition processes have been used by the Centre for Real Property Claims (CRPC), which received over 300,000, or by the Human Rights Chamber, which received over 8,000. Commission for Real Property Claims of Displaced Persons and Refugees. Archived website. Available at http://www.law.kuleuven.be/ipr/eng/CRPC_Bosnia/CRPC/new/en/main.htm; Dakin, B. 2002. The Islamic Community in Bosnia and Herzegovina vs. Republika Srpska: Human rights in a multi-ethnic Bosnia, *Harvard Human Rights Journal*, 15, Spring 2002, 245–66.

21 Mulalić Handan, M. (ed.) 2010. *Report on Assessment of the Architectural and Archaeological Heritage, Bosnia and Herzegovina.* Regional Programme on Cultural and Natural Heritage in South East Europe, Integrated Rehabilitation Project Plan /Survey of the architectural and archaeological heritage (IRPP/SAAH), Sarajevo, March, 5. An inventory of the built heritage conducted by the Republic Statistics Institute of Bosnia and Herzegovina on 31 December 1986. A Republic of BH Law on the Protection of the Cultural, Historical and Natural Heritage had been passed in 1985.

22 International Crisis Group 1999. Is Dayton failing?, *International Crisis Group Report*, 28 October 1999.

23 As noted on the Annex 8 Commission website, 'Once final decisions have been adopted for all the properties on the Provisional List, the Provisional List of National Monuments of Bosnia and Herzegovina will be annulled.' See http://www.kons.gov.ba/main.php?id_struct=108&lang=4.

The initial Provisional List contained 777 prospective National Monuments,[24] and included three general categories of monuments: damaged or destroyed, existing or intact, and movable cultural property (artworks, icons, and so on). Initially there was often a misperception – propagated and fuelled by nationalist politicians and their supporters – that Annex 8 referred *only* to religious buildings or monuments, when in fact, a general review of the Provisional List reveals greater diversity in the initial identification of important sites. Of the total number of potential designated National Monuments, approximately 56 per cent were religious in nature, and can be assigned to the different faiths as detailed in Table 4.1.

Table 4.1 Analysis of religious monuments on the Provisional List

Religious Monuments	RS	FBiH	Total
Muslim	22	103	125
Orthodox	88	27	115
Catholic	100	92	192
Jewish	0	3	3
TOTAL	210	225	435

Looking at these figures, however, it is important to keep in mind the large number of Muslim religious structures that had been completely destroyed during the war leaving little left to preserve and thus affecting the numbers in the provisional exercise:

Table 4.2 provides a rough breakdown of the non-religious potential cultural monuments and buildings included in the Provisional List.

Table 4.2 Analysis of non-religious monuments on the Provisional List

Cultural-Historical Monuments	RS	FBiH	Brčko*	Total
Castles/towers	26	29	1	56
Large governmental, infrastructure, commercial and residential buildings	11	9	13	33
Houses	9	17	2	28

24 Mulalić Handan, M. (ed.) 2010, 7. The final Provisional List can be found on the Commission's website (Bosnian language only): *Privremena lista nacionalnih spomenika Bosne i Hercegovine, Kompletna* at http://www.kons.gov.ba/main.php?id_struct=165&lang=1. The experts on the Annex 8 Commission established in 2002 subsequently reviewed the list and identified errors and duplications. The independent review exercise in this article (developed as part of a 2001 initiative of the European Centre for Minority Issues (ECMI)) is based on the general assessment of the original list as it was available in 2001. However, the general trends are still representative. There was occasionally some difficulty in determining whether a site was Catholic or Orthodox, and in these cases decisions were made based on regional knowledge. The author would like to thank Ermina Porca for her work on this analysis.

Cultural-Historical Monuments	RS	FBiH	Brčko*	Total
Hotels	2	2	0	4
Academies and theatres	3	1	1	5
Museums and libraries	2	4	0	6
Squares and fountains	7	9	2	18
Streets	3	0	0	3
Bridges	5	13	0	18
Socialist era monuments	5	2	0	7
Stećak/necropolis of stećci (medieval tombstones)	34	23	0	57
Other historical/archaeological sites	40	49	1	90
Natural parks	6	3	0	9
TOTAL	157	165	20	342

* Brčko is a special administrative district within Bosnia-Herzegovina that is shared in condominium by both Entities, but is a part of neither Entity.

The Human Rights Chamber Decision

One important development that would affect progress in the implementation of Annex 8 occurred outside the specific mandate of the Annex 8 Commission. In 1996 a case was filed with Bosnia-Herzegovina's Human Rights Chamber (HRC)[25] by the Islamic Community[26] related to the ongoing difficulties and obstructions the Islamic Community had experienced in receiving the necessary permits to begin the reconstruction of 15 mosques in Banja Luka, at least 12 of which had been listed historic monuments before the war and included the sixteenth-century Ferhadija and Arnaudija Mosques.

The HRC case was to bring to the fore the question of totally destroyed monuments and the extent of their protection (or lack of it) under Annex 8, as well the construction vs *re*-construction debate regarding such monuments. In the context of the case, in July 1997 the Islamic Community requested that the site of the Ferhadija Mosque and those of four other totally destroyed Banja Luka mosques be designated National Monuments by the Commission, which was also asked for its comments by the Chamber in October. At its June 1998 meeting the Commission added the site of the Ferhadija mosque to the Provisional List, conferring protection on it. At the same meeting it approved the text of a letter to entity

25 The Human Rights Chamber was created under Annex 6 of the *General Framework Agreement for Peace in Bosnia-Herzegovina,* (Agreement on Human Rights) which incorporated the European Convention for the Protection of Human Rights [European Convention on Human Rights] and other international agreements on human rights and established a Human Rights Chamber. Available at http://www.ohr.int/dpa/default.asp?content_id=374.

26 The Islamic Community = *Islamska Zajednica.* The Islamic Community is a legal entity and official body representing members of the Muslim faith in Bosnia-Herzegovina. *The Islamic Community in Bosnia and Herzegovina v. The Republika Srpska (Banja Luka Mosques), Decision on the Admissibility and Merits, (Delivered on 11 June 1999)*, Case No. CH/96/29, Human Rights Chamber for Bosnia and Herzegovina (HRC).

authorities requesting that they protect the sites of totally destroyed cultural and religious monuments.[27] It informed the Chamber of both these actions.

In its decision, issued on 11 June 1999, the Chamber noted that actions taken by the Commission to Preserve National Monuments established by Annex 8 'did not preclude the Chamber from examining the applicant's grievances.' This was fortuitous, as at that time there was no Annex 8 legislation that could assist in the resolution of grievances and little systematic activity by the Annex 8 Commission. Further, the Commission had apparently decided (but not publicized) that the sites of totally destroyed monuments could *not* be added to the Provisional List – despite the (anomalous) addition of the Ferhadija site to the list.[28]

The Chamber found that:

> the failure of the authorities in Banja Luka to respond to the applicant's request in March 1997 for permission to rebuild seven of the destroyed mosques was an interference with, or a limitation of, the right of the Muslim believers in Banja Luka to freely manifest their religion as guaranteed by Article 9 [of the European Convention for the Protection of Human Rights].[29]

Authorities in Republika Srpska were therefore ordered to cease any obstructions and to issue the permits needed so that reconstruction could begin. This was an important moral and legal victory, and an assertion of basic human rights in Bosnia-Herzegovina.[30] However, in the absence of a consistent and state-wide legislative approach to the issue of reconstruction, there were still no systematic procedures for potential claimants, and no consistent process for parties interested in rebuilding. While the decision began to address the specific judicial issues under consideration, it did not offer a state-wide solution or model for progress.

The Development of Legislation

There were limited initial steps in the development of Annex 8 legislation in the period from 1996–2000. Legislation was needed to supplement the terms of the Annex 8 text and to lay the practical groundwork for subsequent activities and implementation once the Commission's work transferred to domestic authorities. After preliminary work during 1999, in the spring of 2000 UNESCO submitted a draft framework law to begin formally addressing the issues left unresolved yet earmarked for attention by Annex 8. This draft law addressed the matter of National Monument protection at the *state-level*. It included issues of both movable and immovable heritage, suggested the creation of

---

27 HRC Case No. CH/96/29, Nos 7, 45, 77–8. For more on the use of the Human Rights Chamber with regard to historic monuments see Helen Walasek, *Domains of restoration* elsewhere in this volume.

28 For the Commission's discussions of the subject see Goodey, B. and Desimpelaere, W. 2003, Appendix 2, 36.

29 *Annual Report 1999*, Annex G 1 *Summaries of the Decisions on Merits*, Case No. CH/96/29, Human Rights Chamber for Bosnia and Herzegovina (HRC). The report uses the abbreviated 'European Convention on Human Rights.'

30 For more on the human rights aspects of the case see Dakin, B. 2002. The Islamic Community in Bosnia Herzegovina vs. Republika Srpska: Human rights in a multi-ethnic Bosnia, *Harvard Human Rights Journal*, Vol. 15, Spring 2002.

a 30-member 'Heritage Commission' and included procedures for the classification and registration of protected property. Rather than addressing only the issues included in Annex 8 (primarily the designation of National Monuments), this draft provided for a broader structure for cultural heritage protection in Bosnia-Herzegovina. The proposed law was rejected by Republika Srpska, partly in response to the placement of the law at the Bosnia-Herzegovina state-level and the proposed establishment of a 30-member state-level Commission, as RS had been consistently reluctant to cede or share any authority from the entity to the state.

The Council of Europe (COE) had worked closely with issues of cultural heritage management in Bosnia-Herzegovina as well, carrying out a survey of monuments in the country over 1997–1998,[31] and in 2000, in cooperation with related diplomatic initiatives of the COE Special Representative to Bosnia-Herzegovina, developed its own draft law.[32] This draft proposed placing responsibility for implementation of Annex 8 in the cantonal Institutes for the Protection of Monuments in the Federation of Bosnia-Herzegovina, and in the entity Institute in Republika Srpska. It also proposed an approach that expanded on the specific goals of Annex 8, placing a strong emphasis on conservation and protection of heritage, using a broad definition of national culture, and including detailed references concerning archaeological excavations and both movable and immovable monuments. However, this law was neither further developed nor adopted, and failed to be a true catalyst for relevant legislative development.

2001–2002: A New Approach

The nominal transfer of authority from the internationally supervised Commission to a Bosnia-Herzegovina-driven institution took place automatically on 14 December 2000, as set out in Annex 8. The mandate of the UNESCO-led Commission itself ended on 31 March 2001, leaving no subsequent structure and creating a legal void. Early 2001 saw little activity on the issue of Annex 8, despite the clear need for a more comprehensive and consistent approach to site rehabilitation and monument preservation.

Since the end of the war and the beginning of refugee return and related attempts to rebuild, obstructionist local authorities throughout Bosnia-Herzegovina had used a mix of procedural and bureaucratic tactics to stall or deter efforts to rebuild or rehabilitate monuments (particularly religious facilities) that had been damaged or destroyed in the war. Issuance of construction permits, slow and non-transparent decision-making processes, and the application of subjective, locally-developed procedures contributed to slow progress on the community level, and growing frustrations and tensions among community groups seeking to rebuild.

A series of news-making incidents occurred in the spring and summer of 2001, making the issue of Annex 8 increasingly relevant. Cornerstone-laying ceremonies had been scheduled in May to inaugurate the rebuilding of two historic and completely destroyed mosques in Republika Srpska: the Osman-paša and Ferhadija mosques in Trebinje and Banja Luka respectively. In both cases organized opposition to the ceremonies, augmented by

31 *Specific Action Plan for Bosnia-Herzegovina, Preliminary Phase, Final Report*, Technical Cooperation and Consultancy Programme related to the integrated conservation of the cultural heritage, Strasbourg: Cultural Heritage Division, Council of Europe, March 1999.

32 Entitled the *Law on the Protection and Preservation of Cultural Monuments in Bosnia and Herzegovina.*

bussed-in hooligans and fuelled by nationalist chants and alcohol, led to riots and violence. In Banja Luka more than ten people were injured, with one person later dying, and a large group of international observers and dignitaries were trapped for hours during the ordeal.[33] In August 2001, controversy concerning the reconstruction of the Čaršija (or Careva) Mosque[34] in Stolac (in the Federation) further demonstrated the growing need for effective and standardized legal means of both addressing reconstruction as well as preventing or minimizing related civil unrest.[35] Although not as highly visible, debates also continued to surround such sites as the Radimlja stećak necropolis near Stolac, the Žitomislić Orthodox monastery near Mostar and other cultural heritage locales. These incidents together drew the attention of the international community to the issue of cultural heritage reconstruction and the apparent failure of existing procedural mechanisms to effectively implement Annex 8. While UNESCO and the international community in general had spent time and money addressing individual projects and issues on a case-by-case basis, the need for a broader and lasting systemic approach became clear.

In light of these events the Office of the High Representative (OHR), the lead civilian peace implementation agency in Bosnia-Herzegovina, became involved in order to deal with the issue more systematically.[36] The High Representative (then Ambassador Wolfgang Petritsch), unlike the other international organizations that had been working on the question, since 1997 held what became known as the 'Bonn Powers:' the ability to impose legislation and remove politicians assessed to be obstructing Dayton implementation.[37] The involvement of the OHR re-invigorated attention on Annex 8, as a two-track approach was initiated to address the matter. One track involved the development of harmonized, entity-level legislation on the issue (to avoid the resistance of politicians – particularly in Republika Srpska – against state-level legislation not specifically mandated in the DPA), and the other focused on re-constituting the defunct Commission to Preserve National Monuments.

On 10 October 2001, the High Representative requested the Bosnia-Herzegovina state Presidency and the prime ministers of both entities to appoint representatives to serve on a working group to draft harmonized legislation. The harmonized draft laws served as the basis for new entity-level laws in the Federation, Republika Srpska and in Brčko District. Working Group members included representatives from the Federation Ministry of Urban Planning, the Federation Institute for the Protection of Historical Monuments, the Republika Srpska Ministry of Science and Culture, the Republika Srpska Ministry of Urban Planning and the Brčko District government. The first meeting of the Working Group was held on 17 October 2001 in Sarajevo, and additional meetings followed in Banja Luka (29 October 2001) and Brčko (5 November 2001). Legislation was drafted and the draft laws

33 In April 2002, a court in Banja Luka sentenced seven people in connection with the ceremony attacks. Sentences included four months imprisonment or fines of up to $900. See Radio Free Europe/Radio Liberty 2002a, Bosnian Serbs sentenced over mosque incident, *RFERL Newsline*, 29 April 2002.

34 Also referred to as the Carska, Emperor or Imperial Mosque.

35 Unrest continued around the site of the mosque in Stolac, as on 4 December 2001 a small group of extremists tore down a fence surrounding and protecting the building site. See: Herzegovinian Croats vandalize mosque site, *RFERL Newsline*, 5 December 2001. See Amra Hadžimuhamedović in this volume for more on this issue.

36 The OHR had not been previously involved in Annex 8 issues, though some of its decisions were relevant to it. For instance, the 27 May 1999 decision of the High Representative on Socially Owned Land specified that land being used for religious or cultural purposes before 6 April 1992 could be reallocated. See http://www.ohr.int/ohr-dept/hr-rol/thedept/hr-reports/hrcc-hr-rep/98-99-monthly/default.asp?content_id=5091.

37 Bonn Peace Implementation Conference, 10 December 1997.

were presented to the entity governments for consideration and adoption. In December the Federation parliament adopted a law that differed slightly from the draft, including a very broad definition of the monuments to be protected. Republika Srpska government continued the draft law review process into 2002, but did not adopt it.

On 8 February 2002, the High Representative issued two decisions concerning the Annex 8 legislation.[38] One decision amended the law passed in the Federation, changing the definition of a National Monument to reflect those monuments specifically protected under the terms of the DPA and ensuring that permits be issued within 30 days of a request. The second decision imposed the law in Republika Srpska (where the RS National Assembly had failed to adopt it). The goal of the amendments to the Federation law and the imposition of the law in Republika Srpska was the adoption of harmonized legislation and a country-wide, uniform procedure for the implementation of Annex 8.[39] There was little reaction among politicians or the citizenry to the imposition of these laws.

The harmonized laws included a definition of the monuments protected by the legislation, specification of the ministries responsible for implementation[40] and guidelines for the timely issuance of permits for rehabilitation of National Monuments. A key feature of the harmonized laws was the fact that, while the Annex 8 Commission that proclaims National Monuments exists at the *state-level*, responsibility for implementation of these laws rests at the *entity-level*. Previously, during the five years of the original Commission, the laws on administrative procedure placed the issue at the level of municipality and canton, leading to widely varying practices as various regions subjectively employed a panoply of old laws and statutes to address the issue. On 14 August 2002, to ensure continued harmonization of relevant entity-level laws, the High Representative imposed changes to the Republika Srpska Law on Urban Planning, eliminating the requirement for architectural and construction companies working on the rehabilitation of National Monuments to obtain special licences for work in that entity, as this was seen as a potential avenue for unnecessary bureaucracy and potential obstruction.

The harmonized laws continued to use the definition of a National Monument as originally proposed in Article 6 of Annex 8. The laws make a clear point of defining 'rehabilitation' in order to eliminate the construction vs. *re*-construction debate with regard to totally destroyed monuments, and therefore clearly refer to bringing damaged or destroyed property back to its pre-war condition (Article 2). Article 4 stipulates that requests for permits for rehabilitation of National Monuments are to be decided by the appropriate entity ministry and must be issued within 30 days if appropriate documentation is submitted. The laws also reiterated the high levels of protection of National Monuments by the governing bodies, the obligation to co-operate across all levels of governance

38 See OHR Press Release dated 9 February 2002: *High Representative imposes Entity laws supporting the reconstruction of National Monuments*. For full texts of these decisions, see http://www.ohr.int/decisions.

39 On 21 February 2002 the Presidency of Bosnia-Herzegovina issued conclusions to initiate the drafting of documents related to the implementation of the Annex 8 law and the reconstituted Commission.

40 In the Federation, the ministry mandated to address implementation of the law is the Ministry of Urban Planning and Environment. In Republika Srpska the appropriate ministry is the Ministry of Urban Planning, Housing Affairs and Utilities, Construction and Ecology, though there has been some discussion in RS concerning the role of the Ministry for Science and Culture. In Brčko District, the Department of Urbanism, Real Estate Affairs and Economic Development is charged with the implementation of the law.

and basic reference to monitoring, enforcement and inspection.[41] Adoption of these laws provided the needed legislation for progress on Annex 8, and the reconstitution of the Commission would ensure the start of eventual implementation.

The Annex 8 Commission

On 22 December 2001, the Presidency of Bosnia-Herzegovina re-established the Commission to Preserve National Monuments, and five Commission members were selected – once again comprising three BH members and two international members who were all appointed for five years.[42] The Commission decided that the post of the chair of the Commission should rotate among the three domestic Commission members every six months – a common arrangement in official bodies in Bosnia-Herzegovina to ensure equal representation among the three constituent peoples. The Commission's work would be supported by a secretariat funded through the joint institutions of the state of Bosnia-Herzegovina. The first meeting of the reconstituted Commission was held on 4 March 2002.

The Commission decides on the designation of National Monuments in two ways: through acceptance of new petitions and through review of the Provisional List in order to reach a final decision regarding final monument status. Since autumn 2002 the Commission has met regularly (usually every two months), typically holding multi-day sessions at its office in Sarajevo and making visits to selected municipalities to view a range of monuments. As of June 2014 there have been 62 sessions. Between sessions, expert staff members research potential monuments under consideration, gathering the extensive documentation necessary to allow Commission members to determine whether a site should be declared a National Monument.[43]

Since the beginning of the Annex 8 process and the initial development of the Provisional List there have been valid concerns that designations could be made on the basis of political motivations rather than expert heritage considerations. One member of the original UNESCO-led Commission (speaking off the record) noted that it at times seemed like everyone simply wanted to put as many of 'their' monuments on the Provisional List as possible, without critical consideration of their true value. A non-political set of evaluation criteria was necessary to ensure objective and consistent evaluation. Thus, in September 2002, the new Commission agreed upon nine criteria that would be considered in evaluating potential National Monuments (including both movable and immovable objects): time/era, historic value, artistic and aesthetic value, documentary clarity and evidence, symbolic value, townscape/landscape value, authenticity, uniqueness and integrity. Scientific and technical criteria would provide the basis for all decisions, rather than political motivations.

Upon review of the information and documentation, and determination as to whether a site meets the established criteria, a decision by at least four members is necessary to proclaim a site a National Monument. Outside experts may be invited to Commission

41 The issue of movable property is not specifically addressed.

42 The three national Commissioners were (and remain so at time of writing) Amra Hadžimuhamedović, Dubravko Lovrenović and Ljiljana Ševo. The two international Commissioners were Zeynep Ahunbay (Turkey) and Tina Wik (Sweden).

43 See the Commission's website for information on the range of its activities and lists and information on designated National Monuments at www.kons.gov.ba.

meetings to provide additional information and the Commission may (and frequently does) go into the field to visit sites under consideration. If the Commission fails to reach agreement based on the available evidence, the issue must be further studied so that a decision can be made. Once a designation is made, the decision is forwarded by the Commission secretariat to the appropriate municipal and entity authorities, so that they are aware of the decision and of the National Monuments that exist in their communities.[44] Press conferences are held after each session, and the Commission web site updated, to explain the decisions and designations and heighten public awareness of the Annex 8 effort and cultural heritage protection more generally.

The Future of the Commission and Lessons to be Learned

Over the past years the Commission has continued to exercise its function and continually professionalize. At June 2014 772 sites had been designated as National Monuments. In addition to the Provisional List, the Commission has received 1,623 petitions for sites to be considered as potential National Monuments by that date. The Commission has also taken steps to reject certain petitions: petitions for the designation of some newly-constructed mosques (a number built on the site of destroyed mosques) have been found not to have the historical or cultural value necessary to merit designation as a National Monument. For example, at the November 2011 session, 14 sites were designated National Monuments, while the Commission determined that another eight sites did not meet the criteria needed to secure designation.[45]

Every site reviewed for designation as a National Monument is researched by Commission secretariat staff, and much historical and architectural information has been gathered to support this process. All documentation is prepared in the local language as well as in English. The Commission has established a traditional library of resources as well as a data library, and plans to continue to augment these collections; the texts of all designations and much else besides are available on the Commission web site. Since 2001 its annual *Baština / Heritage* has been an important source on Bosnia's monuments and their conservation, as well as on the activities of the Commission.[46] In the absence of any other broader state-wide cultural heritage bodies, the Commission is becoming a rich source of information for scholars and researchers of the cultural heritage of Bosnia-Herzegovina. Historians, architects, archaeologists, anthropologists and others are certain to find information through this documentation, as well as through the individuals and experts working on implementation.

In addition, the Commission has become involved in identifying heritage at risk and undertaking activities to protect sites designated as National Monuments. Throughout the country there are many sites at risk due to illegal nearby construction, vandalism, or simple neglect. A donors' meeting was held in March 2005 as a part of a campaign to protect heritage at risk. Further, the Commission seeks funds from bilateral domestic and

44 *Rules of Procedure for the Work of the Commission to Preserve National Monuments.*

45 Commission to Preserve National Monuments website, http://www.kons.gov.ba/main.php?id_struct=192&lang=1.

46 Papers from *Baština / Heritage, The Annual of the Commission to Preserve National Monuments*, in English and the local language are available from the Central and Eastern European Online Library (CEEOL) www.ceeol.com/.

external donors to support targeted preservation activities at selected sites, and has carried out work at, for instance, the site of the Aladža Mosque and the remains of the Mehmed-paša Kukavica Mosque at Foča (2006), the Šeranić House in Gornji Šeher, near Banja Luka and the Serbian Orthodox churches of St Nicholas at Trijebanj (2007) and Ss Peter and Paul at Ošanići (2009), all with funding from the US Ambassadors Fund for Cultural Preservation (AFCP). The Serbian Orthodox monastery church of the Holy Trinity at Vozuća, near Zavidovići (2007), and the Ali-beg Kapetanović Mosque at Vitina (2009) were both restored by the Commission with significant support from the Federal Republic of Germany and various Federation entity governmental budgets.[47]

The Commission's primary means of action include letters to relevant authorities, public appeals for action and the convening of meetings of relevant authorities and experts to address the problem. The Commission has intervened in such cases as the inappropriate reconstruction of the Hotel Ruža near the Old Bridge (*Stari Most*) at Mostar, as well as on the restoration of the Ćejvan Ćehaja Hamam, also in Mostar, which they judged inadequate.[48] In the case of the Ali-beg Kapetanović Mosque at Vitina, the Islamic Community had begun unsuitable repair work funded by the Embassy of the Islamic Republic of Iran without consulting the heritage authorities. After the Commission intervened, this work was stopped and a new project drawn up in association with the Commission. However, while the Commission tried to put a stop to the construction of a new mosque and minaret on the site of the destroyed Turhan Emin-beg Mosque in Ustikolina (which had resulted in the destruction and removal of the remains of the monument) and despite an order for the minaret's demolition in 2004, the new construction still stands at time of writing.[49]

The Commission's remit includes, crucially, its representation of Bosnia-Herzegovina at state-level internationally in the areas of heritage protection and preservation, including relationships with and participation in the programmes of such bodies as UNESCO, the European Union (EU) and the Council of Europe. One of the most important of these is the Council of Europe's Regional Programme on Cultural and Natural Heritage in South East Europe (RPSEE).[50] The Commission's efforts were key in the successful effort to place the historic city centre of Mostar on the UNESCO World Heritage List, a move aimed at ensuring heightened protection of the area according to UNESCO requirements, as well as boosting tourism. Mostar was awarded this honour in 2005,[51] and the Mehmed-paša Sokolović Bridge in Višegrad (famous from Ivo Andrić's book *Bridge on the Drina*) was added to the list in 2007.[52]

47 For detailed reports on the preservation/restoration of the National Monuments described above at Foča, Trijebanj, Vozuća, Vitina and many others see *Baština / Heritage*, IV/2008.

48 See *Mission Report, Old Bridge Area of the Old City of Mostar (Bosnia and Herzegovina) (C 946rev), 26–28 May 2008*, Paris: UNESCO, 2008 on the Hotel Ruža. See the website of the Commission to Preserve National Monuments on the Ćejvan Ćehaja Hamam, http://www.kons.gov.ba/main.php?mod=vijesti&extra=1075460509&action=view&id_vijesti=271&lang=4.

49 For a full account of the highly-politicized struggle over the reconstruction of the Turhan Emin-beg Mosque and official attempts to have it demolished see Aksamija, A. 2010. Contested identities: Identity politics and contemporary mosques in Bosnia and Herzegovina, in *Mosques in Europe: Why a Solution has Become a Problem*, edited by S. Allievi in cooperation with Ethnobarometer, Network of European Foundations Initiative on Religion and Democracy in Europe. London: Alliance Publishing Trust, 325–32.

50 Council of Europe website http://www.coe.int/t/dg4/cultureheritage/cooperation/see/default_en.asp.

51 *Decisions of the 29th session of the World Heritage Committee (Durban, 2005)*, WHC-05/29.COM/22, Paris: UNESCO, World Heritage Committee, 9 September 2005.

52 *Decisions of the 31st session of the World Heritage Committee (Churstchurch, 2007)*, WHC.07/31.COM /24, Paris: UNESCO, World Heritage Committee, 31 July 2007.

Nevertheless, many obstacles remain for the Commission. Reflecting the fragmentation of the country, virtually everything related to cultural heritage, with the exception of the Annex 8 Commission, remains regulated by entity and cantonal level laws, together with entity or canton-level institutes, with no state-level enforcement or oversight mechanisms. Mirroring the broader political crisis in Bosnia since 2006,[53] there has been little real progress in terms of ensuring state-wide protection of cultural heritage, and in fact the Commission itself has been at risk. There is no state-level law regulating the country's cultural heritage; nor is there a genuine state-level strategy. There is, however, no shortage of domestic legislation or international frameworks, as the exhaustive list of relevant reference texts includes regional planning laws, criminal procedure code and cultural and heritage laws.[54] There has been activity, but few results, due to lack of political will for a more inclusive and expansive BH policy and plan.

For example, in 2004 the Commission, together with experts, drafted a Law for Cultural Heritage which was submitted to the Ministry of Civil Affairs (MoCA)[55] for legislative action. MoCA did not pass the bill to Parliament for consideration or adoption.[56] Instead, there have been efforts to increasingly *disengage* the role of the Commission, and the potential role of the state in cultural heritage affairs, with the entity Ministries for Culture increasingly serving as MoCA counterparts to the exclusion of the Commission. The Commission has struggled to retain and ensure its capacity and role as a state body as the political environment in the country has deteriorated, for example suffering delays as appointments for international Commission members have been stalled through lack of Presidency level attention and confirmation. There are concerns that European Union IPA funds (Instrument for Pre-Accession funds) related to cultural heritage promotion could further validate this disintegrative approach by supporting solely entity-level initiatives, to the exclusion of the Commission, state-level experts, or any broadly integrative perspective. The impact of such an approach became clear in late 2011 as seven state cultural bodies (including the National Gallery, the National Museum (Zemaljski Muzej), and the National Library) which have continued to exist in an unregulated status since the end of the war, and after annual temporary funding fixes, began to close their doors for lack of basic operational funds.[57]

Another example of motion without movement is seen through some other planning documents. In October 2008, the Council of Ministers of BH adopted a document entitled, 'Strategy of Cultural Policy' ('Strategija Kulturne Politike u BiH').[58] This document confirms the decentralized nature of culture in BH, and the minimal coordinating role of the MoCA on basic principles. It reviews a number of issues throughout its 72 pages, including the role of culture in the economy, and aspects of literature, film and music in the culture sector. It reaffirms that some institutions of cultural significance for BH

53 McMahon, P.C. and Western, J. 2009. The death of Dayton: How to stop Bosnia from falling apart, *Foreign Affairs*. September – October 2009, 69–83; Chivvis, C. 2010. Back to the brink in Bosnia? *Survival: Global Politics and Strategy*, Vol. 52, No. 1, February–March 2010, 97–110.

54 Mulalić Handan, M. (ed.) 2010, 9–12.

55 The Ministry of Civil Affairs (MoCA) is the only state-level ministry with some responsibility for culture, largely in the area of co-ordinating Bosnia-Herzegovina's responsibilities and relationships at an international level.

56 Mulalić Handan, M. (ed.) 2010, 12.

57 Zuvela, M. 2012. Bosnia feud threatens end to shared Balkan heritage, *Reuters*, 9 January.

58 *Official Gazette of Bosnia and Herzegovina*, Number 93/08, 64th session of the Council of Ministers, October 2008.

should be supported through the budgets of BH institutions, noting the *Zemaljski Muzej* (National Museum) as one such body that needs such status and recognition. However, rather than define a cultural policy for BH, section 20 of the report is entitled 'Program for *Developing* a Strategy of Cultural Policy in BiH' [emphasis added]. There is also a 2011 'Action Plan for the Implementation of the Strategy of Cultural Policy in Bosnia and Herzegovina.'[59] This document, developed in a classic project management 'logframe' including objectives, indicators and timeframes, with significant project support and funding noted through the United Nations Millennium Development Goals Program, and implemented in a large part by the United Nations Development Programme (UNDP) in BH.[60] While an extensive initiative, including a significant budget (US$8,000,000 from December 2008 through March 2012), the fact remains that culture and cultural policy in Bosnia-Herzegovina remains as fragmented and divided as ever, with little application of culture to bring about forces of reconciliation at either the level of citizens or political elites. And, the status of the seven disputed state institutions remained unresolved at time of writing.

Thus the ultimate protection and preservation of monuments is still an entity obligation, and due to a combination of financial, political and bureaucratic challenges, much of Bosnia-Herzegovina's heritage remains at risk. The entity institutes responsible for heritage protection and Annex 8 implementation suffer from a lack of funding, vision and appropriate expertise. Coordination with the Annex 8 Commission has not been a priority for these entity-level bodies, in spite of the state-wide impact of heritage protection. The lack of a coordinated plan to ensure protection of Bosnia-Herzegovina's heritage is perhaps the biggest challenge facing the Commission today, and it reflects the ongoing political crisis of the structure, competencies and reach of the state *vis-à-vis* its component parts.

The increasing marginalization of the Commission is unfortunate on many levels. As the only state-level body dealing with the issue of cultural heritage, the Commission is increasingly serving as a focal point for issues of cultural importance outside the specific mandate outlined in Annex 8. Additional Commission activities include the launch of a campaign for the protection of endangered monuments, support for the rehabilitation of heritage at risk described above, country-wide awareness-raising projects such as 'My Bosnia and Herzegovina – My Heritage' working with schools and museums; and participation in Europe-wide, regional and other international cultural heritage protection and awareness efforts, as well as scholarly conferences and seminars.[61] The Commission's work supports tourism and related cultural-educational efforts, most visibly through the erection of multilingual signposts and historical information on designated National Monuments – an important step for a country that often pins economic development hopes to its tourism potential. In all of these ways the Annex 8 Commission is filling a gap in Bosnia-Herzegovina, and its efforts were most recently recognized by the European cultural heritage organization Europa Nostra, which in spring 2010 awarded a

59 Ministry of Civil Affairs of Bosnia and Herzegovina (MoCA) website, 10 November 2011, available at http://www.mcp.gov.ba/org_jedinice/sektor_nauka_kultura/pravni_okvir/strategije/Archive.aspx?template_id=19&pageIndex=1.

60 *Culture for Development: Improving Cultural Understanding in Bosnia and Herzegovina*, United Nations Development Programme in Bosnia and Herzegovina (UNDP), available at http://www.undp.ba/index.aspx?PID=21&RID=83.

61 See the Commission's website for an overview of its activities. http://www.kons.gov.ba/index.php?lang=4.

Fig. 4.1 Preparing for a session of the Commission to Preserve National Monuments of Bosnia-Herzegovina at its offices in Sarajevo, 2003. Left to right: Commissioners Ljiljana Ševo and Amra Hadžimuhamedović and Mirela Mulalić Handan, Chief Executive Officer. © Commission to Preserve National Monuments of Bosnia & Herzegovina

European Excellence Award to the Commission for dedicated service in the preservation of cultural heritage.[62]

Very broadly speaking, the implementation progress of Annex 8 in Bosnia-Herzegovina offers an important lesson to the future authors and implementers of other peace plans and post-conflict efforts. The presence of Annex 8 within the Dayton Agreement ensured that cultural heritage was on the radar screen of local and international actors in Bosnia-Herzegovina. If a peace agreement or state-building road map serves as a 'check list' for priorities and necessary tasks, the formal inclusion of cultural heritage preservation can give the issue legitimacy and increase the chances that it will be addressed. In the case of Bosnia-Herzegovina, once legislation was in place, local experts *were* able to successfully begin the implementation process. In the absence of such a mandate in Bosnia-Herzegovina, it is likely that cultural heritage protection would have been ignored by local and international actors alike as attention was devoted to other post-war issues – in spite of the critical role culture can play in post-war reconstruction and reconciliation.

Cultural heritage cannot be limited to, or seen as belonging solely to, a canton, entity or even state. Bosnia-Herzegovina is itself a component of a larger shared European cultural heritage, and as such there should be an interest in a cultural heritage strategy in Bosnia

62 Europa Nostra Laureates 2010, http://www.europanostra.org/projects/30.

that combines the practical principles of subsidiarity with some framework for a state-level strategy and, most importantly, implementation mechanisms. Further, while the issue may be viewed through the expert lens of cultural heritage, it can also be viewed through the prism of Bosnia's post-war reconciliation potential. The experience of the Annex 8 Commission has been useful and note-worthy in the local context and as an example for other post-war countries and could be considered one of Bosnia's post-Dayton quiet successes. However, in light of the tenuous politics in Bosnia-Herzegovina at the onset of 2014, whether culture will ultimately be valued as an integrative tool, or used solely as a disintegrative tool, remains to be seen.

Chapter 5
Domains of Restoration: Actors and Agendas in Post-conflict Bosnia-Herzegovina

Helen Walasek

> And what will happen if the Aladža Mosque … is not rebuilt? They'll say it never existed and children will believe it.[1]

The 1992–1995 Bosnian War was a conflict that shocked the world and galvanized high levels of passionate public polemic. The catastrophic and intentional destruction of a substantial part of the country's cultural heritage (particularly its Ottoman and Islamic heritage) that was a distinguishing feature of the war aroused global condemnation from world leaders and organizations such as the United Nations and UNESCO to the person on the street in London, New York, Istanbul, Cairo and Kuala Lumpur alike. Two destroyed structures more than any others were to become iconic in international perceptions of the destruction: the Old Bridge (*Stari Most*) at Mostar and the National Library[2] (*Vijećnica*) in Sarajevo, though the hundreds of devastated and demolished mosques were potent symbols across the Islamic world.

Thus the restoration of Bosnia-Herzegovina's[3] built heritage after the conflict ended was bound to become a process that brought into play domestic and international actors with widely differing agendas. Yet an examination of this process in the immediate post-war era and over the decade following the signing of the Dayton Peace Agreement shows how these actors sometimes clashed, occasionally worked together, but more frequently had little interaction with each other. In particular, the involvement of the international community in restoration during this crucial period can, for the most part, be characterized by a narrow focus on a small number of high-profile projects coupled with a startling neglect of the wider picture where ethnically cleansed populations returned to reconstruct their communities.

One way of approaching what was a complex scene is to look at what might be called *domains of restoration* – to examine who the actors were in each, how they interacted with other domains (if at all) and how they engaged in what was, after all, a peace process, one that was intended to address a particular feature of the war, a feature which had set in train the attacks on Bosnia's heritage in the first place, and that was ethnic cleansing.[4] There

1 Nihad Čengić, art conservator and staff member of the *Office for the Protection of Cultural Property*, Sarajevo in 1993, quoted in Schiller, B. 1993. Bosnian artists save heritage treasures, *The Toronto Star*, A10, 15 May.

2 The full name of the library is the National and University Library. However, most sources refer to both the building and institution as the 'National Library.'

3 The term Bosnia will be used alternatively with Bosnia-Herzegovina throughout to mean the entire state of Bosnia-Herzegovina

4 The term ethnic cleansing describes the process of the mass expulsion of a specific group which in Bosnia-Herzegovina was accompanied in varying degrees by the forcible expulsion of

were two clearly articulated objectives in the destruction of cultural and religious property that took place as a part of campaigns of ethnic cleansing: first, to obliterate any evidence of the expelled population's historic existence on the territory, and second, to deter the targeted group from ever returning. In other settings, particularly cities like Sarajevo and Mostar, attacks on cultural property were frequently determined attacks on institutions and structures symbolic of Bosnia's diverse shared past.[5]

An overarching aim of the peace agreement that ended the war was to reverse the effects of ethnic cleansing. Thus restoration took place within a particular context and within a particular framework: that of the right of return for refugees and displaced persons under the terms of the Dayton Peace Agreement (DPA)[6] to the homes and communities from which they had been forcibly expelled as a crucial element in the broader attempt to restore Bosnia-Herzegovina's multicultural society within the structure of a democratic state. Yet despite this ambition of the DPA, the treaty's formalization of the division of Bosnia into the two politically powerful entities of Republika Srpska and the Federation[7] was widely seen as rewarding the gains of ethnic cleansing and to reinforce the ethno-national exclusivism that had driven the war.[8] This contradiction was to have its impact in many spheres, not least in the restoration of the built heritage in its role as a marker of identity.

The Framework for Restoration

The destruction of cultural heritage in Bosnia-Herzegovina during the conflict was felt to be of such significance that it was directly addressed in the peace settlement. The most unusual of the Dayton Peace Agreement's 11 annexes – unusual for any peace treaty – was Annex 8 which recognized the role that destruction of the cultural heritage had played during the conflict and aimed at tackling it. Annex 8 established a Commission to Preserve National Monuments (the Commission) whose mandate was to receive petitions to designate property of 'cultural, historical, religious or ethnic importance' as National Monuments and committed entity authorities to protect and make efforts towards the rehabilitation of the designated monuments and to refrain from taking any measures that might damage them – a protection extended to monuments on the interim list awaiting a decision on designation.[9]

---

targeted populations, expropriation of property, destruction of the religious and cultural symbols of the expelled communities, looting, detention in camps, torture, and mass rape and murder. See the *Introduction* for more on the war in Bosnia-Herzegovina and ethnic cleansing.

5 See Helen Walasek, *Destruction of the Cultural Heritage in Bosnia-Herzegovina: An Overview*, in this volume for a fuller account.

6 The peace agreement, more formally The General Framework Agreement on Peace or GFAP, but usually known as the Dayton Peace Agreement or Dayton Accord, was signed in Paris on 14 December 1995. For the full text and Annexes see http://www.ohr.int/dpa/default.asp?content_id=380. Annex 7 on Refugees and Displaced Persons established the right of return.

7 Under the terms of the GFAP Bosnia-Herzegovina was divided into two political entities: the Federation of Bosnia-Herzegovina (sometimes known as the Muslim-Croat or Bosniak-Croat Federation) consisting of ten cantons, dominated either by Muslim or Bosnian Croat majorities and Republika Srpska, a unitary authority dominated by a Bosnian Serb majority. See the *Introduction* for an overview of the war and peace settlement.

8 By its recognition of the concept of Bosnia-Herzegovina's three 'constituent peoples': Bosniaks, Croats and Serbs. See GFAP, Annex 4, Constitution of Bosnia and Herzegovina.

9 See Valery Perry in this volume for a full account of Annex 8 and its implementation. For the text of Annex 8 see http://www.ohr.int/dpa/default.asp?content_id=376.

The Commission was to be run under the aegis of UNESCO for its first five years, then pass to the control of the state of Bosnia-Herzegovina. Two other Annexes of the DPA were also to have an impact on reconstruction of the cultural heritage: Annex 6 on Human Rights, which incorporated the European Convention for the Protection of Human Rights[10] and established a Human Rights Chamber (HRC) whose decisions were final and binding, and Annex 7 on Refugees and Displaced Persons which included the right of return for refugees and displaced persons to live in their pre-war homes.

Actors and Agendas

Implementation of civilian aspects of the Dayton Peace Agreement was to be overseen by a High Representative representing the international community and operating through an Office of the High Representative (OHR) who had wide powers to enforce legislation and dismiss officials. The High Representative and the OHR was to play a key role in the implementation of Annex 8 and thus in the restoration of Bosnia-Herzegovina's destroyed heritage. Military aspects of the treaty were to be supervised by a NATO-led multinational peacekeeping force SFOR (called IFOR during 1996).[11] The DPA also provided for an International Police Task Force (IPTF) to work alongside and monitor local law enforcement agencies.[12] Both SFOR and the IPTF were to become involved in providing a secure environment for returnees.

Already operating in Bosnia-Herzegovina at the end of the war or in its immediate aftermath were the usual array of intergovernmental organizations such as the United Nations Mission in Bosnia-Herzegovina (UNMIBH) and other United Nations bodies (such as UNESCO, UNHCR and UNDP), the Organization for Security and Co-operation in Europe (OSCE), the World Bank, the European Union (EU), the Council of Europe (COE), the Organisation of the Islamic Conference (OIC), national governments (which supported aid projects largely through their emergency relief and international development agencies) and a multitude of non-governmental humanitarian aid agencies of a variety of religious denominations or none (NGOs).

All these actors were to become involved in heritage restoration and reconstruction in one way or another. All had widely differing aims and agendas, from the Swedish government's development agency Sida, and the agencies representing Islamic states like the Saudi High Commission for Relief to Bosnia-Herzegovina, to those representing Austria and Turkey which still maintained vestiges of a post-imperial responsibility for Bosnia. However, unlike the multiplicity of NGOs providing more traditional forms of post-conflict/humanitarian aid, just three worked directly with cultural heritage protection of which only one, the Swedish organization Cultural Heritage without Borders (CHwB), funded almost entirely by Sida, implemented restoration projects across Bosnia-Herzegovina.[13]

---

10 Frequently referred to in a shortened formulation as the European Convention on Human Rights.

11 IFOR = Implementation Force. SFOR = Stabilisation Force from 20 December 1996. EUFOR = European Union Force succeeded SFOR from 2 December 2004.

12 The IPTF was created under Annex 11 of the DPA.

13 The others were University of York's Post-war Reconstruction and Development Unit (PRDU) which developed the initial plans for restoring the historic Ottoman town of Počitelj, http://www.york.ac.uk/politics/centres/prdu/; see below for more on PRDU. Bosnia-Herzegovina Heritage Rescue (BHHR), the only NGO in the field of heritage accredited by UNHCR as a humanitarian aid

Important local actors were to be government authorities at central state and entity levels through to cantonal and municipality levels, which usually operated through their Ministries of Culture[14] and/or the various Institutes for the Protection of Monuments (but which were directed by their political masters drawn from the various ethno-national elites), the Commission to Preserve National Monuments, and the different religious hierarchies. At the time there were no local NGOs involved in historic restoration: Bosnia-Herzegovina was not only a post-conflict country, but also a country in transition from communism where the pre-war centralized socialist heritage hierarchies (and attitudes) still persisted.

There was one other group, however, which was to be a prominent actor in the post-conflict restoration of the cultural heritage. These were the communities which had been ethnically cleansed and were now scattered throughout the country, across the region and around the world.

Domains of Restoration

So what where the *domains of restoration* that engaged these diverse domestic and external actors according to their varying agendas and how was international support and funding for these different domains targeted? And, crucially, how did international involvement in these various domains push forward the peace process?

The domains of restoration in post-conflict Bosnia can be characterized by three core types of sites and structures: the *non-contentious*, the *iconic* and the *contentious*. Sites that can be considered non-contentious were generally in localities where ethnic cleansing had not taken place, or which were regained by expelled populations at the end of the war, and could also include sites in cities like Sarajevo. Here, there was considerable international involvement in heritage restoration, usually in the form of 'value-free' restoration of 'neutral' buildings that housed, for instance, infrastructure entities such as the Austro-Hungarian period Central Post Office building in Sarajevo whose restoration was not perceived as a cultural heritage project but as part of general reconstruction efforts.

It was in this domain, too, particularly in municipalities dominated by Bosniak (Muslim) majorities, that funding from Islamic sources (governments and humanitarian aid agencies) was almost entirely directed, particularly towards mosque construction. But although Islamic donors made substantial investments in mosque rebuilding (and in many new builds), their support almost never went towards restoration or reconstruction of historic Ottoman mosques to their pre-war form.

Iconic sites were those that became hugely symbolic to both the international community and the local population; some iconic sites could also be contentious sites. During the war these sites were the focus of varying levels of worldwide media attention and considerable rhetoric, but as will be seen, their post-conflict fates differed widely. The post-conflict restoration of such sites came to be mobilized as concrete and visible symbols which embodied ideas of reconciliation and the reconstruction of relations between

agency during the war still functioned up to 2000, but carried out only one small restoration project of a fountain in Sarajevo; see Helen Walasek, *Destruction of the Cultural Heritage* in this volume for more on the work of BHHR. See Tina Wik in this volume on the work of CHwB.

14 There is no Ministry of Culture at state level; the Ministry of Civil Affairs (MoCA) has some responsibility for culture and heritage.

Bosnia's ethno-national groups that the international community was keen to promote in the immediate aftermath of the war.

Finally, contentious sites were sites where ethnic cleansing had taken place and where the perpetrators of atrocities remained dominant after the end of the war, many still holding public office where they continued to pursue an exclusivist agenda. Least known and least publicized (except for a handful of cases) it was in this domain, where returned refugees attempted to re-establish destroyed communities in difficult and hostile environments that the most widespread restoration took place.

It is these last two, largely divergent, domains (iconic and contentious) that epitomize all the contradictions of the restoration scene in post-conflict Bosnia-Herzegovina and how restoration of the heritage played out in these spheres shall be examined in more detail.

Iconic Sites

Iconic sites were those that became invested with considerable symbolic value for the international community, for Bosnia's government and – in one way or another – for ordinary Bosnians themselves. Such iconic sites (usually single structures) were widely promoted as 'powerful symbols' of Bosnia's 'multiculturalism and diversity,'[15] of its 'historic tolerance'[16] and 'religious pluralism'[17] whose reconstruction would hopefully be a 'vehicle for reconciliation.'[18]

Destroyed cultural objects or locales that were felt to be of significant beauty, antiquity and/or cultural importance could attain the status of iconic sites. These included the completely demolished sixteenth-century Ferhadija and Aladža Mosques in Banja Luka and Foča respectively (destroyed by Bosnian Serb forces/authorities) and the seventeenth-century Serbian Orthodox Monastery at Žitomislić (destroyed by Bosnian Croat (HVO) forces/authorities)[19] and to a lesser extent the small town of Počitelj with its exceptional ensemble of Ottoman architecture (destroyed by Bosnian Croat (HVO) forces/authorities).

But topping the list of iconic sites were the sixteenth-century Ottoman Old Bridge (*Stari Most*) at Mostar (destroyed by Bosnian Croat (HVO) tank shelling on 9 November 1993) and the late nineteenth-century Austro-Hungarian period National Library (*Vijećnica*, or former Town Hall) in Sarajevo (set alight with incendiary shells fired by Bosnian Serb artillery over the night of 25–26 August 1992).

15 *Reconstruction of three religious monuments – symbols of multiculturalism supported by MDGF Culture and Development Programme*, press release, Sarajevo: United Nations Development Programme in Bosnia-Herzegovina (UNDP), 20 September 2010.

16 *Project Appraisal Document on a proposed credit in the amount of SDR 3.0 million (US$4.0 million equivalent to Bosnia and Herzegovina for a Pilot Cultural Heritage Project*, Report No. 19115-BiH, Washington: World Bank, Infrastructure Sector Unit, Europe and Central Asia Regional Office, 14 July 1999, 2.

17 *Simultaneous reconstruction of religious sites in BiH*, press release, Sarajevo: United Nations Mission in Bosnia-Herzegovina (UNMIBH), 23 May 2000.

18 *Implementation Completion Report on a credit in the amount of US$4.0 million equivalent to the Bosnia and Herzegovina for a Pilot Cultural Heritage Project*, Report No. 32713, Washington: World Bank, Infrastructure Sector Unit, Europe and Central Asia Regional Office, 22 June 2005, 11.

19 *Hrvatsko vijeće obrane* or Croatian Defence Council, the military forces of the self-proclaimed Bosnian Croat nationalist para-state, *Hrvatska zajednica Herceg-Bosna*, or the Croatian Community of Herceg-Bosna.

However, the stakes in the post-war battle for funding the reconstruction and restoration of such 'non-essential' historic monuments (many of them religious structures) were high. While the Bosnian authorities and UNESCO drew up lists of historic buildings requiring restoration to offer to international donors, such projects were certainly not a priority for most outside agencies and governments engaged in reconstruction of the country's shattered infrastructure, housing and manufacturing sectors in the immediate post-war period. Furthermore, non-Islamic or non-denominational donors often saw involvement in the reconstruction of religious structures as problematic, so tied as they were with ethno-national identity, leaving them open to charges of favouring one of the parties in the conflict over the others. Yet these structures had been attacked precisely *because* of their ethno-national affiliation or association with Bosnia's historic diversity.

Thus, such sensitivity could even extend to a secular building like the *Vijećnica*. In an evaluation of Austria's involvement in the building's reconstruction it was suggested that as a symbol of Serb aggression, its restoration might be perceived as a hostile act by the Bosnian Serbs if no restoration of an equivalent important 'Serb' historic monument took place – though quite how such a structure could have been identified is not made clear, seeing that the *Vijećnica* was a building with no ethno-national identity which had served all Bosnia-Herzegovina's citizens regardless of which 'nationality' they belonged to.[20]

An evaluation of UNESCO's work in cultural heritage preservation in Bosnia-Herzegovina from 1996 to 2003 similarly observed as a negative impact 'the predominance of Ottoman monuments restored … with funding imbalances tending to neglect Croat, but especially Serb sites.'[21] And Sweden's Sida initially expressed concern about becoming involved with 'another mosque' when approached by Cultural Heritage without Borders regarding the restoration of the badly damaged seventeenth-century Handanija Mosque in Prusac as they had already funded the restoration of the Kuršumlija Mosque in Maglaj.[22]

The nervousness of external actors with regard to the types of buildings whose reconstruction they were prepared to support could sometimes involve a search for damaged structures belonging to the two other dominant ethno-religious identity groups[23] of Bosnia-Herzegovina in a quest for equivalence – an equivalence hard to attain since the number of important historic mosques and Islamic/Ottoman structures destroyed or badly damaged far outstripped the number of churches and other religious or ethnically-identified structures, whether Catholic/Croat or Orthodox/Serb. Such attempts at even-handedness were to deny the evidence and in effect to buy into the ethno-nationalism that had driven the war. Yet, despite these donor anxieties, the UNESCO evaluation also observed the extreme difficulty of finding international donors to restore Serb monuments.[24]

Thus, the restoration stakes for a major input by international donors narrowed down to two frontrunners: the *Stari Most* (Old Bridge) and the *Vijećnica* (National Library). Neither

20 *Thematic Evaluation: The Relevance of Culture and Cultural Heritage in ADC, Field Report – Bosnia and Herzegovina*. Vienna: Austrian Development Agency (ADC), January 2007, 23.

21 Goodey, B. and Desimpelaere, W. 2003. *External Evaluation of UNESCO's Action in the Preservation of Cultural Heritage Damaged by Conflict: Bosnia-Herzegovina 1996 –2003*, UNESCO Doc. IOS/EVS/PI/24, 23.

22 Personal knowledge. Sida did go on to fund the restoration of the mosque, but only after the support of European experts in Islamic heritage was enlisted, and only after its initial documentation phase was funded by the Barakat Trust. See Tina Wik elsewhere in this volume for more on the restoration of the Handanija Mosque.

23 Bosniak/Muslim, Croat/Catholic and Serb/Orthodox.

24 Goodey, B. and Desimplaere, W. 2003, 19.

were religious structures and both bridge and library were capable of being freighted with considerable symbolic meaning. Equally important, both were located in cities where the international community (and the international media) had their most significant presence. Not only that, while monuments like the Aladža Mosque had been obliterated far from the gaze of the media, the destruction of the *Vijećnica* and the *Stari Most* had been public events, observed by a global constituency. Images of the National Library erupting in flames and the graceful arch of the Mostar bridge crumbling under the final fatal impact of HVO shells had been played out on television screens across the world again and again, and an outpouring of shocked reactions to these brazen 'assassinations' followed over the days, months and years.

The *Stari Most* was already an international icon – one of the most celebrated cultural treasures of the former Yugoslavia, easily accessible from Dubrovnik and the resorts of the Dalmatian coast, it was well-known to many European tourists from happier times. Still, in the months following the end of the war it seemed inconceivable that the devastated National Library, with its cargo of symbolic references and prominent location in the country's capital would not be among the first of Sarajevo's historic monuments to be restored. There had been a UNESCO resolution in 1993 calling for the library's reconstruction and an appeal from the organization's Director-General in 1994,[25] as well as a proliferation of initiatives across Europe and North America.[26] Emotive images of the *Vijećnica* in flames and media articles comparing it with the book burnings of Nazi Germany struck a deep chord. The building had become a shrine to Sarajevo's suffering – and its resilience – after cellist Vedran Smajlović was pictured playing in full evening dress among the shattered columns day after day, bringing renowned conductor Zubin Mehta and opera singers José Carreras and Ruggero Raimondi to perform Mozart's *Requiem* with Sarajevo Philharmonic Orchestra in the ruins and Prince Charles to walk amidst the rubble.

Yet the *Vijećnica* project was doomed from the start. When it came to the crunch, only Bosnia's former colonial ruler Austria stepped forward with funding, and then only because it needed a large visible project in hand to mark the state visit of the Austrian President to Sarajevo in 1996. The funding, too, was only for the first phase of reconstruction to stabilize the structure and restore the vast glass and steel dome.[27] But at the very beginning of the process the issue that was to dog completion of the *Vijećnica* restoration emerged: who owned the building?[28] Though a consultant engaged by the Austrians to assess the project deemed the National Library the legal owner, the Federal Chancellery thought otherwise and entered into a contract with the Canton of Sarajevo, which continually referred to the building as its Town Hall and never as the Library.[29] The ownership dispute rumbled on while an unseemly squabble about the structure's future use began, with the suggestion even

25 *Records of the General Conference, Twenty-seventh Session, Volume 1: Resolutions, 4.8 The situation of the cultural and architectural heritage and of educational and cultural institutions in Bosnia and Herzegovina (Sarajevo National and University Library)*, Paris: UNESCO, 25 October–6 November 1993; *Appeal by Federico Mayor, Director-General of UNESCO for the reconstruction of the National and University Library of Bosnia and Herzegovina in Sarajevo*. Paris: UNESCO, 13 April 1994.

26 There were efforts in the UK, France, Turkey, USA among others.

27 *The Relevance of Culture and Cultural Heritage in ADC, Field Report – Bosnia and Herzegovina*, Austrian Development Agency (ADC), January 2007, 12.

28 The ownership of the *Vijećnica* was only resolved within recent years.

29 *The Relevance of Culture and Cultural Heritage in ADC, Field Report – Bosnia and Herzegovina*, Austrian Development Agency (ADC), January 2007, 22.

raised that it might become headquarters of the SDA,[30] the political party of President Alija Izetbegović whose son Bakir then ran the Sarajevo City Development Planning Institute which coordinated reconstruction efforts in the city and who had dealt with the Austrians.[31]

Though Austria tried to mobilize support for completion of the restoration, no credible partners emerged as international donors took fright. It would not be the first time funding had been given to a reconstruction project in Bosnia only for the real owner of the property to turn up brandishing a title deed and the uncertainty over the building's future use was disturbing.[32] The initial stabilization works funded by Austria were completed in 1997 and restoration ground to a halt.

Now only one of the frontrunners in the restoration race was left and on 30 July 1998 the World Bank, UNESCO and the City of Mostar launched a global appeal for the reconstruction of the *Stari Most*. As a World Bank official succinctly put it:

> The Mostar Bridge is really the symbol of all Bosnia … [Donors] could have chosen the library, and they thought about it and decided the bridge was a better object for symbolic value. If you're doing just that one thing, you have to choose one thing.[33]

A World Bank document also observed that while other sites (like the *Vijećnica*) had some merits, 'none conveys as much symbolism as the Old Bridge.'[34] The *Stari Most* was able to offer symbolic value on many fronts to many constituencies: multiculturalism, cultural diversity, reconciliation, peaceful co-existence and religious tolerance among them, all actualized through the highly visible and concrete process of bridge-building.

So it was in Mostar that the international community was to throw most of its resources in heritage reconstruction for Bosnia to the exclusion of much else. The World Bank in partnership with UNESCO, together with the World Monuments Fund (WMF), the Aga Khan Trust for Culture (AKTC), the European Union (EU), the Council of Europe Development Bank, the Organisation of the Islamic Conference (OIC) via its cultural arm IRCICA, and the governments of Turkey, Italy, Netherlands, Croatia, France, Luxembourg, Saudi Arabia and Germany gathered to fund or participate in projects connected with the restoration of the *Stari Most* and its surroundings in the Old Town of Mostar.

The Old Bridge in Mostar

Mostar was the one place in Bosnia-Herzegovina where the international community had already invested deeply in promoting notions of multiculturalism, co-existence and reconciliation since the end of the so-called Muslim-Croat conflict in early 1994, principally through the European Union Administration of Mostar (EUAM).[35] There the rupture between the Bosniak/Muslim and Croat/Catholic communities appeared (at least at first glance) figuratively clear cut: a divided city split by the Neretva River into the 'Muslim' east bank and the 'Croat' west bank. Thus the reconstruction of the *Stari Most* offered an

30 *Stranka Demokratske Akcije* = Party of Democratic Action.

31 Barry, E. 1999. How the Vijecnica was lost: All the sympathy in the world hasn't rebuilt the Sarajevo Library, *Metropolis*, June 1999.

32 Personal knowledge and see Barry, E. 1999; and Iglesias Kuntz, L. 2002a Whose library is it?, *The New Courier*, UNESCO, May 2002, 17.

33 Quoted in Barry, E, 1999.

34 *Project Appraisal Document*, World Bank, Report No. 19115-BiH, 14 July 1999, 3.

35 See *Introduction* in this volume for an overview of the so-called Muslim-Croat War.

obvious symbol for the story the international community wished to tell about reconnecting communities. The city's Serbs, who had formed almost 19 per cent of Mostar's population in 1991 and who had almost entirely fled in 1992, were left out of this carefully constructed binary pairing.

The Mostar bridge project was also where the idea of the reconstruction of cultural heritage playing an active role in the process of 'reconciliation' between the formerly warring ethno-national groups came actively into play – the most conspicuous example of such a linkage by external actors in Bosnia. The *Stari Most* was characterized as a 'symbol of peace and reconciliation'[36] and in UNESCO's preliminary discussions with the World Bank its Sarajevo office had been keen to ensure a 'reconciled' approach to reconstruction.[37] Thus, the World Bank's stated objective for the bridge project was 'to improve the climate for reconciliation among the peoples in Bosnia-Herzegovina through recognition and rehabilitation of their common cultural heritage in Mostar,' seeing reconciliation as 'a prerequisite for economic revitalization and social cohesion in Bosnia's post-conflict situation.'[38] Whether this could be achieved through cultural heritage restoration was another matter.

Even so, the road to the eventual announcement of the World Bank/UNESCO *Stari Most* project was far from straightforward.[39] In reality, far from reconnecting Mostar's Muslim and Croat communities, the reconstruction of the Old Bridge would actually physically reconnect two 'Muslim' communities. The east bank of the Neretva and a narrow strip along the west bank in the vicinity of the Old Bridge had been held throughout the war by Bosnian government forces and remained largely populated by Bosniaks (Muslims); the rest of the west bank was held by Bosnian Croat nationalist forces (HVO).[40]

This obvious flaw in the scenario was not lost on UNESCO's Head of Office in Bosnia-Herzegovina, Colin Kaiser, when he met a deputation from the well-connected US organization, Council for America's First Freedom in November 1996.[41] The deputation was on a visit to Mostar with the aim of beginning serious fundraising efforts towards the reconstruction of the *Stari Most*, with (as they saw it) its potent symbolism of rejoining two warring religious groups to their former state of harmony. The UNESCO representative was discouraging, telling the group there was little point becoming involved with the bridge's reconstruction as it did not reconnect Muslims and Catholic Croats at all, but only Muslims with Muslims.[42] The deputation was undeterred. Like most international donors and supporters of the project, they were willing to ignore this inconvenient reality in their enthusiasm for the Old Bridge's symbolism and went back to America to raise support for the project from Richard Holbrooke, broker of the Dayton Peace Agreement, and the US State Department.

36 *UNESCO and World Bank to assist City of Mostar in rebuilding its Old Bridge*, Press Release No. 98-163, Paris: UNESCO, 30 July 1998.

37 Goodey, B. and Desimplaere, W. 2003, 47.

38 *Project Appraisal Document*, World Bank, Report No. 19115-BiH, 14 July 1999, 2.

39 See Centar za Mir 2003. *Old Bridge – Monument of Peace*, Exhibition catalogue, Mostar: Centar za Mir, 2004 version, for a useful chronology of events leading to its reconstruction, available at http://www.centarzamir.org.ba/eng/pro/hrono/hronologija.html.

40 See note 18.

41 See the organization's website at http://www.firstfreedom.org/. The first freedom is religious freedom.

42 Personal knowledge.

Apart from its obvious attractions for international donors, the *Stari Most* had been fortunate in having two passionate and indefatigable ambassadors for its reconstruction: the architect Amir Pašić, winner of an Aga Khan Award for Architecture for his conservation plan for Mostar Old Town in 1986, and Safet Oručević, East Mostar's mayor from 1994 to 2001. A week after the bridge's destruction in 1993 the East Mostar authorities had begun making plans for its reconstruction, and the following year a *Protocol on Co-operation for Reconstruction of the Mostar Bridge* was signed in Ankara on 7 February 1994 by the Turkish Foreign Minister Hikmet Cetin and the Bosnian Minister for Education, Science, Culture and Sports, Enes Duraković.[43] Turkey pledged technical co-operation, including financial assistance, and Cetin indicated that while international funding would be sought for the project, Turkey would foot the entire bill if necessary.[44]

Pašić had conceived the *Mostar 2004* workshops with the primary aim of initiating plans for the reconstruction of the Old Town of Mostar and the *Stari Most* and the formation of an international network to champion their reconstruction.[45] IRCICA, the Istanbul-based cultural arm of the Organisation of the Islamic Conference, had taken on the workshops as part of their programmes to help preserve Bosnia-Herzegovina's culture and heritage.[46] The first workshop attracted the support of organizations which would later be deeply involved in Mostar's reconstruction, such as the World Monuments Fund, the Aga Khan Trust for Culture and UNESCO, along with impressive representation from universities and institutions around the world. The *Mostar 2004* workshops were to have the effect of creating an influential corps of professionals and academics committed to the restoration of Mostar and the *Stari Most,* adding to the critical mass behind international support for the project as they lobbied behind the scenes and wrote scores of articles on the restoration of Mostar and the symbolism of the Old Bridge.[47] At the third *Mostar 2004* Workshop in 1997 – the first to actually be held in Mostar – a galaxy of dignitaries appeared for the announcement of the formation of the *Foundation Stari Mostar for the Reconstruction of the Old Bridge*, among them Bosnian President Alija Izetbegović and former US Secretary of State Cyrus Vance, whose ill-thought plan (with British negotiator David Owen) for the division of Bosnia-Herzegovina into mono-ethnic cantons had helped precipitate Croat-Muslim fighting in 1993 and the subsequent destruction of the Old Bridge.

Oručević wanted the reconstruction of the bridge to involve Serbs, Croats and Muslims, but he wanted it to be an international effort as well. Turkish President Suleyman Demirel had offered $1,000,000 towards the bridge's reconstruction as early as June 1996, but Oručević put the offer on hold in search of a more varied assortment of international donors. Demirel later said that Turkey had been prepared to fund the whole restoration but had been given to understand that 'the whole world' should participate in the project.[48] Some felt if

43 Iglesias Kuntz, L. 2002b Mostar, A bridge to peace, *The New Courier*, UNESCO, May 2002, 16–17; *The Situation of the Cultural and Architectural Heritage and of Educational Institutions in Bosnia and Herzegovina, 144th Session of the General Conference. Provisional Agenda*, Annex, Doc. 144 EX/34, Paris: UNESCO, 19 April 1994

44 Bowers, F. 1994. Turkey and Bosnia Agree to Rebuild Historic Landmark, *Christian Science Monitor*, 9 February.

45 The year 2004 was chosen because Pašić optimistically (and presciently) aimed to meet at the restored *Stari Most* in 2004. Mostar 2004 also had wider goals for urban preservation for all of Bosnia-Herzegovina.

46 IRCICA = Research Center for Islamic History, Art and Culture; see http://www.ircica.org.

47 See for example Dodds, J. 1998a and 1998b.

48 Centar za Mir 2003. *Old Bridge – Monument of Peace*, 21.

Turkey was the only donor, once again the restoration would be characterized as a purely 'Muslim' project.

However others who were to be major donors stood in the wings, like Italy which had already given significant support to Mostar's heritage, including the preparation of one of the first (of several) plans for rehabilitating the Old Town,[49] along with the governments of Turkey, France, the Netherlands and Croatia, waiting for a sign that the project would truly be a credible joint international effort. In autumn 1997 Hungarian Army divers attached to SFOR raised the stones of the bridge from the bed of the Neretva[50] to much publicity and discussions about the World Bank's involvement in the *Stari Most* project began. By spring 1998 the bank had still not committed itself to the project – restoring cultural heritage was not a sphere it usually engaged in.[51] But World Bank participation would give the project the political clout and financial accountability such a major effort required.

The Fourth Donors Conference for Bosnia-Herzegovina in Brussels over 7–8 May 1998 offered an opportunity for intensive lobbying of the World Bank by Bosnia's Council of Ministers, who sent a letter requesting the inclusion of the *Stari Most* project in its 1999 lending programme for the country.[52] At the conference US Special Representative Robert Gelbard urged international donors to come forward to participate in this unique opportunity to reconstruct the *Stari Most*, exhorting them to let the bridge 'serve as a metaphor, spanning the differences between diverse ethnic groups' and unite them through working together with a common purpose.[53] Not long afterwards the World Bank commitment finally arrived and with it other donors fell in behind. In 1999 the Bank launched its *Pilot Cultural Heritage Project*, financed via its International Development Association (IDA).[54] Six years later on 23 July 2004, before an audience of international dignitaries, the new Old Bridge was officially opened to worldwide publicity. One year later the Old Bridge and its environs became Bosnia-Herzegovina's first UNESCO World Heritage Site.[55]

49 See for instance UNESCO 1997. *Mostar: Urban Heritage Map and Rehabilitation Plan of Stari Grad/ Mostar: mappa del patrimonio urbanistico e piano di ricostruzione di Stari Grad; Mapa urbanistickog nasljedja i plan rehabilitacije Starog Grada*. Florence: Angelo Pontecorboli.

50 Though most of the stones were ultimately to prove unusable.

51 Although the World Bank had participated in the restoration of the Walled City of Lahore as part of its Lahore Urban Development project from the early 1980s.

52 Centar za Mir 2003. *Old Bridge – Monument of Peace*, 33 and *Project Appraisal Document*, World Bank, Report No. 19115-BiH, 14 July 1999, 50.

53 Centar za Mir 2003. *Old Bridge – Monument of Peace*, 32; Gelbard, R.S. 1998. *Remarks at the Donors' Conference for Bosnia and Herzegovina*. Ambassador Robert S. Gelbard, Special Representative of the President and Secretary of State for the Implementation of the Dayton Peace Accords, 7 May 1998.

54 The World Bank's official partners were listed as the governments of Italy, the Netherlands, Croatia, and the Council of Europe Development Bank, along with UNESCO, the World Monuments Fund, and the Aga Khan Trust for Culture. Contributions from the European Union and the governments of Turkey and France were all made in-kind or outside the financing arrangements for the project. As noted elsewhere in this volume, other countries also contributed to restoration projects in Mostar, namely Saudi Arabia, Luxembourg, Jordan, Sweden, and the United Arab Emirates (via IRCICA). See: *IDA at Work: Bosnia and Herzegovina: Bridges for Peace*, August 2009, http://go.worldbank.org/N8XCPSSIL0 and *Implementation Completion Report on a credit in the amount of US$4.0 million equivalent to the Bosnia and Herzegovina for a Pilot Cultural Heritage Project*, Report No. 32713, Washington: World Bank, Infrastructure Sector Unit, Europe and Central Asia Regional Office, 22 June 2005, 7.

55 The Old Bridge Area of the Old City of Mostar was designated a World Heritage Site on 15 July 2005, a decision that was not without controversy.

Fig. 5.1 **(a)** The new Old Bridge at Mostar nears completion in August 2002. Laying the stones of the span. © Aga Khan Award for Architecture / Tihomir Rozić

Fig. 5.1 **(b)** The new Old Bridge at Mostar was officially reopened on 23 July 2004. In 2005 tourists once again line the bridge. © Tina Wik

The Vijećnica

Meanwhile, 20 years after the beginning of the conflict, the fate of other iconic sites was mixed. By the beginning of 2013 the restoration of the *Vijećnica* was yet to be completed. After the end of the Austrian-supported efforts in 1997, the European Commission (EC) contributed a second tranche of funding in 1999, though work did not begin until 2002, then stopped again in 2004. That the *Vijećnica* was still in a semi-ruinous state ten years after the end of the war and the bickering over ownership and future use continued was now beginning to look not only like a failure of the Bosnian authorities, but of the international community as well, and the Council of Europe's Representative in Bosnia-Herzegovina

Fig. 5.2 **(a)** The Vijećnica, Sarajevo, still boarded up and unrestored in 2009. © Selmer van Alten. **(b)** The Vijećnica in 2011, encased in scaffolding and sheathed with plastic sheeting as reconstruction gathers pace. © Jennifer Boyer

Sonja Moser-Starrach tried to revive international interest in finishing the reconstruction.[56] But no major works resulted and there was considerable public dissatisfaction about how the future use of the building was being decided with little consultation and the exclusion of National Library officials from the process. The Hungarian government provided a small donation, as did national libraries and city governments from across Europe.

Then in 2007 the Spanish government offered €1,000,000 towards completing restoration of the structure's exterior. Still, more was needed and progress was painfully slow. In 2008 the *Vijećnica* appeared on the World Monuments Fund list of the world's 100 most endangered historic sites in the hope of triggering further international support and was named one of Bosnia-Herzegovina's three Consolidated Projects in the EC/COE Ljubljana Process for the rehabilitation of cultural heritage in SE Europe for which major fundraising initiatives were to be undertaken.[57] September 2009 saw the announcement of the third phase of the *Vijećnica*'s reconstruction[58] with funding from Bosnian sources, led by the City of Sarajevo, more than matching the Spanish grant.[59] Now a plan for its use had been hammered out: the building would be shared jointly between the City of Sarajevo (as its Town Hall), the National Library and a museum.[60] An open day was held in mid-September 2010 to present the proposed plans for the building to Sarajevo's citizens. But completion was still some way off, pending receipt of the now-agreed European Union finance coming from Bosnia-Herzegovina's pre-accession funding and the EU's commitments to the Ljubljana Process.[61]

The following year Mayor Alija Behmen declared his conviction that the restoration would be completed by 2014, in time for Sarajevo's hopefully successful application to be a European Capital of Culture – almost 20 years after the end of the war.[62] In April 2011 a tender for the final phase of reconstruction of the *Vijećnica* was announced by the EU Delegation to Bosnia-Herzegovina. The tender was cancelled some months later due to irregularities in the procedure, further delaying completion of the building's reconstruction.[63] March 2012 saw the promised EU pre-accession funding finally coming through, with EU Special Representative to Bosnia-Herzegovina Peter Sørensen committing the EU to the continuing rehabilitation of Bosnia's cultural heritage, commenting that as a way of 'overcoming the wounds of the past, cultural heritage

56 Sonja Moser-Starrach (sometimes known as Sonja Stiegelbauer) founded the society Vijećnica Sarajevo (Rathaus Sarajewo) which acted as a vehicle for fundraising for the project. See: Wiederaufbau der Nationalbibliothek Sarajewo, ÖNB Newsletter. Vienna: Österreichische Nationalbibliothek, No. 3, September 2004, 14.

57 For more on the Ljubljana Process and Consolidated Projects see http://www.coe.int/t/dg4/cultureheritage/cooperation/see/irppsaah/ljubljanaprocess_EN.asp and http://ec.europa.eu/culture/archive/culture_program/pdf/CE_Concept_LJUBLJANA_E.pdf

58 Potpisan ugovor o rekonstrukciji Vijećnice, *Sarajevo-x.com* [online], 24 September 2009. http://www.sarajevo-x.com/.

59 The amount from local sources totalled €1,113,784. See *IPA National Programme 2010 – Bosnia and Herzegovina, Fiche 3 'Cultural Heritage,'* European Commission Enlargement, Instrument for Pre-Accession. http://ec.europa.eu/enlargement/pdf/bosnia_and_herzegovina/ipa/2010/part-1/pf_03_ipa_2010_cultural_heritage_final_eud_en.pdf.

60 The focus of the museum was still undecided in 2012. Suggestions were either to show the history of the Vijećnica throughout its history or to focus on the building during the 1992–1995 war.

61 See *IPA National Programme 2010 – Bosnia and Herzegovina, Fiche 3 'Cultural Heritage.'*

62 *'Sarajevo City Hall Revisited' Exhibition Opens in Vijecnica,'* press release, Sarajevo: FIRMA Project, 14 July 2011.

63 See *Reconstruction works on the Sarajevo and Novi Grad City Halls/Documents/Cancellation*, European Commission, Development and Cooperation – EuropeAid, 19 July 2011.

Fig. 5.3 **(a)** The ruined atrium of the Vijećnica in 1995 showing the damage caused by the intense heat of the shelling of 1992. © Leif Pareli. **(b)** The atrium of the Vijećnica in June 2011 with its pseudo-Moorish arches restored. © Jennifer Boyer

protection and promotion has the potential to be of real support to and a catalyst for better interethnic and intercultural understanding in the future.'[64]

Other Iconic Sites

The Serbian Orthodox monastery at Žitomislić was successfully restored between 2002–2005 by the Republika Srpska Institute for Protection of the Cultural, Historical and Natural Heritage (despite the monastery being located in the Federation) after its restoration had been stalled for many years by the obstruction of the RS and Serbian Orthodox Church authorities (see below). The case of Počitelj was taken up by University of York's Post-war Reconstruction and Development Unit (PRDU) which successfully nominated it to the World Monuments Fund first list of 100 Most Endangered Sites in 1996,[65] receiving funding to prepare an integrated plan for its reconstruction and revitalization.[66] The PRDU plan (uniquely) incorporated cultural heritage restoration

64 *EU and BiH sign Protocol for the reconstruction of the City Hall in Sarajevo and City Hall in Novi Grad*, press release, Sarajevo: Delegation of the European Union to Bosnia and Herzegovina (DELBIH), 7 March 2012. The restored *Vijećnica* reopened on 9 May 2014.

65 World Monuments Fund, WMF Watch, http://www.wmf.org.uk/wmf_watch/.

66 Barakat, S. and Wilson, C. 1997. *The Revitalisation of the Historic Settlement of Počitelj*, York: PRDU, University of York, York. See also Revitalisation Research Award, *Revival*, Iss. 6,

into proposals for the socio-economic reconstruction and development of the town that encouraged the return of the pre-war Bosniak (Muslim) residents, while taking into consideration the needs of the displaced Bosnian Croat families who had been housed there during the conflict. Implementation of at least some aspects of the project by the Federation authorities began in 2001.

More problematically, the reconstruction of two of Bosnia-Herzegovina's most important classical Ottoman mosques, the Aladža in Foča and the Ferhadija in Banja Luka (both in Republika Srpska) remained a long drawn-out process. By December 2012 not a single stone had been laid for the Aladža Mosque. Yet in 2000 the Aladža had been one of four religious sites selected for simultaneous restoration and reconstruction in an ambitious project initiated by Jacques Paul Klein, head of the United Nations Mission in Bosnia-Herzegovina (UNMIBH) in partnership with the Interreligious Council of Bosnia-Herzegovina.[67] The sites, 'of symbolic importance to the community,' whose reconstruction was meant to assist the 'revival of cultural and religious pluralism'[68] were the lightly-damaged Serbian Orthodox Cathedral (1868) in Sarajevo;[69] the ancient Sephardic Jewish cemetery (earliest graves 1630), also in Sarajevo, which had been used as a Bosnian Serb artillery emplacement and had been heavily mined; the Aladža Mosque; and the modern Roman Catholic church of Sveti Juraj Mučenik[70] (1976) in Derventa – these last two both located in Republika Srpska and both completely destroyed.

The Orthodox cathedral was soon repaired with funding from the Greek government and the phased restoration of the Old Jewish cemetery began with support for restoring the chapel from American sources.[71] But the two projects in Republika Srpska languished. The reconstruction of Sveti Juraj was not completed until 2009 in the face of continued obstruction and attacks.[72] The Aladža Mosque fared even worse and its case will be looked at in more detail below.

The reconstruction of the Ferhadija Mosque in Banja Luka, after many setbacks (some of which will be described below), did not even begin until 2005 with preparatory work on fragments of the structure excavated from the municipal landfill site at Ramići and retrieved from a water-filled gravel pit at Karanovac where they had been dumped in 1993.[73] Physical rebuilding of the mosque above foundation level started in 2008 and reconstruction

June 1996, York: Post-war Reconstruction & Development Unit (PRDU), University of York, Institute of Advance Architectural Studies.

67 The Inter-religious Council of Bosnia-Herzegovina is an organization formed in 1997 through the commitments of the heads of Bosnia's four main religious communities. For more information see http://www.mrv.ba/en/about-us.

68 *Simultaneous reconstruction of religious sites in BiH*, press release, UNMIBH, 23 May 2000. For more on this project see Helen Walasek, *Destruction of the Cultural Heritage* elsewhere in this volume.

69 The Serbian Orthodox Cathedral in Sarajevo was damaged by Bosnian Serb Army shelling.

70 Sveti Juraj Mučenik = St George the Martyr.

71 The American Joint Distribution Committee and the US Commission for the Preservation of America's Heritage Abroad contributed to the project.

72 At least the fabric of the church of Sveti Juraj was completed at that time.

73 See the project's website at http://www.ferhadija.ba/. See also Riedlmayer, A. 2008. Banja Luka's Ferhadija mosque rises again, *Bosnian Institute news* [online], 3 March 2008; Hadžihasanović-Katana, M. 2012. The reappearance of Ferhadija Mosque in Banja Luka, in *Protecting Cultural Heritage in Times of Conflict: Contributions from the participants of the International Course on First Aid to Cultural Heritage in Times of Conflict*, edited by S. Lambert and C. Rockwell. Rome: ICCROM, 66–71.

Fig. 5.4 (a) Fragments of the Ferhadija Mosque found at the landfill site at Ramići where they were dumped in 1993, awaiting removal in 2005. © Tina Wik

Fig. 5.4 (b) The foundations of the Ferhadija Mosque in Banja Luka in October 2005 before reconstruction had begun. The Islamic Community building can be seen top left. © Tina Wik

continued to move slowly forward. On 9 July 2013 the first *adhan* called from the minaret of the Ferhadija for 21 years marked the start of Ramadan.

Like the Aladža Mosque, the reconstruction of the Ferhadija attracted little significant international funding in the early stages of rebuilding. This was despite the efforts of the UK-based organization Soul of Europe led by the Anglican priest Donald Reeves, who from 2001 lobbied many potential donors for the project in Europe and the Islamic world.[74] Like other external actors, the Soul of Europe was concerned 'to promote the

74 The Soul of Europe http://www.soulofeurope.org/.

Fig. 5.5 The reconstructed Ferhadija Mosque, Banja Luka, nearly completed in June 2013. © Medžlis IZ Banja Luka

Ferhadija as a sign of unity and reconciliation in Banja Luka'[75] and, not surprisingly given its background, as a focus for interreligious dialogue.[76] The reconstruction of the Ferhadija eventually attracted support from an increasing number of donors, among them Prince Charles,[77] the US Embassy in Sarajevo from the *US Ambassadors Fund for Cultural Preservation* (AFCP),[78] in 2007 from the Republika Srpska authorities themselves, and 300,000 KM from the Ministry for Human Rights and Refugees in 2009 from the budget for sustainable return.

In September 2010 the *United Nations Development Programme (*UNDP) office in Sarajevo announced the reconstruction of three religious sites in Bosnia-Herzegovina in partnership with UNESCO, UNICEF and Bosnian government departments and religious institutions as part of its programme *Culture for Development: Improving Cultural Understanding in Bosnia and Herzegovina* supported by the Millennium Development Goals Achievement Fund (MDGF).[79] Among the chosen sites ('powerful symbols of

75 *Soul of Europe Newsletter*, December 2004. The project also set up a Banja Luka Civic Forum.

76 See: *Soul of Europe Newsletter*, December 2004; Reeves, D. 2005. *The Soul of Europe: Imagination – Confiscated.* Paper delivered at the conference *The Church in the Centre of the City*, Cardiff University 12–15 September 2005.

77 Prince Charles made a donation towards the reconstruction of the Ferhadija Mosque in 2005; personal communication to the author from the Press Officer to TRH The Prince of Wales and The Duchess of Cornwall, 29 September 2011.

78 The donation was made in 2006, *The US Ambassadors Fund* for *Cultural Preservation, Annual Report 06/07*, United States Department of State, 36.

79 *Reconstruction of Three Religious Monuments – Symbols of Multiculturalism Supported by MDGF Culture and Development Programme*, press release, Sarajevo: UNDP, 20 September 2010. Spain gave US$8,000,000 towards the programme.

Fig. 5.6 The neo-Baroque Serbian Orthodox Cathedral of the Holy Trinity (Saborna Crkva) sits above East Mostar in 1981, with the square Ottoman clock tower (before 1636) and the minaret of the sixteenth-century Nesuh-aga Vučijaković Mosque rising below. © William Remsen, 1981 / Aga Khan Visual Archive

multiculturalism and diversity') was the Ferhadija Mosque, along with the Franciscan Monastery in Plehan (church 1902, monastery 1932), both in Republika Srpska, and the Serbian Orthodox Cathedral (*Saborna Crkva*) of the Holy Trinity (*Crkve Svete Trojice* 1863) in Mostar. Yet this appeared to be the recycling of a plan first raised at UNESCO in 2005 for support for the reconstruction of the same three sites,[80] and reaffirmed in March 2008 by its then Director-General Koïchiro Matsuura at Mostar in a speech given on the site of the Orthodox Cathedral.[81]

The massive Serbian Baroque Orthodox cathedral, built towards the end of Ottoman rule (Sultan Abdul Aziz himself had contributed 100,000 groschen towards its construction), had once dominated Mostar's skyline. But from the time of its destruction in June 1992 until 2010 the cathedral was quite literally a heap of rubble. On 2 January 2004 the *Saborna Crkva* was designated a National Monument under Annex 8 of the Dayton Peace Agreement.

80 The plan was announced by Bosnian Presidency Chairman Ivo Miro Jović when he attended the UNESCO General Assembly in Paris in October 2005; see UNESCO to finance reconstruction of Ferhadija mosque, ONASA News Agency, 10 October 2005. The plan was reaffirmed in December 2005 by UNESCO's then Deputy Director Mounir Bouchenaki on a visit to Bosnia; see: FENA 2005. Zamjenik direktora UNESCO-a u Žitomisliću: Najavljena obnova tri poznata vjerska objekta, Oslobođenje, No. 21.158, 20 December, 6.

81 *Address by Mr Koïchiro Matsuura, Director-General of UNESCO, for the Promotion of the Reconstruction of the 3 monuments: The Ferhadija Mosque in Banja Luka, the Orthodox Cathedral of Mostar and the Franciscan Monastery in Plehan*. Mostar: UNESCO 26 March 2008.

Notwithstanding its location in the one city in Bosnia-Herzegovina that had been the focus of most intense international participation in restoration of the cultural heritage, implemented with the overtly declared aim of promoting reconciliation between the formerly warring parties and the return of the country's historic religious pluralism and multicultural identity, the Orthodox cathedral had never acquired the iconic status of other structures and had failed – apparently – to attract the major funding needed for its reconstruction in the early years following the end of the war.

However, the delays in reconstructing the *Saborna Crvka* probably owed more to politics and the relationship of of the Serbian Orthodox Church vis-à-vis the Federation than other factors. Before the war the cathedral was seat of the Bishopric [Eparchy] of Zahum-Herzegovina, which like the Bishopric of Zvornik-Tuzla had been moved in 1992 to Bosnian Serb-controlled territory, in this case Trebinje (now in Republika Srpska) and where it (and the bishop) still remained.

The Orthodox bishop's official residence in Mostar, the Austro-Hungarian period neo-Baroque Metropolitan Palace (*Vladičin dvor* 1910), had been badly damaged during the war and survived only as an empty shell. But the structure was restored by the Project Coordination Unit (PCU) for the Pilot Cultural Heritage Project and returned to the Serbian Orthodox Church for completion by 2005.[82] The reconstruction of the episcopal residence was part of a particular sub-project of the World Bank's restoration efforts in Mostar. Along with the Napredak Cultural Centre (owned by the Croatian Cultural Society and to be used as a cultural centre) and the Vakuf Palace (*Vakufski dvor*, owned by the Islamic Community and also intended for use as a cultural centre), the Metropolitan Palace had been one of the 21 significant historic structures in Mostar chosen by the Aga Khan Trust for Culture and World Monuments Fund to make detailed studies of and offer to prospective donors for rehabilitation in 1999.[83] Of the 21 structures, these three attracted the interest of the World Bank.

While the World Bank's justification for its Pilot Cultural Heritage Project in Mostar decried the tendency to reduce Bosnia's cultural heritage to *ethnic heritage* 'with each of the peoples of Bosnia and Herzegovina using culture – along with language and religion – to reinforce their separate identity, instead of valuing their common heritage while respecting their differences,'[84] paradoxically it was also insistent that, apart from the Old Bridge and its towers, structures restored and whose restoration it would co-finance should include 'monuments that are valuable to each of the peoples of Bosnia and Herzegovina.'[85] Thus these three ethnically-identified buildings were specifically chosen by the World Bank in its plan to rehabilitate as a group 'at least three additional historic monuments that are important to each of the three major peoples'.[86] This element of the World Bank-driven restoration in Mostar was characterized by all concerned as the 'three

82 Ćorić, V. 2005. Obnovljen Vladičin Dvor U Mostaru, *Nezavisne Novine*, 28 April.

83 AKTC/WMF 2004. *Conservation and Revitalisation of Historic Mostar*, Historic Cities Support Programme. Geneva: Aga Khan Trust for Culture/World Monuments Fund, 53.

84 World Bank 1999, 3.

85 World Bank 1999, 4.

86 World Bank 1999. *Project Appraisal Document on a Proposed Credit in the Amount of SDR 3.0 Million (US$4.0 million Equivalent to Bosnia and Herzegovina for a Pilot Cultural Heritage Project.* Project Appraisal Document, Report No. 19115-BiH, Infrastructure Sector Unit, Europe and Central Asia Regional Office, 14 July, 4 and 21.

buildings, three nations' project ('tri objekta tri naroda'), including by the PCU[87] and (often ironically) in the media.[88]

For despite the World Bank's stated desire that the overall project should 'counteract the tendency to reduce cultural heritage to ethnic specificity' through offering the 'opportunity for the three peoples of Bosnia and Herzegovina to cooperate in the rehabilitation of a single site which they all agree is a true example of the multi-cultural heritage of the country,'[89] there could not be a more clear-cut example of international donors' need for equivalence when choosing historic structures to restore. Here, rather than selecting buildings which symbolized a common heritage, the World Bank project instead actively chose to focus on and reinforce the separate ethno-national/ethno-religious identities of Muslims, Croats and Serbs rather than the shared Bosnian identity it claimed to aspire to promote.

As Makas has observed of the 'three buildings, three nations' project: ' … by singling out sites and associating them with distinct groups within the Bosnian population, this kind of "three nation" initiative can also be understood as highlighting their separation,' going on to say that ' … by restoring a project for each of these groups, the international community is indicating that there is no collective Bosnian identity or nation, and no Bosnian collective culture, and that each group must be addressed individually.'[90]

As for the *Saborna Crkva*, there had been a number of attempts to start the reconstruction process. In 2000, inspired by UN Special Representative Jacques Paul Klein's initiative to simultaneously restore four religious sites, the Serbian Orthodox Council (*Srpska Pravoslavna Opština*) of Mostar launched a proposal to reconstruct the cathedral – the final element, as they saw it, in a restored multiconfessional Mostar, affirmed by the minarets and bell towers of Orthodox and Catholic churches that rose above the city.[91] Following Prince Charles' visit to the city for the opening of the rebuilt *Stari Most* in 2004, there was a flurry of talk from the British ambassador Ian Cliff in February 2005 that the prince would contribute to the cathedral's reconstruction[92] (as it was similarly reported he would contribute to the restoration of the Ferhadija Mosque).[93] Later, in June 2005, it was

87 *Prezentacija projekta rekonstrukcije Starog mosta u Mostaru od strane Jedinice za koordinaciju projekta izgradnje Starog mosta u Mostaru*, http://kons.gov.ba/, 1 July 2003.

88 Posle projekta obnove Starog mosta u Mostaru – nastupice projekat 'Tri naroda', *Dnevni Avaz*, 22 September 2003.

89 World Bank 1999, 4.

90 Makas, E. 2007. *Representing Competing Identities: Building and Rebuilding in Postwar Mostar*. Phd thesis [online], 324–9. Available at: http://uncc.academia.edu/EmilyMakas.

91 Ratkušić, E. 2000. Multietničnosti Mostara nema bez munara i zvonik, *Ljiljan*, 4–11 December, 32.

92 See: Outgoing British ambassador to BiH visits Mostar, *ONASA News Agency*, 14 February 2005; and Ćorić, V. 2005. Princ Čarls želi pomoći obnovu Saborne Crkve, *Nezavisne Novine*, No. 2338, 15 February 2005, 9. However, at time of writing no donation to the reconstruction of the *Saborna Crkva* had been given by the Prince Charles; personal communication to the author from the Press Officer to TRH The Prince of Wales and The Duchess of Cornwall, 29 September 2011.

93 The donation to the Ferhadija reconstruction was mentioned on the occasion of the visit of Reis-ul-Ulema Mustafa Ćerić to London to mark the tenth anniversary of the Srebrenica Massacre at the Imperial War Museum in 2005: Burazerović, A. 2005. Princ Čarls obećao ef. Ceriću pomoć u obnovi Ferhadije, *Dnevni Avaz*, No. 3497, 1 July, 2; and as stated by British Ambassador Ian Cliff in 2005 and his successor Matthew Rycroft on a visit to the Ferhadija in 2007, see: Islam is a religion that respects all peoples: UK Ambassador Cliff, *ONASA News Agency*, Sarajevo, 24 February 2005; and V.S. 2007. Ambasador Rajkroft obišao gradilište Princ Čarls spreman pomoći obnovu Ferhadija džamije, *Dnevni Avaz*, No. 4391, 12 December, 7. Prince Charles did make a donation towards the reconstruction of the Ferhadija Mosque in 2005; personal communication to the author from the Press Officer to TRH The Prince of Wales and The Duchess of Cornwall, 29 September 2011.

Fig. 5.7 (a) The ruins of the Serbian Orthodox Cathedral, Mostar in October 2005. A sign announces the building's reconstruction, to be implemented by the government of the Federation of Bosnia-Herzegovina, but little work was to be carried out for several years. © Yuri Stoyanov. **(b)** Progress on the reconstruction of the Serbian Orthodox Cathedral in Mostar in September 2013. © Richard Carlton

announced that the first funds for the reconstruction of the cathedral had been donated by the government of the Federation to the amount of 80,000 KM;[94] a sign was erected at the site announcing the project.

But it was not until spring 2008 that there began to be visible progress on the cathedral project when the Eparchy of Zahum-Herzegovina formally launched a fundraising campaign for its reconstruction with the support of UNESCO and Director-General Koïchiro Matsuura, began preparatory works on the site and held a donors' conference in May attended by a diverse group of from all of Bosnia's communities and religious groups where over US $4,000,000 was pledged, including substantial amounts from governments of both entities.[95] Donations for the cathedral's reconstruction came as well from Muslim citizens of Mostar.[96] Two years later, after extensive preparatory works and the usual delays in receiving the funds pledged, the foundation stone of the reborn *Saborna Crkva* was laid and sanctified on 23 May 2010.[97] In April 2011 the promised funding from the Federation government finally came through – notably, this was only after Grigorije, the Orthodox Bishop of Zahum-Herzegovina, finally moved his episcopal seat back to Mostar from Trebinje in January that year.[98]

Contentious Sites

Least known and least publicized, it was in this domain, where returned refugees and displaced people attempted to re-establish destroyed communities in difficult, often dangerous settings, that the most widespread restoration took place. Yet here there was virtually no involvement from the international community in terms of project participation, support or funding, while contacts with international organizations implementing the peace process were frequently negative. It was here that restoration came up most powerfully against the politics of ethno-national exclusivism that had driven the war and where the implementation of Annexes 6, 7 and 8 of the Dayton Agreement were to be tested.

Here restoration was not a process where there were choices to be made about which group's heritage to restore: there was no equivalence of destruction. This was restoration by the victims of ethnic cleansing in territory now dominated and controlled by perpetrators of the crimes against them. And though this was the domain where the need for heritage restoration was most widespread, it was also the domain where the international community largely feared to tread (particularly in settings dominated by powerful antagonistic ethno-national elites, such as almost all of Republika Srpska and places like Stolac in the Federation), while what contact there was with local heritage institutions and, at times,

94 N.H. 2005. Prva sredstva za radove obezbijedila Vlada FBiH, *Dnevni Avaz*, 20 June.

95 *International Religious Freedom Report 2008 – Bosnia and Herzegovina*, US Department of State, 19 September 2008.

96 Jelin-Dizdar, T. 2008. UNESCO pomaže obnovu Crkve Svete Trojice, *Radio Slobodna Evropa* [online], 26 March.

97 See: Trojičindan u Mostaru, *www.saborna-crkva-mostar.com*, 23 May 2010. See the same website for pictures of the reconstruction. Also see: Jelin-Dizdar, T. 2010. Počela obnova Saborne crkve u Mostaru, *Radio Slobodna Evropa*, 22 October.

98 Potpredsjednik Federacije BiH Mirsad Kebo uručio je danas vladiki zahumsko-hercegovačkom Grigoriju odluku o izdvajanju 1,5 miliona maraka iz budžeta federalne Vlade za obnovu Saborne crkve u Mostaru, *24sata.info*, 23 April 2011, http://www.24sata.info/.

the international officials appointed to implement and oversee the peace process, was frequently unhelpful or even hostile.

The reconstruction of destroyed historic structures (especially mosques) frequently met with determined, often violent, resistance and active obstruction from local authorities, many of whose officials had been active participants in ethnic cleansing. Thus, here restoration was characterized by a quest for justice and human rights – including the right to equality in the public space through the reconstruction of physical markers of identity. It was the restoration of communities ignored for the most part by the international media spotlight that focused almost entirely on Sarajevo and Mostar.

Yet restoration of cultural and religious monuments was to be a key element in promoting the return of refugees and displaced persons to the places from which they had been violently expelled, an essential ingredient in creating an environment for return under the terms of the Dayton Agreement. This is largely a story of Muslim/ Bosniak return to Republika Srpska and Croat-dominated municipalities in the Federation and the battle to rebuild destroyed mosques, though not entirely. The restoration of the Serbian Orthodox monasteries at Zavala and Žitomislić, both in Croat-dominated municipalities in Herzegovina, for instance, were crucial in encouraging Bosnian Serb refugees and displaced people to return to these communities – a challenging task in the face of the political forces operating in post-war Bosnia, where nationalist Bosnian Serb leaders in Republika Srpska actively discouraged the return of refugees and displaced persons to municipalities in the Federation in an attempt to make the effects of ethnic cleansing permanent.[99]

This *refusal to restore* shows how clearly restoration was perceived to be bound up with the return of refugees and displaced persons – an issue many of those responsible for implementing Dayton failed to grasp for some time. When the question of restoring the Serbian Orthodox Monastery of Žitomislić, located not in Republika Srpska but in a Croat-dominated canton in the Federation, was broached in 1998 with a representative from Republika Srpska, they declared that there were no longer any Serbs in Žitomislić and none wanted to return there.[100] It was not until the intervention of the OHR's Senior Advisor on Interreligious and Interethnic issues, Russian diplomat Anwar Azimov, broke the deadlock that agreement on the monastery's reconstruction was finally reached.[101] Displaced Bosnian Serbs did then begin to return to live in the vicinity, citing the reconstruction of the monastery as a crucial factor in their willingness to return.[102]

Restoration of cultural heritage in this domain, then, was literally about 'restoring' communities, of re-establishing their presence on the landscape, reclaiming their right to exist with visible markers that declared their historic roots in a particular locale. It was,

99 For insights on the hostile attitudes of the Republika Srpska authorities regarding the return of Bosnian Serbs to the Federation see ICG 1998. *Return of Bosnian Serb Displaced Persons to Drvar, Bosansko Grahovo and Glamoc*, International Crisis Group, ICG Bosnia Project, Report No. 29, 19 January 1998.

100 See Barrentine, E. *Reconstruction of the Zitomislici Monastery*, quoting Amra Hadžimuhamedović, a member of the Commission to Preserve National Monuments of Bosnia-Herzegovina: 'In 1998, in a meeting with UNESCO representatives, Ms. Hadzimuhamedovic proposed the plan to a representative of the Republika Srpska, who responded that she did not see the need to rebuild the Orthodox monastery as there were no longer any Serbs living in Zitomislici and they did not want to return.'

101 Personal communication from Valery Perry.

102 Personal communication from Valery Perry. See also Tina Wik in this volume regarding the return of displaced Bosnian Serbs to Zavala.

Fig. 5.8 The mosque in Kozaruša near Prijedor in October 2000, the first mosque to be rebuilt in Republika Srpska. The mosque had reopened to the faithful on 26 August 2000. © Richard Carlton

too, about 'restoring' individuals' sense of belonging and identity through the physical restoration of their destroyed heritage, the restoration of a sense of home.[103] This domain is also a narrative of community mobilization where 'absent' citizens played as dynamic a role in the return and reconstruction process as the 'present' citizens who had actually physically returned to their pre-war homes.

The first mosque to be rebuilt in Republika Srpska was at the little village of Kozaruša near Prijedor. It reopened to the faithful on 26 August 2000, with another ceremony held in the presence of high-ranking religious leaders and members of the international community on 30 October 2000. The reconstruction of Kozaruša Mosque was seen by some in the international community as a marker in the movement for return and that the rebuilding of destroyed religious buildings would gather pace. However, this was not to be.

The area of Kozarac where Kozaruša lay was a special case in the return of refugees and displaced people. Kozarac had been the scene of some of the worst ethnic cleansing of the war and the location of its most notorious detention camps at Omarska, Keraterm and Trnopolje. Yet there was a strong well-organized movement for return to the district and a significant number of its former residents now lived not far away in Sanski Most (in the Federation), though many lived much further afield. One of a cluster of factors that enabled the area's diaspora to work towards restoring their physical community was to be the online community created through the medium of a number of websites, most important and long-lasting of which was *www.kozarac.ba*.

103 Hadžimuhamedović, A. 2008. *What happens when the Homeland Cultural Landscapes are Destroyed: A Case from Bosnia*, Sixth International Conference on Science and Technology in Archaeology and Conservation, WATCH, Rome, 8–14 December 2008.

Virtual Bosnia

The crucial role played by the scores of town and village websites like *www.kozarac.ba* created by refugees and internally displaced people during and in the years after the war in constructing a virtual Bosnia which enabled globally-dispersed residents to maintain a vital cohesion as a community has been a little noted phenomenon. With internet usage becoming widespread just as war broke out and large swathes of the country were ethnically cleansed, these websites were to become quite literally sites of memory, heritage and identity, key mechanisms in recreating the localities from which communities had been expelled, keeping alive what had been lost through destruction and absence.

As the process of rewriting the past by those who had carried out the ethnic cleansing began, the 'true' histories of towns and villages were memorialized, documented and made accessible via the internet. Typically sites included galleries of visual images (incorporating pre-war photographs, old postcards, and other illustrative material) of now-destroyed structures and townscapes, people and events from both the recent and more distant past, as well as sections on history and heritage (*baština*) which held transcriptions of written material like scholarly articles, newly-created texts and even, in the exemplary *www.Focaci.org*, such commonplace yet revealing records as pre-war telephone directories.[104] It could be argued that through the medium of these websites communities (particularly its younger members) gained far greater direct access to information on their history and heritage than they had before the war.

After the war, as parts of expelled communities began to return (or in some cases, *before* they began to return), these websites became channels for many 'restorative' and community-sustaining actions, including fundraising for the rebuilding of local mosques and churches. As the years passed and the circumstances of their webmasters and others involved changed, many of such sites have now disappeared from the web, a loss of these irreplaceable and largely unrecorded sites of community memory.[105] However, among those still active in 2013 were *Focaci.org* (www.focaci.org) and *Bileća Online* (www.bileca.org), for former residents of Foča and Bileća respectively (both municipalities now in Republika Srpska where extensive ethnic cleansing had taken place) and the more broadly based *Bošnjaći.Net* (www.bosnjaci.net).

From 2005 *Bošnjaći.Net* promoted a fundraising campaign during Ramadan towards rebuilding devastated or demolished mosques in eastern Herzegovina. The campaign focused on eastern Herzegovina (Republika Srpska) where it was felt the situation of Bosniaks was the worst in the country. Thus projects began to rebuild and restore mosques in Nevesinje, Odžak, Gacko, Bileća, Ljubinje, Trebinje, Gornje Grančarevo, Lastva, Pridvorci and Skočigrm. A text from the campaign explicitly called on Bosniaks not to

---

104 For the telephone directories see http://www.focaci.org/index.php?option=com_content&view=category&id=81&Itemid=116. The directories show the high proportion of Muslims residents in Foča before the war.

105 What may be the only scholarly article on the subject by András Riedlmayer and Stephen Naron came to the author's attention after this section was written. Describing these websites in similar terms as above, it comments as well on the problems of archiving and preserving such ephemeral internet material. Riedlmayer, A. and Naron, S. 2009. From Yizkor Books to Weblogs: Genocide, Grassroots Documentation, and New Technologies, in J.A. Bastian and B. Alexander (eds), *Community Archives: The Shaping of Memory*. London: Facet Publishing, 151–63. 'A town called Kozarac.ba,' a presentation on *www.kozarac.ba* given by Lee Bryant at *Reboot9*, Copenhagen, 2007 can be found at http://www.slideshare.net/leebryant/reboot9-kozaracba-presentation.

Fig. 5.9 The little mosque (mesdžid) at Skočigrm, almost at the Montenegrin border, in August 2012. Although no Bosniaks had returned to live in the village, former residents and their descendants visited to tend the graveyard and it was decided to restore the mosque. © Richard Carlton

abandon their 'cultural and religious traces in the region' and urged them: 'We must be persistent in preserving our heritage and identity' to ensure that the fate of the civilization of Islamic Spain was not repeated.[106]

In the tiny hamlet of Skočigrm, close to the Montenegrin border, it was decided to proceed with reconstruction of the small mosque (or *mesdžid*), despite the absence of any Muslim returnees. In 2009 the campaign page for Skočigrm displayed a list of donations ranging from €10 to US$1,000 from donors dispersed as far afield as New York, San Jose, Toronto, Singapore, Indonesia, Sweden, Denmark, Germany, Belgium and the Netherlands, as well as throughout Bosnia and the region, the majority originally once resident or with family connections to the localities in question.[107]

In August 2010 *Bošnjaći.Net* ran an article listing the achievements of the five years of their fundraising campaign, noting the official openings of the mesdžids at Gornje Garnčarevo and Skočigrm (on 31 July and 7 August 2010 respectively) and the progress on reconstructing mosques in Nevesinje, Odžak, Gacko and Bileća. The rebuilding of the small mosque at Lastva had been funded by a vakuf established by Adamir Jerković whose grandmother had come from Lastva; the opening ceremony there had been held on 17 July 2010.[108]

106 The original text reads: 'Bošnjački je narod protjeran iz dobrog dijela Gacka, Trebinja, Bileće, Ljubinja i Nevesinja, međutim mi ne smijemo odustati od naših kulturnih i vjerskih tragova na tim prostorima. Moramo biti uporni u očuvanju naše baštine i identiteta, kako nam se ne bi ponovila islamska Španija;' from: Ramazanska akcija prikupljanja novca za obnovu džamije u Pridvorcima kod Trebinja, *Bošnjaci.Net*, 14 August 2010.

107 Ramazanska akcija za džamiju u Skočigrmu kod Trebinja, *Bošnjaci.Net*, 21 August 2009.

108 Otvorenje džamije u Lastvi kod Trebinja, *www.rijaset.ba*, 19 July 2010. Adamir Jerković is a former journalist and advisor to President Alija Izetbegović, author and at time of writing director of the Archive of the Federation of Bosnia-Herzegovina.

Fig. 5.10 **(a)** All that remained of the Avdić Mosque at Plana, near Bileća, after it was dynamited in 1993. © Institute for the Protection of the Cultural, Historical and Natural Heritage of Bosnia-Herzegovina. **(b)** The rebuilt but unfinished Avdić Mosque, Plana, with its square campanile style minaret in 2010. The mosque was formally reopened on 3 September 2011. © Richard Carlton

But the website called again for further donations to enable completion of the mosque in Pridvorci.[109] By October 5,369 KM had been collected via the *Bošnjaći.Net* campaign from donors in the United States, Sweden, Singapore, the Netherlands, Slovenia, Canada, France, Belgium, Morocco, Italy, Luxembourg and Croatia.[110] The Pridvorci mosque was finally officially reopened on 23 July 2011, its rebuilding mostly financed by donations, including the results of a concerted fundraising campaign among Bosnian refugees in Denmark.[111]

Campaigns ran on *Bileća Online* for rebuilding the Careva or Obradovića Mosque (1895) at the centre of Bileća[112] and the historic Avdić Mosque in Plana (1617) with its rare 'campanile' minaret.[113] The imposing dressed-stone Careva Mosque had been constructed at the same time as the nearby Serbian Orthodox church of St Sava (1896), both structures receiving contributions from the Austro-Hungarian provincial administration of the time. The Careva Mosque had been dynamited during the 1992–1995 war but survived as a shell, while the Avdić Mosque had been completely destroyed and existed now only as a heap of rubble.

An article by Hasan Eminović from May 2006 published on *Bileća Online* described how with the former residents of Bileća now displaced round the world, the internet would be used to help the reconstruction of the Obradovića Mosque, giving them information on how to donate, how the rebuilding was progressing and when gatherings of Bosniaks from Bileća would be held.[114] By December 2008 the campaign for the Avdić Mosque had raised over US$10,000 from donors in amounts from $20 to over $2,000.[115] Special fundraising campaigns were carried out by the *Udruženje građana Bileće* (*Association of Citizens of Bileća*) in Chicago, but many donors were not originally from the area. However reconstruction of both the Avdić and Careva mosques proceeded in fits and starts as funding became available. In November 2007 the chief imam of Trebinje and Bileća, Husein ef. Hodžić, announced a donation of US $3,500 from the Egyptian Embassy towards the rebuilding of the Careva Mosque. Ef. Hodžić stressed how important this donation was – the first from any embassy or international organization towards restoring a mosque in eastern Herzegovina; none so far, he commented, had even stepped across the entity border (into Republika Srpska).[116]

It was not until 3 September 2011 that the Careva/Obradovića Mosque was officially reopened amid tight security in the still hostile setting of Bileća, where at the time only a dozen Bosniaks had returned. The ceremony was attended not only by representatives and members of the Islamic and Bosniak communities, but also by a small group of Serbs from Bileća, Gacko and Trebinje, representatives from the Serbian Orthodox church and the mayor of Bileća. In the end funds for reconstructing the mosque came not solely from individual donors (among which were members of the Orthodox church), and it was noted how the fact that there were so few Bosniaks returnees had had an impact on fundraising.

109 Ramazanska akcija prikupljanja novca za obnovu džamije u Pridvorcima kod Trebinja, *Bošnjaci.Net*, 14 August 2010.

110 U Akciji Bošnjaci.Net Sakupljeno 5,639 KM, *Bošnjaci.Net*, 19 October 2010.

111 U Pridvorcima otvorena obnovljena džamija, *Trebinje Danas* [online], 23 July 2011; Obavijest iz Medžlisa IZ Trebinje, *www.medzlis-trebinje.ba*, 15 April 2011.

112 Also alternatively known as the Gradska, Carska or Imperial Mosque.

113 For pictures of the Avdić Mosque in its destroyed state see: Pripreme za obnovu Avdića džamije u Planoj, *Bileća Online* (*www.bileca.org*), 20 May 2007.

114 Eminović, H. 2006. Pred obnovu Obradovića džamije u Bileći: Čovjek i džamija, *Bileća Online*. Originally published on *Preporod*, 1 May 2006.

115 Izvještaj o akciji obnove Avdića džamije, *Bileća Online* (*www.bileca.org*), 3 December 2008.

116 Pomoć za obnovu Carske džamije u Bileći, *www.bileca.org*, 3 November 2007.

Thus, significant amounts for the reconstruction had come from the two entity governments (with the Federation Ministry of Displaced Persons and Refugees contributing), from the *Udruženje građana Bileće*, from other non-governmental and citizens' associations, the Mostar Muftijstvo and Sarajevo Centre Municipality.[117] The rebuilding of the minaret was funded by a call for donations after Friday prayers at mosques across Bosnia-Herzegovina.[118] While the main structure of the Avdić Mosque in Plana was rebuilt by 2010, there were not sufficient funds to complete the work and it was not formally reopened until 11 August 2013.[119]

Implementing Dayton

But the restoration of religious buildings in these contentious settings (almost always mosques) were to face a range of degrees of obstruction from local authorities including: extreme delays and other obstacles in issuing building permits;[120] refusal to order the removal of structures built on the sites of destroyed monuments (whether temporary or permanent and some of which had been constructed after the end of the war) or to order the cessation of enterprises carried out on such sites (like markets and car parking);[121] declarations that urban plans had been revised (which included the removal of mosque sites from such plans) and that permitted uses for the site had changed;[122] assertions that the land on which the structure had stood was socially owned and the pre-war owner of the now non-existent building no longer had the right of use;[123] and, finally, the supposed need for archaeological investigations of mosque sites, always on the pretext of testing (usually spurious) claims that a Christian church had existed on the site in pre-Ottoman times. Most contentious and most opposed were the reconstruction of mosques in town and city centres. Yet until 2001 the international community's approach to the issue displayed a lack of attention and a failure to formulate a systematic policy for dealing with this crucial aspect of the return process.

Thus, in this domain, restoration was characterized by a search for justice and human rights, leading some returnee communities and the Islamic Community[124] to take active recourse to the provisions of Annexes 6 and 8 of the Dayton Agreement to ensure that their destroyed and damaged cultural and religious property was reconstructed. These groups made use of the Human Rights Chamber established under Annex 6 and sought the

117 Trebaju nam primjeri živuće tolerancije a ne slatkorječive i puke retorike, *www.medzlis-trebinje.ba*, 15 September 2011.

118 Medzlis Trebinje 2011.

119 Svečano otvaranje Avdića džamije 11.08.2013, *Bileća Online* (*www.bileca.org*), 27 July 2013.

120 As for example for the attempts to reconstruct the Ferhadija Mosque, Banja Luka; the Atik Mosque, Bijeljina; the Kizlaragina Mosque, Mirkonjić Grad.

121 As for example on the site of the Kizlaragina Mosque, Mirkonjić Grad.

122 As in the cases of the Ferhadija Mosque, Banja Luka and the Atik Mosque, Bijeljina.

123 The social ownership obtained from the era of the post-WW2 federal socialist state of Yugoslavia of which the Republic of Bosnia-Herzegovina was a part. On the Atik Mosque in Bijeljina see: *The Islamic Community in Bosnia and Herzegovina v. The Republika Srpska (Bijeljina Mosques), Decision on the Admissibility and Merits (Delivered on 6 December 2000)*, Human Rights Chamber (HRC) Case No. CH/99/2656.

124 The official body and legal entity representing members of the Islamic faith in Bosnia-Herzegovina.

involvement and protection of the OHR, SFOR and the IPTF (International Police Task Force) in carrying out restoration at contentious sites.

These aims were explicitly stated by the active community association of citizens expelled from Stolac and other former residents of the town, the *Society for the Renewal of Civil Trust in the Stolac Municipality*, whose programme sought to systematically restore the devastated Ottoman core of the town in the face of determined obstruction, threats and violence from the hardline Bosnian Croat local administration and its supporters.[125] Restoration of the cultural heritage in Stolac, beginning with the focal structure of the sixteenth-century Čaršija (or Careva) Mosque,[126] was linked unambiguously to testing implementation of the Dayton Peace Agreement (including Annex 8), the return process, justice for returnees and a framework for victims of persecution to exercise their human rights.[127]

On 3 March 1997 the Islamic Community submitted applications to Banja Luka municipality for building permits to reconstruct seven historic mosques in the city, including the Ferhadija Mosque. More than a year later it had yet to receive a response to its application, despite a request from the then High Representative Carlos Westendorp to the city's mayor, Đorđe Umičević, in April 1998 that he allow the reconstruction to proceed. In a rambling open letter Umičević replied that the mosque was 'a monument of the cruel Turkish occupation' and that 'the reconstruction of Ferhadija would be perceived by the Serbian people as the blackest humiliation.'[128]

This refusal was made in the context of a succession of offences against the Ferhadija site and obstructions to the rebuilding of the mosque since the end of the war.[129] Over 16–17 October 1996 the Banja Luka municipal authorities began levelling the site, desecrating graves, destroying a building and uprooting trees. Taking place on the eve of the first meeting of the Republika Srpska Assembly on 19 October, which newly-elected Muslim deputies were due to attend along with high-level international officials, IFOR was quick to react, and with the IPTF established a cordon around the site to prevent any further actions at the site.[130] Both the OHR and the UN Special Representative issued statements condemning the actions of the RS authorities.[131]

More obstruction was to follow. Less than a month later the Islamic Community discovered that the Ferhadija and its adjoining structures did not appear on an official

125 See Amra Hadžimuhamedović in this volume on restoration in Stolac. Also see Mulać, E. (ed.) 2001b. *Written memory against apartheid in the Municipality of Stolac*, The Revival of Stolac. Sarajevo: Association [Society] for the Renewal of Civil Trust in the Stolac Municipality, 8 July 2001. Society for the Renewal of Civil Trust in the Stolac Municipality = *Udruženje za obnovu građanskog povjerenja općini Stolac; Udruženje* is sometimes translated as 'Association.'

126 Sometimes also referred to as the Čarska, Emperor or Imperial Mosque; its reconstruction was completed in 2003.

127 See the opening declaration in Mulać, E. (ed.) 2001 *Crimes in Stolac Municipality (1992–1996)*, The Revival of Stolac, Sarajevo: Association [Society] for the Renewal of Civil Trust in the Stolac Municipality, reprint with addenda.

128 For text of Umičević's letter see: Banja Luka Mayor Umicevic on the Ferhadija mosque, *HAPSBURG Discussion Group*, 9 May 1998.

129 For a good press account of this period see: Husarska, A. 1998. Without A Prayer? Rebuilding a Muslim Temple Is a Test of Bosnia's Peace, *The Washington Post*, 10 May, C01

130 See IFOR AFSOUTH Transcript of Press Briefings, 17–19 October 1996.

131 *Destruction of property on the site of the Mosque Ferhadija in Banja Luka*, OHR Press Releases. Banja Luka: Office of the High Representative, 17 October 1996; and IFOR AFSOUTH Transcript of Press Briefing, 18 October 1996.

property map issued by Banja Luka municipality on 9 November 1996, despite their presence on a similar map issued on 28 May 1996. Thus on 4 December 1996 the Islamic Community filed a case against Republika Srpska at the Human Rights Chamber (HRC) with regard to its 15 mosques in Banja Luka, alleging violation of the freedom of religion and discrimination against members of the Islamic Community and of its property rights in connection with the continuing desecration of the mosque sites and the refusal of the authorities to issue building permits for their reconstruction.[132]

The Chamber found in favour of the Islamic Community in the Banja Luka case in July 1999, ordering the Republika Srpska authorities to take immediate steps to allow the Islamic Community to erect and maintain enclosures around the mosque sites; to refrain from the construction of buildings or objects of any nature on the sites; to refrain from destroying or removing any object remaining on the sites; and to swiftly grant the applicant the necessary permits for reconstruction of seven of the destroyed mosques on their original sites and to comply with the orders by September 1999.[133]

Yet, although the OHR went on to make further statements about the Ferhadija Mosque[134] and while High Representative Wolfgang Petritsch removed Mayor Umičević from office in November 1999, specifically citing his actions and obstruction in relation to the reconstruction of the Ferhadija (including his refusal to implement the HRC decision of July 1999),[135] the Republika Srpska authorities continued to block the rebuilding of the mosque. It was not until the end of March 2001 that the Banja Luka authorities finally issued the building permit that would enable the rebuilding of the Ferhadija Mosque to proceed. Thus, in the absence of any other remedy, up to 2001 the Islamic Community was to bring more cases to the HRC against Republika Srpska with regard to the treatment of its religious facilities and obstructions to their reconstruction at Bijeljina, Zvornik and Mrkonjić Grad.[136] However, as the Islamic Community discovered, it was one thing for the HRC to make a ruling ordering municipal authorities to issue building permits for the reconstruction of various mosques, it was another to make them comply.[137]

The Islamic Community was also to use the Human Rights Chamber process to challenge the UNESCO-run *Commission to Preserve National Monuments* (the Commission) of the immediate post-war period. Thus far the Commission had *not* been accepting petitions for designating properties as National Monuments as directed in Article V of Annex 8, but rather compiling (in secrecy) a list of potential national monuments. On 25 July 1997 the Islamic Community formally requested that the Commission designate the sites of the Ferhadija and

132 *The Islamic Community in Bosnia and Herzegovina v. The Republika Srpska (Banja Luka Mosques), Decision on the Admissibility and Merits, (Delivered on 11 June 1999)*, Human Rights Chamber (HRC), Case No. CH/96/29, Human Rights Chamber for Bosnia and Herzegovina. See also Dakin, B. 2002. Islamic Community in Bosnia and Herzegovina v. The Republika Srpska: Human rights in a multi-ethnic Bosnia, *Harvard Human Rights Journal*, Vol. 15 (Spring 2002), 245–66. See also in this volume: Helen Walasek, *Cultural Heritage, the Search for Justice and Human Rights*.

133 HRC *Banja Luka Mosques*, Case No. CH/96/29, VIII Conclusions.

134 *OHR, OSCE and UNMIBH condemn destruction of mosques in Banja Luka seven years ago*, OHR Press Releases, Sarajevo: OHR / OSCE / UNMIBH, 5 May 2000.

135 *Decision removing Mr. Djordje Umicevic from his position of President of the Municipal Assembly of Banja Luka*, Removals and Suspensions from Office, Sarajevo: Office of the High Representative, 29 November 1999.

136 For the HRC cases regarding mosques in Bijeljina (Case No. CH/99/2656), Zvornik (Case No. CH/01/7701) and Mirkonjić Grad (Case No. CH/98/1062).

137 M.Z. 2001. RS Authority cancels permits for construction of mosques in Bijeljina, *Dnevni Avaz*, 18 August, Bosnian Institute, News and Analysis [online], 9 October 2001.

four other historic mosques in Banja Luka as National Monuments[138] and in October the HRC brought the Commission into the legal proceedings with a request for information.[139] Now the question of the sites of monuments which had been totally razed and which appeared to be without protection under Annex 8 was highlighted and the Commission forced to take a stance. In May 1998 the Commission decided that though totally destroyed monuments could not be listed, their sites *could* be protected, and at its June session added the site of the Ferhadija Mosque to the Provisional List, declaring it a protected site. At the same time the Commission wrote to entity authorities, urging them to rigorously protect sites of historic cultural and religious monuments which had been totally destroyed.[140]

Yet this testing of Dayton implementation by the victims of ethnic cleansing through restoration of their built cultural heritage was often regarded as *provocation* not only by the perpetrators of human rights abuses, but on occasion by those whose duty it was to oversee the peace agreement and support the return process. In 1999 in relation to the Islamic Community's efforts to rebuild the historic Atik Mosque on its city centre site, François Perez, the OHR representative in Bijeljina, stated that the Islamic Community was 'too extreme in their demands,' and that, 'Maybe in time a mosque could be built on the periphery of town.'[141]

Three years later in December 2001 at a workshop on Annex 8 implementation a high-ranking OHR official, Senior Deputy High Representative Ambassador Matthias Sonn, while acknowledging the annex's 'critical importance to peace implementation ... as the issue of national monuments has too often been used to obstruct return and community redevelopment,' also accused 'overzealous minority community leaders' of using Annex 8 to 'push the peace process' and 'make their own political statements,'[142] an accusation apparently directed at Bosniaks and at the Islamic Community, in particular, and its use of legal remedies such as the Human Rights Chamber.

There is no doubt that the rebuilding of mosques and churches became intensely politicized and exploited by religious leaders and politicians from all of Bosnia's three main ethno-national groups. While statements on mosque rebuilding from Serb/Orthodox and Croat/Catholic leaders are mentioned elsewhere in this article, the pronouncements of the leader of the Islamic Community, the Reis-ul-Ulema Mustafa Cerić on the issue also provoked outrage and controversy.[143]

---

138 HRC *Banja Luka Mosques*, Case No. CH/96/29, No. 45.

139 HRC *Banja Luka Mosques*, Case No. CH/96/29, No. 6.

140 HRC *Banja Luka Mosques*, Case No. CH/96/29, Nos 7, 45 and 77–8. The letter asked entity authorities 'to undertake rigorous and complete protection of the sites of cultural and religious monuments of historical value that were totally destroyed. This protection requires that no constructions or use of any other sort encroach on the perimetre of the destroyed monument, or threaten its integrity' and recommended the planting of hedges around such sites. For the Commission's discussions of the subject see Goodey, B. and Desimpelaere, W. 2003, Appendix 2.

141 *Bosnia and Hercegovina, unfinished business: The return of refugees and displaced persons to Bijeljina. Abuses against minorities after the war*, Human Rights Watch, May 2000, Vol. 12, No. 7 (D). Interview with François Perez, Bijeljina, 28 September 1999.

142 For Matthias Sonn's remarks at an ECMI Workshop on Annex 8 implementation see: Perry, V. 2002a *ECMI Civil Society Project in Bosnia and Herzegovina: The Role of Annex 8 in Bridging Communities, Promoting Reconciliation, and Developing a Unique Identity in Bosnia and Herzegovina, The Law Centre, Sarajevo University, 15 December 2001*, Flensburg: European Centre for Minority Issues, Report No. 25, March 2002, 3.

143 Particularly notorious was Cerić's interview with Radio Televizija Republike Srpske (RTRS) on 8 May 2001 following the riots in Banja Luka and Trebinje; see: Transcript of an interview given

Nevertheless, the destruction of cultural and religious property during the conflict had been a fundamental aspect of campaigns of ethnic cleansing and so part of a programme with clear political aims. Many of the actions the international community took in respect of restoring cultural and religious property after the conflict was over were overtly political, as well. Thus to charge returnee and religious communities who sought recourse to the justice and human rights allowed them through the institutions the international community had established under the terms of the Dayton Peace Agreement with 'pushing the peace process' seems disingenuous, particularly as the OHR was about to embark in the same direction itself by its decisions relating to Annex 8.

As discussed elsewhere in this volume, 2001 was a crucial year with regard to the international community's commitment to implementing Annex 8.[144] The disruption of cornerstone-laying ceremonies at the sites of the Osman-paša Mosque in Trebinje and the Ferhadija Mosque in Banja Luka in May by large, violent and well-planned anti-Muslim riots which the Republika Srpska authorities made little effort to control were tremendous embarrassments for the OHR, which realized how it had neglected implementation of Annex 8 and brought to the fore how the issue of National Monuments was being used to obstruct refugee return and community redevelopment,[145] especially as the situation in Stolac regarding the rebuilding of the Čaršija Mosque was now, too, threatening to come to a head.[146]

The UNESCO-led Commission to Preserve National Monuments was widely condemned at the time for its lack of transparency, lack of consultation and politicization of the designation process.[147] The international community, through the OHR, had acquiesced in the Commission's decision not to receive petitions to designate cultural property as National Monuments (as mandated by the terms of the Dayton Peace Agreement) but instead to compile a Provisional List of National Monuments based on the (probably biased) recommendations provided by local heritage preservation institutes and other bodies.

The list was marred by conspicuous omissions and duplications.[148] In Foča, for instance, while the sites of the Aladža and Careva Mosques and the remains of the Mehmed-paša Kukavica Mosque *were* on the list, no other Islamic/Ottoman monuments were included, despite the survival of the substantial shell of the historic Musluk (Atik Ali-paša) Mosque or the town's surviving and then intact eighteenth-century Ottoman *han* and clock tower (*sahat kula*). Not a single site of a destroyed historic mosque was listed for Bijeljina, Prijedor or Nevesinje, although here the totally destroyed Catholic church was included.

---

to RT RS on 8th May by the Reis-ul-Ulema, efendi Mustafa Ceric, *OHR Afternoon Media Reports*, 24 May 2001.

144 See Valery Perry in this volume.

145 See Perry, V. 2002a, 3–4 for OHR officials Matthias Sonn and James McNaught on the issue of Annex 8 and return. The riot in Trebinje took place on 5 May, the one in Banja Luka on 7 May.

146 See Amra Hadžimuhamedović in this volume on the events of June 2001.

147 For a full account of Annex 8 and the Commission to Preserve National Monuments see Valery Perry in this volume.

148 See more on the Provisional List on the website of the Commission to Preserve National Monuments http://www.kons.gov.ba/main.php?id_struct=108&lang=4. For the complete Provisional List adopted on 14 June 2000 (in Bosnian only): *Privremena lista nacionalnih spomenika Bosne i Hercegovine, Kompletna* at the website of the Commission to Preserve National Monuments, http://www.kons.gov.ba/main.php?id_struct=165&lang=1.

Fig. 5.11 The Osman-paša Mosque (originally 1726) in Trebinje, under reconstruction in May 2004. The mosque had been totally destroyed in 1993. Rioting by Bosnian Serb nationalists disrupted the laying of the cornerstone on 5 May 2001, but the rebuilt mosque was eventually reopened on 15 July 2005. © Richard Carlton

In an attempt to defend its secretive practices, unattributed (and implausible) claims emanated from the Commission that it was unwilling to publicize the list for fear that ultra-nationalists might attack the monuments of other ethnic groups.[149] A final blow to the Commission's effectiveness was the appointment of its Chair Léon Pressouyre in 1998 to head the newly-formed UNESCO-led International Committee of Experts for the reconstruction of the *Stari Most*, whereupon, it was noted, the body lost its focus.[150] Though the mandate of the UNESCO-managed Commission ostensibly ran until 31 March 2001, its last meeting in fact took place on 14 June 2000.[151] Thus in 2001 a large void opened in Annex 8 implementation at a critical point in the return process.

But after the crises of mid-2001 over the reconstruction of mosques in Banja Luka and Trebinje, the OHR began to use its extensive powers to speed up the rehabilitation process

149 See Valery Perry in this volume and *Is Dayton Failing?: Bosnia Four Years After the Peace Agreement*. International Crisis Group Report, 28 October 1999.

150 Professor Pressouyre was appointed to head the Mostar Bridge Committee in October 1998. See Goodey, B. and Desimpelaere, W. 2003, 21 on 'fading' leadership of Commission.

151 Authority over the Commission was transferred automatically to the Bosnian government in December 2000. A final meeting was to be held in Trebinje on 16 September but never took place; see Goodey, B. and Desimpelaere, W. 2003, Appendix 2, 36–7. The UNESCO-led Commission was formally established on 31 March 1996 and its five year term ended on 31 March 2001.

and bring an end to obstruction by ensuring that Republika Srpska and the Federation harmonized their legislation in line with Annex 8. High Representative Wolfgang Petritsch's decisions of 8 February 2002 regarding this harmonization explicitly linked Annex 8 implementation with refugee return and creating an environment for return, and clearly defined what was meant by 'rehabilitation,' applying it unambiguously to National Monuments that were totally destroyed.[152] At the same time, the OHR requested the Bosnian government to re-establish the Commission to Preserve National Monuments; the newly reconstituted Commission with five new professionally-qualified members held its first meeting in March 2002.[153] Thereafter, with legislation in place, the OHR was more systematic and assertive in using the terms of Annex 8 to circumvent obstructive local authorities, though such obstructiveness was far from ended.

Archaeology and Restoration

As mentioned above, a particular tactic used as a pretext to halt the reconstruction of mosques temporarily or permanently at a number of important sites where structures had been totally destroyed was the alleged need for archaeological investigations, inevitably proposed in order to test claims that a Christian church had existed on the site before the Ottoman conquest. This ploy was used in relation to the sites of the Atik Mosque in Bijeljina, the Ferhadija Mosque in Banja Luka and the Osman-paša Mosque in Trebinje (all in RS) and the Čaršija Mosque, Stolac and the Šišman Ibrahim-paša Mosque at Počitelj (in the Federation).[154] It was not unknown for those calling for excavations to cite the work of Bosniak historians to back up their claims. Thus in the case of the Čaršija Mosque in Stolac, speculation by the respected Bosniak historian Hivzija Hasandedić in his book *Muslimanska Baština Istočne Hercegovine* regarding the possible location of a church in the town before the Ottoman period was misrepresented by those seeking to delay reconstruction of the mosque (see Hadžimuhamedović in this volume for more on this subject).[155]

The 'morally righteous' obstructionist line was epitomized by the Roman Catholic Bishop of Mostar, Ratko Perić, who wrote in respect of the rebuilding of the Čaršija Mosque in Stolac that it would be 'a fresh act of irreligion, injustice and immorality' to restore a mosque on a former Christian sacred site, introducing the novel concept of

152 For the decision with regard to Republika Srpska see: *Decision imposing the RS Law on Implementation of Decisions of the Commission to Preserve National Monuments Established under Annex 8 of the Dayton Peace Agreement.* With regard to the Federation see: *Decision amending the Federation Law on Preservation of Assets Declared National Monuments of BiH under Decisions of the Commission for Protection of National Monuments.*

153 The first members of the reconstituted Commission were Amra Hadžimuhamedović, Dubravko Lovrenović and Ljiljana Ševo, and two international members, Zeynep Ahunbay (Turkey) and Tina Wik (Sweden).

154 For Amra Hadžimuhamedović on use of archaeology as a tactic to prevent mosques being rebuilt, see Perry, V. 2002a, 12–13. See Hadžimuhamedović also in this volume.

155 Hasandedić, H. 1990. *Muslimanska Baština u Istočnoj Hercegovini.* Sarajevo: El-Kalem, 9. Hasandedić referenced an article published in the periodical *Srđ* in 1902 and which had long been regarded as conjecture and impossible to verify; see the website of the Commission to Preserve National Monuments, Čaršija mosque and Čaršija in Stolac, the Architectural Ensemble, http://kons.gov.ba/main.php?id_struct=50&lang=4&action=view&id=1824. Hasandedić publicly condemned this misuse of his text.

Fig. 5.12 One of the twenty-two pre-Ottoman medieval Christian stećak-shaped tombstones found at the site of the Atik Mosque in Bijeljina in December 2002 when rebuilding of the mosque began. This one has an inscription in Cyrillic. © Commission to Preserve National Monuments of Bosnia & Herzegovina

forcible restoration.[156] In the case of the Atik Mosque, even the Bijeljina rabbi Jozef Atijas was outspoken in his support for the obstructionists, declaring that building a structure of one religion on top of the remains of structures and a cemetery sacred to another religion wouldn't be allowed in Israel.[157]

All these claims were shown to be based on spurious or non-existent historical 'evidence.' Yet they were given a veneer of legitimacy by the active involvement of archaeologists and historians who provided an academic basis for the claims (sometimes propagated through the medium of 'expert' seminars on the subject)[158] which, as Hadžimuhamedović has noted, represented the (criminal) act of destruction as an ultimately beneficial act, by allowing an exploration of the site and thus a deeper exploration of the country's past.[159] This persuasive, apparently rigorously 'scientific' and 'professional,' approach could even co-opt otherwise well-intentioned heritage professionals (and, in the case of the Ferhadija Mosque, the UNESCO representative) into declaring their support for such excavations.[160]

However, with its new tougher stance on Annex 8 implementation, the OHR refused to permit any such archaeological investigations to take place, with the exception of a strictly time-limited excavation at the site of the Atik Mosque in Bijeljina.[161] Here, 22 pre-Ottoman

---

156 Mehmedbašić, S. 2001b. *Second Letter to Bishop Perić, 22 September 2001*, Society for Renewal of Civil Trust in the Stolac Municipality, 5.

157 Đogić, G. 2003. Na početku bi crkva na kraju – knjiga, *Glas Javnosti* [online], 1 April.

158 Such as the academic gathering held at Stolac on 5 October 2001 when a roundtable *Stolac Region in 16th Century, Religious and Political Situation* was held and *Povijest Hrvatskog Počitelja* (*The History of Croatian Počitelj*) held at Počitelj in 1996.

159 Hadžimuhamedović, A. 2005a Cultural memory in Stolac – destruction and reconstruction, in *Human Rights and Destruction of Cultural Memory: The Stolac Case*, edited by A. Hadžimuhamedović. Oslo: Norwegian Helsinki Committee, 278.

160 See Perry, V. 2002a, 15 and Perry, V. 2002b *ECMI Civil Society Project in Bosnia and Herzegovina: Annex 8 of the General Framework Agreement for Peace and the Role of the Preservation of National Monuments in Successful Peace Implementation, Banja Luka, 12 April 2002*, Flensburg: European Centre for Minority Issues, Report No. 29, July 2002, 3 and 4–5 for the discussion on 'multiple layered' sites. For the UNESCO Representative's support for archaeological excavations at the Ferhadija site see Goodey, B. and Desimpelaere, W. 2003, 48.

161 *Bijeljina Atik Mosque Reconstruction*, Coalition Press Information Center, Press Conference, Sarajevo, 11 March 2003, 11.00 Hours, OHR Spokesperson Mario Brkic.

medieval Christian tombstones (*stećci*), as well as a number of unassociated human remains, *had* actually been uncovered in December 2002 while excavating the foundations of the completely destroyed mosque in preparation for its reconstruction, leading to Serb claims that these were the remains of a medieval church and cemetery, the skeletons the bones of long dead Christians.

The discovery of the medieval tombstones with inscriptions in Cyrillic and the human remains had galvanized the local Serb media and opposition to the rebuilding of the mosque, leading to candlelight vigils for the souls of the Christians over whose graves the mosque had allegedly been constructed.[162] A coalition of six local Serb NGOs called for a 'multi-national and multi-religious facility, as a symbol of co-existence' to be constructed on the site, rather than the mosque, threatening to resort to 'unconstitutional measures and activities' if their demands were not met.[163]

This was to be the last of a series of delays and obstructions to the rebuilding of the Atik Mosque, which despite a Human Rights Chamber ruling of 8 March 2001 that the permit for the reconstruction of the mosque should be issued within three months, was not allowed to begin until December 2002. Among the most vociferous objectors to the reconstruction was the Serbian Orthodox Bishop of Zvornik-Tuzla, Vasilije Kačavenda, who in May 1992 had moved the seat of his diocese from Bosnian government-held Tuzla to Bijeljina (in Bosnian Serb-held territory), constructing his new episcopal palace on the site of expropriated and demolished Bosniak (Muslim) houses in full view of the mosque site.[164] In 2009 Kačavenda was still publicly asserting that the Atik Mosque had been built on the site of an Orthodox church, despite evidence to the contrary.[165]

However, the OHR consistently and explicitly cited the HRC ruling regarding reconstruction of the mosque and its protection under Annex 8 as the basis for its decisions.[166] Those who had been prevented from carrying out excavations at the site of the Čaršija Mosque in Stolac were roused to accuse the OHR of double standards in sanctioning the excavations at the Atik Mosque.[167]

The excavations, carried out by a team of archaeologists from the Zemaljski Muzej (the National Museum of Bosnia-Herzegovina) in Sarajevo, Republika Srpska, and from Serbia, conclusively demonstrated that there had been no previous church on the site and that the medieval tombstones had been brought to the site solely to be used as foundation stones when the Atik Mosque was originally constructed in the sixteenth-century. The obstructionists, however, declared that the remains of a church would surely have been found if only they had continued digging.[168] The human remains found were those of Ottoman-period Muslims buried in the cemetery that had once existed beside the mosque.

---

162 The human remains found at the site were actually those of Muslims, buried in a long-disused graveyard.

163 See Bosnian Serb groups warn against rebuilding mosque destroyed during war, *ONASA News Agency* [online], 20 January 2003.

164 Hadžić, H. 2002. Bijeljina: A bastion of apartheid, *Dani*, Sarajevo, 6 December, reprinted in *Bosnia Report*, New Series No: 32–4, December 2002 – July 2003, London: The Bosnian Institute.

165 See Hadžić, H. 2002. and Musli, E. 2009. Vladika Kačavenda širi laži o Atik džamiji, *Dnevni Avaz*, 2 February.

166 See *Bijeljina Atik Mosque Reconstruction*, Coalition Press Information Center, 11 March 2003. The Islamic Community had applied for the site to be designated a National Monument, thus it was afforded protection under Annex 8.

167 See *Construction of Stolac mosque*, OHR, BiH Media Round-up, 9 January 2003, with a summary of an article in *Večernji List*.

168 Đogić, G. 2003.

After a presentation of the results of the excavations on 17 March 2003, reconstruction of the Atik Mosque resumed.

Restoration in Foča

The fate of the Ottoman/Islamic heritage of Foča described in this chapter and elsewhere in this book[169] could be an exemplar of the problems facing the preservation of the rare survivals of such heritage in Republika Srpska in the first post-conflict decade (and beyond), the difficulties faced by willing local actors, and the lack of action on the part of the international community in carrying out or enforcing preservation activities there.

As we have seen, terrible crimes were committed against Foča's Muslim population and almost its entire Ottoman heritage, including its mosques, had been obliterated. For many years after the end of the war Foča was to remain 'a closed, dark place'[170] where hardline Serb nationalists, many responsible for or complicit with war crimes, continued to hold sway with apparent impunity, despite the presence of IFOR/SFOR troops, the IPTF, ECMM and many organizations of the international community such as the OSCE and IRC. Due to its non-cooperation with the International Criminal Tribunal for the former Yugoslavia (ICTY), Foča was among a number of municipalities in Republika Srpska sanctioned under the terms of the so-called Lautenberg Amendment from receiving US government aid and assistance.[171]

While the Aladža Mosque was the most exceptional monument destroyed in Foča, it was perhaps the town's Čaršija, or market area, on the other side of the Ćehotina River which bore the brunt of the widespread destruction. From the medieval period Foča was an important hub on the trade route running north from Dubrovnik. But it was during the Ottoman era that the town flourished as an economic and artisanal centre with its greatest expansion during the sixteenth and seventeenth centuries. The urban plan of the Čaršija, Foča's historic core, with its low shops built along narrow streets had largely been preserved up to 1992. Here stood the sixteenth-century Careva/Sultan Bajezid Valija II and Šehova/ Kadi Osman-efendijina Mosques and the still existing buildings of the *vakuf* (endowment) of Bosnia's mid-eighteenth-century governor Mehmed-paša Kukavica: the mosque and *medresa*, *han* (caravanserai) and *sahat kula* (clock tower) – these last two among the few surviving of their type in Bosnia-Herzegovina. All these monuments and Foča's urban ensemble as a whole had been under state protection before the 1992–1995 war.

The year 2000 saw the start of war crimes trials at the ICTY for atrocities committed in Foča, including on 30 October the prosecution of Milorad Krnojelac, former commandant of the notorious prison KP Dom (Kazneno-Popravni Dom) where Muslim prisoners had been mistreated and killed. Shortly afterwards, on 25 November 2000, after the failure of previous attempts which had been repeatedly blocked by local officials, the first organized return of displaced and refugee Muslims arrived in urban Foča.

169 See *Destruction of the Cultural Heritage in Bosnia-Herzegovina* and *Topography of Destruction* in this volume.

170 The title of the Human Rights Watch report on Foča of July 1998: Human Rights Watch (HRW). July 1998: *Bosnia and Hercegovina "A Closed, Dark Place": Past and Present Human Rights Abuses in Foca*, Vol. 10, No. 6 (D).

171 The so-called Lautenberg Amendment to the US Foreign Operations, Export Financing, And Related Programs Appropriations Act, 2000.

How inextricably the destruction of cultural and religious heritage was bound up with other wartime atrocities could not have been made more evident when on 18 September 2000, in the presence of ICTY investigators, during the excavation of a mass grave near KP Dom where the Federation Commission on Missing Persons was searching for the remains of victims who had been killed there, large recognizable fragments of the Aladža Mosque were uncovered.[172] Among the rubble was an electricity meter later identified as belonging to the mosque. On 18 January 2001 an ICTY criminal investigator testified at length on the discovery and identification of the Aladža Mosque finds at the mass grave site during the Krnojelac trial.[173]

This first discovery of Aladža Mosque fragments in 2000 appears to have received relatively little attention and far more publicity was given to the discovery of further fragments of the mosque in the summer of 2004, again during an excavation at another suspected mass grave site in Foča on the banks of the Drina near the Iron Bridge.[174] However, in 2000 the Sarajevo-based Institute for the Protection of the Cultural, Historical and Natural Heritage of Bosnia-Herzegovina had been notified of the discovery and some of the decorated fragments were taken to Visoko where they were deposited in the town cemetery and later transferred to the Bosniak Institute (*Bošnjački Institut*) in Sarajevo (a private foundation) for safekeeping at the request of the state authorities.[175]

But Republika Srpska had long refused to recognize the once state-wide remit of the Institute and these events took place towards the end of UNESCO's mandate over (and declining interest in) the Commission to Preserve National Monuments.[176] Yet, just a few months earlier in May 2000 the Aladža Mosque had been one of four religious sites 'of symbolic importance to the community' whose reconstruction was meant to assist the 'revival of cultural and religious pluralism' in Bosnia-Herzegovina in a project announced and promoted by the head of the UN Mission in the country, Jacques Paul Klein.[177]

Yet while the Aladža and the majority of Foča's mosques existed only as flattened sites, two *had* survived complete destruction in 1992: the Musluk (Atik Ali-paša) Mosque in the mahala of Donje Polje and the Mehmed-paša Kukavica Mosque in the Prijeka Čaršija. With the first returnees settled in Donje Polje, preparations soon began for the restoration of the Musluk Mosque, though it was not to formally reopen until July 2007.

As we have seen, the Careva and Šehova Mosques had been intentionally destroyed in 1992 and their sites cleared and much of the Čaršija was devastated. Yet despite attempts to blow up the stone-built Mehmed-paša Kukavica Mosque (1751) with dynamite on 28 May 1992, its central domed structure had resisted demolition, although its minaret, medresa and other surrounding buildings were totally destroyed and their remains

172 See *Destruction of the Cultural Heritage in Bosnia-Herzegovina* in this volume on this discovery.

173 *Prosecutor v. Milorad Krnojelac*, Case No. IT-97-25-T, testimony Racine Manas, 18 January 2001, 1861–4.

174 See the Commission to Preserve National Monuments website: *Exhumations team discovers fragments of the Aladža mosque in Foča*, at http://kons.gov.ba/main.php?mod=vijesti&extra=1075460534&action=view&id_vijesti=307&lang=4

175 See the designation of the mosque as a National Monument: *Aladža (Hasan Nazir) mosque, the site and remains of the architectural ensemble*, http://www.kons.gov.ba/main.php?id_struct=50&lang=4&action=view&id=2484.

176 The last meeting of the UNESCO-led Commission took place on 14 June 2000, although its term did not formally end until 31 March 2001. The Commission was reconstituted as a Bosnian state institution on 22 December 2001 and its first meeting was held on 4 March 2002.

177 *Simultaneous reconstruction of religious sites in BiH*, press release, UNMIBH, 23 May 2000.

removed. By the end of the war the Mehmed-paša Kukavica Mosque, although in a partial state, was the only domed mosque structure surviving in all of Republika Srpska. These substantial and completely unprotected remains were seen by the author in October 2000. But where once the shattered mosque had been surrounded by the tightly packed buildings of the Prijeka Čaršija, it now stood isolated in an empty wasteland. However, the *sahat kula* (clock tower) and *han* of Mehmed-paša Kukavica's endowment still stood intact nearby. These were notable survivals of Ottoman architecture in Republika Srpska, and especially in Foča.

The site of the Aladža mosque with its burial ground, the remains of the Mehmed-paša Kukavica Mosque, and the site of the Careva Mosque were all included on the Provisional List of National Monuments compiled under the terms of Annex 8 of the Dayton Agreement by the UNESCO-led Commission to Preserve National Monuments and formally adopted on 14 June 2000. Entry on the Provisional List conferred legal protection on the mosque remains and site and established the responsibility of the entity authorities (in this case, Republika Srpska) to take appropriate measures to ensure its preservation and rehabilitation.

In June 2001 the remains of the Mehmed-paša Kukavica mosque were recorded by Ferhad Mulabegović, then director of the Institute for the Protection of the Cultural, Historical and Natural Heritage of Bosnia-Herzegovina, but given the position of Republika Srpska regarding the Institute, there was little he action he could take. Mulabegović's visit to Foča came at a time when the international community in the form of the Office of the High Representative was belatedly taking an interest in Annex 8 and the issue of National Monuments after the riots the previous month that had accompanied the laying of cornerstones for rebuilding mosques in Republika Srpska at Banja Luka and Trebinje and the OHR's growing realization of the importance of the reconstruction of religious facilities in the return process, no matter how contentious that might be.[178]

Yet the Commission to Preserve National Monuments at that point did not exist: UNESCO's mandate had ended, and the Commission had not yet been re-established by the central state government. Nevertheless, clear responsibility for preserving the mosque already lay with the Republika Srpska entity authorities by virtue of its inclusion on the Provisional List of National Monuments.

But the remains of the Mehmed-paša Kukavica Mosque continued to be left completely unprotected and without stabilization measures taken by either entity or international authorities and during autumn storms in 2002, much of the dome collapsed, bringing with it a large part of the walls. Nevertheless, a significant part of the dome's internal structure and painted wall decorations survived.

On 19 December 2003 the *Medžlis* (Council) of the Islamic Community of Foča (heirs to the Mehmed-paša Kukavica *vakuf*) submitted a petition to the now reconstituted and Bosnian-led Commission to Preserve National Monuments to designate the Mehmed-paša Kukavica Ensemble (including the mosque, *medresa*, *han*, and *sahat kula*) and the Prijeka Čaršija as National Monuments under Annex 8.

In January 2004, concerned about the fragile state of this important Ottoman monument and its continued survival, the Commission sent letters to Republika Srspka prime minister Dragan Mikerević and Borislav Paravac, the Bosnian Serb member of the Presidency, requesting that the entity find funds and take emergency measures to protect the remains

178 See Valery Perry in this volume on this subject and elsewhere in this chapter.

Fig. 5.13 (*facing page*) **(a)** The Mehmed-paša Kukavica vakuf in Foča before the 1992–1995 war. The Mehmed-paša Kukavica Mosque sits at the heart of the historic Prijeka Čaršija, with the great *han* seen to the left of the minaret and the *sahat kula* (clock tower) to the right of the dome. © Institute for the Protection of the Cultural, Historical and Natural Heritage of Bosnia-Herzegovina. **(b)** The remains of the Mehmed-paša Kukavica Mosque in October 2000, its domed central structure still intact. © Richard Carlton. **(c)** The remains of the Mehmed-paša Kukavica Mosque in February 2004, showing the results of the collapse of its dome in 2002, but with some of the internal structure still intact. © Aga Khan Award for Architecture / Emir Softić. **(d)** The remains of the Mehmed-paša Kukavica Mosque in July 2008, now with support and protection carried out by the Commission to Preserve National Monuments in 2006, but little of the structure surviving. © Julian Nitzsche

of the Mehmed-paša Kukavica Mosque.[179] The following month an assessment team from the Commission arrived in Foča to document the structure, recording those parts of the dome and decorations that still survived after the collapse of 2002 and noted the continued absence of protection for the remains.[180]

A further letter from the Commission to prime minister Mikerević, the RS Minister of Education and Culture Gojko Savanović and the RS Minister of Physical Planning, Construction and Ecology Mensur Šehagić on the subject followed on 15 March. Finally on 22 March 2004 members of the Commission met with Savanović and Šehagić in Banja Luka, a meeting also attended by the OHR Deputy High Representative Vadim Kuznetsov. The RS officials were reminded by the Commissioners of the letters and their responsibilities under Annex 8, leading Minister Savanović to indicate that RS was prepared to earmark 10,000 KM for preparation of the project – once it had been completed.[181]

Over 2004–2006 the Mehmed-paša Kukavica mosque, han and clock tower, the sites of the Careva and Šehova Mosques and the entire Čaršija area were designated National Monuments by the Commission to Preserve National Monuments.[182] The Commission began to develop a project to rehabilitate the site of the Mehmed-paša Kukavica Mosque and Medresa, and to document the site and fragments of the Aladža Mosque, not only in the interests of preserving the monuments, but also to preserve the cultural memory of Foča, aiming to raise awareness of the heritage among young people, particularly those in Foča and those in Sarajevo whose families had been displaced from the town, to improve prospects for multiethnic co-existence and to encourage displaced persons and refugees

179 See: Collaboration: 3/22/2004 Meeting in the Building of the Government of Republika Srpska, at the website of the Commission to Preserve National Monuments, http://www.kons.gov.ba/. The letters were sent on 21 and 29 January.

180 The site visit was on16 February 2004. See: *Mosque and medresa of Mehmed-paša Kukavica, the architectural ensemble*, http://www.kons.gov.ba/main.php?id_struct=50&lang=4&action=view&id=2405

181 See note 165.

182 See the website of the Commission to Preserve National Monuments of Bosnia and Herzegovina: www.kons.gov.ba/. The site of the Aladža Mosque and the remains of the Musluk (Atik Ali-paša) Mosque were also designated National Monuments.

(a)

(b)

(c)

(d)

Fig. 5.14 A panorama of what remains of the Mehmed-paša Kukavica vakuf taken by the Commission to Preserve National Monuments in February 2004. The *han* and *sahat kula* survive to the left of the picture, while what remains of the mosque sits in an empty wasteland (compare with Fig. 5.13(a)). © Commission to Preserve National Monuments of Bosnia & Herzegovina

to return. The project was, in addition, to provide studying young professionals with hands-on training in the methodology of preserving a war-damaged historic monument.[183]

In 2005 the Commission received two grants from the United States Ambassadors Fund for Cultural Preservation (AFCP) totalling US $48,825 to support the documentation of the site and the architectural fragments of Aladža Mosque and the preservation of the remains of Mehmed-paša Kukavica Mosque from further deterioration.[184] Work in the Kukavica Mosque project was to include the physical protection of the remains, the methodical clearing of the site and sorting of rubble into useable and unuseable fragments and the thorough documentation and recording of the site and other properties in the Prijeka Čaršija that formed an integral group with the mosque by students from University of Sarajevo's Faculty of Architecture.

In March 2006 the Lautenberg sanctions against Foča were lifted, reflecting the improvement in the situation in the municipality, particularly since the election of Zdravko Krsmanović as mayor in 2004. But by the summer of 2006 when work on protecting the Mehmed-paša Kukavica Mosque from further deterioration was finally able to begin, the shattered dome of the mosque had collapsed completely. All that remained upstanding were two walls and very little else. Thus the only domed mosque in Republika Srpska to have survived the war and which had remained intact up to 2002 and an important historic monument which one might have expected to be the focus of attention by the international community was instead left to decay and collapse before willing local heritage actors were able to intervene.

183 For a detailed account of the mosques and the associated projects see: Lenasi, O. 2008. The architectural ensemble of the Mehmed Pasha Kukavica Mosque and Medresa in Foča / Graditeljska cjelina – džamija i medresa Mehmed-paše Kukavice u Foči, *Baština* (*Heritage*), The Annual of the Commission to Preserve National Monuments, IV/2008, 113–48; and Fočo, M. 2008. Rehabilitation of the Aladža Mosque in Foča: Protection of the remains, *Baština* (*Heritage*), IV/2008, 873–111.

184 See United States Department of State 2006. *The U.S. Ambassadors Fund for Cultural Preservation (AFCP), Annual Report 2005/2006*, 26. http://eca.state.gov/cultural-heritage-center/ambassadors-fund-cultural-preservation/annual-reports. The amounts were US $24,125 for the Aladža Mosque and US $24,700 for the Mehmed-paša Kukavica Mosque. The Ambassadors Fund for Cultural Preservation (AFCP) was established by the US State Department in 2001.

Work on documenting and preserving the site of the Aladža Mosque and its remains went on. Volunteers from the Islamic Community in Foča and returnees participated in clearing the site and excavating fragments of the mosque. The Republika Srpska authorities provided some funding for work to recover fragments from the mass grave sites on the banks of the Drina, though more came from the Federation.[185] However, there were difficulties in finding funds for the actual building works, estimated in December 2009 as almost €3,000,000 for rebuilding the mosque structure alone.[186] By mid-2009 the Islamic Community had collected just over US $145,000 for the mosque's reconstruction, much of which was a further grant of US $98,000 from the US Ambassadors Fund for Cultural Preservation,[187] though Istanbul-based IRCICA had contributed towards the preparation of building plans.

This delay in rebuilding was despite the mosque being the first of Bosnia-Herzegovina's EC/COE Ljubljana Process Consolidated Projects being implemented by the Commission to Preserve National Monuments, the reconstruction of which was intended to 'reinforce the message that justice and human rights will ultimately prevail.'[188] The preparatory phase of the reconstruction of the Aladža Mosque finally began in April 2010 without the greater part of funding needed in place, but in the hope that it would trigger more.[189] In December 2012, the second phase of the reconstruction of the Aladža Mosque was announced, focusing on the conservation of the preserved decorated fragments of the mosque.[190] The Islamic Community in Bosnia-Herzegovina declared 2013 *Year of the Renewal of the Aladža Mosque*.[191]

---

185 Pictures of the fragments recovered from the suspected mass grave sites can be seen at http://www.rijaset.ba/index.php?option=com_content&view=article&id=16239:fotogalerija-ostaci-porusene-aladze-dzamije&catid=259:godina-obnove-aladze&Itemid=457

186 *Business Plan: Preliminary Technical Assessment: Aladza Mosque, Foca, Bosnia and Herzegovina*, Integrated Rehabilitation Project Plan/Survey of the Architectural and Archaeological Heritage, 4 December 2009, 25–6; the total rehabilitation figure given for rebuilding the mosque was 6,050,369 KM (€3,093,505) but includes a figure of 500,000 KM for construction of an imam's house and museum. See also *Preliminary Technical Assessment: Aladza Mosque, Foca, Bosnia and Herzegovina*, 23 June 2008. Both available at (IRPP/SAAH)http://www.coe.int/t/dg4/cultureheritage/cooperation/SEE/Countries/Bosnia_en.asp.

187 See: Odobrena obnova Aladža džamije, nedostaje novac, *Dnevni Avaz*, 1 July 2009; the figure given was 200,000 KM. See also *US Supports Cultural Heritage in Bosnia and Herzegovina* at http://sarajevo.usembassy.gov/press_20090526.html.

188 The third Ljubljana Process Consolidated Project is the important archaeological site of the Hellenistic city of Daorson at Ošanići. See *The Ljubljana Process – funding heritage rehabilitation in South-East Europe, A Joint Action of the European Commission and the Council of Europe*, European Commission (EC) and Council of Europe (COE), 4.

189 See *Bosnia-Herzegovina Intermediary Impact Report 2010*, Integrated Rehabilitation Project Plan/Survey of the Architectural and Archaeological Heritage (IRPP/SAAH), Strasbourg: Council of Europe and: Obnova Aladža džamije u Foči, *Avaz*, 29 April 2010, Bicent/Bošnjački informativni centar [online], 10 May 2010.

190 Fena 2012. Foča: Ozvaničena druga faza rehabilitacije Aladža džamije, *www.vijesti.ba*, 21 December.

191 See: 2013. godina – Godina obnove Aladže džamije, http://www.rijaset.ba/index.php?option=com_content&view=article&id=16226:2013-godina-godina-obnove-aladze-dzamije&catid=259&Itemid=720.

Plans had also been going forward for rebuilding the razed sixteenth-century Careva Mosque after years of preparation and its foundation stone was laid on 22 October 2012.[192] But just as reconstruction was about to begin in early November 2012 the local Serbian Orthodox Church authorities had work on the site stopped contending (as in Stolac) that the mosque had been built on the foundations of an Orthodox church and insisting that archaeological investigations took place to establish the 'facts.'[193] Even though the site of the Careva Mosque had been a designated National Monument since 2004 and the Islamic Community in Foča had been granted planning permission from the Republika Srpska authorities to start rebuilding. It was wondered why it had taken 17 years since the end of the war for the Serbian Orthodox Church to raise this question.

As with the Čaršija Mosque in Stolac, local Orthodox Church officials based their evidence once more on the writings of a Bosniak historian, this time claims made by Alija Bejtić in an article on Foča published in 1956, whose interpretation was dismissed by later historians.[194] Nevertheless, building work was suspended while archaeologists from the Republika Srpska Institute for Protection of the Cultural, Historical and Natural Heritage excavated the site, though the archaeologists stated that whatever the results of the excavations they would not prevent the ultimate reconstruction of the mosque.[195] The excavations were completed by 20 April 2013 and though what were apparently pre-Ottoman foundations were uncovered at the site, there was no evidence that they had belonged to a church and reconstruction of the Careva Mosque resumed.[196]

But the summer of 2013 also saw what remained of Mehmed-paša Kukavica's endowment deteriorate still further. The Kukavica Han, in theory property of the Islamic Community of Foča and apparently restituted to them in 1991 after municipal ownership during the socialist period,[197] was in fact in private ownership. The owner had let the historic building deteriorate to such a state that in June 2013 the Republika Srpska building inspectorate declared the building unsafe and on the verge of collapse and a danger to the public. Despite being a designated National Monument which the entity of Republika Srpska was legally required to protect and preserve, the upper story of the han was quickly demolished.[198] Thus by 2013 little now remained intact of the Mehmed-paša Kukavica

192 Selhanović, S. 2012. Foča: Kurban na temelju Careve džamije, www.Preporod.com, 12 November.

193 See, for instance: Pravoslavna crkva sprečava obnovu Careve džamije u Foči, *www.kalesija.com*, 7 November 2012, http://www.kalesija.com/pravoslavna-crkva-sprecava-obnovu-careve-dzamije-u-foci/.

194 Bejtić, A. 1956. Povijest i umjetnost Foče na Drini (History and art of Foča on the Drina), *Naše starine*, Godišnjak Zemaljskog zavoda za zaštitu spomenika kulture i prirodnih rijetkosti NR Bosne i Hercegovine, Sarajevo, 23–75. See the designation of the Careva Mosque at the website of the Commission to Preserve National Monuments which quotes and discusses this claim and later historians on the issue. http://www.kons.gov.ba/main.php?id_struct=50&lang=4&action=view&id=2483.

195 Arheolozi ne ugrožavaju obnovu hrama, *www.glassrpske.com*, 11 November 2012, http://www.glassrpske.com/drustvo/vijesti/Arheolozi-ne-ugrozavaju-obnovu-hrama/lat/100493.html.

196 See: Arheolozi istražuju lokalitet fočanske Careve džamije, *tuzlalive.ba*, 20 April 2013 at http://tuzlalive.ba/arheolozi-istrazuju-lokalitet-focanske-careve-dzamije-84662/; U Foči ponovno pokrenuti radovi na Carevoj džamiji, Oslobođenje, 21 June 2013, http://www.oslobodjenje.ba/vijesti/bih/u-foci-ponovno-pokrenuti-radovi-na-carevoj-dzamiji.

197 See the designation of the Commission to Preserve National Monuments: *Mehmed-pasha Kukavica han, the historic monument*, referring to Muftić, F. 1997: http://www.kons.gov.ba/main.php?id_struct=50&lang=4&action=view&id=2777.

198 See for instance: Srušen čuveni Han Mehmed-paše Kukavice u Foči, *Dnevni avaz, 17 June 2013; available at* http://www.focaci.org/index.php?option=com_content&view=article&id=2182:han-

vakuf, one of the few groups of important Ottoman built heritage in Republika Srpska to have survived the 1992–1995 war.

The Resulbegović House in Trebinje

The case of the Resulbegović House in Trebinje, Republika Srpska, shows how the restoration of a war-damaged secular structure strongly associated with Bosniak/Muslim identity became a focal point for continuing contention long after the end of the war. Trebinje was the scene of the vicious ethnic cleansing of its Muslim population over 1992 and 1993. The town remained a bastion of hardline Bosnian Serb nationalism after the conflict and, like Foča, was subject to Lautenberg sanctions (lifted in 2005). As described above, in 2001 there were violent anti-Muslim demonstrations in Trebinje as Serb nationalist rioters broke up the cornerstone-laying ceremony for the Osman-paša Mosque.

In 2002, to the horror of Bosniak citizens of the town who had returned there and those still living as refugees abroad, a sports centre began to be erected on the site of the Resulbegović House (Begova kuća/Bey's House). The Resulbegović House was once a beautiful eighteenth-century Ottoman mansion complex and gardens on the banks of the Trebišnjica River which had been an ethnographic museum and cultural centre with a café and outdoor performance area before the war, popular with local people and tourists alike. The idyllic setting of the Begova Kuća and its grounds was fondly remembered by Trebinje's expelled Bosniak citizens (and no doubt by many Bosnian Serbs, too). However, the mansion had been the former summer residence of the powerful Resulbegović family and its construction was credited to Osman-paša Resulbegović, the Ottoman ruler of early eighteenth-century Trebinje, and thus, in the eyes of some nationalist Serbs, a symbol of the Turkish occupation.

But the house had been intentionally attacked during the ethnic cleansing of 1992–1993 and almost entirely destroyed; all that survived were a few crumbling walls.[199] To make matters worse, it was discovered that the remains of the house and its surroundings (which had been socially owned) had in 1996 come into the ownership of the Serbian Orthodox Diocese (Eparchy) of Zahum-Herzegovina, received as a gift from the Municipality of Trebinje.[200] In addition, the construction work was apparently being carried out without a building permit. The house and its surroundings, a protected monument before the war and on the Provisional List of National Monuments of 2000, was declared an endangered National Monument by the Commission to Preserve National Monuments in October 2003 because of the illegal building works taking place. After public opposition the works were suspended.

However, in what appeared to be an attempt to circumvent the complaints, during 2006 the Diocese as a 'co-investor and co-owner'[201] applied to the Municipality of Trebinje for consent to renovate/rebuild the Begova Kuća as a sport, tourist, hotel and restaurant complex, with the restoration to be implemented by the Republika Srpska Institute for

mehmed-pase-kukavice&catid=75:urbicid&Itemid=89. The demolition of a designated National Monument is a criminal offence that carries a prison term of three to five years.

199 For more information see the designation of the site as a National Monument by the Commission to Preserve National Monuments: *Resulbegović family house, the site and remains of the architectural ensemble* at http://kons.gov.ba/.

200 *Resulbegović family house, the site and remains of the architectural ensemble*.

201 *Peticija za obnovu Begove kuće u Trebinju / Petition for the renovation of the cultural-historical monument Bey's House, Trebinje*, 29 April 2006, www.trebinje.se.

Fig. 5.15 The Resulbegović House (Begova kuća / Bey's House) in Trebinje and its outdoor amphitheatre before its destruction during the 1992–1995 war. © Gazi Husrev-beg Library, Sarajevo

Protection of the Cultural, Historical and Natural Heritage – at the time of writing possibly the only instance of the RS heritage preservation institute intending to carry out work on an overtly Ottoman/Bosniak structure.[202]

The Serbian Orthodox Bishop Grigorije began to promote the idea that the restored Resulbegović House would become a centre for interethnic reconciliation and tolerance and claimed to have raised one million KM for its reconstruction.[203] But in April 2006 the Diocese of Zahum-Herzegovina had transferred ownership of the property to the *Foundation Centre for the Development of Physical and Spiritual Culture*,[204] a foundation backed by the Orthodox Church and headed by Dejan Bodiroga, a well-known Serbian basketball player, and of which Bishop Grigorije was a vocal and very public board member. Construction work on the sports centre resumed.

This fresh proposal and the renewed building work caused further outrage in many of Trebinje's Bosniak citizens, a group of whom based in Gothenburg, Sweden, published

202 Duka, V. 2006. Imam Hodžić: Skrnavljenje spomenika; Vladika Grigorije: Obezvređivanje ideje, *Nezavisne Novine* [online], 3 May. See also US Embassy Sarajevo 2006. *Bosnia: Little Progress in Post-sanction Trebinje*, 16 May, Wikileaks: 06SARAJEVO1075_a.

203 See US Embassy Sarajevo 2006. *Bosnia: Little Progress in Post-sanction Trebinje*; Duka, V. 2006; Vladika Grigorije: Obnovom Begove kuće do pomirenja, *Bileća Online* (*www.bileca.org*), 3 May 2006

204 *Fondacija Centar za razvoj fizičke i duhovne kulture.*

a widely circulated petition on 29 April 2006 signed by over 700 people denouncing the plans. The petition declared:

> At the end of 1992 and the beginning of 1993 the Bosniaks of Trebinje were persecuted from their homes and can today be found scattered on all the continents of the world.
>
> Parallel with the persecution of the population, all that which related in any way to their cultural-historical and religious heritage was systematically destroyed. All the mosques and associated buildings were completely destroyed and razed to the ground. Even the cemeteries were desecrated. All the names of buildings, localities and streets whose names reminded of Bosniaks and their culture were changed.
>
> The shameless burning and razing to the ground of the Bey's House, which represented the jewel of the cultural-historical and architectural heritage of the Bosniaks of Trebinje, once more confirmed the intention of destroying all traces of the existence of the Bosniak population on these lands. [205]

The petitioners saw the plans for reconstructing the Resulbegović House as the continuation through new means of the old policy of ethno-nationalist and religious hatred that had brought about the intentional destruction of the house in the first place. The petition called for the Resulbegović House to be restored not by some private foundation with unclear intentions, but for the benefit of the wider community.

There were accusations of 'culturecide' and 'cultural genocide' and open letters from prominent members of the Bosniak community condemning the project.[206] The Bosniak representative for Trebinje to the Republika Srpska Assembly, SDA politician Tarik Sadović, questioned why a foundation whose aims were to organize and gather the Serbian people in Herzegovina and beyond in cooperation with the Serbian Orthodox Church needed to renew and manage a Bosniak National Monument.[207]

The Commission to Preserve National Monuments, too, wrote to the Republika Srpska Ministry of Physical Planning, Construction and Ecology on 26 May 2006 confirming their opinion that the construction works were in contradiction with the original purpose of the property and were endangering the monument. On 2 July the first 'Dejan Bodiroga Basketball Camp' was officially opened on the site by RS prime minister Milorad Dodik. Four days later the Commission designated the Resulbegović House a National Monument; the designation of 6 July 2006 ordered the removal of the illegally-built structures from the site, including the locker rooms and basketball courts.[208]

Press reports over the following years declared that the Republika Srpska authorities had allocated between a half to one million KM for reconstruction of the Begova Kuća,

205 *Peticija za obnovu Begove kuće u Trebinju*, 29 April 2006. See also Peticija za obnovu Begove kuće u Trebinju, *Oslobođenje*, 06 May 2006, available at http://www.idoc.ba.

206 See: Košarkaš Dejan Bodiroga gradi sportski centar u Trebinju (pored Begove kuće), *Bileća Online* (*www.bileca.org*), 31 May. On open letters see for instance Duka, V. 2006, describing the open letter of the Trebinje imam Husein Hodžić to Bishop Grigorije, and *Glas Trebinje* (www.glastb.com), 30 June 2006, iss. 949, Reagovanja predsjednika i potpredsjednika SO Trebinje povodom izgradnje sportskih terena na lokalitetu Begove kuće, contains an open letter from Tarik Sadović, SDA Representative for Trebinje in the RS Assembly.

207 *Glas Trebinje* (www.glastb.com), 30 June 2006, iss. 949, Reagovanja.

208 Kalamujić, A. 2006. Spor oko provodjenja odluke o nacionalnom spomeniku u Trebinju, *Oslobođenje*, 29 November. See the designation: *Resulbegović family house, the site and remains of the architectural ensemble* at http://kons.gov.ba/.

relayed a series of announcements that work on the house had begun or was about to begin, that its reconstruction was a priority for the Trebinje authorities, and published numerous accounts of the myriad alleged financial misdeeds, complicated property speculation and corrupt investments of Bishop Grigorije in which the historic Trebinje property was entangled and asked where the one million KM had gone.[209] The Bishop was credited with making the much-repeated statement: 'Što su Srbi srušili, Srbi moraju i da podignu' ('What the Serbs took down, the Serbs must raise') in relation to the question of why a Serbian-led foundation should reconstruct an Ottoman/Muslim house. The issue of the Resulbegović House became tied, as well, to the campaign for the restitution of property to the Islamic Community (though it had never been owned by them).[210] As of 2013 the Resulbegović House had still not been rebuilt and Bishop Grigorije had still not given up on his plans to restore the structure as a place to promote multiethnicity and multiculturalism.[211]

Restoration, Reconciliation and Return

Thus we have seen how the international community's involvement in heritage restoration in post-conflict Bosnia-Herzegovina in the years following the end of the war was closely tied up with iconic sites, particularly in Mostar, where it invested much of its support and funding to the virtual exclusion of all else. The post-conflict restoration of such iconic sites came to be mobilized as concrete visible symbols which embodied the ideas of reconciliation and the 'reconstruction' of relations between Bosnia's ethno-national groups that the international community was keen to promote in the immediate aftermath of the war.

The frequent claim (and indeed insistence) of the international community was that restoration and reconstruction of historic buildings/cultural heritage destroyed or damaged during the conflict would/should lead to 'reconciliation' between the formerly warring ethno-national groups. The Council of Europe's *Specific Action Plan for Bosnia-Herzegovina*, for instance, had ambitious aims to 'incorporate technical and scientific considerations into an overall scheme of tolerance, reconciliation and reconstruction.'[212] In the case of the Mostar, the international community overtly attempted to use restoration of the heritage as a tool to bring about rapid reconciliation after the end of the war. Yet an observer like Gilles Péqueux, the French engineer responsible for directing the reconstruction of the Old Bridge until February 2003, doubted the wisdom of this approach and whether reconciliation could take place at such a speed.[213] In late 2003 the International Crisis Group (ICG), reporting on the continued attempts of the international community to unite a still very divided

209 See, for instance: Dužničko ropstvo Vladike Grigorija, *Slobodna Bosna*, 27 August 2011, posted on www.trebinje.se; Grigorijevo Rastrošenije: Vladika u bankrot, crkva na doboš, *www.zurnal.info*, 15 August 2010.

210 US Embassy Sarajevo 2006. *Bosnia: Little Progress in Post-sanction Trebinje*.

211 Lagumdžija primio vladiku Grigorija, *Slobodna Bosna* [online], 12 December 2013.

212 *Specific Action Plan for Bosnia-Herzegovina, Preliminary Phase, Final Report*, Technical Cooperation and Consultancy Programme related to the integrated conservation of the cultural heritage, Strasbourg: Council of Europe (COE), Cultural Heritage Division, March 1999, 5.

213 Pequeux, G. 2003. Rencontre avec Gilles Péqueux, ingénieur responsable de la reconstruction du pont de Mostar, *Courrier des Balkans* [online], 1 March; Rossini, A. 2003. Ricostruire il ponte di Mostar seguendone lo 'stato dello spirito', *Osservatorio Balcani e Caucaso* [online], 11 April.

and dysfunctional Mostar, noted how the Old Bridge project had served as 'a species of Potemkin village, designed to create the illusion of interparty, cross-national cooperation, rather than as a manifestation of the real thing.'[214]

In the short term at least, there is little evidence to support the assertion that reconstruction of the cultural heritage led the way towards reconciliation. It is clear that in contentious settings where there had been ethnic cleansing, and where perpetrators of that ethnic cleansing remained dominant, restoration was more frequently a source of conflict and organized aggression against returnees, particularly in the case of religious structures with their function as clear markers of identity – the very reason they had been destroyed in the first place. Yet, there was virtually no linkage by the international community of heritage restoration to the return process as set out in the terms of the Dayton Peace Agreement – certainly none in Republika Srpska. And there was little mention of justice or human rights for the victims of ethnic cleansing in the discourse of external actors on heritage rebuilding.

Though the PRDU plan for Počitelj of 1996 provided an early exemplary model for smaller towns for the integration of heritage restoration with refugee return and socio-economic development, this was not followed elsewhere. External national governments pursued their own agendas, with only Sweden, through Cultural Heritage without Borders (CHwB), supporting restoration projects across Bosnia-Herzegovina. Funding from Islamic sources was only exceptionally applied to the restoration of historic mosques and buildings, with donors preferring to finance the construction of completely new mosques.

International support for restoration projects for war-damaged or destroyed historic structures in Republika Srpska or locations in the Federation where there was determined opposition from hostile ethno-national power structures was all but non-existent as donors feared to become involved in such contentious settings.[215] Here, the restoration of historic structures, crucial elements in rebuilding communities which had suffered and been dispersed by ethnic cleansing, came not through the engagement of the international community, but by the mobilization of the victims themselves. In this other domain returning communities focused on literally restoring their communities, including markers of their identity, as well as on their rights of return and recourse to justice as laid out in the articles of the Dayton Peace Agreement.

It was to take almost a decade after Dayton for restoration of cultural heritage to be seen as an important part of the recovery process in wider aid and development programmes. Even the terminology used by the international community began to evolve and claims for reconciliation achieved through heritage restoration were moderated to express hopes for 'intercultural understanding.'[216]

Perhaps the earliest initiative with this approach was the King Badouin Foundation's *Living Heritage* programme that aimed to promote community development using local

---

214 *Building bridges in Mostar*, International Crisis Group, Europe Report No. 150, Sarajevo/Brussels, 20 November 2003, 5.

215 CHwB's projects in Republika Srpska, undertaken at an early stage of its involvement in Bosnia-Herzegovina, were not of war-damaged buildings, despite the stated aims of the organization. There was a return to these aims after a change of project managers.

216 *EU and BiH Sign Protocol for the Reconstruction of the City Hall in Sarajevo and City Hall in Novi Grad*, press release, Sarajevo: Delegation of the European Union to Bosnia and Herzegovina (DELBIH), 7 March 2012.

heritage and cultural resources.[217] Working in small town and village settings, the programme began implementing projects in 2003 through the recently-formed Bosnian NGO Mozaik in partnership with the Open Society Fund Bosnia & Herzegovina, with major funding also coming from the European Union.[218] While the programme had already begun in other countries of South East Europe, Bosnia's post-conflict situation meant there was a higher proportion of building reconstruction than usual, including of historic monuments such as the eighteenth-century Ottoman clock tower in Donji Vakuf. Most (though not all) of the projects were carried out in the Federation.

With a new emphasis on socio-economic development, the linking of heritage with tourism formed an inevitable part of these programmes as the tourist industry came increasingly to be seen by external funders as a crucial element in Bosnia's post-conflict recovery. Thus, for instance, the British Council's project *Cultural Heritage – Royal Trail*, beginning in 2004, linked tourism development and capacity-building through (among other elements) the preservation of traditional crafts, labelling of historic monuments and monument restoration in the municipalities of Jajce and Šipovo.[219] Jajce is in the Federation, while Šipovo is in Republika Srpska; this was the first time municipalities from the two entities had collaborated since before the war. The project resulted in the restoration of the almost totally destroyed historic Kršlak House in Jajce which now houses the agency for the town's heritage protection and tourism development.[220] As part of its Cluster Competitiveness Activity (CCA) to support tourism development in Bosnia, USAID became involved in carrying out a controversial management plan for Počitelj over 2006–2008.[221]

Another step in this new direction was the Millennium Development Goals Achievement Fund (MDGF) project *Improving Cultural Understanding in Bosnia and Herzegovina* implemented over 2008–2012 by UNDP, UNESCO and UNICEF and funded through a US $8,000,000 grant from the Spanish-financed body.[222] The programme was carried out in partnership with the Ministry of Civil Affairs (MoCA), the entity Ministries of Culture and Education and other state and entity education, culture and heritage institutions, as well as ten core municipalities. It aimed 'to improve cross-cultural understanding' and respect for cultural diversity at community level and promote a focus on Bosnia's shared heritage through social

217 More on the King Badouin Foundation can be found at its website http://www.kbs-frb.be. For the final report on the programme see: Matarasso, F. 2005. *Living Heritage Bosnia and Herzegovina: Community Development Through Cultural Heritage Resources, Programme report 2001–2005*, Brussels: King Badouin Foundation/Mozaik.

218 Mozaik was established in 2002 as the Foundation for the Development of Non-Governmental Organizations (NGOs) specifically to create domestic NGOs that would function after the departure of international organizations from Bosnia; it took the name Mozaik Foundation for Community Development (*Fondacija za razvoj zajednica Mozaik)* in 2004: http://www.mozaik.ba/. Open Society Fund Bosnia & Herzegovina: http://www.osfbih.org.ba/.

219 For more on this project see: *UK Supports Restoration of a National Monument in Bosnia and Herzegovina: From 'Royal Trail' to UNESCO World Heritage Site*, British Council, press release, Sarajevo, 22 March 2007.

220 Agency for Cultural, Historical and Natural Heritage and Development of Tourist Potential of Town Jajce (*Agencija za kulturno-povijesnu i prirodnu baštinu i razvoj turističkih potencijala grada Jajca*), http://www.agencija-jajce.ba.

221 There was much opposition to the plan from heritage professionals and others. See, for instance: Kamenica, E. 2006. Amerikanci hoće koncesiju na Počitelj, *Oslobođenje* [online], No. 21.472, 4, 30 October.

222 See the MDGF website for various documents on the programme http://www.mdgfund.org/program/improvingculturalunderstandingbosniaandherzegovina. For a short film on the programme see http://www.youtube.com/watch?v=Fe6x0nwSMdM.

and economic development. Projects were carried out in both the Federation and Republika Srpska. Revitalization of the cultural heritage was seen as a crucial factor in restoring inter-community relations, as well as creating assets for cultural development and education and a number of historic monuments were conserved or restored as part of the project.[223]

Generally, there was increasing support for restoring war-damaged and destroyed cultural heritage from domestic authorities. The Federation entity government, for instance, between 2003–2008 through its Ministry of Culture and Sports allocated 34,000,000 KM[224] for restoring cultural heritage, much of it war-damaged.[225] This was in contrast to its budget of just 200,000 KM for both culture and cultural heritage protection in 2002.

There was growing recognition, too, of the need to restore religious, cultural and other public buildings (which might be historic monuments) as part of programmes for sustainable return and reintegration of refugees and displaced persons and financial support for such projects became available. From 1 January 2004 the Bosnian state authorities, under a Ministry for Human Rights and Refugees (MHRR),[226] became responsible for implementation of Annex 7 of the Dayton Agreement governing the return of refugees and displaced persons to their places of origin. Programmes were financed by a central state Return Fund, as well as funds from entity and other levels of government and international donors. In a report of December 2005 the MHRR noted that in the context of full sustainability of return and reintegration, special attention needed to be paid not only to the reconstruction of social, health and cultural infrastructure, but also to the reconstruction of mosques, churches and other places of worship, which was still a priority for the Ministry of Human Rights and Refugees.[227]

Ministries of refugees and displaced persons were established at entity-level in both the Federation and Republika Srpska[228] to support sustainable return and included in their remits funding for the restoration of religious and other cultural buildings. The RS Ministry, for instance, described one of its main activities from 2006 as 'participation in the restoration, reconstruction and construction of housing, infrastructure, cultural, religious and public buildings,'[229] reporting that in 2011 it had allocated funds for the restoration of 40 religious buildings in both Republika Srpska and the Federation.[230]

One of the projects to receive funds from the RS Ministry in 2011 was the reconstruction of the totally destroyed Careva/Sultan Bayezid Mosque in Nevesinje to which it gave 10,000 KM as part of its support for sustainable return to the municipality, the first funding the

223 Travers, R. 2012. *MDGF Final Evaluation: Improving Cultural Understanding in Bosnia and Herzegovina*. Sarajevo: United Nations Office of the Resident Coordinator, June 2012, 10–11, 50.

224 The sum equals US $24,199,194, Euros €17,383,924, GBP £16,557,469.

225 Hamidović, M., Mićić, L. and Petrović, A. (eds) 2010 *Zaštita kulturne i graditeljske baštine: 2003–2009*. Sarajevo: Federalno ministarstvo kulture i sporta, 5.

226 *Ministarstvo za ljudska prava i izbjeglice BiH*. See: *RRTF Closure and Transfer of Responsibities [sic] to BiH Authrorities [sic]*, http://www.ohr.int/ohr-dept/presso/pic/rrtf-resp-w-bih-authorities/.

227 MHRR 2005. *Uporedna analiza pristupa pravima izbjeglica i raseljenih osoba*. Sarajevo: Ministarstvo za ljudska prava i izbjeglice BiH, December, 307.

228 *Federalno ministarstvo raseljenih osoba i izbjeglica / Federal Ministry of Displaced Persons and Refugees* (FMROI) and *Ministarstvo za izbjeglice i raseljena lica Republike Srpske / Ministry for Refugees and Displaced Persons of Republika Srpska*.

229 *Ministarstvo za izbjeglice i raseljena lica Republike Srpske / Djelatnosti*, http://www.vladars.net/sr-SP-Cyrl/Vlada/Ministarstva/mirl/DJEL/Pages/Splash.aspx.

230 *Najznačajniji rezultati rada Ministarstva za izbjeglice i raseljena lica RS u 2011. godini*, http://www.vladars.net/sr-SP-Cyrl/Vlada/Ministarstva/mirl/DJEL/Pages/Dostignuca.aspx.

Islamic Community had received to rebuild what was a designated National Monument.[231] The Federation Ministry also allocated a high number of points to the reconstruction of the Careva Mosque under the heading 'objects of religious and sacred character' on its July 2012 list of potential beneficiaries for assistance for sustainable return for the construction and rehabilitation of infrastructure and buildings of importance to the wider community in areas of return in Republika Srpska.[232]

231 Obnavlja se Careva Džamija u Nevesinju, *Lokoportal Trebinje* (*http://lokoportal.com*), 3 August 2011.

232 'Objekti vjerskog i sakralnog karaktera.' See *Odluku o utvrđivanju konačne rang liste potencijalnih korisnika*, July 2012, http://www.fmroi.gov.ba/bosanski/tenderi/odluke/2012/juli/Odluka RC Mostar-infrastruktura.pdf.

Chapter 6
The Built Heritage in the Post-War Reconstruction of Stolac

Amra Hadžimuhamedović

Introduction

Stolac, a small town in southern Bosnia-Herzegovina, had one of the greatest concentrations and diversity of cultural heritage in the region until it was systematically destroyed in the late twentieth century, and the rehabilitation of its heritage has been an integral part of the complex process of the return of refugees and displaced persons, civic reconciliation, the exercise of human rights, the establishment of the rule of law and defining post-war development plans. The state of the heritage in Stolac since the 1992–1995 war in Bosnia and people's attitudes towards its rehabilitation might serve as a model for understanding the role cultural heritage plays in the complex social processes taking place in war-torn communities.

The post-war rehabilitation of the built heritage in Stolac has been based on the relation between those who were displaced and their home town as the central premise in the process of overcoming the chaos generated by the ravages of war. Almost all non-(Bosnian) Croats[1] in Stolac were held in concentration camps or forced to leave the town before the destruction of its historic built heritage began. Clearing an area of the 'other' in the process known as 'ethnic cleansing' to allow for the unimpeded, systematic destruction of all physical evidence of their connection with the area, is clear evidence of the familiar, tried and tested model of destroying a society or community while at the same time the cultural memory attesting to their history, art, religion, customs and laws – to their very existence – is wiped out. Milan Kundera has described how this is done:

> You begin to liquidate a people by taking away its memory. You destroy its books, its culture, its history. And then others write new books for it, give another culture to it, invent another history for it. Then the people slowly begin to forget what it is and what it was.[2]

In the summer of 1993 around 62 per cent of the population of Stolac[3] was either incarcerated in concentration camps, from which most of the men who survived were

1 That is Bosnian Muslims/Bosniaks and Bosnian Serbs. References to 'Croats' or 'Serbs' in this chapter will mean Bosnian Croats (rather than Croats from Croatia) and Bosnian Serbs (rather than Serbs from Serbia) unless otherwise indicated. Bosniak is the term many Bosnian Muslims choose to call themselves.

2 Kundera, M. 1999. *The Book of Laughter and Forgetting*. New York: Harper Perennial Modern Classics, 218.

3 According to the census in Bosnia-Herzegovina conducted in 1991, Stolac Municipality had a population of 18,681 of which 43.36 per cent per cent were Muslim (Bosniak), 33.12 per cent Bosnian Croat, 20.96 per cent Bosnian Serb, 1.64 per cent Yugoslavs and 0.92 per cent others. The population

forced to seek refuge in third countries, or in the case of women and children, forced to become internally displaced elsewhere in Bosnia; the town's Serb residents had largely fled in 1992. Stolac's non-Croat community – about 80 per cent of the population – thus became a 'displaced community.'

Kundera's model for the destruction of a people suggests that a refugee or displaced person is a person whose crystallization of the past and evidence of identity has been consigned to oblivion by being physically cut off from his or her cultural heritage and the destruction of that heritage ensures that the effects will be permanent. Yet the processes leading to Stolac's post-war reconstruction could be said to have all the features of a model showing that the destruction of a community by consigning its cultural memory to oblivion cannot be done easily or quickly, illustrating (paradoxically) how the destroyed cultural heritage can become a factor that *strengthens* a sense of identity. In the case of Stolac it can be shown that following the destruction of the cultural heritage, the awareness of those displaced from the town that their survival depended on the strength of their cultural memory was heightened, taking centre stage in their efforts to ensure that they could 'truly' return home.

One should note, however, that the case of Stolac is perhaps not a typical case, but one in which the value of the heritage, the bond between the people and the heritage[4] and the extent of its destruction are all very pronounced. Thus the conclusions which might be drawn from the case of the post-war reconstruction of Stolac may not universally be applicable, but still are a significant indicator of the potential for integrating the heritage into post-war reconstruction and the complete restoration of the war-ravaged townscape.

The Heritage of Stolac and its Destruction

In Stolac and its hinterland can be found the legacy of human presence reaching through the Palaeolithic, Bronze Age, Hellenistic, Roman and Medieval periods. But it was the Ottoman era that gave the town of Stolac its distinctive character, along with structures from the start of Austro-Hungarian rule to the twentieth century. Though the town on the banks of the Bregava River was not a centre of major architectural and artistic activity, and despite the loss (as elsewhere in Bosnia-Herzegovina) of a significant part of its built heritage, particularly in the years following World War II, up to 1993 it had largely retained the harmony of its uniquely complex appearance.

The chronological and stylistic wealth and typological diversity of Stolac's built heritage included architecturally unique examples of both public edifices: mosques (the oldest dating from the early sixteenth and the newest from the early twentieth century), Orthodox and Catholic churches, a *havra* (Jewish religious structure[5]), a seventeenth-century *hamam* (public baths), a sixteenth-century *musafirhana* (guesthouse), a *tepica* (seventeenth-century: a market weighhouse) and bridges; and private residential and commercial buildings, including four types of traditional house, flour and fulling mills.

of the town itself (5,530) was 61.95 per cent Muslim (Bosniak), 20.09 per cent Bosnian Serb, 11.80 per cent Bosnian Croat, 4.32 per cent Yugoslavs and 1.84 per cent others. See 'Nacionalni sastav stanovništva – Rezultati za Republiku po opštinama i naseljenim mjestima 1991,' *Statistički bilten 234*, Sarajevo. National Statistics Institute of the Republic of Bosnia and Herzegovina.

4 See for instance, Kolind, T. 2008. *Post-war Identification: Everyday Muslim Counterdiscourse in Bosnia Herzegovina*. Aarhus: Aarhus University Press, 149.

5 The *havra* is located at the tomb of the legendary Sarajevo rabbi, Moše Danon.

Fig. 6.1 **(a)** A prewar photo of Begovina, the most famous of the great Ottoman residential complexes in Stolac that lined the banks of the Bregava River. © Gazi Husrev-beg Library, Sarajevo. **(b)** A prewar photograph of the sixteenth-century Čaršija Mosque, Stolac, a quintessential Mediterranean scene. © Gazi Husrev-beg Library, Sarajevo

The čaršijas (bazaars) took shape from the sixteenth to the nineteenth century, while the grand residential complexes of elite Muslim Slav families of the Ottoman period, such as Serdarevina, Behmenluk, Ada and Begovina, are among the most important examples of this type in Bosnia-Herzegovina. The rich heritage of the town and its surroundings is felt to be of such significance that the natural and architectural ensemble of Stolac is included on the tentative list of UNESCO World Heritage Sites.[6]

Thus Stolac is a town in which the heritage was the most striking feature of the townscape, and one in which, by comparison with others elsewhere in Bosnia, its residents were unusually aware of the value of that heritage. It is a small town in which every change is noted and is significant to the townscape as a whole. From 1992 to 1996, but especially in the summer of 1993, Stolac was subjected to systematic destruction, aimed

6 See UNESCO World Heritage Centre website http://whc.unesco.org/en/tentativelists/5282/.

Fig. 6.2 Aerial view of the centre of Stolac showing the extent of the destruction of July 1993. © Amra Hadžimuhamedović

largely at the houses, business premises and public buildings dating from the sixteenth to the twentieth century that were regarded as non-Croat/Catholic, that is Muslim/Bosniak/Ottoman or Serb/Orthodox.

At the outbreak of war in April 1992 Stolac was already occupied by JNA (Yugoslav People's Army) forces and Serb and Montenegrin militias and the majority of the town's Croat residents had fled. In July Bosnian Croat nationalist troops (HVO[7]) and the Croatian Army regained control of the town and now the town's Serb population for the most part departed as the JNA withdrew; shortly after the Orthodox Church of the Holy Assumption in Stolac was devastated.[8] By spring 1993 the HVO increased their efforts to establish their hold over territory where they hoped to create an ethnically pure Croat para-state of Herceg-Bosna and embarked on the persecution of Bosniak intellectuals and leaders. In early July 1993, supported by units of the Croatian Army, the HVO carried out the ethnic cleansing of Stolac's Bosniak citizens. The systematic and planned destruction of Stolac's non-Croat heritage by HVO troops by dynamite and fire then began.

At the very heart of the town, the Čaršija mosque and the old market area (*tepa*) with its shops, *musafirhana* and other public buildings, as well as the burial ground forming part of its *harem* were dynamited. Parts of the *mihrab* wall and the *šadrvan* (fountain) survived the destruction, but the entire area was later cleared and the rubble dumped in the Bregava and Radimlja rivers or taken to landfill sites. The historic urban core was thus transformed into a blank weed-strewn space, and Stolac became a town without a centre which no longer

7 HVO = *Hrvatsko vijeće obrane* = Croatian Defence Council.

8 It was destroyed by HVO troops.

had the nature of a place. In the historic city area 33 monuments and valuable housing complexes and their contents were either destroyed or badly damaged. While the remains of public buildings were removed and their sites levelled, residential structures were left as crumbling rat-infested ruins. The removal of the remains took place out of public view over early autumn 1993 – autumn 1996 during the period when Stolac was emptied of its non-Croat citizens and it was impossible for anyone to approach the town without authorization from the HVO.

Following the removal of most of the non-Croat inhabitants of western Herzegovina and the destruction of their built heritage, as part of their attempt to 'Croatize' the region, the Herceg-Bosna/HVO leadership began the transfer of Croats from central Bosnia into Stolac and its hinterland, a process that continued even after the end of the war.[9] The Washington Agreement in March 1994 marked the official end of hostilities between Bosniaks and Croats.[10] However, Stolac remained a place of notorious hardline Croat nationalism promoted by the municipal authorities with the support of the local police force, even after the signing of the Dayton Peace Agreement in December 1995 that ended the war. Two of its mayors were to be removed by the Office of the High Representative (OHR)[11] for actively obstructing implementation of the peace agreement, particularly in relation to the return of refugees.[12] It was in this environment that Stolac's expelled residents began their attempts to return to their home town.

The Stolac Case

Many of those violently displaced from the town during the 1992–1995 war were fully aware of the importance of the heritage for their sustainable return to Stolac. The role of individuals and groups who identified Stolac as their home town in defining the significance of the heritage for restoring civic trust in a place where atrocities had been so recently committed was of crucial importance, all the more so since the perpetrators and those who incited and encouraged them, their witnesses and their victims still encounter and recognize each other, but do not all have equal access to the heritage.[13]

The approach of those involved in defining a strategy for the post-war reconstruction of the heritage in Stolac has been based on evaluation, judging each case on its own merits, not on reaching agreement between victims and perpetrators. It is an approach that requires great commitment and faces many serious obstacles since the municipal authorities in Stolac are prone to justifying the destruction of the cultural heritage and to contesting the right to

---

9 Williams, C.J. 1993b. On Bosnia's chessboard, refugees feel like pawns, *Los Angeles Times*, 2 August; and Roberts, D. 1996. Bosnia / Croats, *Voice of America*, transcript, 28 August.

10 The agreement signed on 18 March 1994 brought an end to the so-called Muslim-Croat War which began in 1993. The agreement included the abolition of the para-state of Herceg-Bosna and the dissolution of the camps of which people from Stolac had been among the inmates.

11 The Office of the High Representative (OHR) is the *ad hoc* international institution responsible for overseeing implementation of civilian aspects of the Dayton Peace Agreement (formally the General Framework Agreement for Peace or GFAP), the international peace agreement negotiated in Dayton, Ohio and signed in Paris on 14 December 1995, which marked the end of the 1992–1995 war in Bosnia-Herzegovina.

12 Pero Raguz in 1998 and Pero Pazin in 1999.

13 Hadžimuhamedović, A. 2008a. The meaning of homeland: Heritage and uprootedness, *Forum Bosnae* 44/08, 328–46.

its restoration, a stance exacerbated by the bureaucracy of the international community, most recently by the European Union (see below), which prefers to 'reach agreement between democratically elected government representatives' and apparently firmly rejects any kind of commitment to the principles laid down in fundamental international conventions and resolutions of the Council of Europe, the European Parliament and the European Commission.

The case of Stolac has played an emblematic role in the challenges facing the post-war restoration of Bosnia-Herzegovina's built heritage. It was a crucial model for the implementation of Annex 8 of the Dayton Peace Agreement (DPA, formally the General Framework Agreement for Peace or GFAP) which aimed at addressing the destruction of cultural heritage during the 1992–1995 war. At the time it was adopted Annex 8 was a unique and certainly a pioneering document for the world as a whole, evidence of the undoubted role of the cultural heritage as a factor for peace and reconciliation in our time.[14] Yet almost six years after it came into force it was the only part of the DPA that had not a single specific result to show for its implementation.

The date of 1 June 2001 could be taken as marking the start of the physical post-war reconstruction of Stolac when a meeting was held in the headquarters of the OHR in Mostar between representatives of returnees to Stolac, the Islamic Community,[15] the Stolac municipal authorities, the Catholic Church, the Ministry of Regional Planning of the Federation of Bosnia-Herzegovina and international military and civilian officials.[16] The meeting was prompted by steps being taken by the municipal authorities, the Croat nationalist HDZ political party,[17] the Catholic Church (both the bishopric in Mostar and the parish church in Stolac), the Institute for the Protection of the Cultural-Historical Heritage of the (illegal) para-state of Herceg-Bosna and the Ministry of Spatial Planning of Herzegovina-Neretva Canton with the intention of blocking the reconstruction of Stolac's main mosque, the sixteenth-century Čaršija (or Careva) Mosque,[18] by beginning to erect a solid fence and barring access to the mosque site, at the same time as announcing plans to conduct archaeological excavations there to investigate claims of the former existence of a church on the site.

That day, however, could also be said to clearly mark the beginning of the concrete implementation of Annex 8. The meeting took place in the aftermath of the violent demonstrations that had accompanied the ceremonies laying cornerstones for the rebuilding of the totally destroyed Ferhadija and Osman-paša Mosques in Banja Luka and Trebinje in May – events that prompted the international community to consider how it had underestimated the importance of Annex 8 for the return process. In Stolac, too, plans had been made by the Islamic Community to start the reconstruction of the Čaršija Mosque. Following the meeting, chaired by Colin Munro, Deputy High Representative and head

14 *Agreement on Commission to Preserve National Monuments*. See Valery Perry in this volume for a detailed account of Annex 8. For text of Annex 8 see http://www.ohr.int/dpa/default.asp?content_id=376.

15 The official body and legal entity representing members of the Muslim faith in Bosnia-Herzegovina.

16 See *Joint Statement on Stolac*, OHR Press Release, 1 June 2001.

17 *Hrvatska demokratska zajednica* = Croatian Democratic Union, the Croat nationalist political party.

18 Mosques in Bosnia-Herzegovina are frequently known by a variety of names. Čaršija = bazaar or market; Careva = Emperor's. The mosque is also sometimes referred to as the Čaršijska or Sultan Selim Mosque, and among other local names are the Old Masjid, the Carska or Imperial mosque, the mosque with eleven pillars and the mosque in the Tepa.

of the OHR in Mostar, the principles of Annex 8 were explicitly reaffirmed in a statement issued by the OHR and responsibility for their implementation was identified in every document regarding the reconstruction and protection of historic monuments subsequently issued by the Office of the High Representative.[19] This was particularly so up to the end of 2001, when all these principles were stipulated in the texts of the harmonized entity laws requested by the OHR on implementation of decisions of the *Commission to Preserve National Monuments* established under Annex 8.[20]

Everything that ensued concerning the reconstruction of the Čaršija Mosque in Stolac – from the formal context of issuing a permit for its rehabilitation to research, investigation and reconstruction – took place just as responsibility for the operations of the *Commission to Preserve National Monuments* was being transferred to the state of Bosnia-Herzegovina (after being under the aegis of UNESCO 1995–2000) and a law on the implementation of decisions rendered by the Commission was being enacted, and (with a number of other significant cases)[21] had a marked impact on determining the shape to be taken by the legislative and institutional framework as well as the overall strategy for the implementation of Annex 8.[22] The decision to ensure the transfer of authority for the implementation of Annex 8 to a professional, independent central state institution of Bosnia-Herzegovina was taken at a meeting at the OHR in Sarajevo in late June 2001,[23] convened to try to find a solution for the Stolac case, and was implemented on 21 December that year by resolution of the Presidency of Bosnia-Herzegovina.[24]

The case of the reconstruction of Stolac's built heritage could be said to be an ongoing model for testing possible solutions within the framework provided by Annex 8. If the reconstruction of the Old Bridge (*Stari Most*) in Mostar can be said to have been a kind of

19 In letters from High Representative Wolfgang Petritsch, Senior Deputy High Representative Matthias Sonn (27 June 2001), and in a communiqué issued by the Peace Implementation Council (*Communique by the PIC Steering Board, Brussels, PIC SB Political Directors, 30 October 2001*).

20 The preamble of High Representative Wolfgang Petritsch's decision of 8 February 2002 imposing the harmonization law on Republika Srpska reaffirms the importance of commitment of the authorities to implementing Annex 8 and it firmly stresses the link between the reconciliation process and cultural heritage. For more on the harmonization of the entity laws see Valery Perry in this volume. For the text of the decision see http://www.ohr.int/decisions/plipdec/default.asp?content_id=6888.

21 The Ferhadija mosque in Banja Luka, the mosques in Bijeljina and the Orthodox monastery in Žitomislić were others.

22 For more on Annex 8, its meaning, implementation, and the work of the *Commission to Preserve National Monuments* see Hadžimuhamedović, A.: 2007 Reconstruct or forget: European history and Bosnian reality, in *Unity and Plurality in Europe, Forum Bosnae*, 38, 222–44; 2008b Sedam godina rada Komisije za očuvanje nacionalnih spomenika/Seven years of work of the Commission to Preserve National Monuments, *Baština/Heritage*, Annual Journal of Commission to Preserve National Monuments 4/08; 2009 Naslijedje u miru nakon rata – slučaj Bosne/Heritage in post-war peacetime – the case of Bosnia, *Baština/Heritage*, Annual Journal of Commission to Preserve National Monuments 4/08; 2010 Integriranje naslijeđa u proces poslijeratne obnove/Integrating the heritage into post-war revival, *Baština/Heritage*, Annual Journal of the Commission to Preserve National Monuments, 5(10), 23–63.

23 Personal knowledge. The author was present at the meeting as well as at all other meetings organized either by OHR or by Presidency of Bosnia-Herzegovina at which the strategy of Annex 8 implementation was drafted.

24 The *Commission to Preserve National Monuments of Bosnia-Herzegovina* officially began operations in March 2002. The Commission is undoubtedly one of the country's most successful state institutions, recognized in 2010 when it received the European Heritage/Europa Nostra Award for dedicated service.

test of the international community's political, technical, professional and formal will to take a position and proceed accordingly in cases of atrocities against the cultural heritage, then the case of Stolac can be described as the same kind of test on the institutional, legal and political stage in Bosnia-Herzegovina. In another analysis, beyond the scope of this chapter, it can be shown that neither the international community nor the national authorities of Bosnia-Herzegovina have demonstrated unambiguous resolve, and that both lack a systematic approach to the heritage following the ravages of war.

In other spheres the destruction of Stolac's heritage featured in the ground-breaking indictments for war crimes and crimes against humanity at the *International Criminal Tribunal for the former Yugoslavia* (ICTY) in the *Prlić, et al.* case[25] and was the subject of the conference held in partnership with the Helsinki Human Rights Committees of Bosnia-Herzegovina and Norway (Sarajevo, April 2004) where the *Declaration on Human Rights and the Destruction of Cultural Memory: the Case of Stolac*[26] called for the establishment of an international legal framework in which human rights and the cultural heritage are intimately linked and the destruction of the cultural heritage is seen as a violation of human rights.

The Heritage and the Return of Refugees and Displaced Persons

One's homeland is reflected in the heritage – evidence of permanence and a kind of 'immortality.' Displaced persons cannot become 'true' returnees until their past is recreated in the image they carry with them as the last sight of their home town before they were expelled. That nostalgia which is generated by the absence of visible signs of belonging to a particular place creates this relationship between displaced persons and their sometimes imagined heritage which, if it is to symbolize their home town, must take physical shape, at times by way of reincarnation.

Boym's non-technical definition of restoration offers an understanding of the link between human rights and fundamental freedoms and the destruction and rehabilitation of the built heritage by seeking to recreate familiar surroundings, a process of liberation from the chaotic, unrecognizable image of the home town generated by its destruction:

> Restoration (from re-staure, re-establishment) signifies a return to the original stasis, to the prelapsarian moment. The past for the restorative nostalgic is a value for the present; the past is not a duration, but a perfect snapshot. Moreover, the past is not supposed to reveal any signs of decay; it has to be freshly painted in its 'original image' and remain eternally young.[27]

The displaced people of Stolac derived their right to reconstruct their historic centre as a first step towards restoring the townscape in which their past is authentically expressed from

---

25 *Prosecutor v. Jadranko Prlić, Bruno Stojić, Slobodan Praljak, Milivoj Petković, Valentin Ćorić and Berislav Pušić, Second Amended Indictment*, ICTY Case No. IT-04-74-Pt, *Stolac Municipality*, paras 162–6, describing the destruction of property, religious buildings and the heritage as a violation of the provisions of international humanitarian law and of the laws and customs of war.

26 Hadžimuhamedović, A. (ed.) 2005b *Human Rights and Destruction of Cultural Memory – the Stolac Case*. Sarajevo: Helsinki Committee for Human Rights in Bosnia and Herzegovina and Norwegian Helsinki Committee. See also http://humanrightshouse.org/Articles/5519.html.

27 Boym, S. 2001. *The Future of Nostalgia*. New York: Basic Books, 49.

Fig. 6.3 The ruins of great Ottoman housing complex of Behmenluk in Stolac, with a journalist from *Feral Tribune*, published in the magazine on 2 May 1994: the first documentary record of the destruction. © Amra Hadžimuhamedović / Feral Tribune

their sense of nostalgia for the (violently-destroyed) right to experience gradual changes to their surroundings. Thus the aim of integrating the heritage into the post-war reconstruction of Stolac was always framed in such a way that displaced persons who are to become returnees were always central to the process. The conservation professionals involved, however, were to find themselves providing a critical balance between those wanting their past to be more real than their present reality and the need to maintain the authenticity and integrity of the heritage.

Efforts to include the cultural heritage of Stolac as part of the solution to the fate of the town's refugees and displaced persons began in the autumn of 1993 soon after its destruction, at a time when not even United Nations observers or international journalists could enter the ravaged town. At the first meeting of displaced citizens of Stolac living in Sarajevo it was agreed that one of the issues to be at the forefront was the fate of the cultural heritage which was a matter of surmise only at the time, based on the unreliable accounts of a few young people who had managed to evade capture and find their way into Sarajevo, itself under siege. Few other towns laid waste during the 1992–1995 war in Bosnia were so cut off, so inaccessible.[28]

28 In December 1993 the author met with Vinko Puljić (Archbishop of Vrhbosna since 1990; Cardinal since 1994) and his secretary in the Archbishop's Palace, Sarajevo, in order to incorporate their information on the destruction of Catholic property in Vrhbosna Archbishopric into an official report of the Information Office of the Government of Bosnia-Herzegovina. Following the official business Archbishop Puljić was asked whether it was true that all of Stolac's mosques had been razed to the

The first published report on the state of the devastation in Stolac was compiled by journalists from the Split-based magazine *Feral Tribune* who managed to enter the town in the spring of 1994 by following the vehicle carrying Ratko Perić, Roman Catholic Bishop of Mostar,[29] on a visit to Stolac to hold a confirmation service in the Catholic church of St Elijah, the town's only religious structure still standing. The Croatian journalists produced a dramatic photographic report, the first evidence of the destruction of Stolac's built heritage. The report constitutes a unique documentary record, since the remains of most destroyed monuments were later removed.[30] The photographs revealed the extent of the damage, resembling that of the European cities that had been carpet-bombed during World War II.

Archive research, opinion polls and gathering material for a study of the structure and form of Stolac before its destruction began at that time in Sarajevo under the auspices of an informal group called the *Society for Renewal of Civil Trust in the Stolac Municipality* (*Udruženje za obnovu građanskog povjerenja općini Stolac)*.[31] For many members of the Society, mainly displaced persons from Stolac, discussions about the built heritage based on the collated materials were a kind of virtual return to their home town: collecting photographs and family reminiscences and making sketches from memory of architectural details were an attempt not to forget, a way of healing the malady of 'placelessness'[32] from which they were suffering. By 1996 extensive research had been conducted and a study titled *Crimes in Stolac Municipality* had been published by the Society, setting out (along with other war crimes) the scale of the destruction of Stolac's immovable and movable heritage on the basis of the information available at the time.[33] This study became the starting point for drafting post-war reconstruction plans and identifying priorities for action.

The dedicated remembrance of their home town by members of the *Society for Renewal of Civil Trust in the Stolac Municipality* at their meetings prompted the introduction of opinion polling to assist in drafting a strategy for the post-war reconstruction of Stolac.[34] The survey, conducted during 2000 and 2001 among a random sample of 146 returnees, refugees, displaced persons, and natives of Stolac now living elsewhere, was intended to test the hypotheses that the heritage is vital to the return of those who were forced from their homes, that re-establishing the sense of belonging to a place is more important than 'elementary' needs and that restoring the lost forms of the destroyed town is the equivalent of restoring the recognizable cultural landscape, and completes the return to the homeland.

---

ground, to which he replied that he did not know, but that he would not be surprised, since in 1992 certain Muslims from Stolac had welcomed the arrival of JNA (Yugoslav People's Army) tanks in Stolac.

29 Ratko Perić has been Bishop of Mostar and Duvno since May 1992. He is well known for his remarks relativizing the crime of destruction of the non-Catholic heritage in his diocese. Bishop Perić's political actions and his considerable influence on the authorities in Stolac are also plain to see in his published speeches, letters and other writings.

30 Malić, G. 1994. Aladin više ne stanuje ovdje: Stolac grad bez prošlosti. *Feral Tribune*, 2 May, 14.

31 *Udruženje* is sometimes translated as 'Association.'

32 The term 'placelessness' was coined by Eduard Relph in Relph, E. 1976. *Place and Placelessness*. London: Pion.

33 Hadžimuhamedović, A. 2000. *Zločin u Općini Stolac/Crimes in Stolac Municipality (1992–1996)*, edited by E Mulać. Mostar, second revised and enlarged edition 1996, Sarajevo: DID. Given the prevailing insecurity at the time the author's name was not included in the published edition of the book.

34 The aim, contents and findings of the surveys and the conclusions drawn from the findings were presented in Hadžimuhamedović, A. 2008a. The meaning of homeland: Heritage and uprootedness, *Forum Bosnae* 44/08, 328–46.

Sections were designed to assess the relationship between cultural memory, exile and return; to seek respondents' views on the importance of the heritage to the process of return and the restoration of a town in which nearly every building of cultural and historic significance had been totally eliminated or was in ruins; to discover the priority they accorded restoration of the heritage in their notion of sustainable return, and a question designed to indicate the community's views on the central issue: how to restore? Questions asking how the most important religious buildings within the structure of the town might be restored offered choices intended to resolve the dilemma of whether to reconstruct them or to erect modern buildings.

Even after the signing of the Washington Agreement in 1994, Stolac remained equally cut off and no less devastated than before, seemingly beyond the reach of the laws and authorities of Bosnia-Herzegovina. With the war ended in December 1995, in early 1996 UNHCR, the United Nations refugee agency, backed by the European Commission's Humanitarian Aid & Civil Protection (ECHO), launched a pilot project for the return of Bosniak and Bosnian Serb displaced persons to their home town of Stolac.[35] The project, which included repairs to one hundred houses that had been damaged, was forcibly opposed: the returnees and construction workers were subjected to physical violence and the restored houses were dynamited and set on fire on more than one occasion.[36] In this climate of hostility and aggression returns did not begin until June 1997, but the numbers of returnees initially remained small in the face of the continued intransigence of the local authorities and the lack of security.

It was only in 2000, after four years of effort, that substantial numbers of displaced persons were finally able to return.[37] Large parts of the town they had been forced to leave seven years earlier had become completely unrecognizable. Where the old buildings and streetscapes that constituted their familiar homeland had once stood were now empty spaces and overgrown ruins, rubbish dumps or car parks. At the town's heart, the site of the Čaršija Mosque was a blank gravel plain on which from 1998 the local Catholic priest, in association with the municipal authorities, placed a nativity scene over the Christmas period – a pointed message to the returnees of the transformed identity of Stolac. In this changed townscape, compelled to live in one area of the town that had all the features of a ghetto,[38] those taking part in this pilot project were not true returnees. In such a context, the

35 As part of the *Dayton Pilot Projects* agreed at Dayton on 2 November 1995. Stolac was one of four towns chosen as the first programme to target so-called 'minority' returns (returnees had often formed the pre-war majority in the towns targeted). The other towns were Bugojno, Travnik and Jajce. With the agreement between the 'non-warring' signatories of the Washington Agreement, returns were intended to take place by the end of 1995. See *Bosnia Peace Operation: Pace of Implementing Dayton Accelerated as International Involvement Increased,* US General Accounting Office (USGAO), GAO/NSIAD-98-138, 1998, 116, and: Refugees in B&H Do right solutions exist?, *AIM Sarajevo*, 23 June 1997.

36 See for instance International Community Joint Press Briefings 3, 5 and 7 February 1997, at http://www.nato.int/SFOR/trans/trans-prev.htm and *OHR Human Rights Report, May 1998*, 15 May 1998; *Report on the situation of human rights in Bosnia and Herzegovina, the Republic of Croatia and the Federal Republic of Yugoslavia (Serbia and Montenegro) submitted by the Special Rapporteur of the Commission on Human Rights*, to the General Assembly of the United Nations, 24 October 1999, A/54/396, S/1999/1000.

37 Cousens, E.M. and Cater, C.K. 2001. *Toward Peace in Bosnia: Implementing the Dayton Accords*. Colorado: Lynne Rienner Publishers, 77.

38 This characterization of ghettoization was made by many outside observers. See for instance: United States Committee for Refugees and Immigrants, *U.S. Committee for Refugees World Refugee Survey 1998 – Bosnia and Herzegovina*, 1 January 1998.

need to establish the image of the homeland by reconstructing the most significant features of the built heritage becomes the most vociferous demand of displaced persons.

By 1998 UNHCR and ECHO's implementing partner, the Danish Refugee Council, had restored the proposed one hundred houses, most of which were distinctive traditional stone-built houses with stone-clad roofs. However, the houses' special features were completely ignored in the rebuilding: most houses were increased in height, reinforced concrete was used, window openings were enlarged and fitted with metal or PVC frames instead of wood, and the roofs clad with red tiles or corrugated iron, so that this humanitarian action in fact constituted a lasting obstacle to the restoration of the destroyed cultural memory.[39]

The *Society for Renewal of Civil Trust in the Stolac Municipality* adopted a statement of principles for the restoration of the destroyed built heritage on 3 March 2000.[40] The starting point consisted of a triple definition of the town through image, structure and form.[41] The first principle specified that the structure of the town must be re-established by rebuilding its landmarks and nodes – the town's central points, the places imbued by its citizens with the most powerful meaning. Discussions on how to reconstruct historic buildings and on the methodological approach were conducted in the light of the debate over authenticity and integrity, the value of the material and intangible features of the built heritage, and the symbolic, social and townscape role of historic monuments.

Crucial to the form ultimately adopted were the views of returnees, as expressed in the surveys undertaken during the opinion polling described above.[42] At the time the survey was undertaken in 2000–2001 there were three proposals for the post-war reorganization of Stolac town centre:

- the regulatory plan proposed by the Bosnian Croat nationalist-dominated municipal authorities for a large, central open space to be known as Ante Starčević Square[43] where the Čaršija Mosque had stood.
- a scheme proposed by the Minister of Education, Science, Culture and Sport of Federation of Bosnia-Herzegovina and designed by the architect Zlatko Ugljen (whose Hotel Bregava of 1979 was among the buildings devastated) for the reconstruction of the destroyed buildings of the Čaršija Mosque and surrounding structures to new, modern designs.
- the solution developed by the *Society for Renewal of Civil Trust in the Stolac Municipality* proposing reconstruction of Čaršija Mosque and its surroundings as they were prior to their destruction.

The first proposal was a continuation of the plan to wipe out the cultural memory in Stolac. The void at the heart of the town was filled temporarily with elements designed to reshape the historical context – crosses, car parks and Christmas cribs – backed by the politically dominant factions in Stolac municipality and with financial support from the authorities. Thanks to the memory of the people of Stolac who were driven from their homes, however,

39 See Tina Wik on this subject in this volume.

40 Personal knowledge. The author was present at the meeting. The document with the title *Letter of Intention* is in the author's private archive.

41 This definition may be derived, for example, from the theoretical postulates set out in works such as Kevin, L. 1960. *The Image of the City*, Cambridge: The MIT Press; and Norberg-Schulz, C. 1971. *Existence, Space and Architecture*, New York: Praeger Publishers.

42 See note 33.

43 Ante Starčević (1823–1896) was a Croat nationalist politician and writer.

the old layers that had been 'scraped off' yielded a recognizable place in every image, every account and every vision of their future.

The second proposal was heavily promoted by the Federation authorities through an exhibition and publications and was ready to be backed by Turkish donor funds.[44] In 1994 Zlatko Ugljen had presented a modern replacement design for the destroyed Ćuprijska Mosque in Stolac,[45] and in 2000–2001 he was commissioned by the then Minister of Education, Science, Culture and Sport of Federation of Bosnia-Herzegovina, Fahrudin Rizvanbegović, to produce a new design for the Čaršija Mosque. Rizvanbegović strongly advocated Ugljen's design at meetings with reis ul-ulema Mustafa Cerić (head of the Islamic Community in Bosnia-Herzegovina) in June 2001 and resolutely opposed reconstruction of the mosque to its original form.[46]

But the opinion poll was to show that this attempt to use modern architecture to conceal the trauma of wounds that were still raw was wholly unacceptable. It revealed, rather, that the preferred option for returnees, refugees, displaced persons, and natives of Stolac living elsewhere was the third proposal for restoration to the town's pre-war state developed by the *Society for Renewal of Civil Trust in the Stolac Municipality*. Indeed, there was almost complete unanimity on the desire for the restoration of Stolac's built heritage and its importance in speeding refugee return.

The Reconstruction of Stolac's Heritage

Thus in the case of Stolac, the integration of the heritage into post-war reconstruction aimed to preserve its unique historic townscape, to recreate the setting required to reconnect returning refugees and displaced persons with their home town and to ensure that the heritage became one of the key factors in its sustainable development. Among the critical starting-points identified and developed by the author through an interactive relationship with those involved in return to Stolac was the need for comprehensive preservation of the structure of the historic urban centre in terms of both its physical layout and the associated evocation of meaning generated by physical structures, respecting their last known state prior to destruction. This reconstruction has been conducted systematically and largely extra-institutionally as part of the activities of civil society.

The reconstruction of Stolac's heritage began in 2001 with the rebuilding of the town's central and oldest mosque, the sixteenth-century Čaršija Mosque, for which

44 Alija Rizvanbegović, pre-war president of the SDA party in Stolac (who had been a refugee in Istanbul during the war) called Nedžad Čerkez in August 2001 while the author was present, during the period when the works clearing the site of the Čaršija Mosque were in progress, and informed him that donors from Turkey were ready to provide the full budget for the construction of a new modern mosque at the site of the Čaršija.

45 The design was presented at an exhibition in 1996; see Ugljen, Z. 1996. *Izložba: Džamija Hadži Alije Hadžisalihovića; Vjersko i kulturno sjedište Plehan*, Zagreb. The mosque is also known as the Hadži Alija Hadžisalihović Mosque. Ugljen's design was not used in this case, either, and the Ćuprija Mosque has been reconstructed to its pre-war form.

46 The author was present at the meeting. Ugljen's design is presented in Bernik, S. 2002. *Arhitekt/Architect Zlatko Ugljen,* Međunarodna galerija portreta, Tuzla. Both Ugljen's Ćuprija and Čaršija Mosque designs can be found in Tomšič Čerkez, B. 2010. Arhitektura med gradnjo in rušenjem identitete / Architecture between building and destroying identity. *Arhitektura, raziskave / Architecture, Research*, 2010/2, 23–30. At the same time, while being minister, Rizvanbegović committed 300,000 KM for the reconstruction of his own historic house in Stolac, Begovina.

Fig. 6.4 (*facing page*) **(a)** The historic urban core was transformed into a blank weed-strewn space, and Stolac became a town without a centre: the site of the Čaršija Mosque during the winter of 1995–1996, before Bosniaks had returned to the town. © Kemal Zukić / Centra za Islamsku Arhitekturu, Sarajevo. **(b)** Work begins on the reconstruction of the Čaršija Mosque, August 2001: discovering fragments of mosque at the site where its remains were dumped. © Amra Hadžimuhamedović. **(c)** Tombstones from the graveyard of the Čaršija Mosque after numbering and recording. © Amra Hadžimuhamedović

a preliminary project had been drawn up following eight years of research. As the only surviving documentation for the mosque was a single pre-war architectural survey, apart from consulting the usual array of available records, research also included analysis of such finds as a video recording of a 1980s television documentary on the mosques of Stolac found among rubbish dumped on the site of a destroyed house in the town. This was the first instance of one of Bosnia- Herzegovina's totally destroyed monuments, the remains of which had been removed from the site, being fully reconstructed through anastylosis, with repristination where anastylosis could not be applied.

It was a complex project carried out in extremely difficult circumstances with a very limited budget, lacking the necessary equipment, and without institutional support. The project management, supervision, research, documentation, decoration and so on, and the greater part of the physical reconstruction work was carried out by volunteers, both experts and local people, with a professional construction company brought in only as building works became more complex. Despite these challenges, as work proceeded a sizeable team of professionals from various backgrounds was formed, with the author of this chapter as responsible conservator and project manager, and a methodology was evolved which was to become a model for dealing with other destroyed cultural heritage properties in Bosnia.[47]

Most of the funding for the restoration was contributed by the people of Stolac, as well as by supporters from the region and around the world. The *Society for Renewal of Civil Trust in the Stolac Municipality* produced widely disseminated leaflets with information on the reconstruction strategy, plans and activities, as well as a breakdown of costs, while lists of contributors and their donations were regularly published, together with reports on activities and expenditures. The Society also issued a call for volunteers, and anyone could participate in the physical work of reconstruction under the guidance of the conservation architect. The investigation of the remains of the mosque and its reconstruction was to become a factor of wide social significance, involving both returnees to the town and their many supporters in Bosnia and abroad.[48]

Stolac returnees began the reconstruction process on 22 August 2001 by clearing garbage and earth from the site to uncover the structure's foundations.[49] By the end of

47 The author has been a member of the Commission to Preserve National Monuments since 2001 and has influenced both the policy of integration of cultural heritage into post-war recovery and the practical projects that Commission has been responsible for across Bosnia-Herzegovina.

48 For the history of the mosque, its qualities and the conservation approach to its rehabilitation, see Hadžimuhamedović, A. 2005c Redefinicija primjene metoda zaštite i prezentacije graditeljskog naslijeđa u procesu poslijeratne obnove – slučaj Čaršijske džamije u Stocu/Redefinition of protection and presentation methods of the architectural heritage in the process of post-war rehabilitation – the case of the Čaršija Mosque in Stolac, *Baština/Heritage, Annual Journal of the Commission to Preserve National Monuments*, 1(05), 71–117.

49 Officially through the Board for Reconstruction of Stolac Mosques, composed of returnees.

(a)

(b)

(c)

Fig. 6.5 **(a)** Aerial view of the centre of Stolac and the site of the Čaršija Mosque in September 2001 after clearing and archaeological research. The foundations of the mosque, a stack of recovered material and the protective wooden fence surrounding the site can be clearly seen. © Amra Hadžimuhamedović. **(b)** Bosniak returnees and other volunteers working on the rebuilding of the Čaršija Mosque in August–September 2001.The ruined eighteenth-century Šarić house (the Branko Šotra Gallery) can be seen in the background. © Amra Hadžimuhamedović

Fig. 6.6 **(a)** Stolac, September 2002. Behind the wooden fence erected to protect the site, the minaret of the Čaršija Mosque nears completion. © Helen Walasek. **(b)** First prayers in the restored Čaršija Mosque, August 2003. Worshippers spill out onto the porch of the mosque. © Amra Hadžimuhamedović

August the foundations and parts of the wall of the mosque had been found intact, as well as those of the minaret, the *šadrvan* (fountain), the *minber* (pulpit), inner pillars, *čatrnja* (well), the right half of the porch and the paving under the porch. Four days after work began local Croat residents who had witnessed the destruction informed the Islamic Community of the location of the remains of the mosque which had been dumped on the banks of the Bregava River. The reconstruction of the Čaršija Mosque in Stolac was thus also a crucial contribution to the first signs of restoring civil trust. Serb residents, who had returned in small numbers to Stolac up to 2001, participated actively in the reconstruction process; one of them, Bogdan Nikolić, was responsible for the restoration of iron elements of the mosque.

The Čaršija Mosque was officially reopened in August 2003 on the completion of the greater part of the building works.[50] Its reconstruction brought the town back its centre and symbolizes the demand for the human right to participate in public life. The weed-covered dead space at the heart of Stolac became again a key point of the town on which all activities converge, a place of public gatherings, coffee houses and shops. An old Serb inhabitant watched the process of reconstruction from his house in the centre of Stolac after eight years of exile. When the minaret was erected he said: 'Now I feel I am at home.'

By the end of the Čaršija Mosque reconstruction project several hundred Stolac residents had worked on a voluntary basis every day for two years in the physical

50 Although some of the interior works in the Čaršija mosque, including the finishing touches to the mihrab and the reconstruction of the stone minber, remain incomplete at time of writing.

investigative and building works; several hundred more helped by organizing meals for the workers, providing accommodation for non-local experts and holding exhibitions, lectures and presentations for children and adults on the cultural memory of Stolac. The masons, carpenters and roofers who worked on the project were all from Stolac. The work was also assisted by members of the International Police Task Force (IPTF)[51] and the international military Stabilization Force (SFOR),[52] who carried out daily accurate and detailed video and photographic recordings on the ground and from the air of all the finds and every stage of the work. Bosniaks, Serbs, Croats, French, British, Americans and Italians, Muslims, Jews, Catholics, Orthodox Christians, Anglicans and others – all participated in the reconstruction of the Čaršija Mosque.

The process of integrating restoration of the built heritage into the return of refugees and displaced persons and the exercise of their human rights as a prerequisite of a sustainable peace has been ongoing in Stolac and its surrounding area since 2001. The projects that have been carried out at the time of writing include four mosques in the town centre, ten traditional houses, three traditional Islamic primary schools (*mekteb*, Koranic school), the *havra*, three Orthodox churches (one in Stolac town centre, and the others in Ošanići[53] and Trijebanj), one *konak* (guest house), a number of shops in the čaršija, Orthodox cemeteries in Poljuba and Bokulja, two necropolises with mediaeval *stećak* tombstones, and the fort, with its many historical strata from the Roman to the Austro-Hungarian period. Most monuments and groups of buildings have been rehabilitated thanks to the efforts of non-governmental organizations.[54] As a consequence of such activities more and more owners of private historic properties are using traditional materials and seeking expert assistance when making their damaged houses habitable again. Thus the old silvery-grey stone roofs, unrendered stone walls, cobbles and woodwork are being restored to the Stolac townscape.

Obstructions to the Integration of the Heritage into Post-war Reconstruction in Stolac

However, not one of the projects described above, carried out to protect and preserve the cultural heritage of Stolac, enjoyed the support of the local authorities who continued to pursue their policy of cultural and ethnic exclusivism that during the war had taken the form of ethnic cleansing and destruction of the cultural heritage. The authorities not only failed to welcome the reconstruction of destroyed monuments but fiercely contested the entire process, placing every obstacle they could in its way, a tactic they were still employing in 2012.

51 The IPTF or International Police Force was one of the major organizations within the United Nations Mission in Bosnia-Herzegovina.

52 The Stabilisation Force (SFOR) was the NATO-led multinational peacekeeping force in Bosnia which was tasked with upholding the Dayton Peace Agreement from 1996 to 2004.

53 Sometimes referred to as Ošanjići.

54 The organizations include International Forum Bosnia, Lëtzebuerger Guiden a Scouten, Stolac Youth Forum, Society for the Restoration of Civil Trust in the Stolac Municipality, Committee for Reconstruction of the Stolac Mosques, Church-Parish Council in Stolac, Scouts' Union of Bosnia and Herzegovina, Croatian Helsinki Committee, as well as the International Summer School of Religion and Public Life.The Commission for Protection of National Monuments of Bosnia and Herzegovina has also participated in the activities.

The municipal authorities persisted in their refusal to issue a permit for the reconstruction of the Čaršija Mosque to the Islamic Community, despite a warning letter from OHR Senior Deputy High Representative Matthias Sonn on 27 June 2001 requesting that approval be issued immediately in accordance with Annex 8.[55] A permit was eventually issued by the Federation Ministry of Regional Planning and the Environment on 27 October 2001, a permit Željko Obradović, the mayor of Stolac, declared 'illegal.' In December 2001 Obradović went on to announce his decision to suspend the reconstruction, shortly after a Croat nationalist mob had destroyed part of the fence surrounding the site.[56] Work nevertheless continued, but violence and opposition continued to accompany the mosque's restoration in the form of protests, arson, physical attacks and threatening behaviour.[57]

In their efforts to preserve the effects of the wartime destruction, the local authorities had the full backing of the Roman Catholic Church, in the persons of both the Stolac parish priest, Don Rajko Marković and the Bishop of Mostar, Ratko Perić. The *Society for Renewal of Civil Trust in the Stolac Municipality* actively addressed this resistance to the reconstruction of the destroyed Čaršija Mosque, the role of Bishop Perić and other Catholic priests in justifying these atrocities and persisting in their exclusivist attitude towards non-Catholics, human rights, justice and the significance of the cultural heritage. The six open letters sent to Bishop Ratko Perić by Senad Mehmedbašić, president of the Society and Bishop Perić's responses are valuable documents setting out the philosophical basis for advocating reconstruction and the opposition to it, and attitudes towards issues of continuity and discontinuity, identity and rights to the cultural heritage.[58]

When the reconstruction of the historic centre of Stolac began in 2001, marking the start of the genuine return of refugees and displaced persons, asserting their rights to their heritage and identity, it was accompanied by a post-war 'make-over' of cultural memory – new books were written, new cultures proposed, new histories concocted by those who had supported the ethnic cleansing.[59] It is significant that this aspect of the cultural 'extinction' of the non-Croat Stolac community was not put into effect, or at least not systematically and intensively, until displaced persons began to return to the town and work started on the reconstruction of the destroyed heritage.

From spring 2001 Don Rajko Marković and Bishop Perić began to promote the idea of the existence of an earlier Christian structure on site of the Čaršija Mosque and the importance of archaeologically investigating the site, based on the 'evidence' of traditional stories and a reference by the respected Bosniak historian Hivzija Hasandedić in his book on Islamic heritage in eastern Herzegovina; Hasandedić later issued a public condemnation of

---

55 See *Sonn requests rebuilding of Careva Mosque in Stolac*, OHR BiH Media Round-up, 2 July 2001 http://www.ohr.int/ohr-dept/presso/bh-media-rep/round-ups/default.asp?content_id=533.

56 OHR BiH Media Round-up, 15 November 2001, http://www.ohr.int/ohr-dept/presso/bh-media-rep/round-ups/default.asp?content_id=6349. On Obradović's decision to suspend the reconstruction see OHR Press Office Chronology/Monthly Tracker 2001.

57 See NATO/SFOR Joint Press Briefing 4 December 2001, http://www.nato.int/SFOR/trans/2001/t011204a.htm and 6 December 2001 http://www.nato.int/SFOR/trans/2001/t011206a.htm.

58 See Mehmedbašić, S. 2001a. *Letter to Bishop Perić: 9 August 2001*; 2001b. *Second Letter to Bishop Perić: 22 September 2001*; 2001c. *Third letter to Bishop Perić: 8 December 8 2001*; 2002. *Fourth Letter to Bishop Perić: 28 January 2002*, all Society for Renewal of Civil Trust in the Stolac Municipality, Stolac.

59 Kuzman, Z. (ed.) 2002 *Stolac mjesto spora umjesto razgovora. Radovi s Okruglog stola Stolac, 5 listopada 2001,* (*Papers from a seminar on Stolac, 5 October 2001*), Stolac: Municipal Authorities of Stolac.

this wilful misinterpretation of his text.[60] While Marković applied for planning permission to construct a church on the site of the mosque,[61] Bishop Perić, who had never condemned the destruction of the mosques in Stolac, declared that it would be a 'crime' to 'forcibly' rebuild the mosque on the site of a church.[62]

As described above, on 1 June 2001 a high solid fence was erected by the Stolac authorities around the site of the Čaršija Mosque barring access to it, while a sign announced that archaeological excavations were to be conducted on the site by the Institute for the Protection of the Cultural Heritage of Herceg-Bosna, funded by the Municipality of Stolac and the Ministry of Culture of Herzegovina-Neretva Canton. It was clear the excavations were aimed at proving that a church had existed on the site of the mosque and were solely to act as justification for preventing its reconstruction. Rumours grew in the Bosniak community that the remains of a medieval church taken from a museum in Zadar were to be planted as 'evidence' at the site. This action led to the meeting called by the OHR Deputy High Representative and the agreement that the fence would instead serve to protect the site of the Čaršija Mosque and that no archaeological or other works would take place there except in accordance with Annex 8.[63]

But those in the process of rewriting cultural memory and intent on blocking the reconstruction of the mosque did not stop there. On 5 October 2001 a roundtable was held in the town titled *Stolac Region in the 16th Century, Religious and Political Situation*, based on a 'scientific' framework provided by the archaeologists Boris Graljuk and Snježana Vasilj and academics Mladen Ančić and Ante Škegro, who confidently asserted the former existence of a church on the site of the Čaršija Mosque.[64] A further call for archaeological investigations to take place at the site was made at the roundtable, which was hosted by the then Mayor Zdravko Kuzman, and attended by, among others, prominent Catholic churchmen including the Vicar General of Herzegovina bishopric, Don Luka Pavlović.

It never crossed the mind of anyone from the Stolac local authorities or their ideologists to raise the issue of archaeological excavations on the sites of destroyed mosques until work began on their reconstruction. Only when the returnee community launched the process of rehabilitation of the heritage that had been the subject of destruction did the local authorities propose archaeological excavations as an alternative. The advocates of justifying the destruction proposed, as an alternative to the familiar image of the home town, an unfamiliar, dubious and more remote history based on speculation, not for the sake of the truth or of a more profound knowledge of the complex, multi-layered cultural heritage, but with the explicit aim of erasing the stratum of 'the other' from the town's complex chronology. Thus archaeology became the tool of ideologues and the re-writers of cultural memory, with archaeologists and historians joined by amateurs and quasi-experts to power the machinery of ethnic exclusivism.

The assertion that the remains of thus far unknown churches lie beneath the town's mosques and that it was vital to use archaeology to prevent them being covered over

60 Hasandedić, H. 1990. *Muslimanska Baština u Istočnoj Hercegovini*, Sarajevo: El-Kalem, 9. Hasandedić condemned the misuse of his text. See Hadžimuhamedović, A. (ed.) 2005a, 278.

61 Letters from Senad Mehmedbašić to Archbishop Leanza and Cardinal Puljić, 1 July 2002. See http://stolac.info/2002/07/letters-to-archbishop-leanza-and-cardinal-puljic/

62 Mehmedbašić, S. 2001b, 6.

63 OHR 2001. *Joint Statement on Stolac*, press release dated 1 June 2001, Sarajevo: Office of the High Representative.

64 Hadžimuhamedović, A. 2005b Cultural memory in Stolac – Destruction and reconstruction, in Hadžimuhamedović, A. (ed.) 2005a, 277.

by mosques was used once again to justify or relativize the crime of the destruction of cultural memory – about which more and more was studied and written as the process of reconstruction continued – hoping to encourage the Catholic/Croat population to oppose the reconstruction of mosques as threatening the destruction of churches. The roundtable described above was extensively promoted and discussed in the local media and its proceedings later published by the municipality.[65] For added credibility, the putative church on the Čaršija Mosque site was even given a name – the church of St Ann.[66] At the same time, crosses began springing up to mark public spaces in Stolac, again with the dual purpose of conveying the message to non-Catholics that they had no right to the public space and to arouse the sentiments of Catholic/Croats with a view to gaining their support for the opposition to the reconstruction of destroyed monuments.[67]

The opposition to the rehabilitation of the heritage destroyed during the war that began with the reconstruction of the Čaršija Mosque had a high-profile strategy clearly revealing the municipal authorities' views that cultural properties in the historic centre of Stolac should not be designated as National Monuments as provided for under Annex 8 (since this would constrain the authorities' decision-making capacity to transform the town), that cultural properties damaged or destroyed during the war should not be restored and that streets and sites in Stolac should be renamed as part of the rewriting of cultural memory.[68]

The Role of the International Community in the Integration of the Heritage into Post-war Reconstruction in Stolac

Between the destruction of Stolac in 1993 and 2001, the international community[69] took no steps to inform the wider world of the state of the devastation in Stolac. Unlike with Mostar and Sarajevo, the destruction of the heritage in Stolac was not the subject of official reports or condemnation, and its rehabilitation formed no part of the international community's plans.

One of the first significant operations by the international community in Stolac took place in 1999 and bore the code name *Operation Westar*.[70] The aim of *Operation Westar* was 'to disrupt the anti-Dayton activities of the Bosnian Croat secret service (the SNS) and to discover illegal activities of the Croatian National Intelligence Service' in the parts

65 See Kuzman, Z. (ed.), 2002. On the meeting see also OHR BiH Media Round-up 10/10/2001, *Dnevni List: Round Table Takes Place on the Stolac and Stolac Area in 16th century* at http://www.ohr.int/ohr-dept/presso/bh-media-rep/round-ups/default.asp?content_id=6033 and Dragić, Z. 2001. Croat scientists appealed for protection of cultural-historical heritage: Their Bosniak colleagues ignored gathering!, *Hrvatska Rijec*, 29 October.

66 Dragić, Z. 2001.

67 See Sells, M. 2003. Crosses of blood: Sacred space, religion and violence in Bosnia and Herzegovina, *Sociology of Religion*, 64(3), 309–31.

68 The views are presented in numerous letters from the mayor of Stolac to the Commission to Preserve National Monuments, OHR and the other authorities.

69 Using the term to denote the representatives of the inter-governmental bodies responsible for the implementation of the Dayton Peace Agreement including Annex 8, such as the Office of the High Representative, the United Nations, the Council of Europe, the European Union, UNESCO and so on.

70 For more on Operation Westar see Operation Westar at SFOR Online at http://www.nato.int/sfor/sfor-at-work/opwestar/indexsaw.htm and Warner, M. et al. 2000. *SFOR Lessons Learned in Creating a Secure Environment with Respect for the Rule of Law, based on a study of Bosnia, May 2000*, US Army Peacekeeping Institute.

of Bosnia under Croat control, particularly in Stolac and West Mostar. The materials confiscated during *Operation Westar* revealed the depth of infiltration by the Bosnian Croat Intelligence Agency into the staff of international organizations and could suggest one of the reasons for the international community's lack of success in ensuring that the heritage was protected and for its failure to apply the law, international conventions and the principles set forth in the resolutions of various international bodies.

The attitude of the international community, which could be described as ineffective and unconcerned with changing the status quo, was a betrayal of the principles set out in international documents such as the preamble to the Hague Convention,[71] or such eloquent statements as Resolution No. 4 on conservation in situations of conflict adopted at the Third European Conference of Ministers Responsible for Cultural Heritage (Malta, 16–17 January 1992), which not only condemned the destruction of the heritage in Europe, but called on international bodies to take part 'in concerted action to repair the damage done.'[72]

In 2001 when the *Society for Renewal of Civil Trust in the Stolac Municipality* began operating in the very heart of Stolac the international community (principally the OHR, SFOR and the IPTF) monitored its activities closely, but did nothing to hinder it. Indeed, it provided the necessary military and police protection for returnees to Stolac who had begun reconstruction by clearing the sites of destroyed monuments and assembling surviving fragments; without this protection reconstruction would have been impossible. This was a new task for members of the IPTF and SFOR and its importance and justification were described in an article by a 1st lieutenant in the Spanish SFOR Battalion: 'As stated in the General Framework Agreement for Peace (GFAP), Annex 8, SFOR troops are ready to provide support to international organizations and local authorities also in all matters related to national monuments.'[73] SFOR and IPTF personnel also actively assisted the reconstruction process with aerial and ground photography, as described above.

Beginning the reconstruction of the Čaršija Mosque in Stolac was a crucial factor in prompting the international community, particularly the OHR, to ensure the establishment of the *Commission to Preserve National Monuments* as a central state institution of Bosnia-Herzegovina, to insist on its powers and to give it authority over all decisions concerning the manner of dealing with the destroyed or damaged heritage. This was the most significant step made by the international community (along with a number of public condemnations) with regard to upholding international heritage protection principles in Stolac.

However, until 2007 the question of the heritage of Stolac was not considered in the light of the obligations assumed by the Conference of European Ministers of Culture when adopting Resolution No. 4 in Malta. Faced with increasingly high-profile reports on the condition and importance of the cultural heritage in the Western Balkans for the process of European integration, on 13 December 2007 the European Parliament approved the *Pilot Project for Preserving and Restoring Cultural Heritage in Conflict Areas*,[74] a plan

71 *Convention for the Protection of Cultural Property in the Event of Armed Conflict* 1954: '*Considering* that the preservation of the cultural heritage is of great importance for all peoples of the world and that it is important that this heritage should receive international protection.'

72 *3rd European Conference of Ministers Responsible for the Cultural Heritage*: *Proceedings, Malta, 16–17 January 1992*, Strasbourg: Council of Europe (COE), 1993, 60.

73 Sanchez, L. 2001. SFOR helps to preserve national monuments, *SFOR Informer Online*, 115, 13 June.

74 See Chapter 22 02 – *Enlargement process and strategy (cont'd) 22 02 08 Pilot project for preserving and restoring cultural heritage in conflict areas*; and 15. 22 02 08 *Pilot project for preserving and restoring cultural heritage in conflict areas, at eur-lex.europa.eu/.*

for implementing pilot schemes for preserving and restoring cultural heritage in conflict areas in the Western Balkans, for which a budget of €3 million was earmarked for heritage protection projects in Bosnia-Herzegovina and Kosovo.

The criteria set by the Delegation of the European Union to Bosnia-Herzegovina (DELBIH) were that the properties to be restored should be designated National Monuments, that they should have been damaged or destroyed during the war, and that they were not religious properties. The main aim of the project with respect to Bosnia-Herzegovina (as set out by the European Parliament) was to contribute to reconciliation between the different communities by means of the cultural, social and economic restoration of the Stolac community. The project, called *ARCH Stolac: Heritage as a bridge between communities*,[75] was designed to stimulate cross-cultural dialogue within the community and, in particular, to encourage refugees and displaced persons to return to their home town.

Following lengthy discussions between the *Commission to Preserve National Monuments* (which was chosen to implement the project) and the European Union Delegation it was decided that the project objectives could be achieved by reversing the damage to the cultural heritage caused by war and ensuring an inclusive approach to the method of evaluation based on current laws and standards as a means of ensuring justice. The heritage of 'others' that had been the subject of hatred and destruction should be reconstructed not only to affirm the right to be different, but also to demonstrate that the universal values of the cultural heritage belong to everyone.

The properties selected to be partly rehabilitated in Stolac were destroyed or damaged during the war and were all monuments of national importance: the seventeenth-century *hamam*, the sixteenth-century *musafirhana*, two houses in Behmenluk (an architectural ensemble that took shape from the sixteenth to the late nineteenth century) and the eighteenth-century flour mills. Work was to be conducted in full transparency so as to involve all citizens of Stolac and the region by holding workshops, lectures, visits to the various sites and 'open days' when people would take part in activities relating to the implementation of the project.

However, despite the agreement reached, as yet unexplained efforts began in spring 2010 to prevent the implementation of the project or to radically alter it. At a meeting on 8 April 2010 at the Ministry of Culture and Sport of the Federation of Bosnia-Herzegovina, the Delegation of the European Union in Bosnia and Herzegovina, together with the Ministry, altered the agreed list of monuments without consultation with the *Commission to Preserve National Monuments*. Radical changes to the aims of the project renounced the idea of rehabilitating properties damaged or destroyed by the war and instead proposed working on 'neutral' monuments like the bridges over the River Bregava and the archaeological site of the old fort of Vidoški (where the cross erected illegally in 2002 still stood with a view to bestowing on it an exclusively Croat/Catholic identity).

After having completed and agreed on the project proposal, which was grounded entirely on the basis of the law, European standards, international conventions and universal principles, the European Union Delegation in Sarajevo reopened the process. Attempts to understand the reasons for the alteration of the original approach set out by the European Parliament are based on the assumption that the mayor of Stolac Municipality, Stjepan Bošković, and his fellow party members holding ministerial posts at entity and national

75 The full name of the project is *ARCH Stolac: Heritage as a bridge between communities. Reconstruction of Architectural Heritage in Stolac – Sustainable Revival and Development of a war-torn community.*

level – Dragan Vrankić, the Minister of Finance and Treasury of Bosnia-Herzegovina (who for two years refused to sign the funding agreement because of Mayor Bošković's stance), as well as Gavrilo Grahovac, Minister of Culture and Sport for the Federation of Bosnia-Herzegovina – are not in favour of rehabilitating structures with 'Muslim' architectural features like the *hamam*, *musafirhana* and the Ottoman houses.

However, after suddenly changing the criteria for selecting monuments after two years of negotiations the EC Delegation indicated that the reason for the alteration was because the EU did not want to fund the reconstruction of buildings in private ownership – which the *musafirhana* and *hamam* were not. As the Croatian Helsinki Committee for Human Rights commented in their protest to the changes: 'These are the monuments that were destroyed and, whether in private or public ownership, it was they that reflected the diversity that was the object of destruction and 'ethnic cleansing' in Stolac, the victims of which were the owners of these historic monuments.'[76] In any case, every one of the original monuments was designated as a National Monument under Annex 8 and the decisions of the *Commission to Preserve National Monuments* require that such properties be restored to the state they were in prior to their damage or destruction during the war.[77]

Despite this, the European Union Delegation abruptly substituted as a method of evaluation the unprincipled method of 'agreement by all stake holders.' The heritage thus once again became not a tool for dialogue and the expression of universal values, but particularized within the dangerous ideological and nationalist model of agreement by power-holders in which the majority outvote the minority, the strong outweigh the weak, and the universal value of the heritage is lost to the horse-trading between simplistic, exclusivist identities: 'we'll agree to the restoration of your monument if ours is restored too.' This principle paves the way for endless areas of conflict, divisions, and a widening gulf of misunderstanding driving a wedge through a diverse Bosnian identity. Evidence of this can be seen in the reaction to the proposed changes by NGOs in Stolac involved in heritage protection, the Croatian Helsinki Committee for Human Rights and numerous individuals and groups in their letters of protest sent to officials of the European Union and the authorities blocking the rehabilitation of the Stolac heritage destroyed during the war.[78]

---

76 Croatian Helsinki Committee HHO (20 May 2010): *Letter concerning Stolac*, etc. See http://www.bosniaks.net/prilog.php?pid=38058.

77 See decisions of the Commission to Preserve National Monuments, http://kons.gov.ba, relating to the architectural ensemble of the Čaršija Mosque and Čaršija in Stolac, No. 08.1-6-915/03; 6 May 2003; the site and remains of the historic building of the Hammam on the Ćuprija (bridge) in Stolac, No. 09-2-123/06-3, 6 July 2006; and the residential architectural ensemble of the Behmen family house in Behmenluk mahala in Stolac, No. 07.1-02-75/07-12, 12 September 2008.

78 Such as *Stolac Youth Forum* and *Society for the Renewal of Civil Trust*: letter to the Prime Minister of the Federation of Bosnia-Herzegovina (cc. members of the Presidency, the Council of Ministers, the Ministry of Foreign Trade, the Federal Ministry of Culture and Sport, the *Commission to Preserve National Monuments*, OHR, the European Union in Bosnia-Herzegovina, the US Ambassador, the mayor of Stolac and the media): an account of the project and a call to ascertain who is blocking and relativizing the Pilot Project; *Caucus of Bosniak Councillors of Stolac* to the Prime Minister of the Federation of BH (15 April 2010): view on the ARCH-Stolac project, supporting the project proposed by the *Commission to Preserve National Monuments*; Hector Gullan to High Representative Valentin Inzko (25 April 2010): (cc. *Commission to Preserve National Monuments* (01.2-35.11-69/10-14), *Stolac Youth Forum*, *Society for Renewal of Civil Trust*, *Helsinki Committee in BH*, *International Forum Bosnia*, UK Ministry of Defence/War Crimes Office); *Croatian Helsinki Committee* HHO (20 May 2010): Letter concerning Stolac, etc.

At the time of writing the case of the *Pilot Project for Preserving and Restoring Cultural Heritage in Conflict Areas* remained unresolved. On 29 September 2010 the European Union Delegation announced the amendment of the project and the properties for restoration and that the project had been 'revised in order to reach consensus with all competent government authorities on the foreseen activities.'[79] The European Union administration has ignored the fact that the *Commission to Preserve National Monuments*, Bosnia-Herzegovina's central state authority for cultural heritage protection (with responsibility for implementing the project), strongly objected to the amendment of the project, as did numerous civil society organizations and returnees. Stolac once again became a test case for a range of political, legal and conservation-related principles.

Young People and the Heritage in the Post-war Reconstruction of Stolac

Integrating the heritage into education in Stolac is a significant process since education can serve to develop mutual respect and reconciliation by demonstrating the historically diverse patterns of community in the town, upheld and attested to by its heritage, while raising awareness of that heritage and its importance among young people and so ensuring its sustainability. Teaching about the heritage in Stolac's schools is particularly complicated given that pupils are rigidly segregated on the basis of ethno-religious affiliation.

The annual *Youth and Heritage* (Ćuprija) summer camps[80] have become one of the more important ways of integrating the heritage into post-war reconstruction in Stolac. From 2005 the programme, organized by the NGO *International Forum Bosnia* with its partners,[81] has brought young people from the region and around the world to spend three weeks at the camp beside the River Bregava, participating in heritage protection work, learning about the region's heritage, and taking part in workshops on co-existence, cultural tolerance and the heritage. Since their inception the camps have involved the young people of Stolac of different ethnicities, allowing them to exchange views with their peers and to understand their role in the protection of Stolac's heritage, while helping to break down the nationalist exclusivity forced upon them by the war. From 2008 the camps have been organized by the *Youth and Heritage International Centre,* a more permanent tool for integrating the heritage and its protection into youth education.

Conclusion

The reconstruction of the Čaršija Mosque in Stolac was not only an outcome of technical research and investigation, but also a crucial contribution to the first signs of restoring civil trust. Returnees to Stolac described themselves as living in a ghetto or a state of apartheid and began to demand that it be changed and since 2001 the heritage has played a key role as they begin to exercise their collective and individual human rights.

---

79 See *European Union support to the preservation of cultural heritage in Stolac* at *European Union support to the preservation of cultural heritage in Stolac* at http://www.delbih.ec.europa.eu. The Branko Šotra Gallery (the 18th century Šarić House) was added to the final list. See the 2011 project details on *Annual Work Programme for Grants 2011 – DG ELARG* at http://ec.europa.eu/enlargement/pdf/how_does_it_work/grants_tenders/grants/2011/list_of_grants2011_02-20110927.pdf.

80 See the camp's website at http://www.cuprija.ba/ for more information about its activities.

81 See the website of International Forum Bosnia at http://www.ifbosna.org.ba.

Even as the current authorities in Stolac continue to obstruct the implementation of Annex 8, and in the absence of systematic domestic or international institutional support for its implementation, nevertheless by 2012 the methodical rehabilitation of the heritage was in its eleventh year. Stolac is a fascinating case for this very reason: here heritage rehabilitation does not reflect the town's dominant power, as one might stereotypically characterize post-war reconstruction. Rather, the informal sector of civil society has evolved a strategy and systematic approach to the post-war reconstruction of the structure, image and form of the town, and its strategy is being implemented. The rehabilitation of the heritage in Stolac could be seen an example of the strength of meaning of the cultural heritage as a means of exercising human rights.

Integrating the heritage in the post-war reconstruction of Stolac as an undeniable fact that attests to the historic mingling of influences from different times and places, and all the meanings imbued in the heritage that generate its value and are the reason for its protection or for its denial and destruction, along with its reshaping and false interpretation, is one of Bosnia's most significant contemporary social phenomena.

Chapter 7
Restoring War-Damaged Built Cultural Heritage in Bosnia-Herzegovina: An International Perspective

Tina Wik

Introduction

Following the news from the former Yugoslavia during the 1990s was an unexpectedly shocking experience, particularly from a Swedish perspective where there had been no wars on its territory since 1809. When the Yugoslav wars of 1991–1995 erupted into widespread ethnic cleansing in Bosnia-Herzegovina and the systematic elimination of the symbols of peoples who had been forcibly expelled, this seemed an act of unbelievable violence and violation in a part of the world we considered 'civilized' western society. A group of Swedish heritage bodies, including ICOMOS Sweden, the Central Board of National Antiquities (Swedish National Heritage Board) and the Swedish National Commission for UNESCO, reacted to these events by convening an expert seminar in Stockholm in June 1994 on the protection of cultural property during armed conflict, looking particularly at the former Yugoslavia and considering what responses could be made to the destruction there.[1] The discussions at the seminar and the desire of a group of architecture, heritage preservation, and media professionals to take concrete action through emergency rescue aid to cultural heritage under attack, beginning in Bosnia-Herzegovina, led to the founding of the non-governmental organization (NGO) Cultural Heritage without Borders (CHwB) in April 1995.[2]

Cultural Heritage without Borders' mission is governed by the principles of the preamble to the 1954 Hague Convention for the Protection of Cultural Property in the Event of Armed Conflict[3] and its founding aspirations were to integrate conservation rescue work into standard emergency aid to preserve cultural heritage threatened by war, natural disasters, neglect, poverty or political and social conflicts. From the start CHwB found support at the highest levels of Sweden's cultural and governmental structures and several state and national organizations concerned with cultural heritage protection and preservation were institutionally linked to CHwB, with the National Heritage Board, the

---

1 *Information as an Instrument for Protection against War Damages to the Cultural Heritage. Report from a Seminar, June 1994*, Stockholm: Svenska Unescoradets skriftserie, 4/1994; the seminar was a follow-up to the 27th Session of the General Conference of UNESCO held in Paris 25 October to 16 November 1993.

2 CHwB's founding group was architect Jan-Krister Boman, heritage professionals Margareta Biörnstad and Bengt O H Johansson and journalist Ingamaj Beck. For more on CHwB and its activities in Bosnia-Herzegovina, Kosovo and elsewhere, and the projects described in this chapter see the CHwB and CHwB Regional Balkan websites at http://www.chwb.org/ and http://www.chwb.org/regional/index.php.

3 That 'damage to cultural property belonging to any people whatsoever means damage to the cultural heritage of all mankind, since each people makes its contribution to the culture of the world.'

National Committees for ICOM and ICOMOS, and the Swedish Association of Architects appointing six members of its nine-person board.

With a deteriorating security situation in Bosnia over the months following its foundation, CHwB was not able to begin work in the field until after the signing of the Dayton Peace Agreement in December 1995 and the end of the Bosnian War. In the meantime, it concentrated on building its network of contacts and potential sources of funding, in particular, the Swedish International Development Cooperation Agency (Sida). A ChwB board member accompanied the Swedish Minister of Culture an official visit to Sarajevo in February 1996 when contacts were made with the Zemaljski Muzej, the National Museum of Bosnia-Herzegovina – the start of a long-term relationship with the institution. However, the organization's first mission to the country to assess where assistance could best be targeted did not take place until April 1996, a year after the organization was founded.

CHwB entered into a dialogue with Sida and proposed that as part of its massive post-conflict aid programme in Bosnia-Herzegovina it begin to finance the conservation and restoration of public monuments in war-damaged areas, both as a human right of those who had been ethnically cleansed and as a peacekeeping strategy. They pointed out, too, the provisions for preserving monuments in Annex 8 of the Dayton Peace Agreement.[4] Restoring cultural heritage was not the kind of humanitarian aid Sida normally dealt with and this was a novel concept for them. But to accept this was to grasp the specificities of the conflict in Bosnia and to understand the central importance of restoring symbolic buildings for expelled populations as a vital ingredient in creating an environment that would enable them to re-establish themselves in their homes as a part of a return process.

The development of CHwB was also a novel endeavour for those involved, with few models to follow. At the time there were virtually no NGOs working internationally in the field of disaster aid to cultural heritage – which remained the case at time of writing.[5] CHwB was to be the only NGO carrying out restoration of historic buildings across Bosnia-Herzegovina and its work broadened into training and professional development, as well as support for museums and archives. CHwB's work later expanded beyond Bosnia-Herzegovina, but has remained so far almost exclusively in the Western Balkans.[6]

The Work in Bosnia Begins

Among those who founded CHwB were some of the most well-established heritage professionals in Sweden, notably Margareta Biörnstad, a former Director-General of the

4 The General Framework Agreement on Peace or GFAP, more usually known as the Dayton Peace Agreement or Dayton Accord, 14 December 1995. See *Annex 8: Agreement on Commission To Preserve National Monuments*; the full text of Annex 8 is available at http://www.ohr.int/dpa/default.asp?content_id=376. For more on Annex 8 see Valery Perry elsewhere in this volume.

5 Bosnia-Herzegovina Heritage Rescue (BHHR) had been working in Bosnia since 1992 and its director Marian Wenzel was a speaker at the 1994 Stockholm seminar. BHHR's work was an inspiration to the founders of CHwB. See Helen Walasek, *Destruction of the Cultural Heritage, etc.* in this volume for more on the work of BHHR. The French NGO Patrimoine sans Frontières was founded in 1992, again in response to the cultural heritage destruction in Bosnia, but did not carry out projects in the country.

6 The Western Balkan countries CHwB has worked in are Bosnia-Herzegovina, Kosovo, Montenegro, Serbia and Albania. CHwB's only involvement outside this region at time of writing was support for the restoration of the Beri Monastery, Gandze, Sichuan, China, implemented by the Tibet Heritage Fund.

Swedish National Heritage Board and Bengt O H Johansson, then a department head at the same body. These connections within the governmental framework were undoubtedly vital in successfully obtaining funding and support from Sida in 1996 for CHwB's pilot restoration projects in Maglaj, central Bosnia: the sixteenth-century *Kuršumlija Mosque* and the nineteenth-century *Uzeirbegović Konak*, both severely damaged by shelling. With the peace still fragile, Sida had wanted CHwB to begin work in an area where Swedish peacekeeping troops were deployed and Maglaj fitted this criteria. Work on an increasing number of projects followed,[7] and in spring 2000 I was appointed CHwB's project manager to implement their programme in Bosnia, a role I continued in until 2007. In this chapter I will outline the work, priorities and approach of CHwB during this developmental phase of its activities, describing its collaborations with local heritage bodies and communities where projects took place and some of the issues that arose from working in a post-conflict environment, as well as a number of key projects that embodied its aims at that time.

After 2005, under the influence of Sida, CHwB's approach in Bosnia-Herzegovina was to change markedly. Despite CHwB's original ambitions to find financial support from a variety of sources, up to the time of writing its work has been financed almost entirely by Sida, with only minor additional contributions coming from other donors. Thus, in order to maintain this relatively secure funding stream CHwB's project proposals have tended to follow Sida's humanitarian and development aid priorities, increasingly so as its relationship with Sida deepened. However, conservation projects carried out between 1996 and 2005 during the initial phase of CHwB's work in Bosnia-Herzegovina aimed at restoring the identity of place and so improving conditions for returnees, at ensuring that the majority of work was implemented by local partners using (as far as possible in this post-conflict setting) local expertise and contractors and, through the conservation process, helping to strengthen state organizations by educating and training their young professionals through a 'hands-on' method according to the philosophy of *learning by doing*.

Thus the *Institute for the Protection of the Cultural-Historical and Natural Heritage of Bosnia-Herzegovina*, at the time a central state institution, was initially chosen as CHwB's principle local collaborating partner. However, the political division of Bosnia-Herzegovina under the Dayton Peace Agreement into two entities (the Federation of Bosnia-Herzegovina and Republika Srpska) was to have a detrimental impact on CHwB's first local partner. Although the Institute had continued in its pre-war nationwide role throughout the conflict and into the immediate post-war period, the refusal of Republika Srpska to accept its state-wide remit led to the downgrading of its status and responsibilities in the years following Dayton until it became the heritage preservation body for the Federation alone (the *Federal Institute*). Nevertheless, in the evolving post-war conditions of 1996 there was no other body which might have been expected to obtain this role in the future except for the Institute, and it was judged crucial to protect the continued existence of a body that represented the collective knowledge of the built heritage of Bosnia-Herzegovina and which held the greater part of the surviving documentation of that heritage in its archives.

In this chaotic transitional environment CHwB chose to focus on developing the professional skills of the younger members of the Institute's staff whose careers had been severely disrupted by the conflict. As investing in individuals rather than institutions can have potential drawbacks in terms of long-term benefit to a country, CHwB later spread its support to other heritage preservation institutes, to young local architects who

7 The first two CHwB project managers in Bosnia were Tuula Rauma, followed by Pål Anders Stensson.

expressed interest in working in restoration in Bosnia-Herzegovina and to the masters' degree programme in restoration at Sarajevo University's Faculty of Architecture. So while projects were chosen primarily from a conservation point of view, they also needed to act as suitable educational projects for young professionals where they could solve practical problems through analysis based on thorough research.

CHwB's objectives at the time were to establish a solid base of restoration architects familiar with international charters relating to cultural heritage and historic buildings and able to convert these charters into their practical work. The restored buildings themselves could be viewed as an additional outcome of CHwB's hands-on training philosophy of *learning by doing*, becoming in effect a document for future study. It became increasingly evident, however, that CHwB's programmes should integrate training and professional development into its project work, as well as to make efforts to reconnect professionals whose relationships were broken by the war and post-war division of Bosnia.

Projects were chosen in collaboration with the Federal Institute or other regional Institutes for Protection of the Cultural-Historical Heritage and, as well as the training opportunities with the interesting restoration problems they offered, CHwB's criteria for choosing monuments were to save an important cultural heritage object for the country, strengthen local identity and help heal wounds of the war. However, the size of the project was also an important consideration as it had to fit within CHwB's relatively small-scale budget and be within the capacity of what could be handled by one project leader from Sweden.[8] CHwB demanded high professional standards in projects as one of its main objectives and it was the responsibility of the project leader to achieve this. Proposals for viable projects came almost entirely from the various preservation institutes, though occasionally there were suggestions from other directions.

However, CHwB's customary policy was not to work on complete reconstructions but to work on structures where the greater part survived (see below). Conservation practice has often been perceived as a local matter, but the Bosnia-Herzegovina case is a good example of its evolution into an international issue. Sharing knowledge, discussing solutions to both technical and philosophical problems, such as how to apply international charters in practical conservation projects where local praxis or political events demand contradictory solutions, was actualized in the post-conflict environment of Bosnia-Herzegovina. CHwB's approach was to follow international charters as a general guideline and to choose objects whose authenticity could be saved by urgent action. Even if the question of and the desire for complete reconstructions of historic structures was frequently raised by Bosnian professionals as a result of the large number of buildings which had been totally destroyed during the war, CHwB always worked where intervention would preserve the existing remnants of a structure and prevent the authenticity they represented from being lost.

Inevitably, in many cases due to the extensive war damage some parts of structures needed to be completely reconstructed and solutions to save a structure's authenticity and how lost parts should or could be rebuilt were considered and discussed. Where advanced technical and materials solutions were required CHwB held workshops to discuss methodology before work began, bringing together local, regional and international experts, such as that held in April 2000 before the projected restoration of the *Despić House*

8 CHwB kept its head-office operations deliberately small so that funding could focus on projects. Board members contributed much voluntary work, and while an administrator was eventually employed, it had no director (Secretary-General) until October 2002, more than seven years after its foundation.

(1700–1900), a Serbian merchant's house belonging to Sarajevo City Museum. Other discussions rose from considering the identity of a place and the role the project played in re-establishing the dignity of people and their environment and CHwB always worked closely with local inhabitants and property owners.

In this post-conflict context conservation issues included practical limitations like the availability (or lack) of equipment, materials and building skills and occasionally these had to be imported, while materials samples were sent to Sweden for analysis as there was nowhere in Bosnia-Herzegovina which could carry out such tests. When necessary, external specialists were brought in during both the project and construction phases. Work was often carried out in environments where destroyed infrastructure and the presence of landmines were additional challenges. Unexpected factors could impinge on project implementation. The presence of a displaced family living in the *Despić House*, for instance, delayed the start of restoration work; once the cantonal minister of culture understood that work would not begin until the family was relocated, the local authorities quickly found them a new home.

After 2001 the work of CHwB in Bosnia-Herzegovina grew as Sida extended their funding. This resulted in a more collaborative interaction between CHwB and Sida when choosing projects, with the additional objective of supporting other Sida programmes, such as its extensive returnee project, or working in areas where other Sida projects were taking place. The Jajce project was a direct outcome of this new direction.

Jajce: The Historic City Core 2001–2005

Historically one of Sida's two largest projects worldwide was its *Integrated Area Programmes* (IAP) in Bosnia-Herzegovina which helped at least 50,000 people return to their homes.[9] Several NGOs implemented this project, among them the Swedish organization Cross Roads International[10] which operated in Jajce. Cross Roads assisted refugees and displaced people to rebuild their homes through an approach Sida described as 'assisted self-help,' where eligible returnees were given enough materials to repair their houses to a basic standard.[11] To be chosen for the programme families had to fulfil certain criteria: to be a low-income family, own only the house they wanted to return to, be able to participate in its reconstruction, guarantee that they would return and not sell the house, and that the reconstructed house would be the same size as the destroyed one. Traditionally the historic Old Town of Jajce had been a predominately Muslim area and the majority of houses in the neighbourhood were destroyed or heavily damaged during the war. Thus several families who had been expelled from the district satisfied the IAP conditions for assistance from Cross Roads.

Due to its location close to the borders of Hungary, Croatia and later the Habsburg and Ottoman Empires, Jajce has played an important role in Bosnia-Herzegovina throughout its long history, coming under the influence of many cultures. This role has given Jajce its main character and the medieval fort and city walls still dominate the town as a constant reminder of this history which is of significant importance to newly independent Bosnia-Herzegovina as the final part of Bosnia to fall to the Ottomans. Even the urban plan,

9 Čukur, M., Magnusson, K., Molander, J. and Skotte, H. 2005. *Returning Home: An Evaluation of Sida's Integrated Area Programmes in Bosnia and Herzegovina*, Sida Evaluation 05/18.

10 The name used in Bosnia-Herzegovina by PMU Interlife, http://www.pmu.se/.

11 Čukur, M. et al. 2005, 6.

Fig. 7.1 Restoration of a historic panorama. Jajce's fortress and Old Town with its steep-roofed traditional houses after CHwB's restoration work. The shiny metal roofs of two houses reconstructed by CHwB can be seen left, with the entrance gate to the fortress (Sahat Kula) far left. The wooden minaret of the restored Sinan-beg Mosque is seen right, below the tower of the fortress. © Tina Wik

concentrated within the medieval walls, is more or less the same as over past centuries, while the Old Town (*Stari Grad*) on the southern slope of the hill crowned by a fortress, is famous for its panoramic view. Jajce's other attractions include a fourth-century Mithraic temple, a medieval catacomb complex, the building where the formation of a federal socialist Yugoslavia was decided in 1943 (the AVNOJ Museum)[12] and a spectacular waterfall.

The predominate house type in the Old Town, the *Jajce House*, is famous for its distinctive features developed through an interaction with several cultures, especially the Ottoman and the Central European. Typically situated on a steep slope, it has a stone basement for storage entered from the garden, and a residential floor above, entered from street level, of plastered timber-frame construction using local *sedra* stone as filler. The residential floor consists of a series of rooms with panoramic windows towards the south and a splendid view over the landscape – an important feature of these houses. The roofs are steeply hipped and traditionally were covered with wooden shingles. However, while shingles remained on the roofs of publicly-owned buildings, most private houses had lost

12 AVNOJ = Anti-Fascist Council of the People's Liberation of Yugoslavia (*Antifašističko Vijeće Narodnog Oslobođenja Jugoslavije*)

this covering. The most common replacement roofing material from the 1950s onwards had been unpainted metal sheets, now rather rusty in appearance.

Before the conflict new houses in this historic area were being built in modern designs similar in appearance to traditional structures. However, after the war the plans of new houses were uncontrolled and very divergent designs were being constructed. This damaging trend was speeded up by international humanitarian post-war aid, where agencies distributed poor quality standardized building materials, expecting residents to reconstruct their houses themselves, with no consideration of the historic nature of the destroyed structures. While there is a strong tradition in Bosnia-Herzegovina of self-building homes, these complex houses were impossible to be rebuilt by their owners. Thus, international aid was helping to destroy the history and identity of the place it had come to help, continuing the destruction of the war. This was far from an isolated phenomenon and could be found across Bosnia as refugee return gathered pace.[13]

The house form in Jajce had influenced public buildings on the hillside, unifying the neighbourhood into a harmonious whole and the panorama of structures with their steep hipped roofs scattered across the slope below the fortress had become Jajce's trademark view. It was vital not to lose this important historical feature when so many of the buildings were destroyed or badly damaged in the war. These damaging international interventions had to be brought into focus and stopped. Both non-professionals and professionals reacted to this detrimental trend. One was Janne Haasitalo, head of Cross Roads, another was Ferhad Mulabegović, director of the Federal Institute. The Institute and CHwB prepared an application to Sida describing the urgency of the situation in Jajce's historic core, where its heritage was severely at risk, proposing as well a partnership with Cross Roads in the restoration of some of its historic houses. We received a quick positive answer both to the restorations and the collaboration with Cross Roads, since this met Sida's humanitarian aid objectives as well.[14]

During the project Cross Roads worked as usual, finding families and houses which fulfilled their criteria, while CHwB and the Federal Institute prepared an inventory of houses of cultural-historical importance. The final selection of six houses to be restored needed to satisfy the objectives of all partners. However, finding structures which satisfied the all the project requirements was not without complications. CHwB had wanted to restore one of the most important of these Jajce houses, but were unable to do so as the owners did not want to move back to Jajce. On the other hand, one homeowner who otherwise fulfilled the programme criteria had illegally extended his home before the war towards the medieval city walls. When CHwB offered to rebuild the house following local planning regulations and placing the extension in a different direction, the owner refused to agree, later making complaints that only a few returnees were receiving help from the programme. Another owner tried to acquire a larger house via the project; but as CHwB would only consider rebuilding structures to their pre-war footprint, the property owner withdrew.

CHwB's aim was to show that it was possible to conserve these traditional buildings in an economically realistic way, partly through the use of modern materials. At the same time, an important consideration was to make the houses attractive places to live, meeting contemporary living standards with modern kitchens and bathrooms, and ensuring as well

13 The phenomenon is described in Calame, J. and Sechler, K. 2004. Is preservation missing the point, in *Future Anterior*, Columbia University, Vol. 1, No. 1, Spring 2004, 57–63. See also Amra Hadžimuhamedović in this volume.

14 Additional funding for the Jajce projects was also received from a private donor Amanda Sebestyen, and the British Embassy, Sarajevo.

that they were easy for their low-income owners to maintain. Thus, while the original roof covering would have been wood shingles, these were difficult and expensive to obtain and install, and we chose instead to use the metal sheets of the pre-war era.

Jajce had been attacked, then occupied by the Bosnian Serb Army in October 1992, resulting in the destruction of many of its buildings, particularly in the historic quarter under the fortress, and all the town's mosques and Catholic churches. The town's Serbian Orthodox church had been dynamited by Bosnian Croat extremists just weeks before. Thus all religious buildings in Jajce had been eliminated except for two small mosques on the hillside from which it would have been too difficult to carry away the stones of their remaining shells: the tiny *Dizdar Mosque* (1812) and the *Sinan-beg Mosque* (1689). These two mosques, both with interior masonry domes, stood open to the weather and needed urgent intervention. They could still be saved with their authentic shape and stone construction in place, of crucial importance as the only remaining historic religious buildings in the area. With quick action required, it was decided to begin with the restoration of the mosques rather than the houses which had been demolished to their foundations.

Although Muslims had formed a slight majority in pre-war Jajce,[15] the post-war town has a Bosnian Croat majority. However, it is the traditionally Muslim neighbourhood on the slopes below the fortress that characterizes the town and its historic presence. Though the Croat[16] mayor was excited about the prospect of help rebuilding the town's historic structures, he was adamantly opposed to the restoration of the mosques. Instead, he tried to persuade us to undertake the conservation of St Mary's Church or its bell tower (St Luke's Tower), two historic buildings from the medieval period that symbolized the Catholic presence. The church had been converted into a mosque in the sixteenth century, but had been a roofless shell since 1833; neither structure had suffered war damage. The mayor refused to issue CHwB building permits on the grounds that he could not guarantee our security if we began the conservation of a mosque. After a full year of negotiations, the mayor finally understood that there would be no conservation projects at all if we were not able to start with the mosques, and he provided the permits that allowed us to begin.

Along with the two mosques, CHwB also conserved the wooden shingle roof of the entrance gate to the fortress (*Sahat kula*) and began planning the reconstruction of six timber-frame *Jajce Houses*. The houses had been almost completely destroyed and we started from the remnants of the basement or foundation walls using what we could of the authentic material. Homeowners participated in the reconstruction as much as they were able, but most were elderly and unable to contribute much in this respect. In the end four houses were rebuilt, in parallel with the public buildings. Working on a group of buildings simultaneously revitalized the area as a whole, aiming to encourage returnees by the restoration of the mosques as well as domestic structures. It was very evident how regaining the religious structures so dramatically lost, as well as the restoration of other monuments, restored as well feelings of security and a base for existence, especially for minority groups.

The seven buildings CHwB completed on the hillside made a significant difference to the appearance of the town and was a start in regaining pre-war Jajce which had been a

15 In 1991 Muslims formed 38.61 per cent of the population, Croats 35.13 per cent, Serbs 19.24 per cent, Yugoslavs 5.54 per cent and Others 1.48 per cent .

16 I will refer to Bosnian Croats and Bosnian Serbs (as distinct from Croats/Croatians in Croatia or Serbs/Serbians in Serbia) as Croats and Serbs for brevity after the first mentioning. Bosnian Muslims frequently refer to themselves as Bosniaks.

Fig. 7.2 **(a)** The ruins of House Number 17 in Jajce's historic Old Town before rebuilding. **(b)** House Number 17 almost completed, with its steep Jajce roof in place. © Tina Wik

major tourist destination in the former Yugoslavia. Re-establishment of the tourist industry was one of the main hopes for the future of the town which (as elsewhere in Bosnia) has extremely high unemployment. This motivated CHwB to continue conservation work there, looking at Jajce as a historic urban landscape where tourism needed support with an information centre, as well as improved urban spaces. However, although CHwB wanted to co-operate with the local authorities in a more direct way and the municipality was willing, due to a lack of trained staff at the municipality it proved difficult to establish a working partnership and negotiations ended instead with two final projects: the seventeenth-century *Omerbegova House*,[17] intended to house a tourist information centre and small museum, and section of the medieval city walls including one of the towers (*Papaz kula*) as part of a larger conservation programme.

A number of issues arose during CHwB's work in Jajce. The selection of a small number of houses for restoration caused contention among returnees as it was perceived that a only few house owners benefited from the project, while others did not and Sida received letters of complaint from some who felt excluded. Already unhappy with the idea of financing the restoration of private houses and believing the cost had been too high, Sida was unwilling to fund similar schemes again, and this was the only project of its type CHwB carried out in Bosnia.

In the end CHwB implemented nine restoration projects in Jajce, aiming to help the town regain its historic panorama and identity, offer security for Muslim returnees, and demonstrate how it was possible for humanitarian aid agencies to respect local building traditions and the features of the houses that form such an important part of the famous panorama and cultural-historical ensemble of Jajce. While CHwB was not able to share its contribution equally with each family that had suffered loss, everyone can enjoy the panoramic view that is now a little more recaptured. And by this step towards rehabilitation one additional step was taken towards improved tourism prospects and potential economic benefits for the town.

17 Unfortunately the *Omerbegova House* became caught in new legislation regarding the ownership of local historic buildings and was not used as originally planned.

Fig. 7.3 **(a)** The seventeenth-century Handanija Mosque in Prusac as restoration began, full of shell holes, its original roof and internal wooden dome destroyed and an emergency roof in place. **(b)** The Handanija Mosque in Prusac after restoration. Both © Tina Wik

Prusac: The Handanija Mosque 2003–2005

The village of Prusac in central Bosnia has an important complex of historic buildings, including several mosques (one with a Sufi *tekke*), a medieval fortress and an Ottoman clock tower. At the end of the sixteenth–early seventeenth century the village flourished as a major administrative, legal, scholarly and religious centre. Today Prusac holds an important place in Bosnia's religious and cultural life due to the famous pilgrimage to the spring of Ajvatovica which takes place every summer, starting from the *Handanija Mosque*.[18] The largest Islamic pilgrimage in Europe, Ajvatovica was banned under communism and revived in 1990. Pilgrims come in their thousands, often wearing traditional folk costumes and many riding horses, which play a central part in the processions.

Yet the stone-built *Handanija Mosque* (1617) which plays such a crucial role in the pilgrimage was in 2002 the only mosque in Prusac yet to be restored of its severe wartime damage. Architecturally important, its two notable features were an arcaded porch and interior wooden dome. It was the only mosque in the village still in its original form and one of the few historic country mosques left in Bosnia since the 1992–1995 war. The *Handanija*'s association with Ajvatovica was one of the reasons the imams from Donji Vakuf and Prusac contacted me, urging CHwB to help with the conservation of the mosque, as did Ferhad Mulabegović of the Federal Institute and Helen Walasek of the Bosnian Institute, London.

The mosque survived, but in a very poor state. During the war Prusac had been shelled continuously by Bosnian Serb troops entrenched on a ridge above the village. The *Handanija* had caught fire so that its whole roof was lost, while the main structure and its stone minaret were punctured by shell holes. The condition of the unroofed structure worsened each year from the heavy rains and snow common in the area. In 1999 the Federal Institute undertook emergency action, erecting a temporary roof and dismantling the top of the minaret which was at risk of collapse.[19]

CHwB decided to apply for funding from Sida for the *Handanija*'s restoration as well as to set up a team of young architects who would work on the project as a part of their master's degree programme. However, initially Sida was unwilling to support restoration of another religious structure since the religious communities in Bosnia seemed well able to finance plenty of new buildings. It was pointed out that while there was considerable funding coming from Islamic sources for erecting new mosques, there was little for the restoration of historic Ottoman ones. In the meantime, via the Bosnian Institute, CHwB successfully applied to the Barakat Trust for support for the documentation phase of the project.[20] With input from two noted scholars, Machiel Kiel and Harry T Norris, and the Barakat funding in place, we finally convinced Sida this was important heritage at risk with

18 Where, according to legend, the saintly Ajvaz-dedo found water for the population of Prusac from a rock which miraculously split into two halves.

19 Under the Italian-funded Progetto Atlante / Project Atlas, a programme linking Italian local authorities with those in Bosnia-Herzegovina (in this case Donji Vakuf municipality) aimed at promoting decentralized cooperation for human development. Funding was provided by Italian Cooperation and the programme was coordinated by the UNDP and implemented by the UNOPS and Italian Consortium of Solidarity (ICS), in collaboration with the WHO and the International Decade for Natural Disaster Reduction (IDNDR). http://sustainabledevelopment.un.org/content/dsd/dsd_aofw_mg/mg_success_stories/csd6/atlas_hd.htm.

20 The Barakat Trust was established in 1987 for the study, promotion and understanding of Islamic Art History, Architecture and Archaeology. See http://www.barakat.org.

a contemporary social relevance.[21] With a co-financer in the picture, Sida was willing to back the project, even though their contribution would be the larger.

Also problematic was reaching agreement on integrating the work into the Sarajevo University Faculty of Architecture's master of restoration programme. CHwB's practice-based approach was not considered sufficiently 'scientific' enough by the faculty's professors who expected students only to produce papers on theoretical themes, though the conservation process in fact had many interesting scientific issues to investigate. A solution was eventually reached, but it was obvious these two different approaches clashed and there was little continuing support from the Faculty.

The most difficult discussions with Bosnian colleagues concerned how to reconstruct the lost wooden dome and hipped roof, particularly as all surviving similar mosques had new roof constructions erected after World War II. CHwB suggested a hanging beam construction, a common medieval type, but the Sarajevo Faculty of Architecture professors reacted negatively to a solution they had never seen before. Discussions went on for months with CHwB accused of trying to implement Swedish solutions in a Bosnian mosque! When neither the professors nor local structural engineers could offer alternatives, we proceeded with the hanging beam construction; we later felt justified in our choice when I discovered similar structures in wooden buildings in Istanbul.

The *Handanija Mosque* was at the centre of Prusac life and its restoration was of great importance for the whole community and the local imam, Sakib ef. Began, who supported our work wholeheartedly throughout a long project. One of the most satisfying moments was when the roof structure was being erected and the repair of the mosque became a reality for the villagers. As a lime-making village, Prusac residents produced all the lime used in the building – the same lime type used in the original construction as we discovered from samples analysed. We tried to finish the project in time for Ajvatovica in June 2005. Unfortunately the works were delayed, but the re-inauguration in October was a festive occasion presided over by the Reis-ul-ulema of Bosnia-Herzegovina, Dr Mustafa ef Čerić.

Zavala: The Church of the Presentation of the Virgin 2001–2006

The Serbian Orthodox monastery *Church of the Presentation of the Virgin* (*Crkve Vavedenja Bogorodice*) at Zavala, just 32 kilometres from Dubrovnik, lies set into the side of a steep cliff overlooking the plain of Popovo Polje in Herzegovina. The oldest records for the church date from 1514 and in 1619 it was decorated by the renowned fresco painter Georgije Mitrofanović of Hilandar Monastery on Mount Athos. Two other churches in Bosnia-Herzegovina have frescoes by Mitrofanović, but those at Zavala are the best preserved, making this one of the most important historic Orthodox churches in the country.

The first sign of the coming war was the destruction of the neighbouring Bosnian Croat village of Ravno by the Yugoslav People's Army (JNA) in October 1991. Zavala, a predominately Bosnian Serb village, was later completely devastated during 1992. All the buildings belonging to the monastery were badly damaged or destroyed, but the tiny

21 For a full description of the restoration project see Wik, T., Hadžić, L. and Walasek, H. 2006. *The Handanija Mosque Restoration*, report series No. 3/2006, Cultural Heritage without Borders (CHwB) Regional Office, Sarajevo.

Fig. 7.4 Church of the Presentation of the Virgin (Crkve Vavedenja Bogorodice) of the Serbian Orthodox monastery at Zavala in Herzegovina, overlooking the plain of Popovo Polje. The church was restored and conserved 2001–2003. © Tina Wik

church with its beautiful frescoes miraculously survived.[22] CHwB was asked to undertake the church's restoration by the *Institute for Protection of Cultural, Historical and Natural Heritage of Republika Srpska* (although Zavala is located in the Federation).

When I came to Zavala for the first time I had difficulty finding the church. It was covered in plastic sheets to prevent water from entering wide cracks in its walls and as the church's north wall is the cliff face itself, the building is concealed under a large overhanging rock. My first experience of the place was an unforgettable moment when I finally found this hidden, covered church above the ruined and empty village full of landmines and at the same time imbued with a magical solitude. The contrast between the small church and the grandiose large-scale landscape was striking.

The church, an irregular barrel-vaulted basilica, was in very bad condition, with large cracks and the eastern vault at risk of collapse. Our immediate concern was to save the fabric of the church and to stabilize the south-east corner. The Swedish structural engineer Krister Berggren had solved the problem of structural stability by designing support for the church's shallow foundations with a micro-piling system that would be inserted into the ground until the piles reached solid rock. While the west part of the church stands on solid rock, the east stands on packed clay 13 metres deep. The

22 Its moveable heritage also survived. Information from the Commission to Preserve National Monuments of Bosnia-Herzegovina indicates that its collections spent the war in the Art Gallery of Dubrovnik where it had been taken, probably by Croatian Army forces. Thirty-five items were returned to the church in Zavala on 11 September 2009.

Fig. 7.5 Workers from Swedish firm Stabilator training a church volunteer in the micropiling process during restoration of the Church of the Presentation of the Virgin at Zavala. © Tina Wik

probable cause of the soil movement that was producing the cracks in the church was poor packing of the clay after an intervention in the 1960s. The arch of vault above the apse was about to give way, so we had to take action urgently. The project was to be a collaboration between the Republika Srpska and Federation entity heritage preservation institutes under my supervision. This proposed collaboration between the two institutes which had barely communicated until that time was a further reason for Sida agreeing to fund the project.

As at the time we began the work in 2001 Zavala was a depopulated Bosnian Serb village in a Bosnian Croat-dominated canton with no electricity supply, our first challenge was how to transport an electricity generator over war-damaged roads with all nearby bridges blown up. Furthermore, we needed to use a hydraulic press for the micro-piles since an ordinary drill would cause vibrations that might further damage the church, already at risk of collapse. The micro-piling system was unknown in Bosnia-Herzegovina, so besides importing materials and equipment, we needed to import two workers from Sweden who would work with church volunteers.

Before the micro-piling, however, the frescoes needed to be conserved and protected. After the micro-piling, the complex process began of temporarily removing the frescoes from the damaged parts of the building that needed to be taken down and re-erected. This was followed by dismantling the stone slate roof and remaining plasterwork, marking and dismantling the stones of the collapsing vault, and finally re-erecting everything and replacing the frescoes which then underwent further conservation – a process that took altogether three years. We continued work on the church and its surroundings, repairing

Fig. 7.6 Conserving the seventeenth century frescoes by Georgije Mitrofanović at the Orthodox Church of the Presentation of the Virgin at Zavala. © Tina Wik

and conserving the bell tower as well as the vaulted entrance tunnel to the church that was in danger of collapse. One problem that could not be completely 'built away' was the water continually seeping from the rock and surface water during periods of heavy rain, though we carried out some preventive work to minimize the amount of water into the church by leading the ground water in another direction. Maintenance guidelines were prepared to deal with this humidity but they demand constant supervision from the priest in charge and relied on his commitment to preserve the authenticity of the structure.

We worked six years on this project, which not only preserved one of Bosnia-Herzegovina's most important historic monuments, but crucially helped begin the return process to Zavala. Father Vasilije, the Orthodox priest who had come to re-establish the monastery, was an important participant in the restoration process. The Serb residents of Zavala began moving back, first to help Father Vasilije restore the monastery and later, as security and living conditions improved, simply to re-establish their lives in the locality. During the first years of CHwB's work a Croatian flag had been raised in the middle of the empty village and Bosnian Croat police regularly visited the monastery. Father Vasilije handled these visits with admirable patience and diplomacy, which along with the return of Serb families, resulted in the restoration of electricity and other services to the village and the arrival of de-mining teams, encouraging more villagers to return and start rebuilding their homes.

Fig. 7.7 The archaeological pavilion of the Zemaljski Muzej (the National Museum of Bosnia-Herzegovina), Sarajevo, after restoration by CHwB. © Tina Wik

Sarajevo: Zemaljski Muzej (The National Museum of Bosnia-Herzegovina) 2000–2004

The Zemaljski Muzej was the largest project undertaken by CHwB in Bosnia-Herzegovina, involving not only conservation of its buildings, but also support for the institution's museological work. The *Zemaljski Muzej*, the National Museum of Bosnia-Herzegovina, was founded in 1888 by Bosnia-Herzegovina's Austro-Hungarian administration and is the oldest of the country's modern cultural and scientific institutions. When the museum outgrew its original premises, plans for a new building were initiated in 1908. The architect Karel Pařík[23] designed a complex of four pavilions in a rectangle around a botanical garden. This large complex could not fit in the small-scale winding streets of Ottoman Sarajevo and, like most other large Austro-Hungarian institutional buildings, it was built in a new grid-patterned extension to the city. In the years following the end of the 1992–1995 war the restoration of these structures which stood like dark and sad giants in the townscape was to be crucial in revitalizing Sarajevo.

The museum's main building contains the prehistory and archaeology collections, the rear pavilion the natural history collections, while the smaller side pavilions house the ethnographic collections and the library. Exteriors are Neoclassical in style, the interiors Secessionist and use was made of modern materials for the time such as concrete. The new complex was inaugurated in 1913 and documents show that the final stage of works was speeded up so that the building would be ready in time for its official opening. The *Zemaljski Muzej* was the first purpose-built museum in the Balkans and is still the second largest museum in the region.

The *Zemaljski Muzej*, facing the wide boulevard of Zmaja od Bosne, known as Sniper's Alley during the 1992–1995 war, saw heavy fighting at the start of the conflict, while the frontline was directly behind it across the Miljacka River. Thus it suffered considerable

23 Sometimes called Karl or Karlo Paržik. Pařík was a Czech who made his career in Bosnia-Herzegovina during the Austro-Hungarian occupation/annexation of Bosnia-Herzegovina from 1878 to 1918, designing many major buildings.

Fig. 7.8 (a) The master craftsman in the workshop at the Zemaljski Muzej where missing or damaged neoclassical details from the exterior were recreated. **(b)** Detail of neoclassical decorations on the facade of the Zemaljski Muzej after restoration. Both © Tina Wik

damage from shrapnel and shell impacts. Tank shells had pierced exterior walls and the glass roofs over the main exhibition halls were broken and open to the elements for much of the war. The many holes in the facades caused continuous damage to the fabric of the structures, aggravated by the 'restoration' carried out before the 1984 Winter Olympics on many public buildings in Sarajevo when structures were sprayed with cement plaster and painted with acrylic paint, trapping humidity in the walls and ceilings. Sculptures on the main façade lacked parts, as did the balustrades crowning the building. Still, this obvious damage was not as detrimental to the structures as the many small bullet and shrapnel holes peppering the plastered walls.

In usual conservation practice damaged and lost parts means a loss of authenticity, especially with regard to materials and patina and how to deal with these parts would normally have been discussed far more than we did. But following a conflict, one of which's aim was to destroy the historic identity of Bosnia-Herzegovina, our priority was to restore and put back in place the country's foremost heritage institution rather than to discuss the value of the material authenticity of every damaged part. Since we had reliable evidence in the remaining balustrades and other classical details of how the missing parts looked and what they had been made of, we repaired this damage in the same way as we did the roofs and the facades, by using the same materials the original builders had used. We could and we did, however, preserve the authenticity of the design.

Samples of each detail and different layers of plaster were sent for testing in Sweden – one of our greatest challenges was to stop workers from rushing to complete tasks, but to wait for test results and investigations based on authentic documents. Our Bosnian contractors established a workshop on site where all the missing or damaged exterior neoclassical details were reproduced in moulds by an older master craftsman using the same material as the originals: the balustrades and statues in concrete, the mutules and other small Neoclassical ornaments in gypsum. He became a tutor for younger workers who learned skills that had almost been lost and were reintroduced during the restoration process. The moulds for casting the lost features were made by one of the museum's conservators, who also repaired and conserved the sculptures on the facades as well as the fresco paintings in the entrance hall.

Through the restoration we aimed to show how historical evidence, both from archives and the building itself, could be used as guidelines in such work. Equally important was to show how the decision to restore this important state institution symbolized respect for and the rehabilitation of the multicultural identity of Bosnia-Herzegovina. This symbolic dimension was later to cause most of the problems we faced in this project. CHwB had originally received a request from the Federation government to help with the conservation of the *Zemaljski Muzej*. This was a much larger project than CHwB normally undertook and initially we intended only to repair the roof of the Natural History Department, but it quickly grew to include the facades of the same building and eventually to the other pavilions. Since the work could be easily divided it was possible to proceed in discrete phases as funding became available.

One conservation dilemma concerned the colours of the facade. We had no information on the now completely removed original colours and no documentation on the subject – possibly as a result of the rush to complete the buildings in 1913. With only black and white photographs to go on we could draw no firm conclusions and had long discussions with our Bosnian collaborator, the Federal Institute. They proposed changing the colour from yellow to grey based on their interpretation of the old pictures, but this seemed too uncertain. Two other factors also came into play. We were unsure whether Sida would finance further work on the museum, thus the prospect was raised if funding was not forthcoming of one pavilion being grey and the rest a sharp yellow! Another consideration was what colour would be acceptable to the citizens of Sarajevo who were used to the yellow facades. In the end a toned down yellow palette was used with different parts of the façade decoration emphasized: the pilasters and the cornices in greyish white and the walls a light yellow, a common Neoclassical colour scheme where the structural elements are emphasized.

CHwB was able to restore all the pavilion exteriors with Sida providing funding except for that of the Ethnology pavilion which was financed by the European Union. In 2004, the exterior work finished, we continued on the interiors, starting with the entrance hall. The toilets were run down and the main dome had been hit by a shell impact that had been poorly repaired. Here we wanted to conserve the last part of the museum's interiors not yet been brought back into use: the prehistory department. Humidity had caused large parts of its beautiful Pompeian-style frescoes to detach and fall from the ceilings, and the rooms were badly in need of repair. In addition, we discussed the inclusion of a small café as well as a bookshop to generate some income for the museum.

Everyone was excited by these plans and CHwB applied for funding from Sida as usual. By now they had supported our projects at the *Zemaljski Muzej* for several years with a sum that added up to more than US$1,800,000.[24] Sida was supportive of this last intervention. However, since the museum had constant problems with its running costs, with no money for heating, electricity and salaries, Sida decided to require a political

24 The sum was more than 12,000,000 Swedish crowns.

guarantee that funding for the institution's overheads would in future be provided by the state of Bosnia-Herzegovina. Due to the devolvement of considerable powers to the two entities under the Dayton Peace Agreement there is no state ministry of culture and the museum had not had its legal status resolved since the end of the war, existing in a budgetary limbo.[25] The Canton of Sarajevo and the Federation had so far provided financing for the museum, but each year this was an exception since they believed the museum was a state institution and should be legally recognized and financed by the state. However, the Republika Srpska authorities worked actively against the recognition of state-level institutions, particularly those based in Sarajevo.

In October 2004 the museum closed to the public and staff went on strike in an attempt to bring the issue to public attention. Museum staff had remained at their posts throughout the war and the museum, which had not closed to the public since its opening in 1888, had become a symbol of cultural survival and its closure was widely reported in the local and international media.[26] Despite the intervention of the then High Representative, Paddy Ashdown, a one-off first time grant from the central state government, the return of heating and the museum's subsequent re-opening, Republika Srpska continued to refuse to acknowledge its status as a state-level institution.

CHwB was given an 18-month time-limit within which the proposed works had to be completed if they were to be financed by Sida. By then, to meet Sida's conditions on providing funding for the final phase of the project, the question of the museum's status had to be resolved, including who would finance its future maintenance and running costs. But despite many meetings at all levels of government, it proved impossible to overturn the objections of Republika Srpska, CHwB's final planned intervention was never carried out and the funding was lost. In 2012 the prehistory department remained unrestored, the museum's legal and funding problems unresolved and on 4 October the Zemaljski Muzej closed its doors to the public.[27]

Closing Comments

Implementing the conservation of historic buildings in a post-conflict, post-socialist and recently-independent country, where destruction of cultural heritage had been a key war aim, inevitably included challenges apart from the purely operational. Events that took place centuries ago continued to have currency and were still debated and a cause of conflict, while power struggles in evolving political structures or commitment to a cause made some experts misuse historic material. Archival material was too often considered a source of power and kept hidden by experts in charge of documentation, hoping for assignments to restore or conserve important monuments. The collegial sharing of information was frequently a battle. This has potentially disastrous consequences for

25 Six other state level institutions were in the same position: the Historical Museum of Bosnia-Herzegovina, the Art Gallery of Bosnia-Herzegovina, the Museum of Literature and Theatre, the National and University Library, the Library for Blind and Sight-Impaired Persons, and the Film Library.

26 See for example Sito-Sucic, D. 2004. Post-war squabbles threaten Bosnia's museums, *Reuters*, 12 November; and Plazonic, M. 2005. A culture served cold, *Transitions Online*, 9 February.

27 See, for instance, Zuvela, M. 2012. Bosnia feud threatens end to shared Balkan heritage, *Reuters*, 9 January. IWPR Staff Sarajevo 2013. Sarajevo Museum in Peril, *IWPR, Institute for War & Peace Reporting*, 22 February, http://iwpr.net/.

a depleted post-war society, so badly in need of collaboration in rebuilding the country. Such in-fighting steals the focus from the actual conservation or rehabilitation project which risks being poorly executed. The political fighting thus continues after the war and prevents professionals from collaborating and learning from the international scene and from stepping out into the globalized world where sharing information is an inevitable part of professional improvement.

One issue thrown up by the war which Bosnian heritage preservation professionals were not prepared to debate was a serious discussion of the complete reconstruction of totally destroyed historic buildings. While the Venice Charter of 1964 regards such reconstructions justifiable only if they have an educational aim, the charter itself has been much debated in recent years and the question of complete or near-complete reconstructions have been put into another light following recent conflicts where a feature of the warfare has been the elimination of heritage with the intention of humiliating or destroying particular groups.[28] This was recognized by the Council of Europe's 2005 Faro Convention with its declaration that the right to enjoyment of one's cultural heritage is a human right.[29]

The Bosnian case demonstrates clearly the importance of historic monuments and their role as symbols of identity and approaches to complete reconstructions must necessarily be different when loss is caused by violent acts of destruction. Still, such complete reconstructions should fulfil certain conditions of integrity based on knowledge of the lost building, its design and materials. These conditions were rarely thoroughly discussed in Bosnia and even less controlled in ongoing projects, resulting in reconstructions where poor quality was justified by a lack of information, time or money because it was judged more important to quickly regain a symbolic building in an area where ethnic cleansing had taken place.

Finally, how should one assess the work and aspirations of CHwB and heritage restoration as a form of humanitarian aid more generally? In 2007 CHwB's activities were evaluated by Torsten Kälvemark.[30] Kälvemark raised several important issues of which I will comment on two. One is a critique of conservation professionals for their lack of involvement in social development work.[31] This is an interesting argument, especially seen from the perspective of another issue raised: the role of cultural heritage restoration in promoting reconciliation in a post-conflict society. Reconciliation, Kälvemark claims, is encouraged by conservation projects. Even UNESCO asserts this role for cultural heritage and Kälvemark quotes a former Director-General's statement that 'the cultural heritage is … an instrument for peace and reconciliation.'[32] CHwB declares contributing to the reconciliation process as one of its key aims[33] and Kälvemark's report, along with an impact

28 *The Venice Charter*, ICOMOS, 1964, see http://www.icomos.org/venice_charter.html.

29 *Council of Europe Framework Convention on the Value of Cultural Heritage for Society (Faro Convention)*, Faro, 27 October 2005. For the text of the convention see http://conventions.coe.int/Treaty/EN/Treaties/Html/199.htm.

30 Kälvemark, T. 2007. *Cultural Heritage for Peace and Reconciliation, An evaluation of Cultural Heritage without Borders (CHwB)*, Stockholm: Cultural Heritage without Borders (CHwB),.

31 Kälvemark, T. 2007, 16. He cites Calame, J. and Sechler, K. 2004.

32 The Director-General of UNESCO's message at the start of the United Nations Year for Cultural Heritage (2002), quoted in Kälvemark, 2007, 15.

33 See the CHwB information brochure: Johansson, B.O.H. et al. 2005: *Cultural Heritage: Reconciliation, Restoration, Hopes for the Future*, Stockholm: Cultural Heritage without Borders (CHwB).

assessment study carried out for the organization by a Sarajevo-based public opinion research company in 2006,[34] appear to find its work in Bosnia had achieved this aim.

Unfortunately, I saw no signs that this wishful but unproven thinking was true in Bosnia-Herzegovina – at least in the short term. There were no indications of diminished political and professional tensions being brought about by the execution of conservation projects. Indeed, as was shown in incidents across Bosnia-Herzegovina, heritage restorations can trigger increased political tensions and even violence. One of the worst episodes was the demonstration at Trebinje in May 2001 accompanying the laying of the cornerstone for the reconstruction of the totally destroyed Osman-paša Mosque. On the other hand, the reaction to the violence *did* speed up reconstruction of the mosque which had seemed a political impossibility at the time.[35]

What I did see, on the other hand, was how the rehabilitation of symbolic buildings important to the identity of a place and a particular group contributed to the return process. The inclusion of a greater awareness of social context in the work of conservation experts could perhaps improve the prospects for reconciliation reached through the conservation process. However, in conflicted and war-torn environments the social stability this kind of collaboration requires can be as difficult to achieve.

Kälvemark points out that a particular strength of CHwB was their 'hands-on' philosophy – *learning by doing* – giving young professionals practical experience in implementing projects. It was a method appreciated by many professionals in Bosnia-Herzegovina as one of the best learning processes that no seminars or lectures could replicate. Yet this philosophy became a victim of what I believe was a disadvantage for CHwB: its almost total reliance on Sida as a funder, forcing the organization to react to Sida's changing political priorities, policies and definitions – such as what constitutes 'capacity-building' – rather than developing a strategy and methodology of its own. Thus, while it was evident that implementing restoration projects was indeed a form of capacity-building, developing a skilled base of architects and heritage professionals, contractors and building workers, with the additional outcome of a restored monument, this was not how Sida saw it. Consequently, in order to maintain its Sida funding after 2005, CHwB was required to stop its restoration activities in Bosnia-Herzegovina and focus instead on developing regional networks, at the same time as beginning to plan its exit strategy from the country.

Sida's policies and political considerations and CHwB's lack of financial independence from them had an impact as well on project implementation. One example was Sida's reluctance to fund the Jajce model of restoring groups of buildings and environments (including private houses) elsewhere in Bosnia due to its negative perceptions of the project. Another was tying unachievable political conditions to the release of funding for completing the restoration of the Zemaljski Muzej. Furthermore, CHwB's directors (Secretary-Generals) have all come from political or development backgrounds – the organization's board judging a good relationship with Sida to be more important than having a director with a background and real interest in building conservation.

34 CHwB 2007 *The Restoration Process in Eight Cities, Towns and Villages in Bosnia-Herzegovina, Impact Assessment Study*, carried out by market research/polling agency Puls, Cultural Heritage without Borders (CHwB), Sarajevo, January 2007, 8.

35 See Valery Perry elsewhere in this volume for more on the violence surrounding heritage reconstruction in 2001.

Ultimately, however, with its small organizational structure, active board, and the relatively secure support of Sida, CHwB's achievements in Bosnia-Herzegovina were considerable – probably far greater than its founders imagined in 1995. Its work was acknowledged by the Federation of Bosnia-Herzegovina with an award in 2004, while its activities in Bosnia-Herzegovina were recognized internationally in 2006 with a European Union Prize for Cultural Heritage/Europa Nostra award.

Chapter 8
Cultural Heritage, the Search for Justice and Human Rights

Helen Walasek

The search for justice for victims of the horrific events of the 1992–1995 Bosnian War was to become an important testing ground of international humanitarian law and international human rights law, not only with regard to more commonly recognized violations, but also in respect of the protection and preservation of cultural and religious property, the right to a people's enjoyment of their cultural heritage and the development of concepts of cultural heritage and identity. Key in this were the ground-breaking legal precedents established by the International Criminal Tribunal for the former Yugoslavia (ICTY), though there were other significant articulations such as the decisions of the International Court of Justice (ICJ) and the Human Rights Chamber for Bosnia and Herzegovina (HRC) and domestic courts in Bosnia-Herzegovina. This brief overview highlights some of the most important cases in these different venues and refers to other publications which treat this subject in more depth.

Cultural property has long been given particular protection in times of war and armed conflict on the basis of its universal significance to mankind, most notably in the post-World War II era through *The Convention for the Protection of Cultural Property in the Event of Armed Conflict* of 1954 (*The Hague Convention*).[1] The destruction of cultural heritage during the Second Gulf War had revived concerns about the Convention's inadequacy with regard to the definition of 'military necessity' and its conformity with current international humanitarian law.[2] By September 1992 the continued destruction of cultural property in the ongoing conflicts in Croatia and Bosnia-Herzegovina (and in particular the bombardments of Dubrovnik and the historic centres of Sarajevo and Mostar – the first a World Heritage Site, the others on the tentative list) prompted UNESCO to place a long overdue review of the Convention at the top of its agenda.[3] The review by Professor Patrick Boylan[4] was to lead to the adoption of the Second Protocol to the Hague Convention of 1954 in 1999.[5]

1 See Vrdoljak, A.F. 2009. Cultural Heritage in Human Rights and Humanitarian Law, in *Human Rights and International Humanitarian Law*, edited by O. Ben-Naftali. Oxford: Oxford University Press for a lucid overview. See also Vrdoljak, A.F. 2007. Intentional Destruction of Cultural Heritage and International Law, in *Thesaurus Acroasium* XXXV (2007): 377–96.

2 Hladík, J. 1999. The 1954 Hague Convention for the Protection of Cultural Property in the Event of Armed Conflict and the notion of military necessity, *International Review of the Red Cross* No. 835, 30 September 1999, 621–35.

3 *Review of the application of the Convention for the Protection of Cultural Property in the Event of Armed Conflict (The Hague, 1954)*, UNESCO Doc. 140 EX/26, Paris, 11 September 1992. See III, 3 for references to Dubrovnik, Sarajevo and Mostar.

4 Boylan P.J. 1993. *Review of the Convention for the Protection of Cultural Property in the Event of Armed Conflict (The Hague Convention of 1954)*, (1993), UNESCO Doc.CLT-93/WS/12.

5 *Second Protocol to the Hague Convention of 1954 for the Protection of Cultural Property in the Event of Armed Conflict The Hague, 26 March 1999.*

However, the work of the ICTY and other bodies in relation to the destruction of cultural property in Bosnia-Herzegovina was to move beyond the 'classical' model of the UNESCO review. Van Boven has noted the 'historic evolution' of human rights as 'an integral and dynamic part of international law' in the decades following World War II,[6] an evolution which led to the beginning of a process at the United Nations in 1989 to attempt to address the need for justice for victims of human rights abuses, their right to 'redress and reparation,' and to 'combat impunity'[7] for those who committed such abuses, which led ultimately to the UN General Assembly Resolution A/RES/60/147 on 16 December 2005: *The United Nations basic principles and guidelines on the right to a remedy and reparation for victims of gross violations of international human rights law and serious violations of international humanitarian law*.[8]

Part of this burgeoning movement was the 'exponential growth'[9] of international humanitarian law relating to cultural property. This movement incorporated an increasing linkage of cultural heritage with the enjoyment of human rights that advanced a more complex 'recalibration'[10] of the rationale for the protection of cultural heritage (transcending the concept of universal significance) with its attendant implications for safeguarding and promoting cultural pluralism and diversity.[11]

The development of these trends and the seminal role played by the precedent-setting case law of the ICTY have been explored by Abtahi, Bhat, Francioni and Vrdoljak. Vrdoljak observes that until the establishment of the ICTY in 1993 there was little jurisprudence on war crimes relating to cultural property since the post-World War II Nuremberg Trials,[12] while Bhat remarks on the conspicuous development during the 1990s of the criminalization of wrongs against cultural property under customary international law – and that such wrongs would be dealt with accordingly.[13] Abtahi notes the achievements of the ICTY, first through the inclusion in its Statute of crimes relating to cultural property, 'a major step toward strengthening previous international instruments' protection of cultural property in times of armed conflict' and through its prosecutions and subsequent convictions that demonstrate the importance given to such crimes.[14] The most extensive overview and assessment of the statute and case law of the ICTY in relation to crimes against cultural property to date can be found in Petrovic (2012).[15]

---

6 van Boven, T. 2010. *The United Nations basic principles and guidelines on the right to a remedy and reparation for victims of gross violations of international human rights law and serious violations of international humanitarian law*, 2. United Nations Audiovisual Library of International Law [online], an introduction to UN Resolution A/RES/60/147, 16 December 2005 of the same name.

7 van Boven, T. 2010, 1.

8 GA Res. 60/147, UN Doc. A/RES/60/147 (16 December 2005).

9 Francioni, F. 2004. Beyond State Sovereignty: The Protection of Cultural Heritage as a Shared Interest of Humanity, in: *Michigan Journal of International Law*, Vol. 25, Iss. 4, 1209–28.

10 The term is used by Vrdoljak, A. 2009, 4.

11 Bhat, P.I. 2001. Protection of cultural property under international humanitarian law: Some emerging trends. *ISIL Year Book of International Humanitarian and Refugee Law* (2001), Vol. 1, 47–71.

12 Vrdoljak, A. 2009, 44–5.

13 Bhat, P.I. 2001. 9.

14 Abtahi, H. 2001. The protection of cultural property in times of armed conflict: The practice of the International Criminal Tribunal for the former Yugoslavia. *Harvard Human Rights Journal*, 14 (Spring 2001), 30–31.

15 Petrovic, J. 2012. *The Old Bridge of Mostar and Increasing Respect for Cultural Property in Armed Conflict*. Leiden: Martinus Nijhoff.

The International Criminal Tribunal for the former Yugoslavia (ICTY)[16]

The International Criminal Tribunal for the former Yugoslavia (ICTY) came into being after a series of United Nations Security Council resolutions over the course of the summer and autumn of 1992 as the war in Bosnia-Herzegovina raged. In response to accounts of widespread and systematic atrocities, violations of human rights, and other grave breaches of the Geneva Conventions and international humanitarian law taking place in Bosnia-Herzegovina, in particular the practice of ethnic cleansing, a resolution of 13 July 1992 (resolution 764 (1992))[17] reminded all parties in the conflict of their obligations under international humanitarian law, affirming the principle of individual responsibility for any crimes committed under these laws. A further resolution of 13 August 1992 (resolution 771 (1992)) condemned the persistent violations, particularly that of ethnic cleansing, giving notice that failure to comply with the resolution would result in further measures being taken under the Charter of the United Nations.[18]

With continued reports of non-compliance, a Security Council resolution of 6 October 1992 (resolution 780 (1992))[19] called for the formation of a Commission of Experts to gather and assess evidence of grave breaches of the Geneva Conventions of 1949 and other international humanitarian law committed on the territory of the former Yugoslavia (in practice, initially Bosnia-Herzegovina and Croatia), evidence that ultimately could lead to the prosecution of those responsible. On 18 December 1992 a UN General Assembly resolution (A/Res./47/121) added to the momentum to when it urged the Security Council to consider recommending the establishment of an *ad hoc* criminal tribunal to try and punish those who had committed war crimes in Bosnia-Herzegovina.[20] Resolution A/Res./47/121 further declared that ethnic cleansing was not a consequence of the war, but rather its goal and went on to categorize ethnic cleansing as a form of genocide.

Among the evidence of substantial and severe violations, the Commission of Experts' Interim Report submitted to the Security Council on 10 February 1993 found widespread 'destruction of cultural and religious property.'[21] The report of the Security Council's Human Rights Special Rapporteur, Tadeusz Mazowiecki, of the same date, also confirmed 'the systematic destruction and profanation of mosques, Catholic churches and other places of worship, as well as other sites of cultural heritage.'[22] The report clearly identified the deliberate and methodical targeting of historic and religious structures as a fundamental aspect of ethnic cleansing. Defining all these many violations of international humanitarian

16 See the ICTY website for more on the organization and its work: http://www.icty.org/.

17 *United Nations Security Council Resolution 764 (1992.) Adopted by the Security Council at its 3093rd meeting, on 13 July 1992*, UN Doc. S/RES/764 (1992).

18 *United Nations Security Council Resolution 771 (1992) Adopted by the Security Council at its 3106th meeting, on 13 August 1992*, UN Doc. S/RES/771 (1992).

19 *United Nations Security Council Resolution 780 (1992). Adopted by the Security Council at its 3119th meeting, on 6 October 1992*, UN Doc. S/RES/780 (1992).

20 *The situation in Bosnia and Herzegovina*, UN Doc. A/Res./47/121, 91st Plenary Meeting of the General Assembly, 18 December 1992, No. 10.

21 *Interim Report of the Commission of Experts Established Pursuant to Security Council Resolution 780 (1992)*, UN Doc. S/25274, 10 February 1993, No. 27.

22 *Report on the situation of human rights in the territory of the former Yugoslavia submitted by Mr. Tadeusz Mazowiecki, Special Rapporteur of the Commission on Human Rights, pursuant to Commission resolution 1992/S-1/1 of 14 August 1992*, UN Doc. E/CN.4/1993/50, 10 February 1993, No. 106.

law as war crimes and crimes against humanity, both reports recommended the establishment of an *ad hoc* international tribunal by the United Nations.[23]

The Commission of Experts had gathered an overwhelming amount of information on destruction of cultural property which it did not have the resources to consider, let alone investigate. Thus, it concentrated on preparing in-depth studies of two important cases: the attacks on the World Heritage Site of Dubrovnik in Croatia in 1991 and the destruction of the Old Bridge (*Stari Most*) in Bosnia-Herzegovina in 1993. The studies were included in the Commission of Experts' Final Report of 27 May 1994 in a section on destruction of cultural property,[24] which evolved into a more detailed account as *Annex IX* (*Destruction of cultural property report*) to the Commission's Final Report presented to the Security Council on 28 December 1994.[25]

Thus, Security Council Resolution 827 (1993) of 25 May 1993[26] established the International Criminal Tribunal for the former Yugoslavia, or ICTY, the first international war crimes tribunal since the Nuremberg and Tokyo tribunals that followed World War II and the first established by the United Nations.[27] The ICTY was provided with a mandate and authority to prosecute persons responsible for serious violations of international humanitarian law committed in the territory of the former Yugoslavia since 1991, specifically: grave breaches of the Geneva Conventions of 1949 (Article 2), violations of the laws or customs of war (Article 3), genocide (Article 4), and crimes against humanity (Article 5).[28] As the ICTY website declares: 'Since its establishment in 1993 it has irreversibly changed the landscape of international humanitarian law and provided victims an opportunity to voice the horrors they witnessed and experienced.'[29]

The ICTY Statute afforded protection to cultural property indirectly under all its Articles and directly under *Article 3 (Violations of the laws or customs of war)*, Clause (d): seizure of, destruction or wilful damage done to institutions dedicated to religion, charity and education, the arts and sciences, historic monuments and works of art and science.[30] However, it was through the inclusion of the destruction of cultural and religious property in indictments made under *Article 5 Clause (h): (Crimes against humanity)* persecutions on political, racial and religious grounds, and *Article 4 (Genocide)* that the landmark

23 *Interim Report*, S/25274, No. 27, Nos 47–50 and 74 and *Report on the situation of human rights,* UN Doc. E/CN.4/1993/50, No. 269.4.

24 *Final report of the United Nations Commission of Experts established pursuant to Security Council resolution 780 (1992)*, UN Doc. S/1994/674, 27 May 1994. IV.J, *Destruction of cultural property*, 66, Nos 285–97.

25 *Final Report of the United Nations Commission of Experts Established Pursuant to Security Council Resolution 780 (1992), Annex XI (Destruction of Cultural Property Report)*, UN Doc. S/1994/674/Add.2 (Vol. V) 28 December 1994.

26 *United Nations Security Council Resolution 827 (1993) Adopted by the Security Council at its 3217th meeting, on 25 May 1993*, 25 May 1993, UN Doc. S/RES/827 (1993).

27 Pocar, F. 2008. *Statute of the International Criminal Tribunal for the Former Yugoslavia*, United Nations Audiovisual Library of International Law, http://untreaty.un.org/cod/avl/pdf/ha/icty/icty_e.pdf and ICTY website 'Establishment' http://www.icty.org/sid/319.

28 Updated *Statute of the International Criminal Tribunal for the Former Yugoslavia*, September 2009, available http://www.icty.org/x/file/Legal%20Library/Statute/statute_sept09_en.pdf.

29 *About the ICTY* available at http://www.icty.org/sections/AbouttheICTY. The role of the ICTY in providing justice for victims of war crimes and crimes against humanity and its impact in the countries of the former Yugoslavia has, of course, been the subject of much analysis since its formation; however this phrase could still be considered essentially true.

30 Abtahi, H. 2001 and see *Updated Statute*, September 2009.

judgements of the Tribunal were to make a significant contribution to the developments described above, setting precedent with regard to the destruction of cultural property both as a crime in itself and as a manifestation of persecution.

Notable early judgements were those on the *Lašva Valley* cases relating to the ethnic cleansing of the Lašva Valley in central Bosnia from May 1992 to April 1993 (the *Blaškić* case (2000) and *Kordić and Čerkez* cases (2001)).[31] Tihomir Blaškić, a senior officer and later Commander of the Croatian Defence Council (HVO),[32] was found guilty on counts of persecution and crimes against humanity, citing (among many others) 5(h) (the destruction and plunder of property and, in particular, of institutions dedicated to religion or education) and Article 3(d) (Destruction or wilful damage done to institutions dedicated to religion or education).[33] The judgement on the cases against Dario Kordić (a leading Bosnian Croat nationalist politician) and Mario Čerkez (formerly Commander of the HVO Vitez Brigade) found them guilty on similar counts, but went further in explicitly linking the destruction and wilful damage of institutions dedicated to Muslim religion or education with persecution, seeing such acts (when coupled with discriminatory intent) as 'a nearly pure expression of the notion of "crimes against humanity."'[34]

The *Brđanin* case[35] indicted Radoslav Brđanin, President of the Crisis Staff of the so-called Autonomous Region of Krajina (ARK) on charges related to the ethnic cleansing of non-Serb populations in north-west Bosnia (the Bosnian Krajina)[36] between April–December 1992 and included counts of 'the destruction or wilful damage to Bosnian Muslim and Bosnian Croat religious institutions' under Article 3(d).[37] The 2004 judgement, which found Brđanin guilty of these and other war crimes, observed that the ICTY Trial Chamber was 'satisfied beyond reasonable doubt that there was willful damage done to both Muslim and Roman Catholic religious buildings and institutions in the relevant municipalities by Bosnian Serb forces' and that the devastation during the summer months of 1992 was 'targeted, controlled and deliberate.'[38] The judgement goes on to list such destruction in 13 municipalities, describing the mining, firing and razing of Roman Catholic churches and mosques (with particular attention devoted to ensuring the collapse of minarets) and in many cases, the subsequent removal of their remains.

The judgement in the *Krstić* case (2001), in which Radislav Krstić, former Deputy Commander/Commander of the Drina Corps of the Bosnian Serb Army, was found guilty of genocide, persecutions and murder in relation to the fall of Srebrenica in spring–summer 1995, includes important discussions on interpretations of the 1948 Genocide Convention and definitions of genocide under customary international law and whether they include the destruction of the cultural property of a group, remarking:

> The physical destruction of a group is the most obvious method, but one may also conceive of destroying a group through purposeful eradication of its culture and identity

31 *Prosecutor v. Tihomir Blaškić 'Lašva Valley,'* ICTY Case No. IT-1995-14-T; *Prosecutor v. Dario Kordić and Mario Čerkez 'Lašva Valley,'* ICTY Case No. IT-95-14/2-T.

32 HVO = *Hrvatsko vijeće obrane.*

33 *Blaškić, Judgment*, Case No. IT-1995-14-T, 3 March 2000.

34 *Kordić and Čerkez, Judgment*, Case No. IT-95-14/2-T, 26 February 2001, No. 207.

35 *Prosecutor v. Radoslav Brđanin*, ICTY Case No. IT-99-36-T. Brđanin is sometimes spelled Brdjanin.

36 In Bosnian the region is known as Bosanska Krajina.

37 *Brđanin, Sixth Amended Indictment*, Case No. IT-99-36-T, 9 December 2003, Nos 61–64, Count 12.

38 *Brđanin, Judgement*, Case No. IT-99-36-T, 1 September 2004, Nos 640, 642 and 658.

> resulting in the eventual extinction of the group as an entity distinct from the remainder of the community.[39]

The Trial Chamber acknowledged that while customary international law currently limited the definition of genocide to those acts seeking the physical or biological destruction of all or part of the group, it also pointed out that the United Nations General Assembly had characterized ethnic cleansing as a form of genocide in 1992[40] and:

> … that where there is physical or biological destruction there are often simultaneous attacks on the cultural and religious property and symbols of the targeted group as well, attacks which may legitimately be considered as evidence of an intent to physically destroy the group. In this case, the Trial Chamber will thus take into account as evidence of intent to destroy the group the deliberate destruction of mosques and houses belonging to members of the group.[41]

The ICTY began to take a more systematic approach to collecting evidence relating to the deliberate destruction of cultural property in Bosnia-Herzegovina when the Office of the Prosecutor (OTP) commissioned two expert consultants to compile reports, both focused on the destruction of Bosnian Muslim and Bosnian Croat heritage in selected municipalities (almost entirely ones held by Bosnian Serb forces/authorities during the war). The first, by Colin Kaiser,[42] carried out from summer 2001 to early 2002 and based on a combination of field trips and examination of documentation already collected by ICTY investigators, assessed damage and destruction only of sacral buildings; the field trips to all municipalities requested appear not to have been completed however. Kaiser had testified in 1998 on destruction of (mainly Islamic) sacral and cultural monuments in the Lašva Valley during the Blaškić case and in 2004 was an expert witness in the case against Pavle Strugar and Miodrag Jokić for the attacks on Dubrovnik (see below).[43]

In April 2002 András Riedlmayer,[44] who had already carried out a survey of deliberately damaged and destroyed cultural heritage sites in Kosovo in 1999 to assist the work of

---

39 *Prosecutor v. Radislav Krstić, Trial Judgment*, ICTY Case No. IT-98-33, 2 August 2001, No. 574.

40 *The situation in Bosnia and Herzegovina*, UN, AG/Res./47/121.

41 *Krstić, Trial Judgment*, Case No. IT-98-33, No. 580.

42 Kaiser, C. 2002. (*Report on Destruction of Cultural Property* 09-Jul-02, Exhibit P906, *Krajišnik* Case No. IT-00-39) *Report on the Damaging and Destruction of Islamic and Roman Catholic Sacral Buildings in the Municipalities of Bosanski Novi, Donji Vakuf, Kljuc, Kotor Varos, Prijedor and Sanski Most in 1992 (April to December) and Preliminary Remarks on Damaging and Destruction of Islamic and Roman Catholic Sacral Buildings in the Municipalities of Bijeljina, Brcko, Bratunac, Cajnice, Doboj, Foca, Ilijas, Rogatica, Visegrad and Zvornik*, 9 July 2002. Dr Kaiser (a former director of ICOMOS) had been consultant expert to the Parliament Assembly of the Council of Europe (COE) reporting on war damage to the cultural heritage in Bosnia-Herzegovina and Croatia from 1992 on and was later UNESCO Representative in Bosnia-Herzegovina; see elsewhere in this volume for more on his activities in Bosnia.

43 *Blaškić* Case No. IT-1995-14-T, Kaiser testimony, 16 July 1998; *'Dubrovnik' Strugar*, Case No. IT-01-42-T and *Jokić*, Case No. IT-01-42/1-S, Kaiser testimony, 12–13, 16–18 February 2004.

44 András Riedlmayer has directed the Documentation Center for Islamic Art and Architecture at the Fine Arts Library of Harvard University since 1985. He was a co-founder of the Bosnian Manuscript Ingathering Project and has written and spoken widely on the destruction of cultural heritage in Bosnia-Herzegovina and Kosovo.

the ICTY,[45] was commissioned by the OTP to carry a further field investigation assessing both sacral and non-sacral sites in 19 municipalities (nearly identical to those in the Kaiser report, and with the addition of central Sarajevo).[46]

These cultural heritage reports (along with the testimonies of the experts involved) were used in more than one ICTY prosecution, including those of Momčilo Krajišnik, Slobodan Milošević, Radovan Karadžić and Vojislav Šešelj.[47] Riedlmayer went on to be the principal expert witness in cultural heritage for the ICTY on the destruction of non-Serb religious and cultural property in Bosnia-Herzegovina and carried out further field investigations for the OTP in relation to specific prosecutions, such as for instance, visiting the Bosnian Serb municipalities of Bileća, Gacko, Teslić and Vlasenica in 2009 with respect to the Mico Stanišić and Stojan Župljanin case.[48] Over the course of his work for the ICTY Riedlmayer documented 36 municipalities, reporting on 534 individual sites, some of which were composed of multiple separate elements.[49] He made his ninth and final appearance as an expert witness in the ICTY Trial Chamber in October 2013 at the trial of Ratko Mladić.

It is notable that the first ICTY indictments that followed the 'classical' interpretation of protection for cultural property (on the basis of its universal significance to mankind) were not issued until 2001 against former senior Yugoslav military officers, Pavle Strugar and Miodrag Jokić, relating to the shelling of the Old Town of Dubrovnik in Croatia during October–December 1991 by the Yugoslav People's Army (JNA) and Yugoslav Navy forces over which they had command. The indictment cited 'destruction or wilful damage' to historic monuments and emphasized the Old Town's status as a UNESCO World Heritage Site.[50] Jokić and Strugar were subsequently convicted for this and other war crimes on 18 March 2004 and 31 January 2005 respectively, Jokić's sentencing judgement stating that the 'shelling attack on the Old Town was an attack not only against the history and heritage of the region, but also against the cultural heritage of humankind.'[51]

An act equally as iconic in the destruction of cultural heritage during the Wars of Yugoslav Succession as the shelling of Dubrovnik was the destruction of the sixteenth-century *Stari Most* (Old Bridge) at Mostar on 9 November 1993. Both the Dubrovnik and the *Stari Most* cases had featured in *Annex XI (Destruction of cultural property report)* of the United Nations Commission of Experts *Final Report* of December 1994, the only two examples selected to represent such grave breaches of the Geneva Conventions and international humanitarian law committed in respect of cultural property in the former Yugoslavia.[52]

---

45 The survey was carried out with Andrew Herscher. See Herscher, A. and Riedlmayer, A. 2000. *Architectural Heritage in Kosovo: A Post-War Report.*

46 Riedlmayer, A. 2002. *Destruction of Cultural Heritage in Bosnia-Herzegovina, 1992–1996: A Post-war Survey of Selected Municipalities*, (*Milošević* Case No. IT-02-54, Exhibit P486, Date: 08/07/2003, and *Krajišnik* Case No. IT-00-39, Exhibit P732, Date: 23/05/2005).

47 *Prosecutor v. Momčilo Krajišnik 'Bosnia and Herzegovina'* (Case No. IT-00-39), *Prosecutor v. Slobodan Milošević* (Case No. IT-02-54); *Prosecutor v. Radovan Karadžić* (Case No. IT-95-5/18-I); *Prosecutor v. Vojislav Šešelj* (Case No. IT-03-67). Riedlmayer also testified in cases related to the Kosovo conflict.

48 See: Destroying all traces of mosques and churches, *Sense Tribunal* [online], 2 June 2010.

49 Personal communication with András Riedlmayer 22 October 2013.

50 *Prosecutor v. Pavle Strugar, Judgment*, ICTY Case No. IT-01-42-T, 31 January 2005 and *Prosecutor v. Miodrag Jokić, Judgment*, ICTY Case No. IT-01-42/1-S, 18 March 2004.

51 *Prosecutor v. Miodrag Jokić, Judgment*, Case No. IT-01-42/1-S, No. 51.

52 *Final Report of the United Nations Commission of Experts*, UN Doc. S/1994/674/Add.2 (Vol. V), 28 December 1994.

There had been many calls to bring to justice those who had destroyed or ordered the destruction of the Stari Most. An investigation by the Mostar District Military Court in late 1993 of the tank crew which had fired on the bridge found that the three HVO soldiers had acted on their own initiative without orders from their commanding officers.[53] These proceedings were widely regarded as an attempt to find convenient scapegoats for the attack after the international outcry that followed the demolition of the bridge. A further investigation of the three soldiers at the Mostar Cantonal Court in 2001 reached the same conclusion and the case was referred to the ICTY.[54] The whereabouts of the three soldiers was unknown time of writing; two are apparently dead.[55]

ICTY war crimes charges for the attack on the Stari Most came on 2 March 2004 with the indictment of Jadranko Prlić, Bruno Stojić, Slobodan Praljak, Milivoj Petković, Valentin Ćorić and Berislav Pušić (*Prlić et al*),[56] coincidentally just as the rebuilt Stari Most neared completion.[57] All were formerly high-ranking political or military officials connected with the wartime Bosnian Croat para-state, the Croatian Republic of Herceg-Bosna and the Croatian Defence Council (HVO); Praljak had been commander of HVO forces at the time of the final attack on the Stari Most.[58] All were charged with participating in a *joint criminal enterprise* ' … to politically and militarily subjugate, permanently remove and ethnically cleanse Bosnian Muslims and other non-Croats who lived in areas on the territory of the Republic of Bosnia and Herzegovina which were claimed to be part of the Croatian Community (and later Republic) of Herceg-Bosna … '[59] and detailed through a long list of charges which included the destruction of the Stari Most.

However, the treatment of the Stari Most in the *Prlić et al* indictment differed noticeably from that of Dubrovnik, despite it featuring with Dubrovnik in the United Nations Commission of Experts' *Final Report* of December 1994 in its annex on destruction of cultural property. Though described as an 'international landmark,' the Stari Most was not accorded the special status of historic monument and only featured in the indictment at the end of a list of destroyed religious properties (many of them historic monuments themselves), whose destruction was cited as solely 'destruction or wilful damage to institutions dedicated to religion or education.'[60] This decision of the Prosecution was to have significant implications in the final judgement in respect of the case of the Stari Most.

Yet, once the trial began the destruction of the Stari Most and the command responsibility of Slobodan Praljak and other defendants for the act featured prominently in the proceedings with many prosecution and defence witnesses called to give evidence,[61]

---

53 *Fourth Information Report on war damage to the cultural heritage in Croatia and Bosnia-Herzegovina*, Parliamentary Assembly of the Council of Europe (COE), Strasbourg, Doc. 6999, 19 January 1994, No. 72; Croats questioned over destruction of Mostar bridge, *The Independent*, 23 December 1993.

54 Petrovic, J. 2012, 207.

55 According to Slobodan Praljak in 2009. *Prlić et al*, Case No. IT-04-74-T, testimony Slobodan Praljak, 8 June 2009, 41281–2.

56 *Prosecutor v. Jadranko Prlić, Bruno Stojić, Slobodan Praljak, Milivoj Petković, Valentin Ćorić and Berislav Pušić, Indictment*, ICTY Case No.IT-04-74-T. The indictment was made public on 2 April 2004. See ICTY *Prlić et al*, Case No. IT-04-74-T, *Case Information Sheet*.

57 The new *Stari Most* was officially opened on 23 July 2004.

58 *Prlić et al, Initial Indictment*, Case No. IT-04-74-T, 2 March 2004, paras 116 and 118.

59 *Prlić et al, Initial Indictment*, Case No. IT-04-74-T, para. 15.

60 *Prlić et al, Initial Indictment*, Case No. IT-04-74-T, para. 116. The *Stari Most* was not a UNESCO World Heritage Site at the time.

61 For a detailed account of the ICTY proceedings in relation to the *Stari Most*, see Petrovic, 2012, *The Old Bridge of Mostar*.

so much so that President of the Trial Chamber Judge Jean-Claude Antonetti remarked that the issue of the Old Bridge had been discussed 'for hours on end.'[62] Antonetti also noted that there was ' … not even one sentence on the destruction of the Old Bridge in the pre-trial brief, although the bridge is something extremely relevant for all the communities in Mostar.'[63] By the time the trial ended in March 2011, the focus of the case in relation to the charges of 'destruction or wilful damage to institutions dedicated to religion or education' was almost entirely on the destruction of the Stari Most.

While no one disputed the cultural-historical and symbolic importance of the bridge, the trial proceedings examined such questions as who was actually responsible for its destruction (with Slobodan Praljak's defence seeking to prove that it had been explosives set on the east bank of the Neretva controlled by the Bosnian Army that caused the final collapse of the Stari Most), how badly it had been damaged during the earlier JNA/Bosnian Serb Army siege of the city and whether that could have contributed to its collapse, and whether it could be considered a legitimate military target, since the Bosnian Army had used the bridge to move men and military supplies to a small Muslim-held enclave on the west bank of the river (as well as supplying the civilian population remaining there).[64]

On 29 May 2013 the ICTY Trial Chamber delivered its judgement and found (by a majority) Praljak and his co-defendants guilty of participating in a joint criminal enterprise. In respect of the charge relating to the destruction of the Stari Most, the Chamber found by a majority that although the bridge was used by the Bosnian Army and thus (in its judgement) constituted a legitimate military target for the HVO, its destruction caused disproportionate damage to the Muslim civilian population of Mostar.[65]

The judgement noted the Stari Most's immense cultural, historical and symbolic character for Muslims in particular, of which the HVO was well aware and found that the HVO command intended to destroy the Old Bridge in Mostar and thereby destroy the morale of the Muslim population of Mostar. The Chamber found, by a majority, that the HVO armed forces destroyed the Old Bridge of Mostar, thereby committing the crime of wanton destruction of cities or villages, or devastation not justified by military requirements.[66] Judge Antonetti dissented from these findings.[67] One early legal commentator on the judgement with respect to the Stari Most noted that ' … questions on proportionality when attacking dual use objects … and whether the threshold for (important) cultural property to lose its protected status when used for military purposes is different from a regular civilian object, are still open to debate.'[68]

62 *Prlić et al,* Case No. IT-04-74-T, testimony Slobodan Praljak, 8 June 2009, 41279.

63 *Prlić et al,* Case No. IT-04-74-T, testimony Slobodan Praljak, 8 June 2009, 41279.

64 See *Prlić et al, Jugement, Vol. 2,* Case No. IT-04-74-T, 29 May 2013, 367–86. See also: *Prlić et al,* Case No. IT-04-74-T, *Final Trial Brief* of the Prosecution,1 April 2011, *XI.4 Praljak and the HVO Main Staff approved the destruction of the Old Bridge (Stari Most)*, Nos 821–38, 259–66; and *Final Trial Brief* of Slobodan Praljak, 31 March 2011, *X.g Destruction of the Old Bridge, 9 November 1993*, Nos 316–57, 112–15.

65 *Prlić et al, Judgement Summary,* Case No. IT-04-74-T, 29 May 2013.

66 *Prlić et al, Jugement, Vol. 3,* Case No. IT-04-74-T, 29 May 2013, Nos 1582–7.

67 *Prlić et al, Jugement, Vol. 6, Opinion Individuelle Séparée et Partiellement Dissidente du Juge Jean-Claude Antonetti Président de la Chambre*, Case No. IT-04-74-T, 29 May 2013, 386–87. In his long separate dissenting opinion, in relation to the Stari Most, Antonetti declared that he considered that who exactly had destroyed the bridge had not been established, that it had already been damaged during the conflict with the Serbs, and that as the bridge was considered a military objective, there could be no consideration of proportionality.

68 Bartels, R. 2013. Prlić et al.: The Destruction of the Old Bridge of Mostar and Proportionality. *www.ejiltalk.org*, 31 July. http://www.ejiltalk.org/prlic-et-al-the-destruction-of-the-old-bridge-of-mostar-and-proportionality/.

However, the Chamber noted the inclusion of the destruction of the Stari Most in Count 21: 'destruction or willful damage done to institutions dedicated to religion or education' and that the prosecution chose to retain the Old Bridge in this count. The ICTY judges clearly found this decision inexplicable. As Judge Antonetti noted in his separate opinion, he found it surprising that this internationally important monument (by that time listed by UNESCO as a World Heritage Site),[69] one of the most symbolic cultural sites affected by the wars in the former Yugoslavia, did not merit a separate paragraph in the indictment, especially as it was clearly neither a religious nor an educational structure.[70]

The Chamber noted that the Prosecution did not choose to include the destruction of the Stari Most in the indictment in its status as a historic monument as provided for by Article 3 d) of the ICTY Statute (seizure of, destruction or wilful damage done to institutions dedicated to religion, charity and education, the arts and sciences, historic monuments and works of art and science). This negligence made it impossible for the Chamber to be able to take into account the destruction of the Old Bridge as a historic monument, with major historical and symbolic value (especially in the eyes of the Muslim community) within the terms of Count 21 which applies only to destruction or wilful damage to institutions dedicated to religion or education. This lapse also meant that the Chamber had no option but to consider the destruction of the Old Bridge in the case against Slobodan Praljak under the terms of Count 19 (extensive destruction of property, not justified by military necessity and carried out unlawfully and wantonly) and Count 20 (wanton destruction of cities, towns or villages, or devastation not justified by military necessity).[71]

Other ICTY prosecutions that include counts of the deliberate destruction of cultural and religious property awaiting verdicts and judgements at time of writing include those against Radovan Karadžić and Vojislav Šešelj.[72]

The War Crimes Chamber (WCC) of the Court of Bosnia and Herzegovina[73]

As part of the ICTY's completion strategy, a number of its cases were transferred to be tried in Bosnia at the War Crimes Chamber (WCC) of the Court of Bosnia and Herzegovina.[74] The Court of Bosnia and Herzegovina, whose panels of judges include national and (at time of writing) international members, established a War Crimes Chamber in 2005 and prosecutes the most serious of such cases, drawing on ICTY and other jurisprudence and legal opinion. One of the ICTY cases transferred to Bosnia in 2006 was that of Paško Ljubičić, former Commander of the Fourth Military Police Battalion of the HVO, in

69 Old Bridge Area of the Old City of Mostar was inscribed as a World Heritage Site in July 2005.

70 *Prlić et al, Jugement, Vol. 6, Opinion Individuelle Séparée et Partiellement Dissidente du Juge Jean-Claude Antonetti Président de la Chambre*, Case No. IT-04-74-T, 29 May 2013, 359.

71 *Prlić et al, Jugement, Vol. 3*, Case No. IT-04-74-T, 29 May 2013, No. 1611.

72 *Karadžić*, Case No. IT-95-5/18-I; *Šešelj*, Case No. IT-03-67.

73 The Court of Bosnia and Herzegovina was established in 2002. For more information see the Court's website at http://www.sudbih.gov.ba and for a good overview of the war crimes chamber see Ivanišević, B. 2008. *The War Crimes Chamber in Bosnia and Herzegovina: From Hybrid to Domestic Court*. New York: International Center for Transitional Justice. http://ictj.org/publication/war-crimes-chamber-bosnia-and-herzegovina-hybrid-domestic-court.

74 Under the Tribunal's Rule 11 *bis*. See *Rules of Procedure and Evidence*. Rule 11 *bis* (Referral of the Indictment to Another Court). IT/32/Rev. 43, 24 July 2009, The Hague: International Criminal Tribunal for the former Yugoslavia.

relation to war crimes committed in the Lašva Valley in 1993 which included charges for the 'destruction or wilful damage to institutions dedicated to religion or education.'[75] At the WCC Ljubičić's indictment included a specific charge of 'destruction or deliberate damaging of establishments devoted to for religious purposes (destruction of mosques)' under Article 179 of the Criminal Code of Bosnia-Herzegovina.[76] However, following a Plea Agreement by Ljubičić, his amended indictment did not include this charge, although he was found guilty of War Crimes Against Civilians, including the destruction of property, on 28 April 2008.

Other WCC cases which have included charges for destruction of cultural and religious property were those against Zijad Kurtović,[77] a military police commander in the Independent Drežnica Battalion of the 4th Corps of the Army of Bosnia-Herzegovina, and Predrag Kujundžić, former commander of *Predini vukovi* (Predo's Wolves), a paramilitary unit attached to the Bosnian Serb Army.[78]

Kurtović was accused of a series of war crimes during the conflict between the HVO and the ARBiH, including the physical and mental torture of Bosnian Croat civilians and prisoners of war in the Roman Catholic church of All Saints in the village of Donja Drežnića in October 1993. He was further charged with desecrating and, vandalizing the interior of the church and destroying the religious objects there, including using crosses and statues of the saints to beat detainees and forcing detainees to eat pages from the Bible and other holy books.[79]

On 30 April 2008 the first instance verdict of the Court of Bosnia and Herzegovina found Zijad Kurtović guilty of the criminal offences of War Crimes against Civilians, War Crimes against Prisoners of War, and Violation of Laws and Practices of Warfare and sentenced him to 11 years imprisonment. However, on 25 March 2009, the Appellate Panel of the War Crimes Chamber modified the first instance verdict in terms of the legal assessment and qualification of the offence. The second instance verdict found Kurtović guilty of the criminal offence of War Crimes against Civilians and the criminal offence of Destruction of Cultural and Historic Monuments of the Criminal Code of SFRY.[80] This was the first time the Court of Bosnia and Herzegovina found a defendant guilty specifically for damage to cultural property as a war crime. Kurtović's sentence remained unchanged.

The case against Predrag Kujundžić once again clearly linked the destruction of religious property with persecution. In its First Instance Verdict of 30 October 2009, citing judgements of the Nuremberg Trials, ICTY jurisprudence and the 1991 Report of the International Law Commission (IRC) which found that the destruction of religious objects was 'an unequivocal example of crimes against humanity' and when 'committed with a necessary discriminatory intent, this offense represents an attack against the very religious identity of a people,' the Court concluding 'that the mosque in Bukovačke Čivčije was destroyed just because it was intended for the practice of Islamic religion, and thus its destruction represents an attack on the religious identity of the Bosniak population with

75 *Prosecutor vs Paško Ljubičić, Corrected Amended Indictment,* ICTY Case No. IT-00-41-PT, 2 April 2002, No. 53.

76 *Case No. X-KR-06/241 – Ljubičić Paško,* Court of Bosnia and Herzegovina, 2006.

77 *Case No. X-KRŽ-06/299 – Kurtović Zijad,* Court of Bosnia and Herzegovina, 2007.

78 *Case No. X-KRŽ-07/442, Kujundžić Predrag,* Court of Bosnia and Herzegovina, 2008.

79 *Case No. X-KRŽ-06/299, Kurtović Zijad, First Instance Verdict,* Court of Bosnia and Herzegovina, 30 April 2008.

80 *Case No. X-KRŽ-06/299, Kurtović Zijad, Second Instance Verdict,* Court of Bosnia and Herzegovina, 25 March 2009.

a clearly expressed discriminatory intent which constitutes the basis of persecution.'[81] Kujundžić was found guilty on all counts against him.

The International Court of Justice (ICJ)[82]

A further important legal step in testing the link between the destruction of a group's cultural heritage and genocide began in 1993 when Bosnia-Herzegovina filed a case of genocide at the International Court of Justice (ICJ) against Yugoslavia (later Serbia and Montenegro) for violations of its obligations under the 1948 Genocide Convention.[83] The charges were brought under Article II(c) of the Convention concerning the deliberate infliction on a group of conditions of life calculated to bring about its physical destruction in whole or in part, and made three sets of claims against Serbia and Montenegro with regard to its treatment of non-Serb populations: encirclement, shelling and starvation; deportation and expulsion; and destruction of historical, religious and cultural property, claiming 'that Bosnian Serb forces attempted to eradicate all traces of the culture of the protected group through the destruction of historical, religious and cultural property.'[84]

Thus the linkage of genocide with the destruction of cultural and religious heritage was crucial to the case of Bosnia-Herzegovina and András Riedlmayer, the ICTY's key consultant expert in the field, was one of the only two witnesses it offered for its submission during the Merits stage, presenting testimony on the destruction of the cultural and religious heritage of Bosnia-Herzegovina.[85] Like the ICTY, in its judgement the ICJ examined legal definitions of genocide, but unlike ICTY, it rejected arguments that the destruction of a group's cultural heritage would bring about the physical destruction of the group,[86] finding that although there was ' ... conclusive evidence of the deliberate destruction of the historical, cultural and religious heritage of the protected group ... ,' in the Court's view:

> the destruction of historical, cultural and religious heritage cannot be considered to constitute the deliberate infliction of conditions of life calculated to bring about the physical destruction of the group. Although such destruction may be highly significant inasmuch as it is directed to the elimination of all traces of the cultural or religious presence of a group, and contrary to other legal norms, it does not fall within the categories of acts of genocide set out in Article II of the Convention.[87]

---

81 *Case No. X-KRŽ-07/442, Kujundžić Predrag, First Instance Verdict*, Court of Bosnia and Herzegovina, 30 October 2009, 320–21.

82 For more on the International Court of Justice (ICJ) see its website: http://www.icj-cij.org/.

83 *Application of the Convention on the Prevention and Punishment of the Crime of Genocide (Bosnia and Herzegovina v. Yugoslavia (Serbia and Montenegro))*, ICJ, 20 March 1993. See also *United Nations Convention on the Prevention and Punishment of the Crime of Genocide*, Paris, 9 December 1948.

84 *Application of the Convention on the Prevention and Punishment of the Crime of Genocide (Bosnia and Herzegovina v. Serbia and Montenegro), Judgement*, ICJ, 26 February 2007, 320–22.

85 Vrdoljak, A. 2009, 66. See Riedlmayer's testimony in *Application of the Convention on the Prevention and Punishment of the Crime of Genocide*, ICJ CR 2006/22, 17 March 2006. See also ICJ *Judgement*, 26 February 2007, *Destruction of historical, religious and cultural property*, paras 335–44.

86 *Judgement*, ICJ, 26 February 2007, 335–44.

87 *Judgement*, ICJ, 26 February 2007, para. 344.

The Human Rights Chamber for Bosnia and Herzegovina (HRC)[88]

The Human Rights Chamber for Bosnia and Herzegovina (HRC) became a further setting where the association between the protection and preservation of cultural heritage and the enjoyment of human rights under international humanitarian law was articulated. One of the institutions established under the terms of the Dayton Peace Agreement (Annex 6),[89] the HRC was a court with national and international members whose mandate ran from March 1996 to 31 December 2003.[90] Annex 6 included in its articles the European Convention for the Protection of Human Rights[91] and many other international agreements relating to human rights, all of which were incorporated in the country's new constitution. Dakin has noted the importance of the Human Rights Chamber's contribution to the development of international human rights law[92] as part of a trend towards 'judicial globalization,'[93] calling it 'a third voice in the ongoing dialogue between the European Court of Human Rights and national jurisdictions in Europe about individual and group rights and the relationship between citizens and the state.'[94]

In 1996 the Islamic Community in Bosnia-Herzegovina[95] filed a case against Republika Srpska at the Human Rights Chamber with regard to the events surrounding the destruction of 15 mosques in Banja Luka in 1993, the desecration of their sites and the subsequent refusal in the post-war period of the Republika Srpska and Banja Luka authorities to issue permits for the rebuilding of the destroyed mosques.[96] At least 12 of the mosques were listed as historic monuments in the pre-war era and included the sixteenth-century Ferhadija and Arnaudija Mosques. The case was brought under Article 9 of the European Convention for the Protection of Human Rights and Article 1 of Protocol No. 1 to the Convention, alleging violation of the freedom of religion of Islamic Community members and of its property rights, as well as discrimination against the Islamic Community on the grounds of the religion and national origin of its members.

The HRC found in favour of the Islamic Community, its decision of 11 June 1999 noting that the 'refusal to allow reconstruction of any mosques is clearly aimed at preventing the applicant from providing its members in the Banja Luka area with adequate premises for the manifestation of their religion and ethnic identity.'[97] It noted as well that the destroyed

88 For more information see the website of the Human Rights Chamber (HRC) at http://www.hrc.ba/english/default.htm.

89 Formally the *General Framework Agreement for Peace* or GFAP. Annex 6, Agreement on Human Rights, available at http://www.ohr.int/dpa/default.asp?content_id=374.

90 From 1 January 2004 human rights violations formerly considered by the HRC came under the remit of the Constitutional Court of Bosnia and Herzegovina.

91 Frequently referred to as the European Convention on Human Rights, but formally the *European Convention for the Protection of Human Rights and Fundamental Freedoms and its Protocols.*

92 Dakin, B. 2002. Islamic Community in Bosnia and Herzegovina v. The Republika Srpska: Human rights in a multi-ethnic Bosnia, *Harvard Human Rights Journal*, Vol, 15 (Spring 2002).

93 A term used by Anne-Marie Slaughter and cited in Dakin, B. 2002, Note 131.

94 Dakin, B. 2002, 264.

95 A legal body established in 1882 during the Austro-Hungarian occupation of Bosnia-Herzegovina.

96 *The Islamic Community in Bosnia and Herzegovina v. The Republika Srpska, (Banja Luka Mosques), Decision on the Admissibility and Merits, (Delivered on 11 June 1999)*, HRC Case No. CH/96/29. See also Dakin, B. 2002 for a full account of the case.

97 *The Islamic Community (Banja Luka Mosques), Decision, 11 June 1999*, HRC Case No. CH/96/29, No. 206.

mosques had 'held significant religious and cultural importance not just for the applicant and its members' and 'formed part of the cultural-historic heritage of Bosnia and Herzegovina.'[98] The Islamic Community brought several similar suits to the HRC against Republika Srpska with regard to the treatment of the sites of its religious facilities and the non-issuance of building permits for their reconstruction in Bijeljina, Zvornik and Mrkonjić Grad.[99]

The Islamic Community Reparations Case

The Islamic Community was to seek redress and reparations for the destruction of its mosques in other legal arenas. On 27 October 2000 it filed an action at the Basic Court (Osnovi Sud) in Banja Luka against Republika Srpska and the Municipality of Banja Luka for compensation for the destruction of its religious facilities in Banja Luka in 1993. Crucial factors in the case were the cultural-historical values of the mosques (most important of which were the Ferhadija and Arnaudija) and the status of most as listed historic monuments (declared as such by the competent pre-war heritage authorities of both the Republic of Bosnia-Herzegovina and the City of Banja Luka), the legal obligations of the RS and Banja Luka authorities to protect them, and their failure to do so. The cultural-historical values of the structures formed the basis of the calculations for the amount claimed by the Islamic Community in compensation.[100] The subsequent designation of the sites of historic mosques as National Monuments under Annex 8 of the Dayton Agreement in the post-war period was also noted. A key argument in the case was that there were no war operations ongoing in the Municipality of Banja Luka and that the destruction of the mosques took place during the hours of curfew when the movement of people was restricted to those in official positions.

However, the suit was not to come to trial for almost seven years. It took a Constitutional Court of Bosnia and Herzegovina ruling in 2006 that the delays in proceeding with the case represented a violation of the right to a fair trial in relation to the right of access to court under Article II(3)(e) of the Constitution of Bosnia-Herzegovina and Article 6, paragraph 1 of the European Convention for the Protection of Human Rights for the Banja Luka court to begin to take action on the claim in 2007.[101]

Remarkably, on 20 February 2009 the Basic Court found in favour of the Islamic Community, ordering the Republika Srpska and Banja Luka authorities to pay total compensation of more than 64.7 million KM[102] (approximately US$42 million, or €33

98 *The Islamic Community (Banja Luka Mosques), Decision, 11 June 1999*, HRC Case No. CH/96/29, No. 202.

99 Regarding mosques in Zvornik see: HRC Case No. CH/01/7701, (9 November 2000), in Bijeljina: HRC Case No. CH/99/2656 (6 December 2000), and in Mirkonjić Grad: HRC Case No. CH/98/1062 (22 December 2003).

100 The Basic Court judgement (in Serbian), *Presuda Islamska zajednica BiH protiv Republike Srpske*, Osnova Sud Banjaluka, No. P-2757/00, 20 February 2009, with details of the case can be found at the website of the Islamic Community's lawyer: Advokat Esad Hrvačić, Aktuelna sudska praksa, at http://www.advokathrvacic.com.ba.

101 *Thirty-sixth session of the Grand Chamber, 13 June 2006*, Constitutional Court of Bosnia and Herzegovina.

102 KM = *konvertibilna marka*. The currency of Bosnia-Herzegovina is also given the code BAM = Bosnian convertible marka.

million), which included 32.4 million KM for the demolition of the Ferhadija Mosque, 15.1 million KM for the Arnaudija Mosque and 17.7 million KM for other demolished religious buildings.[103] While hailed as a landmark judgement, the decision was soon overturned on appeal by Republika Srpska later in 2009 at the District Court in Banja Luka (Okružni Sud), which cited the statute of limitations in pursuing compensation and ordered the Islamic Community to pay court costs.[104] Yet, crucially, the District Court confirmed Republika Srpska's and Banja Luka's responsibility for failing to prevent the destruction of the mosques, leading Esad Hrvačić, the lawyer representing the Islamic Community, to describe the verdicts of both courts as of historic significance.[105]

The Islamic Community's legal battle continued as it went to the Republika Srpska Supreme Court in an attempt to overturn the District Court ruling. Though the RS Supreme Court's judgement in February 2012 upheld the District Court decision, once again it did not contest the responsibility of the Republika Srpska and Banja Luka authorities for the mosques' destruction. Following the RS Supreme Court judgment, the Islamic Community announced its intention to lodge an appeal with the Constitutional Court of Bosnia and Herzegovina and to pursue the case as far as the European Court of Human Rights in Strasbourg if they did not receive justice in Bosnia-Herzegovina.[106]

Esad Hrvačič asserted that the destruction of the Banja Luka mosques had been a war crime and thus not subject to any statute of limitations,[107] and cited the precedent set by the ICTY judgements in the Brđanin case with regard to destruction of religious property.[108] The General Secretary of the Islamic Community, Muhamed Salkić, declared pursuing the case was 'not about money, but about the destruction of historic monuments that occurred without reason,'[109] remarking that the destruction was a huge loss, not just for Muslims, but for Banja Luka as a city with a common heritage of living together and co-existence.[110]

103 See many articles in the news media, for instance: BIRN, 2009, Islamic Community to Receive Around 33 Million Euros in Compensation, *Justice Report, Balkan Investigative Reporting Network (BIRN)* [online], 20 February 2009, and: Serbs ordered to pay for mosques, *BBC News* [online], 20 February 2009. See also *Presuda Islamska zajednica BiH protiv Republike Srpske*, Osnova Sud Banjaluka, No. P-2757/00, 20 February 2009. The values given are those calculated at the time of the judgement.

104 For the decision of the District Court (in Serbian) see: Okružni Sud u Banjaluci, *Ferhadija*, Case No. : 011-0-Gž-09-000328, 30 September 2009.

105 Mackic, E. 2009. Historic Decisions by Banja Luka Court, Justice Report. *Balkan Investigative Reporting Network (BIRN)*, 13 November.

106 Remiković, D. 2012. Islamska zajednica u BiH žalit će se na sudsku presudu u RS po pitanju džamija, *Southeast European Times*, 16 February; Bjelajac, M. 2012. Sud odlučio: RS neće isplatiti odštetu za srušene džamije, *Radio Slobodna Evropa*, 3 February.

107 See *Resolution adopted by the General Assembly: The United Nations basic principles and guidelines on the right to a remedy and reparation for victims of gross violations of international human rights law and serious violations of international humanitarian law*, UN Doc. A/RES/60/147 on statutes of limitations (Article IV).

108 Bjelajac, M. 2012.

109 Remiković, D. 2012, Bjelajac, M. 2012.

110 Bjelajac, M. 2012.

Fig. 8.1 András Riedlmayer testifying at the International Criminal Tribunal for the Former Yugoslavia on 27 May 2008 as an expert witness on cultural heritage at the trial of Vojislav Šešelj. © Sense News Agency

Conclusion

Thus the search for justice for victims of the human rights violations of the 1992–1995 Bosnian War brought into focus the importance of cultural heritage as a crucial feature of group identity, as well as the importance of the protection and preservation of cultural and religious property in safeguarding pluralism and diversity. Charges for attacks on cultural and religious property formed part of war crimes prosecutions at both the ICTY and later in domestic war crimes courts in Bosnia-Herzegovina, offering a form of retributive justice. The Islamic Community in Bosnia-Herzegovina has been unique so far in its quest for reparative justice and demands for the enjoyment of human rights on behalf of its members in their pursuit of cases against Republika Srpska and municipal authorities within the entity, both in domestic courts and at the Human Rights Chamber of Bosnia and Herzegovina.

Appendix
A Brief Introduction to the Cultural Heritage of Bosnia-Herzegovina

Helen Walasek

At the start of the Bosnian War in 1992 (and even now) few outside the region knew much of Bosnia-Herzegovina's history – except that World War One famously had started in Sarajevo with the assassination of Archduke Franz Ferdinand, heir to the Austro-Hungarian throne and (perhaps) that it had once been part of the Ottoman Empire. Little, if anything, was known of what its cultural heritage consisted. As the war went on knowledge grew of a handful of its historic monuments, largely through reports of their destruction, particularly the National Library in Sarajevo (an Austro-Hungarian structure), the Old Bridge at Mostar (an Ottoman structure), as well as the hundreds of mosques and scores of Catholic and Orthodox churches which had been attacked. But not much else was known, apart from an awareness that before the war the country had had a diverse, 'multiethnic' history and cultural identity. Thus this appendix seeks to give a very brief introduction to Bosnia-Herzegovina's historical-cultural patrimony, focusing particularly on the material heritage context at the outbreak of war in 1992.

Located on the fracture between Eastern and Western Christendom, then at the interface of the Ottoman and Habsburg Empires for over 400 years, as well as being the site of a distinctive medieval kingdom, Bosnia-Herzegovina has a complex heritage interweaving Mediterranean, Byzantine, Ottoman and Central European cultural elements. The Una, Sana, and Drina rivers have formed natural borders on the north-west, north and east for much of its history (and still do), while most of Bosnia's terrain is dominated by the mountainous landscapes of the Dinaric Alps which descend in the north towards the country's only significant region of rich agricultural lands along the Sava River basin. Landscapes and climates vary dramatically from the green wooded hills and rolling pastures of central and north-western Bosnia, the heavily forested mountains of eastern Bosnia to the rugged barren limestone mountains of Herzegovina as they run down towards the Adriatic. Brilliant green and turquoise rivers are striking features in this varied landscape.

Much of Bosnia's ancient past awaits investigation, but modern humans have lived in present day Bosnia-Herzegovina for tens of millennia, leaving traces like the rock engravings in Badanj Cave near Stolac, dated to 14–12000 BC, considered the oldest Paleolithic monument in South East Europe. During the Neolithic period human groups began to settle in permanent villages in the region from around the middle of the sixth millennium BC and the remains of rich cultures have been found. The important late Neolithic site at Butmir near Sarajevo revealed fine pottery decorated with geometrical motifs including dramatic moulded and incised spirals, as well as unique tiny clay figurines. Recent excavations at Okolište near Visoko have uncovered an exceptionally large and densely occupied settlement of the sixth and fifth millennia BC.

A federation of Illyrian tribes flourished from around 1,000 BC until their final conquest by the Romans in 9 AD. The Bronze Age is marked by the production of metal items decorated with engraved geometric motifs; the characteristic hill-top settlements and burial grounds of stone tumuli continued into the Iron Age. These settlements reached their most developed expression in the megalithic hillfort of Daorson at Ošanići near Stolac, where jewellery moulds with sophisticated motifs show the spread of Hellenistic culture into Herzegovina from Greek colonies along the Adriatic coast. In the north-west, centred on the Una River valley, the Japodes established a culture distinguished particularly by low relief stone carving and decorated metalwork from at least the ninth century BC until well into the Roman period.

The Romans began to develop an enduring source of Bosnia's wealth, its abundant mineral deposits – especially silver – and Roman military, civilian and industrial settlements were established throughout the country. In 2008 the remains of a settlement and harbour as well as sunken boats laden with amphorae were discovered in Hutovo Blato, near Mostar, indicating that the site was an important Roman-Illyrian trading post from the first century BC. Among the Roman and Late Antique remains found across Bosnia are a significant number of reliefs dedicated to the god Mithras, most notably at Jajce where one such relief is still in place in the remnants of a Mithraic temple from the early fourth century AD; elsewhere, several Early Christian basilicas with distinctive plans have been found, like that at Cim, near Mostar, dating to 500–600 AD.

Medieval Bosnia

Following the disintegration of the Roman Empire, Bosnia's next milestone was the arrival of the Slavs (from whom most Bosnians descend) in the early seventh century. Bosnia's position at the interface of the Western and Eastern Roman Empires and their respective churches meant their successors continued to exert influence on the area. But after centuries of fluctuating control by the Byzantines, Serbs, Croats and Hungarians, by the 1180s Bosnia was effectively an independent state. It was during the medieval period that Bosnia's unique cultural identity began to evolve. The earliest of a trio of able rulers was Ban Kulin (1180–1204), while under Ban Stjepan Kotromanić II (1322–1353) Herzegovina became part of the Bosnian kingdom and by the end of King Tvrtko I Kotromanić's reign (1353–1391) Bosnia was the most powerful state in the Western Balkans. But the period was often politically unstable as feudal lords vied for power, and there was the ever-present threat of a papal crusade against 'heretical' Bosnia, egged on by the Hungarians who had ambitions to acquire its territories. Bosnia's alleged 'heresy' was the most exceptional feature of its medieval history – an indigenous Bosnian Church whose beliefs are still debated. Co-existing with Roman Catholicism and Eastern Orthodoxy, for many Bosnians the Bosnian Church symbolizes Bosnia's pluralism and tolerance. Bosnia proper was predominately Catholic, eastern Herzegovina largely Orthodox; but far from organized church structures, religious identities were fluid.

Bosnia's medieval elites were widely literate and alongside other alphabets a local idiosyncratic form of Cyrillic script called *Bosančica* evolved. One of the earliest surviving documents in *Bosančica* is the legendary Charter of Ban Kulin of 1189 to the citizens of Dubrovnik encouraging its traders into his territories – considered a defining document of Bosnian statehood. The Romanesque and Gothic styles arrived via Central Europe, Byzantine influences through Serbia and the Adriatic. Manuscript painting flourished, as did metalworking using ores from the country's rich mines. Bosnian metalwork was prized

Fig. A.1 A decorated *stećak*, one of the many in the famous necropolis of stećci at Radimlja near Stolac, here carved with a sword and grape vines. © Yuri Stoyanov

across Europe; the fifteenth-century silver bowl at the Franciscan monastery in Fojnica with its heraldic beasts is a superb example.

But relatively little of Bosnia's medieval material culture survived dispersal and destruction during the Ottoman onslaught. Of a number of extant important manuscripts, like the early fifteenth-century Hval Codex (now in Bologna), only one, the Čajniče Gospel, is still in Bosnia-Herzegovina. Few medieval churches survive, and almost all are a small number converted into mosques, such as the Fethija Mosque at Bihać and the ruined church of St Mary and the adjoining intact bell tower of St Luke at Jajce. Many movable treasures were gathered in Franciscan and Orthodox monasteries and churches. The Franciscans had first arrived in the country in 1291 and a Bosnian vicariate was established in 1340. Over the centuries, however, collections and libraries were destroyed by conflict and disastrous fires; nevertheless Franciscan monasteries still hold some of the country's greatest cultural treasures.

What does survive from the medieval period are tens of thousands of Bosnia's most iconic historic monuments, the massive stone tombstones known as *stećci* (singular *stećak*), many of them decorated and inscribed, which lie scattered singly or in groups across the countryside, most famously at the necropolis at Radimlja near Stolac. Research has shown that stećci can mark Bosnian Church, Catholic and Orthodox burials alike. A remarkable feature is the occurrence of stećak necropolises intermingling with and overlapping earlier prehistoric burial tumuli (such as at Veliki and Mali Han near Livno) or side by side with later Islamic burials with *nišan* tombstones (such as at Lavšići near Olovo). Some of the best examples of stećci from across the country were brought to Sarajevo to be displayed in the botanical garden of the Zemaljski Muzej, the National Museum of Bosnia-Herzegovina.

Also surviving are the nearly two hundred castles which still command river basins from the Drina and Buna in the east to the Una in the west and along the Sana in the north. Constructed mainly during the twelfth to fifteenth centuries (like the *stećci*), some castles were actually fortified towns, such as Jajce, and even now are called *stari grad* or old town. Jajce, with its massive fortress and town walls, alone among Bosnian towns still retains its medieval urban plan. In Herzegovina, the mighty castle of Stjepangrad looms above the

Fig. A.2 Srebrenik Castle, in north-east Bosnia, dates from at least 1333 and sits high above the River Tinja guarding the basin of the River Bosna. © Wikimedia Commons

Ottoman dervish *tekke* at the source of the Buna river that gushes from the base of a steep cliff at Blagaj. The royal seat of Bobovac, however, was destroyed by the invading Ottoman Army in 1463, but fragments of vivid Gothic wall paintings and finely carved slabs of red Hungarian limestone that covered the graves of three Bosnian kings, was well as other artefacts excavated from the site and now at the Zemaljski Muzej (the National Museum of Bosnia-Herzegovina) give a glimmering of the high Central European material culture of the last kings of Bosnia and the feudal aristocracy.

Ottoman Bosnia 1463–1878

The Ottomans had been relentlessly encroaching on Bosnia since the 1380s. Faced with the might of Sultan Mehmet II the Conqueror, Stjepan Tomašević, the last king of Bosnia, desperately sought help from his Christian allies – help which never arrived. Stjepan Tomašević was captured and beheaded outside Jajce in May 1463 and Bosnia became the Ottoman Empire's frontier with Western Europe, though complete domination over Bosnia-Herzegovina was not achieved until 1592 with the capture of Bihać. Now another element was added to Bosnia's religious mix: Islam. The Islamization of Bosnia took place gradually over 150 years; by 1624 half the population was Muslim, the high conversion rate most likely a result of Bosnians' weak loyalty to any particular form of Christianity. Sephardic Jews expelled from Spain found refuge in Ottoman lands and Sarajevo's once

Fig. A.3 **(a)** The Franciscan monastery of the Holy Spirit at Fojnica in central Bosnia holds some of the country's most important cultural treasures. © Institute for the Protection of the Cultural, Historical and Natural Heritage of Bosnia-Herzegovina. **(b)** The iconostasis and royal doors of the Old Orthodox Church of the Archangels Michael and Gabriel in Sarajevo. © Jennifer Boyer

Fig. A.4 **(a)** A view of the beautifully situated Ottoman village of Počitelj, south of Mostar which was an artists' colony before the 1992–1995 war. © Courtesy of the Architect / Aga Khan Trust for Culture (AKTC). **(b)** The dervish *tekija* (tekke) at Blagaj at the source of the River Buna. © Sean MacEntee / Wikimedia Commons substantial Jewish community dates back to at least 1565. A celebrated symbol of Bosnia's cultural diversity, the fourteenth-century Spanish-Jewish manuscript known as the Sarajevo Haggadah came to the city via Italy, although it was unknown to the wider world until 1894. Following the Ottoman occupation, the Franciscan order remained in the country after a legendary meeting between Anđeo Živković, head of the order, and Sultan Mehmet II in May 1463. The outcome was the Ahdnama charter, guaranteeing the Franciscans and the Catholics of Bosnia the right to follow their faith without harassment, a decree sometimes more honoured in the breach than its observance. The Franciscans were to remain the only Catholic clergy in Bosnia until 1881. For the Orthodox church the early Ottoman period was notable as an era of growth and expansion during the sixteenth and seventeenth centuries.

The system of *devşirme*, where Christian boys were taken to Istanbul, converted to Islam and trained as members of the janissary corps or the imperial administration during approximately the first century and a half of Ottoman rule helped to form an expatriate Muslim elite who could still retain strong ties with their Christian families. Thus, many of the Ottoman grand viziers of the sixteenth and seventeenth centuries came originally from Bosnia. Most eminent was the former *devşirme* boy Mehmed-paša Sokolović (Sokollu Mehmed Pasha, 1505–79), Grand Vizier to Suleyman the Magnificent and builder of the famous bridge on the Drina at Višegrad. Sokolović was instrumental in the restoration of the defunct Orthodox patriarchate of Peć in 1557 and his relative (possibly brother) Makarije Sokolović was elected Archbishop. The patriarchate had jurisdiction over a large area in what is now Serbia, Bosnia-Herzegovina and Croatia, which had important implications for the position of Orthodox Christians in the region.

The Ottomans rapidly urbanized Bosnia-Herzegovina, frequently around the fortified towns of the medieval period. But Sarajevo was an Ottoman creation, established around 1462 as capital of the newly-conquered territories, growing rapidly to become the third largest city in the Ottoman Empire's European provinces. The dynamic growth of Sarajevo and other towns in Bosnia was propelled by the Islamic charitable foundation, the *vakuf*. Endowed by wealthy individuals and financed by income-producing enterprises such as shops, *vakufi* funded the building of mosques (*džamija*), religious schools (*medresa*), tekke or dervish lodges (*tekija*) and other facilities for the general good like public fountains (česma), baths (*hamam*), covered markets (*bezistan*), caravanserai (*han*), inns (*musafirhana*), clock towers (*sahat kula*) and bridges (*most*) in settings where the flowing water of streams and rivers formed an essential element in the pattern. Structures in classical Ottoman styles began to be built from the sixteenth century on, most famously the domed Ferhadija Mosque in Banja Luka (1579), the Aladža Mosque in Foča (1550–1), the Old Bridge (*Stari Most*) at Mostar (1566) and Mehmed-paša Sokolović's Bridge on the Drina (1571–77). The greatest of Sarajevo's benefactors was its mid-sixteenth-century governor, Gazi Husrev-beg, whose monumental mosque, library, *medresa*, *bezistan* and *imaret* (charitable kitchen) still dominate central Sarajevo. Bosnia's metalworking tradition continued in the new Islamic/Ottoman environment, as did manuscript painting and calligraphy.

The arts of the other faiths thrived, too, during periods of good relations with the Turkish authorities. The sixteenth and seventeenth centuries saw a flowering of arts in Orthodox monasteries like Lomnica: icon and fresco painting and goldsmiths' work. The Orthodox monastery at Zavala, not far from Dubrovnik and the Adriatic, was functioning as early as 1514 and the frescoes in its church of the Presentation of the Virgin by the renowned artist Georgije Mitrofanović date from 1619. Important collections of icons, largely by Serbian and Italo-Cretan painters, were formed in Orthodox churches during the Ottoman period, many donated by the wealthy Orthodox merchant families who had trading connections up and down the Dalmatian coast.

Vernacular buildings had wide regional variations from the Mediterranean-style stone structures of Herzegovina roofed with stone tiles, to the wood and wood frame buildings of the highlands of central Bosnia with their steeply pitched wood-shingled roofs, designed to cope with the heavy snows of the region, to the red clay fish scale roof tiles and whitewashed stucco walls of farm buildings in the agricultural lands to the north. Distinctively Bosnian are the small mosques with wooden minarets found across central Bosnia and tucked away in Sarajevo's *mahalas* (neighbourhoods), and a type of Herzegovinian mosque built by Catholic stonemasons from Dalmatia, many with square church tower-like minarets, such as the Avdić Mosque at Plana.

Austro-Hungarian Bosnia 1878–1918

For Bosnia-Herzegovina the next dramatic transformation in its history began in 1878 when at the Congress of Berlin, convened by the Great Powers to deal with the aftermath of the brutal repression of Christian peasant risings in the moribund Ottoman Empire's European provinces which had led to the Russo-Turkish War of 1877–1878 and the unsatisfactory Treaty of San Stefano, Austria-Hungary was given the mandate to occupy and administer its territories, with formal annexation following in 1908. Austria-Hungary entered its new colonial project with great energy and Bosnia-Herzegovina was propelled into the modern world. Economic development and industrialization were seen as the key to 'civilizing' the newly-acquired territory and miles of new roads and railways were constructed and industrial enterprises begun. The *Zemaljski Muzej* or *Landesmuseum* in Sarajevo, now the National Museum of Bosnia-Herzegovina, established in 1888, was the most important cultural institution founded by the Austro-Hungarian administration, and was for instance, responsible for the archaeological investigations of the Neolithic site at Butmir after its discovery in 1893. While rural Bosnia and a still feudal way of life changed little for thousands of impoverished peasants, for many, particularly educated elites, the new regime opened access to western ideas and technological advances and cultural life blossomed, as artists like Gabrijel Jurkić made their way to Vienna for training.

The Austro-Hungarian administration dramatically transformed Bosnian cities and towns. Central European-style buildings designed by Vienna and Prague-trained architects followed the latest fashions from Neo-Renaissance to Secessionism and the exotic pseudo-Moorish style (like Sarajevo's Town Hall, the *Vijećnica*) was promoted by the authorities who believed it evoked typical 'Oriental' architecture. It is also argued that a distinctive Bosnian style (*Bosanski Slog*//*Bosanski Stil*) influenced by Bosnia's traditional architecture evolved among architects who spent the greater part of their careers in Bosnia-Herzegovina.[1]

Sarajevo, the centrepiece of Austria's modernization plans, was seat of the imperial administration and its newly-created institutions and headquarters for scores of new enterprises. A new Central European-style city was laid out to the west of the old Ottoman centre and Sarajevo began to acquire its distinctive fusion of styles. There were wide residential streets of apartment blocks and villas, very different from the narrow winding

1 See, for instance, Kurto, N. 1998 *Arhitektura Bosne I Hercegovine: Razvoj Bosanskog Stila*, Sarajevo: Sarajevo Publishing / Međunarodni centar za mir; Krzović, I. 2004. *Arhitektura Secesije u Bosni i Hercegovini. Sarajevo:* Sarajevo Publishing.

Fig. A.5 One of the many splendid Austro-Hungarian buildings in Sarajevo. Built as the Palace of Justice (1914), it was designed by Viennese architects Hans Glaser and Alfred Kraupa and built by the prolific Czech architect Karl Pařik (Karlo Paržik) who spent much of his career in Sarajevo. © Dayou_X

streets of the Ottoman *mahalas*. With Catholicism the dominant religion of Bosnia's new rulers, and with thousands of migrants arriving to administer the territory – or simply seek their fortune – there was a flurry of church building and many of Bosnia's Catholic churches date from this era.

1918–1992

Sarajevo has notoriously been labelled as the place where World War One began with the assassination there of the heir to the Austrian throne Archduke Franz Ferdinand on 28 June 1914 by the young Bosnian Serb nationalist Gavrilo Princip. The war brought the collapse of the Austro-Hungarian Empire and Bosnia-Herzegovina became part of one of the states created out of its territories in 1918: the Kingdom of Serbs, Croats and Slovenes (the Kingdom of Yugoslavia from 1929). Between the world wars Bosnia was somewhat of a backwater, but nevertheless young architects and artists like painter Vojo Dimitrijević studied or worked abroad, absorbing trends from Western Europe.

World War II came in 1941 when Axis forces invaded Yugoslavia and Bosnia-Herzegovina was incorporated into the Nazi puppet state, the Independent State of Croatia (NDH) and split into German and Italian zones of occupation. The mountains of Bosnia became the heartland of the Partisan resistance to Axis occupation, led by Josip Broz 'Tito.' It was at Jajce in 1943 at the Anti-Fascist Council of the People's Liberation of Yugoslavia (AVNOJ), that Bosnia-Herzegovina was re-established as a political unit as one of the six constituent republics that would form a post-war federal Socialist Yugoslavia.

In post-WW2 socialist Yugoslavia, Bosnia-Herzegovina remained relatively open to western trends in the arts and architecture and Sarajevo was known for its vibrant cultural life. From the 1950s painting and graphics flourished in particular and the work of Bosnian artists was recognized internationally through such names as Mersad Berber, Dževad Hozo, Bekir Misirlić, Edin Numenkadić and others. Public buildings in a restrained Modernist style were a feature of the decades immediately following the war; among the most influential architects were Juraj Neidhardt (Najdhart) and later, Zlatko Ugljen.

However, socialism's most spectacular legacy in the built environment are perhaps the massive abstract memorials by well-known sculptors and architects commemorating the battles of WW2 constructed during the 1960s and 1970s, like the complex dedicated to the fallen of Battle of Kozara near Banja Luka designed by sculptor Dušan Džamonija with its huge columnar monument and walls carrying the names of thousands of Partisans (still a popular visiting place) and Bogdan Bogdanović's beautiful but now sadly neglected Partisan Memorial Cemetery in Mostar.

By 1992, as elsewhere in Europe, much of the historic built environment in Bosnia-Herzegovina had been swept away by conflict, changing political regimes and the pressures of modernization and urban development. Yet many cities, towns and villages still retained that distinctive mix of the Ottoman and the Austro-Hungarian, nowhere more so than Sarajevo and Mostar, while places like Stolac and Počitelj in Herzegovina felt both Ottoman and Mediterranean in character. Fortresses still dominated many towns and cities. Examples of the medieval tombstones – the stećci – were found at every turn – in front of municipal offices or local museums, in parks, at protected historic sites, along roadsides or hidden in woodland. Also typically Bosnian was how many small towns had at least one mosque, a Catholic church and an Orthodox church, often in within metres of each other.

Glossary

abdesthana	Place for ritual washing next to a mosque, often associated with cisterns.
beg/bey	Lord or landowner in the Ottoman period.
bezistan	Covered market of the Ottoman period.
Brčko/Brčko District	A district in north-east Bosnia whose status remained undecided at the signing of the Dayton Peace Agreement due to its geographical position. It was not part of either entity of Bosnia-Herzegovina and its affairs were overseen by an internationally-appointed supervisor (post suspended but not terminated in 2012).
čaršija	Bazaar.
čardak	Free-standing agrarian building sometimes used as a dwelling.
čatrnja	Well, or cistern.
česma	Public water fountain, often associated with cisterns and often built by a public benefactor as part of a vakuf.
crkva	Church.
džamija	Mosque.
Federation	The Federation of Bosnia-Herzegovina (*Federacije Bosne i Hercegovine)*, the so-called Muslim-Croat Federation. One of the two main political entities of which Bosnia-Herzegovina is formed according to the terms of the Dayton Peace Agreement. Created under the Washington Agreement of March 1994, it is formed of ten cantons.
groblje	Graveyard, cemetery.
gusulhana	Mortuary house for ritual bathing, usually part of a mosque complex.
hamam	Public Turkish baths.
han	Caravanserai.
harem	Muslim burial ground within the precincts of a mosque.
imaret	Charitable kitchen for the poor.
Islamic Community	An official body and legal entity representing members of the Muslim faith in Bosnia-Herzegovina.
Herceg-Bosna	Unrecognized and illegal para-state formed by Bosnian Croat ethno-nationalists, in existence 1991–1994. Sometimes referred to as the Croatian Community (later Republic) of Herzeg-Bosnia.
mahala	A neighbourhood or district in an urban area.
manastir	Usually an Orthodox monastery.
medresa	Muslim secondary school.
medžlis	Local council of the Islamic Community.
mekteb/mejtef	Muslim religious primary school.
mesdžid	Mosque without a minber (pulpit) and normally without a minaret.
mezaristan	Muslim graveyard, separate from a mosque.
musafirhana	Guesthouse or inn, often associated with a religious institution.

most/pl. mostovi	Bridge.
muftijstvo/muftistvo	Offices of the chief mufti of a region.
nišan/pl. nišani	Muslim gravestones.
Republika Srpska	One of the two main political entities of which Bosnia-Herzegovina is formed according to the terms of the Dayton Peace Agreement. The misleading English translation 'Serbian Republic' is now almost never used. A para-state formed by secessionist Bosnian Serbs, it broke ties with Bosnia-Herzegovina in April 1992. The Dayton Peace Agreement formalized the existence of Republika Srpska as one of two entities within Bosnia-Herzegovina.
šadrvan	Fountain with a basin for ritual washing in the precincts of a mosque.
sahat kula	Clock tower, usually of the Ottoman era.
samostan	Usually a Catholic monastery.
šebilj	Drinking water fountain.
stari grad	Literally 'old town.' Often applied to the historic core of a town, but also to a medieval castle/fortress or the precincts of a castle.
stećci/sing. stećak	Monumental medieval tombstones, sometimes decorated with motifs and/or inscriptions.
tekija	Tekke, or dervish meeting house.
turbe/pl. turbeta	Mausoleum.
vakuf	Islamic charitable endowment (waqf) for religious purposes and/or for public benefaction.

Acknowledgement

The Bosnian Institute, London supported the early stages of the preparation of this book and funded the fieldtrips of Helen Walasek and Richard Carlton to Bosnia-Herzegovina in 2000 and 2001.

Illustration Acknowledgements

Finding the right illustrations seemed at times more challenging than writing and editing the book itself. Several people went well beyond the call of duty to help with sourcing photographs. First and foremost was András Riedlmayer of Harvard University Fine Arts Library and principal expert prosecution witness on cultural heritage destruction at the ICTY. He is followed by Bedrudin Gušić, head of the Islamic Community of Banja Luka during the 1992–1995 Bosnian War. Particular thanks are also due to Mirsad Sijarić of the Zemaljski Muzej in Sarajevo for liaising with Bosnian sources; Šejla Holland of Green Cube Gallery, Laguna Beach, California; Dražen Ravlić for permission to reproduce the photographs taken by his late father Aleksander Aco Ravlić; staff of the Commission to Preserve National Monuments in Sarajevo (especially Azra Delalić and Almedina Salihagić); and staff of the Medžlis of the Islamic Community of Banja Luka. Andrea Schuler of Massachusetts Institute of Technology (MIT) Libraries and staff at the Press Office of the International Criminal Tribunal for the former Yugoslavia (ICTY) were always ready to help. Special thanks to Tina Wik and Richard Carlton (who also drew the map). Thanks for support in the quest to Robert Bevan, Amra Hadžimuhamedović, Medina Haždihasanović, Mirko Klarin, Peter Lippmann, Alojz Čakarić, Fiona Lloyd-Davies, Leif Pareli, Edward Serotta at Centropa, Yuri Stoyanov, and staff of the Center for Civil Courage/Centar za građansku hrabrost, Zagreb. Finally, Ivo Žanić, Elša Turkušić and Graham O'Hare helped enthusiastically in the early stages of the picture search, as did Marko Attila Hoare (who also kindly reviewed the historical introduction of the final text).

For scanning and/or providing or giving permission to use the photos belonging to them or their institution, thanks are due to the Commission to Preserve National Monuments of Bosnia & Herzegovina, Sarajevo; Dr Mustafa Jahić, director of the Gazi Husrev-beg Library, Sarajevo and staff; Kemal Zukić, former director of the Center for Islamic Architecture, Sarajevo and staff; Ferhat Mulabegović and Džihad Pašić, former directors of the (now defunct) Institute for the Protection of the Cultural, Historical and Natural Heritage of Bosnia-Herzegovina and staff; the International Criminal Tribunal for the former Yugoslavia (ICTY); the Aga Khan Trust for Culture; the Aga Khan Visual Archive at MIT; the Medžlis of the Islamic Community of Banja Luka; Dražen Ravlić; András Riedlmayer; Richard Carlton; Tina Wik; Leif Pareli; Kemal Hadžić; Robert Bevan; Edward Serotta; Yuri Stoyanov; and Peter Lippmann.

Figures 1.1a, 1.2e, 1.2f, 1.2g, 1.2h, 1.3, 1.5d, 1.7a, 1.12a, 1.12b 1.13c, 1.14b 1.17b 1.17c, 1.25a, 1.27 were lodged as prosecution evidence at the International Criminal Tribunal for the former Yugoslavia (ICTY) and are thus in the public domain. Of these, figures 1.1a, 1.2e, 1.2f, 1.3, 1.5d, 1.12a, 1.12b, 1.17b 1.17c, and 1.25a were provided courtesy of András Riedlmayer, and 1.14.b courtesy of Robert Bevan.

Figures 1.1b and 1.1c by Aleksander Aco Ravlić are courtesy of the Medžlis of the Islamic Community of Banja Luka.

Figures 1.9c, 1.10c, 1.15a, 5.2a, 5.2b 5.2d, 5.10d, A.2a, A.3, A.4b and A.5 are reproduced on Creative Commons licenses and were sourced via Wikimedia Commons and Flikr.

Bibliography

Publications on Damage, Destruction and Reconstruction of Cultural Property in Bosnia-Herzegovina

Balić, S. 1993. Culture under fire, in *Why Bosnia? Writings on the Balkan War*, edited by R. Ali and L. Lifschutz. Stony Creek, CT: Pamphleteer's Press, 75–83.

—— 1994. Destruction of the Bosnian architectural heritage: An interim report. *Journal of Islamic Studies*, 5 (1994), 2, 268–73.

Barakat, S. and Wilson, C. 1997. *The Revitalisation of the Historic Settlement of Počitelj*. York: Post-war Reconstruction and Development Unit (PRDU), University of York.

—— 1998. *Challenges and Dilemmas Facing the Reconstruction of the Cultural Heritage: The Case Study of Počitelj, Bosnia-Herzegovina*. Paper to World Archaeological Congress: The Destruction and Conservation of Cultural Property, Brać, Croatia, 3–7 May 1998.

Barakat, S., Wilson, C., Sankovic Simcic, V. and Kojakovic, M. 2001. Challenges and dilemmas facing the reconstruction of war-damaged cultural heritage: The case study of Pocitelj, Bosnia-Herzegovina, in *Destruction and Conservation of Cultural Property*, edited by R. Layton, P. Stone and J.S. Thomas. London: Routledge, 168–81.

Begić, A. 1995. The fate of moveable cultural heritage in Sarajevo and Central Bosnia – an insider's observations. *Museum Management and Curatorship*, 14:1, (March 1995), 80–83.

Calame, J. and Pasic, A. 2009. Post-conflict reconstruction in Mostar: Cart before the horse. *Divided Cities/Contested States, Working Paper No. 7, 2009*, Conflict in Cities and the Contested State: Everyday life and the possibilities for transformation in Belfast, Jerusalem and other divided cities, UK Economic and Social Research Council Large Grants Scheme, RES-060-25-0015, 2007–2012.

Čamdžić, E. ef. 2000. *Sa Džamijama i Povratak*. Banja Luka: Banja Luka Muftistvo.

Čengić, A. and Duraković, F. (eds) 1995. *Cultural Institutions and Monuments in Sarajevo*. Budapest: Open Society Institute, Open Society Project of Sarajevo, March 1995.

Centar za Mir 2003. *Old Bridge – Monument of Peace*. Exhibition catalogue. Mostar: Centar za Mir, 2004 version. Available at: http://www.centarzamir.org.ba/eng/pro/hrono/hronologija.html.

Chapman, J. 1994. Destruction of a common heritage: The archaeology of war in Croatia, Bosnia and Hercegovina. *Antiquity*, 68, 120–26.

Ćorović, A. 2008. Šeranića kuća u Gornjem Šehera kod Banka Luka/The Šeranić House in Gornji Šeher near Banja Luka. *Baština* (*Heritage*), The Annual of the Commission to Preserve National Monuments, IV/2008, 199–218.

Ćurić, B. et al. 1994. *Warchitecture – Urbicide Sarajevo*. Exhibition catalogue. Originally Sarajevo: Association of Architects of Bosnia-Herzegovina (DAS-SABIH), October 1993, reprint Munich, with the assistance of Baumeister, May 1994.

Detling, K.J. 1993. Eternal silence: The destruction of cultural property in Yugoslavia. *Maryland Journal of International Law and Trade*, 17:1, Spring, 41–75.

Dodds, J.D. 1998. Bridge over the Neretva. *Archaeology*, 51:1, January/February 1998, 48–53.
Donia, R.J. 2004. Archives and cultural memory under fire: Destruction and the post-war nationalist transformation. Paper at conference: *15th International Congress on Archives: Archives, Memory and Knowledge*, Vienna, 23–29 August 2004. Available at: http://www.arhivsa.ba/ica2004/robert.htm.
Fahy, M. and Mogul, J. 1995. An interview with Lidija Fekeza: An archaeologist in Sarajevo: Culture under siege, *The Journal of the International Institute* [online], 3:1 (Fall).
Fočo, M. 2005. National monuments endangered for lack of measures to save them from further dilapidation/Nacionalni spomenici ugroženi nepreduzimanjem mjera zaštite od daljeg propadanja. *Baština/Heritage*, The Annual of the Commission to Preserve National Monuments, I/2005, 29–45.
Fočo, M. 2008. Rehabilitation of the Aladža Mosque in Foča: Protection of the remains, *Baština* (*Heritage*), The Annual of the Commission to Preserve National Monuments IV/2008, 873–111.
Hadžihasanović-Katana, M. 2012. The reappearance of Ferhadija Mosque in Banja Luka, in *Protecting Cultural Heritage in Times of Conflict: Contributions from the Participants of the International Course on First Aid to Cultural Heritage in Times of Conflict*, edited by S. Lambert and C. Rockwell. Rome: ICCROM, 66–71.
Hadžimuhamedović, A. 2005a. Cultural memory in Stolac – destruction and reconstruction, in *Human Rights and Destruction of Cultural Memory: The Stolac Case*, edited by A. Hadžimuhamedović. Oslo: Norwegian Helsinki Committee.
—— (ed.) 2005b. *Human Rights and Destruction of Cultural Memory: The Stolac Case*, edited by A. Hadžimuhamedović. Oslo: Norwegian Helsinki Committee.
—— 2005c. Redefinition of protection and presentation methods of the architectural heritage in the process of post-war rehabilitation – The case of the Čaršija Mosque in Stolac/Redefinicija primjene metoda zaštite i prezentacije graditeljskog naslijeđa u procesu poslijeratne obnove – slučaj Čaršijske džamije u Stocu. *Baština /Heritage*, The Annual of the Commission to Preserve National Monuments, I/2005, 71–117.
—— 2008a. *What Happens When the Homeland Cultural Landscapes are Destroyed: A Case from Bosnia*. Paper to: The Sixth International Conference on Science and Technology in Archaeology and Conservation, WATCH, Rome, 8–14 December 2008.
—— 2008b. Sedam godina rada Komisije za očuvanje nacionalnih spomenika/Seven years of work of the Commission to Preserve National Monuments, *Baština (Heritage)*, The Annual of the Commission to Preserve National Monuments, IV/2008, 11–17.
Ivanović, V. and Vlahušić, A. and Croatian Community of Herzeg-Bosnia 1993. *The Ethnic Cleansing of Croats in Bosnia and Herzegovina, 1991–1993*. Mostar: Office of President of the Croatian Community of Herzeg-Bosnia, August 1993.
Ihsanoglu, E. (ed.) 1993. Bosnia-Hercegovina: History, culture, heritage. *Research Center for Islamic History, Art and Culture Newsletter*, 31, April 1993. Istanbul: IRCICA.
Jovanović, D., Bundalo, G. and Govedarica, M. 1994. *Iskorenjivanje Srba u Bosni i Hercegovini, 1992–1993* [*The Eradication of Serbs in Bosnia and Herzegovina, 1992–1993*], '*Destruction of Orthodox Religious Structures*.' Belgrade: Rad, 205–20.
Kapetanović, E., Šarančić Logo, A., Numanagić, H. and Vesković, E. 2008. The Ali Bey Kapetanović Mosque in Vitina: Restoration of the national monument/Džamija Ali-bega Kapetanovića u Vitini: Restauracija nacionalnog spomenika, *Baština* (*Heritage*), The Annual of the Commission to Preserve National Monuments, IV/2008, 335–64.

Komarica, F. 2000. *In Defence of the Rightless/U Obrani Obespravljenih*. Zagreb: Targa. Available at: http://www.biskupija-banjaluka.org/izdanja_/obespravljeni_/english_/index.htm.

Kovačević, M. 1996. War damage suffered by the State Archive of Bosnia and Herzegovina. *Memory of the World at Risk, Archives Destroyed, Archives Reconstituted, Archivum, International Review on Archives*, XLII, International Council on Archives.

Kožar, A. 1999. *War Destruction of the Archival Materials in Bosnia & Herzegovina*. Unpublished paper cited in Riedlmayer, 2001.

Lenasi, O. 2008. The architectural ensemble of the Mehmed Pasha Kukavica Mosque and Medresa in Foča/Graditeljska cjelina – džamija i medresa Mehmed-paše Kukavice u Foči, *Baština (Heritage)*, The Annual of the Commission to Preserve National Monuments, IV/2008, 113–48.

Makas, E. 2007. *Representing Competing Identities: Building and Rebuilding in Postwar Mostar*. Phd thesis [online]. Available at: http://uncc.academia.edu/EmilyMakas.

Mehmedbašić, S. 2001a. *First Letter to Bishop Perić, 9 August 2001*. Society for Renewal of Civil Trust in Stolac Municipality.

—— *Second Letter to Bishop Perić, 22 September 2001*. Society for Renewal of Civil Trust in Stolac Municipality.

—— 2001c. *Third Letter to Bishop Perić: 8 December 8 2001*. Society for Renewal of Civil Trust in the Stolac Municipality, Stolac.

—— 2002. *Fourth Letter to Bishop Perić: 28 January 2002*. Society for Renewal of Civil Trust in the Stolac Municipality, Stolac.

Mileusnić, S. 1997. *Spiritual Genocide 1991–1995*. Belgrade: Museum of the Serbian Orthodox Church.

Mulać, E. (ed.) 1996. *Crimes in Stolac Municipality (1992–1994)*. The Revival of Stolac, *Iss. No. 1*. Sarajevo: Association [Society] for Renewal of Civil Trust in the Stolac Municipality.

—— (ed.) 2001. *Crimes in Stolac Municipality (1992–1996)*. The Revival of Stolac. Sarajevo: Association [Society] for the Renewal of Civil Trust in the Stolac Municipality, reprint with addenda.

—— (ed.) 2001b. *Written Memory against Apartheid in the Municipality of Stolac*. The Revival of Stolac. Sarajevo: Association [Society] for the Renewal of Civil Trust in the Stolac Municipality, 8 July.

—— (ed.) 2002. *Destruction of the Opijač House in the Ada Residential Ensemble in Stolac*. The Revival of Stolac. Sarajevo: Association [Society] for the Renewal of Civil Trust in the Stolac Municipality, Iss. No. 17, March.

Obratil, S. 1996. In memoriam: Dr Rizo Sijarić. *Glasnik Zemaljskog muzeja Bosne i Hercegovine*, new series, 47, 441–7.

Omerdić, M. 1999. *Prilozi Izučavanju Genocida nad Bosnjačima 1992–1995*. Sarajevo: El-Kalem.

Pavić, V. (ed.) 2005. *War Damage to Museums and Galleries in Croatia/Ratne štete na muzejima i muzejskoj građi u Hrvatskoj*. Zagreb: Museums Documentation Center (MDC). Available at: http://www.mdc.hr/RatneStete/eng/fs-glavni.html.

Pašić, A. 1994. *Islamic Architecture in Bosnia and Herzegovina*. Istanbul: IRCICA.

Perry, V. 2002a. *ECMI Civil Society Project in Bosnia and Herzegovina: The Role of Annex 8 in Bridging Communities, Promoting Reconciliation, and Developing a Unique Identity in Bosnia and Herzegovina*, The Law Centre, Sarajevo University, 15 December 2001. Flensburg: European Centre for Minority Issues, Report No. 25, March 2002.

—— 2002b. *ECMI Civil Society Project in Bosnia and Herzegovina: Annex 8 of the General Framework Agreement for Peace and the Role of the Preservation of National Monuments in Successful Peace Implementation*, Banja Luka, 12 April 2002. Flensburg: European Centre for Minority Issues, Report No. 29, July 2002.

—— 2003. Reversing ethnic and cultural cleansing: The role of Annex 8 in reclaiming history, promoting reconciliation, and developing a unique multiethnic identity in Bosnia and Herzegovina. *The European Yearbook of Minority Issues*, 2, 2002/2003. Leiden: Martinus Nijhoff Publishers, 197–226.

Petrovic, J. 2012. *The Old Bridge of Mostar and Increasing Respect for Cultural Property in Armed Conflict*. Leiden: Martinus Nijhoff.

Raguž, J. 2013. Sudbina posmrtnih ostataka kralja Stjepana Tomaševića 1463.–1888.–1992.–1999. (The Fate of Posthumous Remains of King Stjepan Tomašević, 1463 – 1888 – 1992 – 1999) in *Stjepan Tomašević (1461.–1463.) – slom srednjovjekovnoga Bosanskog Kraljevstva: zbornik radova sa Znanstvenog skupa održanog 11. i 12. studenoga 2011. godine u Jajcu*, edited by A. Birin. Zagreb and Sarajevo: Hrvatski institut za povijest u Zagrebu : Katolički bogoslovni fakultet u Sarajevu, 235–82.

Rakic, S. 2006a. 'Migrating icons:' politics and Serbian cultural heritage in Bosnia and Herzegovina before and after 1992. *Mediterranean Quarterly*, Summer, 17:3, 43–54.

—— 2006b. *Migrating Icons: Politics and Serbian Cultural Heritage in Bosnia and Herzegovina before and after 1992*. Presentation at Woodrow Wilson International Center for Scholars, 8 November 2006, Meeting Report 329.

Ravlić, A.A. 1996. *Banjalučka Ferhadija: ljepotica koju su ubili/Banja Luka's Ferhadija Mosque: A Murdered Beauty*. Rijeka: AARis.

Ribarević-Nikolić, I. and Jurić, Ž. (eds) 1992. *Mostar '92 – Urbicid*. Mostar: Hrvatsko vijeće obrane općine Mostar/Javno poduzeće za obnvu izgradnju Mostara/IPD Općinskog Stožera Mostar.

Riedlmayer, A. 1995. Erasing the past: The destruction of libraries and archives in Bosnia-Herzegovina. *Middle East Studies Association Bulletin*, 29 I, July 1995, 7–11.

—— 1996. *Libraries Are Not for Burning: International Librarianship and the Recovery of the Destroyed Heritage of Bosnia and Herzegovina*. Paper to: The 61st IFLA General Conference, Istanbul, August 20–25, 1995.

—— 2001. Convivencia under fire: Genocide and book-burning in Bosnia, in *The Holocaust and the Book: Destruction and Preservation*, edited by J. Rose. Amherst: University of Massachusetts Press, Studies in Print Culture and the History of the Book, 266–91.

—— 2002b. From the ashes: The past and future of Bosnia's cultural heritage, in *Islam and Bosnia: Conflict Resolution and Foreign Policy in Multi-Ethnic States*, edited by M. Shatzmiller. Montreal: McGill-Queens University Press, 98–135.

—— 2007. Crimes of war, crimes of peace: Destruction of libraries during and after the Balkan Wars of the 1990s. *Library Trends*, 56:1, Summer, 107–32.

Rizvanbegović, F. 1999. *Begovina – Stolac*. Sarajevo: Federation of Bosnia and Herzegovina, Ministry of Education, Science, Culture and Sports.

Serdarević, M. 1993. Postscript. *Museum Management and Curatorship*, 12:2, June 1993.

—— 1997. *Pravna zaštita kulturno-historijskog naslijeđa BiH sa posebnim osvrtom na staro urbano jezgro Sarajeva: (nastajanje, očuvanje, destrukcija)*. Sarajevo: Međunarodni centar za mir.

Slavitt, L.D. 1994. *Report on the Destruction of Cultural Heritage, Property and Structures as a Result of War in Croatia and Bosnia-Herzegovina 1992–1994*. Chicago: International Human Rights Law Institute.

Sijarić, R. 1993. Update on the Zemaljski Muzej, Sarajevo. *Museum Management and Curatorship*, 12:2, June, 195–9.

—— 1996. Zemaljski Muzej Bosne i Hercegovine u ratu 1992/93 godine. *Glasnik Zemaljskog muzeja Bosne i Hercegovine*, new series, 47, 7–39.

Tucaković, Š. 1998. *Aladža Džamija – Ubijeni Monument*. Sarajevo: Institut za istraživanje zločina protiv čovječnosti i međunarodnog prava.

Ugljen, Z. 1996. *Izložba: Džamija Hadži Alije Hadžisalihovića; Vjersko i kulturno sjedište Plehan*, Zagreb.

Wenzel, M. 1993a. The Zemaljski Muzej. *Museum Management and Curatorship*, 12:1, March.

Wenzel, M. 1994. Obituary: Dr. Rizo Sijaric, Director of the Zemaljski Muzej, Sarajevo. Killed in Sarajevo, 10 December 1993. *Museum Management and Curatorship*, 13:1, 79–80.

Wik, T., Hadžić, L. and Walasek, H. 2006. *The Handanija Mosque Restoration*, report series No. 3/2006. Sarajevo: Cultural Heritage without Borders (CHwB) Regional Office.

World Monuments Fund 1999. *Reclaiming Historic Mostar: Opportunities for Revitalization*. New York: World Monuments Fund.

Živković, I. (ed.) 1997. *Raspeta Crkva u Bosni i Hercegovini: Uništavanje Katoličkih Sakralnih Objekata u Bosni i Hercegovini (1991–1996)*. Banja Luka, Mostar, Sarajevo: Biskupska Konferencija Bosne i Hercegovine and Zagreb: Hrvatska Matica Iseljenika Bosne i Hercegovine and Hrvatski Informativni Centar.

Zukić, K. 1999. *Slike Zločina: Rušenje Islamskih Vjerskih Objekata u Bosni i Hercegovini*. Sarajevo: Centar za islamsku arhitekturu.

Books and Articles: Selected Historical Background

Andrić, I. 1945. *Na Drini ćuprija*. Belgrade: Prosveta.

Ayverdi, E.H. 1981. *Avrupa'da Osmanli Mimâri Eserleri, Yugoslavya, Vol. II, Part 3*. Istanbul: Istanbul Fetih Cemiyeti.

Basler, Đ., Begić, A., Čelić, D., Kajmaković, Z. and Vartabedijan, M. 1987. *The Art Treasures of Bosnia and Herzegovina*. Sarajevo: Svjetlost.

Bernik, S. 2002. *Arhitekt/Architect Zlatko Ugljen*. Tuzla: Međunarodna galerija portreta.

Bećirbegović, M. 1990. *Džamije sa Drvenom Munarom u Bosni i Hercegovini*. Sarajevo: Biblioteka Kulturno Nasljeđe, Veselin Masleša.

Carlton, R. and Rushworth, A. 2009. The Krajina Project: Exploring the Ottoman–Hapsburg borderland, in *The Frontiers of the Ottoman World*, edited by A.C.S. Peacock. London: British Academy.

Čepalo, H. 1990. *Prusac – Istorijski Spomenici*. Prusac: Mala Biblioteka.

Donia, R.J. 2006. *Sarajevo: A Biography*. Ann Arbor, MI: University of Michigan Press.

Donia, R.J. and Fine Jr, J.V.A. 1994. Bosnia and Hercegovina: A Tradition Betrayed. New York: Columbia University Press.

Fine, J.V.A. 2007. The Bosnian Church: Its Place in State and Society from the Thirteenth to the Fifteenth Century. London: Saqi Books and the Bosnian Institute.

Handžić, A. 1996. *A Survey of Islamic Cultural Monuments until the End of the 19th Century in Bosnia*. Istanbul: IRCICA.

Hasandedić, H. 2000. *Mostarski Vakifi i njihov Vakufi*. Mostar: Medžlis Islamske Zajednice Mostar.

—— 1999. *Muslimanska Baština Bošnjaka Herceg-Novi Okolina, etc, Vol. II*. Mostar: Islamski Centar Mostar.

—— 1997. *Muslimanska Baština Bošnjaka u Južnoj (Srednjo) Hercegovini, Vol. I.* Mostar: Islamski Centar Mostar.
—— 1990. *Muslimanska Baština u Istočnoj Hercegovini.* Sarajevo: El-Kalem.
Hoare, M.A. 2007. *The History of Bosnia: From the Middle Ages to the Present Day.* London: Saqi Books.
Horozović, I. and Vukmanović, M. 1984. *Banjaluka.* Zagreb: Turistcomerc.
Kiel, M. 2011. The campanile minarets of southern Herzegovina, in *Centres and Peripheries in Ottoman Architecture: Rediscovering a Balkan Heritage*, edited by M. Hartmuth. Sarajevo: Cultural Heritage without Borders, 60–79.
Krzović, I. 2004. *Arhitektura Secesije u Bosni i Hercegovini, Sarajevo:* Sarajevo Publishing.
Kudĕla, J., Dimitrijević, B. and Vacík, I. 2007. *Arhitekt/Architect/The Architect Karel Pařík 1857–1942.* Sarajevo: Embassy of the Czech Republic in Bosnia and Herzegovina.
Kurto, N. 1998. *Arhitektura Bosne I Hercegovine: Razvoj Bosanskog Stila*, Sarajevo: Sarajevo Publishing:Međunarodni centar za mir.
Lovrenović, I. 2001. *Bosnia: A Cultural History.* London: Saqi Books/The Bosnian Institute.
Malcolm, N. 2002. *Bosnia: A Short History.* London: Pan Books.
Mannergren Selimovic, J. 2010. *Remembering and Forgetting after War. Narratives of Truth, Justice and Reconciliation in a Bosnian Town*, unpublished PhD Dissertation in Peace and Development Research, School of Global Studies, University of Gothenburg.
Microbook 1998. *Mostar – 99 Pictures.* Mostar: Microbook.
Mujezinović, M. 1974. *Islamska Epigrafika u Bosne i Hercegovine, Vol. I – Sarajevo.* Sarajevo: Biblioteka Kulturno Nasjleđe, Veselin Masleša.
—— 1977. *Islamska Epigrafika u Bosne i Hercegovine, Vol. II – Istočna i Centralna Bosna.* Sarajevo: Biblioteka Kulturno Nasjleđe, Veselin Masleša.
—— 1982. *Islamska Epigrafia BiH, Vol. III – Bosanska Krajina, Zapadna Bosna i Hercegovina.* Sarajevo: Biblioteka Kulturno Nasjleđe, Veselin Masleša.
Rakic, S. 2000. *Serbian Icons from Bosnia-Herzegovina: Sixteenth to Eighteenth Century.* New York: A. Pankovich Publishers.
Sells, M.A. 1996. *The Bridge Betrayed: Religion and Genocide in Bosnia.* Berkeley, CA: University of California Press.
Sells, M.A. 2003. Crosses of blood: Sacred space, religion and violence in Bosnia and Herzegovina. *Sociology of Religion*, 64:3, 309–31.
Selman, M. 2006. *Banjaluka – Uspomena i Stvarnost sa Mejdana.* Available at: www.bosnafolk.com, October 2006, http://www.bosnafolk.com/pdf/mejdan_screen.pdf.
Sujoldić, E. 1999. *Travnicke Sahat-Kule [The Clock Towers of Travnik].* Travnik: Sediment – bibliotaka odabranih djela.
—— 1998. *Džamije Travniku.* Travnik: Sediment – biblioteka odabranih djela, Zavičajne Muzej Travnik.
Tomašević, N. et al. (eds) 1980. *Treasures of Yugoslavia.* Belgrade: Yugoslaviapublic.
Truhelka, Ć. 1904. *Naši Gradovi. Opis Najljepsih Sredovjecnih Gradova Bosne i Hercegovine.* Sarajevo: Bosanska Posta.

Other Books, Articles and Conference Papers

Abdulhak, T. and Busatlić, M. 2007. *Court of Bosnia and Herzegovina.* Paper to: Conference on International Criminal Justice, Turin, 14–18 May 2007. Available at: http://www.torinoconference.com/conferencepapers.shtml.

Abtahi, H. 2001. The protection of cultural property in times of armed conflict: the practice of the International Criminal Tribunal for the Former Yugoslavia. *Harvard Human Rights Journal*, 14 (Spring 2001), 1–31.

Aksamija, A. 2010. Contested identities: Identity politics and contemporary mosques in Bosnia and Herzegovina, in *Mosques in Europe: Why a Solution has Become a Problem*, edited by S. Allievi in cooperation with Ethnobarometer, Network of European Foundations Initiative on Religion and Democracy in Europe. London: Alliance Publishing Trust, 325–32.

Antić, M. 2003. Living in the shadow of the bridge: Ivo Andric's 'The Bridge on the Drina' and western imaginings of Bosnia. *Spaces of Identity*, The Canadian Center for German and European Studies, 3:2. Available at: http://www.yorku.ca/soi/

Bala, S. 2000. The Opposition in Croatia, in *Burn This House: The Making and Unmaking of Yugoslavia*, edited by J. Udovicki and J. Ridgeway. Durham, NC: Duke University Press, 267–80.

Barakat, S. 1999. *A Culture of War – A Heritage of Destruction*. Paper to the conference: The Idea of Heritage: Past, Present and Future, London Guildhall University, London, 7–9 September 1999.

—— 2007. Postwar reconstruction and the recovery of cultural heritage: Critical lessons from the last fifteen years, in *Cultural Heritage in Postwar Recovery, Papers from the ICCROM Forum held in Rome 4–6 October 2005*, edited by N. Stanley-Price. Rome: ICCROM, 26–39.

Bartels, R. 2013. Prlić et al.: The Destruction of the Old Bridge of Mostar and Proportionality. Available at: http://www.ejiltalk.org, 31 July. Available at: http://www.ejiltalk.org/prlic-et-al-the-destruction-of-the-old-bridge-of-mostar-and-proportionality.

Baumann, R.F., Gawrych, G.W., Kretchik, W.E. 2004. *Armed Peacekeepers in Bosnia*. Fort Leavenworth, KS: Combat Studies Institute Press.

Bevan, R. 2006. *The Destruction of Memory: Architecture at War*. London: Reaktion Books.

Bhat, P.I. 2001. Protection of cultural property under international humanitarian law: Some emerging trends. *ISIL Year Book of International Humanitarian and Refugee Law 2001*, 1, 47–71.

Bogdanović, B. 1994. *Grad i Smrt*. Belgrade: Beogradski krug.

Bose, S. 2002. *Bosnia after Dayton: Nationalist Partition and International Intervention*. London: Hurst & Company.

Boylan, P.J. 2006. *Implementing the 1954 Hague Convention and its Protocols: Legal and Practical Implications*, Paper to the conference: *Protecting Cultural Heritage: International Law after the War in Iraq*, Cultural Policy Center, University of Chicago, 3 February 2006.

Bugnion, F. 2004. The origins and development of the legal protection of cultural property in the event of armed conflict, *International Review of the Red Cross, Special Issue: Protection of Cultural Property in Armed Conflict*, 854, 11 November 2004.

Burg, S.L. and Shoup, P.S. 2000. *The War in Bosnia-Herzegovina: Ethnic Conflict and International Intervention*. London: M.E. Sharpe.

Calame, J. and Sechler, K. 2004. Is preservation missing the point? Cultural heritage in the service of social development. *Future Anterior: Journal of Historic Preservation History, Theory and Criticism*, 1:1, Spring, 57–63.

Campbell, D. 1998. *National Deconstruction: Violence, Identity and Justice in Bosnia*. Minneapolis, MN: University of Minnesota Press.

Chivvis, C. 2010. Back to the brink in Bosnia?. *Survival: Global Politics and Strategy*. 52:1, February–March 2010, 97–110.

Chollet, D. 2005. *The Secret History of Dayton, US Diplomacy and the Bosnia Peace Process 1995*. National Security Archive Electronic Briefing Book No. 171, George Washington University [online], 21 November 2005. Originally a classified document published as: *The Road to Dayton: US Diplomacy and the Bosnia Peace Process May–December 1995*, United State Department of State, May 1997. Available at: http://www.gwu.edu/~nsarchiv/NSAEBB/NSAEBB171/index.htm.

Conversi, D. 1996. Moral Relativism and Equidistance in British Attitudes to the War in the Former Yugoslavia, in T. Cushman and S. Mestrovic (eds), *This Time We Knew: Western Responses to Genocide in Bosnia*. New York and London: New York University Press, 244–81.

Cousens, E.M. and Cater, C.K. 2001. *Toward Peace in Bosnia: Implementing the Dayton Accords*. Colorado: Lynne Rienner Publishers.

Coward, M. 2001. *Community as Heterogeneous Ensemble: Mostar and Multiculturalism*, ISA Annual Convention, Chicago, 21–24 February 2001

—— 2004. Urbicide in Bosnia, in *Cities, War, and Terrorism: Towards an Urban Geopolitics*, edited by S. Graham. Oxford: Blackwell, 154–71.

—— 2008. *Urbicide*. Abingdon: Routledge.

Cushman, T. 2004. Anthropology and genocide in the Balkans: An analysis of conceptual practices of power. *Anthropological Theory*, March 2004, Vol. 4, No. 15–28.

Dahlman, C.T. and Ó Tuathail, G. (G. Toal) 2005. The legacy of ethnic cleansing: The international community and the returns process in post-Dayton Bosnia-Herzegovina. *Political Geography*, 24:5, 569–99.

Dakin, B. 2002. Islamic Community in Bosnia and Herzegovina v. The Republika Srpska: Human rights in a multi-ethnic Bosnia. *Harvard Human Rights Journal*, 15 (Spring 2002), 245–67.

Dutli, M.T. (ed.) 2002. *Protection of Cultural Property in the Event of Armed Conflict; Report on the Meeting of Experts (Geneva, 5–6 October 2000)*. Geneva: International Committee of the Red Cross, Advisory Service on International Humanitarian Law.

Eyre, Lt Col. D. 1998. Working with NGOs: What Every SOF Soldier Should Know, *Special Warfare*, Spring 1998, 15.

Francioni, F. 2004. Beyond state sovereignty: The protection of cultural heritage as a shared interest of humanity. *Michigan Journal of International Law*, 25:4, 1209–28.

Friesendorf, C. 2012. *International Intervention and the Use of Force: Military and Police Roles*, SSR Paper 4, Geneva: The Geneva Centre for the Democratic Control of Armed Forces (DCAF).

Glaurdic, J. 2011. *The Hour of Europe: Western Powers and the Breakup of Yugoslavia*. New Haven, CT and London: Yale University Press.

Gordy, E.D. 1999. *The Culture of Power in Serbia: Nationalism and the Destruction of Alternatives*. University Park, PA: Pennsylvania State University Press.

Gutman, R. and Rieff, D. (eds) 1999. *Crimes of War: What the Public Should Know*. New York and London: Crimes of War Project, W.W. Norton & Co.

Henckaerts, J-M. 1999. New rules for the protection of cultural property in armed conflict, *International Review of the Red Cross*, 835, 30 September 1999.

Herscher, A. and Riedlmayer, A. 2000. *Architectural Heritage in Kosovo: A Post-War Report*. First published in US/ICOMOS Newsletter 4 (July-August 2000).

—— 2001. Monument and crime: The destruction of historic architecture in Kosovo. *Grey Room*, MIT Press Journals, 01, Fall 2000, 108–22.

Hladík, J. 1999. The 1954 Hague Convention for the Protection of Cultural Property in the Event of Armed Conflict and the notion of military necessity. *International Review of the Red Cross*, No. 835, 30 September 1999, 621–35.

Hoare, M.A. 2004. *How Bosnia Armed.* London: Saqi.

—— 2010. The War of Yugoslav Succession, in *Central and Southeast European Politics Since 1989*, edited by S.P. Ramet. Cambridge: Cambridge University Press, 111–35.

Hoare, Q. and Malcolm, N. (eds) 1999. *Books on Bosnia*. London: The Bosnian Institute.

Hodge, C. 1999. *The Serb Lobby in the United Kingdom*. Seattle: University of Washington, Donald Treadgold Papers, No. 22, September 1999.

ICTY Outreach 2005. View from The Hague: Facts about Foča, in *Justice in Transition (Pravda u tranziciji*), Vol. 2, December 2005. Belgrade: Office of the War Crimes Prosecutor of Serbia. Available at: http://www.tuzilastvorz.org.rs/html_trz/%28CASOPIS%29/ENG/ENG02/540.htm.

Ivanišević, B. 2008. *The War Crimes Chamber in Bosnia and Herzegovina: From Hybrid to Domestic Court*. New York: International Center for Transitional Justice. Available at: http://ictj.org/publication/war-crimes-chamber-bosnia-and-herzegovina-hybrid-domestic-court.

Jogan, S. 2002. The procedure for marking cultural property in Slovenia, in *Protection of Cultural Property in the Event of Armed Conflict; Report on the Meeting of Experts (Geneva, 5–6 October 2000)*, edited by M.T. Dutli. Geneva: International Committee of the Red Cross, Advisory Service on International Humanitarian Law.

Kent, G. 2004. Framing and problem definition: British responses to the war against Bosnia, in *Media in Security and Governance: The Role of the News Media in Security*, edited by M. Caparini. Bonn: Nomos/Bonn International Center for Conversion (BICC) and Geneva: Geneva Centre for the Democratic Control of Armed Forces (DCAF), 161–77.

Kalender, F. 1997. Stanje fondova Nacionalna i univerzitetske biblioteke prije i poslije katastrofalnog pozara. *Bibliotekarstvo: Godišnjak društva bibliotekara Bosne i Hercegovine*, Vol. 37–41 (1992–1996), 26–9.

Kila, J.D. 2012. *Heritage under Siege: Military Implementation of Cultural Property Protection Following the 1954 Hague Convention*. Leiden: Brill.

Kolind, T. 2008. *Post-war identification. Everyday Muslim Counterdiscourse in Bosnia-Herzegovina*. Aarhus: Aarhus University Press.

Kostić, R. 2007. *Ambivalent Peace: External Nation-building, Threatened Identity and Reconciliation in Bosnia and Herzegovina*. Dissertation. Uppsala: Uppsala University.

—— 2009. *Reconciling the Past and the Present: Evaluating the Dayton Peace Agreement 1995*. Uppsala: Uppsala University and New York: The Mediation Support Unit, Department of Political Affairs, United Nations.

Kurspahic, K. 2003. *Prime Time Crime: Balkan Media in War and Peace*. Washington, DC: United States Institute of Peace Press.

Kuzman, Z. (ed.) 2002. *Stolac mjesto spora umjesto razgovora. Radovi s Okruglog stola Stolac, 5 listopada 2001* [*Papers from a Seminar on Stolac, 5 October 2001*]. Stolac: Municipal Authorities of Stolac.

Landry, Lt-Col R. 1999. *The European Community Monitor Mission (ECMM) in Former Yugoslavia: Lessons Learned for OAU Civilian Missions*. ACCORD Occasional Paper: Peacekeeping, ACCORD (The African Centre for the Constructive Resolution of Disputes), No. 5/99.

Mahmutćehajić, R. 2000. *Bosnia the Good: Tolerance and Tradition*. Budapest: Central European University Press.

McDowall, L. (ed.) 2009. *Bridging the Gap Between the ICTY and Communities in Bosnia and Herzegovina, Conference Proceedings, Foča, 9 October 2004*. The Hague: International Criminal Tribunal for the Former Yugoslavia.

O'Kane, M. 1996. *The Role of the Media in Bosnia*. Balkan Papers/Interview Series, Australia Region Media (armedia.net.au), May 1996. Available at: http://ics-www.leeds.ac.uk/papers/pmt/exhibits/807/okane.pdf.

O'Keefe, R. 2006. *The Protection of Cultural Property in Armed Conflict*. Cambridge: Cambridge University Press.

O'Shea B. 1998. *Crisis at Bihac: Bosnia's Bloody Battlefield*. Phoenix Mill: Sutton Publishing.

Peruaca, B. 2003. *Can Faith-Based NGOs Advance Interfaith Reconciliation? The Case of Bosnia and Herzegovina*. Washington DC: United States Institute of Peace, Special Report 103, March 2003.

Pesic, M. 1994. *Manipulations on Television Belgrade*, MA thesis, Department of Journalism, City University, London, September 1994.

Phuong, C. 2000. 'Freely to return': Reversing ethnic cleansing in Bosnia-Herzegovina, *Journal of Refugee Studies*, 13:2, 165–80.

Pocar, F. 2008. Statute of the International Criminal Tribunal for the Former Yugoslavia, Security Council Resolution 827 (1993), 25 May 1993. *United Nations Audiovisual Library of International Law.*

Popov, N. (ed.) 2000. *The Road to War in Serbia: Trauma and Catharsis*. Budapest: Central European University Press.

Praljak, S. 2006. *Kako je srušen Stari most* [*How the Old Bridge Was Destroyed*]. Zagreb: Oktavijan.

Ramet, S.P. 2006. *The Three Yugoslavias: State-Building and Legitimation, 1918–2005*. Bloomington, IN: Indiana University Press.

Ramos, R. 2003. Imagining Peace, *Trickster's Way*: Vol. 2, Iss. 2, Article 4. Available at: http://digitalcommons.trinity.edu/trickstersway/vol2/iss2/4

Reeves, D. 2005. *The Soul of Europe: Imagination – Confiscated.* Paper to the conference: *The Church in the Centre of the City,* Cardiff University 12–15 September 2005.

Riedlmayer, A. and Naron, S. 2009. From Yizkor Books to Weblogs: Genocide, Grassroots Documentation, and New Technologies, in *Community Archives: The Shaping of Memory*, edited by J.A. Bastian and B. Alexander. London: Facet Publishing, 151–63.

Roethlisberger, E. 2002. The importance of respect for international humanitarian law and the activities of the ICRC, in *Protection of Cultural Property in the Event of Armed Conflict; Report on the Meeting of Experts (Geneva, 5–6 October 2000)*, edited by M.T. Dutli. Geneva: International Committee of the Red Cross, Advisory Service on International Humanitarian Law.

Sandoz, Y. 1998. The International Committee of the Red Cross as guardian of international humanitarian law, *ICRC Resource Centre* [online], 31 December 1998.

—— 2002. Competing priorities: Placing cultural property on the humanitarian law agenda, in *Protection of Cultural Property in the Event of Armed Conflict; Report on the Meeting of Experts (Geneva, 5–6 October 2000)*, edited by M.T. Dutli. Geneva: International Committee of the Red Cross, Advisory Service on International Humanitarian Law, 21–25. Special Advisor to the ICRC.

Simms, B. 2002. *Unfinest Hour: Britain and the Destruction of Bosnia*. London: Penguin.

—— 2003. The end of the 'official doctrine': The new consensus on Britain and Bosnia, *Civil Wars*, 6:2, 53–69.

Stanley-Price, N. (ed.) 2007. *Cultural Heritage in Postwar Recovery, Papers from the ICCROM Forum held in Rome 4–6 October 2005*. Rome: ICCROM.

—— 2007. The thread of continuity: Cultural heritage in postwar recovery, *Cultural Heritage in Postwar Recovery, Papers from the ICCROM Forum held in Rome 4–6 October 2005*, edited by N. Stanley-Price. Rome: ICCROM, 1–16.

Stovel, H. 1998. *Risk Preparedness: A Management Manual for World Cultural Heritage*, Rome: ICCROM, UNESCO, ICOMOS, WHC.

Thompson, M. 1999. *Forging War: The Media in Serbia, Croatia & Bosnia-Hercegovina*. Luton: University of Luton Press.

Toal, G. and Dahlman, C.T. 2011. *Bosnia Remade: Ethnic Cleansing and its Reversal*. New York: Oxford University Press.

Torov, I. 2000. The Resistance in Serbia, in J. Udovicki and J. Ridgeway, James (eds), *Burn This House: The Making and Unmaking of Yugoslavia*. Durham, NC: Duke University Press, 247–66.

Tomšič Čerkez, B. 2010. Arhitektura med gradnjo in rušenjem identitete/Architecture between building and destroying identity. *Arhitektura, raziskave/Architecture, Research*, 2010/2, 23–30.

Udovicki, J. and Ridgeway, J. (eds) 2000. *Burn This House: The Making and Unmaking of Yugoslavia*. Durham, NC: Duke University Press.

van Boven, T. 2010. *The United Nations Basic Principles and Guidelines on the Right to a Remedy and Reparation for Victims of Gross Violations of International Human Rights Law and Serious Violations of International Humanitarian Law (Introductory Note)*. United Nations Audiovisual Library of International Law [online]. Available at: http://untreaty.un.org/cod/avl/ha/ga_60-147/ga_60-147.html.

Verdery, K. 2000. *The Political Lives of Dead Bodies*. New York: Columbia University Press.

Vrdoljak, A.F. 2007. Intentional Destruction of Cultural Heritage and International Law, in *Thesaurus Acroasium* XXXV (2007): 377–96.

—— 2009. Cultural heritage in human rights and humanitarian law, in *Human Rights and International Humanitarian Law*, edited by O. Ben-Naftali. Oxford: Oxford University Press.

—— 2010. Genocide and restitution: Ensuring each group's contribution to humanity. *European Journal of International Law*, Vo. 22, Iss. 1, 17–47.

Warner, M. et al. 2000. SFOR Lessons Learned in Creating a Secure Environment with Respect for the Rule of Law, based on a study of Bosnia, May 2000. Carlisle, PA: US Army Peacekeeping Institute.

Wentz, L. 1998. *Lessons from Bosnia: The IFOR Experience*. Washington, DC: Command and Control Research Program (CCRP).

Yarwood, J. 1999. *Rebuilding Mostar: Urban Reconstruction in a War Zone*. Liverpool: Liverpool University Press.

Young, K. 2001. UNHCR and ICRC in the former Yugoslavia: Bosnia-Herzegovina. *International Review of the Red Cross*, September 2001, 83:843, 781–805.

Websites

Balkan Witness. Available at: http://balkanwitness.glypx.com. Extensive archive of material on the wars and conflicts in the former Yugoslavia (including Bosnia), with material on destruction of monuments.

Bosnian Manuscript Ingathering Project. Available at: http://bmip.info. Full information on the BIM Project and many other articles relating to the destruction of cultural heritage and libraries.

Commission to Preserve National Monuments of Bosnia and Herzegovina. Available at: http://www.kons.gov.ba. Information on the activities of the Commission with Decisions of designation of properties as national monument.

COE/EC Integrated Rehabilitation Project Plan/Survey of the Architectural and Archaeological Heritage (IRPP/SAAH), Bosnia and Herzegovina. Information on the programme in Bosnia-Herzegovina with links to surveys and technical assessments. Available at: http://www.coe.int/t/dg4/cultureheritage/cooperation/SEE/Countries/Bosnia_en.asp.

Michael Sells Pages. Extensive archive of internet articles to supplement *A Bridge Betrayed* with much information on monument destruction. Formerly available at: www.haverford.edu/relg/sells/; now archived at http://balkanwitness.glypx.com/sells-reports.htm.

Kantonalni zavod za zaštitu kulturno-historijskog i prirodnog naslijeđa Sarajevo, Ratne štete [*War Damage*]. Document with damage lists and photographs of damage monuments in Sarajevo Canton. Available at: http://www.spomenici-sa.ba/kulturna-i-prirodna-bastina-sarajeva/item/61.

Newspaper, Magazine and Web Articles, Press Releases.

24SI 2011. Potpredsjednik Federacije BiH Mirsad Kebo uručio je danas vladiki zahumsko-hercegovačkom Grigoriju odluku o izdvajanju 1,5 miliona maraka iz budžeta federalne Vlade za obnovu Saborne crkve u Mostaru. *24sata.info* [online]. 23 April. Available at: http://www.24sata.info/vijesti/bosna-i-hercegovina/mostar/61868-Jedan-simbola-grada-Milion-pol-obnovu-Saborne-crkve-Mostaru.html.

Ahmetašević, N. 2002. Kada je Klein stigao i BiH, Hagada je, srećom već bila spašena!. *Slobodna Bosna*, Sarajevo, 12 December, 31.

AIM 1997. Refugees in B&H Do right solutions exist? *AIM Press* [online]. *Alternativna Informativna Mreža (Alternative Information Network) Sarajevo*, 23 June. Available at: http://www.aimpress.ch/dyn/trae/archive/data/199706/70629-004-trae-sar.htm.

Anđelić, S. 2000. Arhivsku građu iz Jasenovaca prebacio u Muzej holokausta u Washingtonu!. *Slobodna Bosna*, 28 December, 35–7.

Arnautovic, M. 2010. Bosnia: The village where hate never triumphed. *Institute for War and Peace Reporting* [online] 10 April, TRI Iss. 642. Available at: http://iwpr.net/report-news/bosnia-village-where-hate-never-triumphed.

Art Newspaper 1992. Bosnia's written history in flames?. *The Art Newspaper*, 3, No. 21, 1 October, 1.

Associated Press 1993. Croats besieged in Bosnia monastery get lift out from British. *Los Angeles Times* [online], 10 June, http://articles.latimes.com/1993-06-10/news/mn-1614_1_croats.

—— 1992. Report: Serbs would change name of Sarajevo, *Associated Press* [online], 10 November. Available at: http://www.apnewsarchive.com/1992/Report-Serbs-Would-Change-Name-of-Sarajevo-With-AM-Yugoslavia/id-33f5d010e6ee75c81c8668b74d8bdb12

Avaz 2010. Obnova Aladža džamije u Foči. *Avaz*, 29 April. Bicent/Bošnjački informativni centar [online], 10 May 2010. Available at: http://www.bicent.com/kultura/obnova-aladza-dzamije-u-foci.

Barrentine, E. *Reconstruction of the Zitomislici Monastery*. Available at: http://bartik.brynmawr.edu/students/jkatz/reconstruction#path.

Barry, E. 1999. How the Vijecnica was lost: All the sympathy in the world hasn't rebuilt the Sarajevo Library. *Metropolis* [Online], June 1999. Available at: http://www.metropolismag.com/html/content_0699/ju99howt.htm.

BBC News 2009. Serbs ordered to pay for mosques. *BBC News* [online], 20 February. Available at: http://news.bbc.co.uk/1/hi/7902929.stm.

BBC Summary of World Broadcasts 1993. Sarajevo, Zagreb radios report destruction of Mostar's Old Bridge, (a) *Radio Bosnia-Hercegovina*, Sarajevo, in Serbo-Croat 14.00 GMT 9 November 1993, (b) *Croatian Radio*, Zagreb, in Croatian 14.00 GMT, 9 November 1993. *BBC Summary of World Broadcasts*, EE/1843/C, 11 November. Accessed on LexisNexis.

Beelman, M.S. 1993. Muslims win fierce battle for key Bosnian town. *Seattle Times* [online], 8 June. Available at: http://community.seattletimes.nwsource.com/archive/?date=19930608&slug=1705505

Bensman, T. 1993. Mechanic's skills useful in Bosnia war. *St Petersburg Times*, 10 January. Accessed at NewsBank.

BH Press news agency 2001. Bosnian Serb official criticizes former government for transfer of archives. *BH Press News Agency*, 19 January. *BBC Worldwide Monitoring*, 19 January. Accessed on LexisNexis.

Bileća Online 2006a. Vladika Grigorije: Obnovom Begove kuće do pomirenja, *Bileća Online* (www.bileca.org), 3 May. Available at: http://www.bileca.org/modules.php?name=News&file=article&sid=465

—— 2006b. Košarkaš Dejan Bodiroga gradi sportski centar u Trebinju (pored Begove kuće), *Bileća Online* (www.bileca.org), 31 May. Available at: http://www.bileca.org/modules.php?name=News&file=article&sid=488

—— 2007a. Pripreme za obnovu Avdića džamije u Planoj, *Bileća Online* (www.bileca.org), 20 May 2007. Available at: http://www.bileca.org/modules.php?name=News&file=article&sid=695

—— 2007b. Pomoć za obnovu Carske džamije u Bileći, www.bileca.org, 3 November. Available at: http://www.bileca.org/modules.php?name=News&file=article&thold=-1&mode=flat&order=0&sid=868

—— 2008. Izvještaj o akciji obnove Avdića džamije, *Bileća Online* (www.bileca.org), 3 December 2008. Available at: http://www.bileca.org/modules.php?file=article&name=News&sid=1329

—— 2013. Svečano otvaranje Avdića džamije 11.08.2013, *Bileća Online* (www.bileca.org), 27 July 2013. Available at: http://www.bileca.org/modules.php?name=News&file=article&sid=2402

BIRN 2009. Islamic Community to receive around 33 million Euros in compensation. *Balkan Investigative Reporting Network (BIRN), Justice Report* [online], 20 February. Available at: http://www.bim.ba/en/154/10/16877.

—— 2011. Karadzic: Destruction of mosques and Catholic churches. *Balkan Investigative Reporting Network (BIRN), Justice Report* [online], 9 December]. Available at: http://www.bim.ba/en/299/10/33999.

Bjelajac, M. 2012. Sud odlučio: RS neće isplatiti odštetu za srušene džamije. *Radio Slobodna Evropa* [online], 3 February. Available at: http://www.slobodnaevropa.org/content/sud_odlucio_rs_nece_isplatiti_odstetu_za_srusene_dzamije/24472881.html.

Block, R. 1993. Croatian death squad talks tough around the pooltable. *The Independent*, 6 September.

Bogdanovic, B. 1993. *Murder of the City*. *New York Review of Books*, 40:10, 27 May.

—— 1992. Der rituelle Städtemord. *Die Zeit* (Hamburg), 18 September 1992. Available at: http://www.zeit.de/1992/39/der-rituelle-staedtemord.

—— 1992. Ritualno ubijanje grada. *Borba* (Belgrade), 28 April.

Bollag, B. 1995. 'We must save what is left'. *The Independent*, 20 April. Available at: http://www.independent.co.uk/arts-entertainment/we-must-save-what-is-left-1616365.html#

Bošnjaci.Net 2009. Ramazanska akcija za džamiju u Skočigrmu kod Trebinja. *Bošnjaci.Net*, 21 August. Available at: http://www.bosnjaci.net/prilog.php?pid=35123.

—— 2010a. Otvorena Džamija Adamira Jerkovića, *Bošnjaci.Net*, 18 July. Available at: http://bosnjaci.net/prilog.php?pid=38489

—— 2010b. Ramazanska akcija prikupljanja novca za obnovu džamije u Pridvorcima kod Trebinja. *Bošnjaci.Net*, 14 August. Available at: http://www.bosnjaci.net/prilog.php?pid=38585.

—— 2010c. U Akciji Bošnjaci.Net Sakupljeno 5,639 KM. *Bošnjaci.Net*, 19 October. Available at: http://www.bosnjaci.net/prilog.php?pid=39282

Bowers, F. 1994. Turkey and Bosnia agree to rebuild historic landmark. *Christian Science Monitor* [online], 9 February. Available at: http://www.csmonitor.com/1994/0209/09032.html.

Boyes, R. 1992. This is cultural genocide. *The Times*, 28 August. Accessed at NewsBank.

Burazerović, A. 2005. Princ Čarls obećao ef. Ceriću pomoć u obnovi Ferhadije. *Dnevni Avaz*, No. 3497, 01 July, 2. Accessed at Inforbiro, www.idoconline.info.

Buric, N. 1993. Croatian daily suggests Croats destroyed 16th century bridge. *The Associated Press*, 10 November. Accessed at LexisNexis.

Burns, J.F. 1992. Cease-fire or no, Serbian shells fall. *New York Times*, 24 November. Available at: http://www.nytimes.com/1992/11/24/world/cease-fire-or-no-serbian-shells-fall.html.

Chronicle News Services 1995. Serb homes, Croatian church leveled in Bosnia – Orthodox Church condemns violence against Catholics. *The San Francisco Chronicle*, 20 May. Accessed at NewsBank.

Cohen, R. 1995a. CIA report on Bosnia blames Serbs for 90% of the war crimes. *New York Times*, 9 March. Available at: http://www.nytimes.com/1995/03/09/world/cia-report-on-bosnia-blames-serbs-for-90-of-the-war-crimes.html.

Cohen, R. 1995b. Bosnia Jews glimpse book and hope, *New York Times*, 16 April. Available at: http://www.nytimes.com/1995/04/16/world/bosnia-jews-glimpse-book-and-hope.html.

Commission to Preserve National Monuments 2003. *Prezentacija projekta rekonstrukcije Starog mosta u Mostaru od strane Jedinice za koordinaciju projekta izgradnje Starog mosta u Mostaru*, http://kons.gov.ba/, 1 July.

Community of Bosnia Foundation *A Sister-community in Bosnia: Foca-on-the-Drina*. Community of Bosnia Foundation [Online]. Archived at: http://web.archive.org/web/20050708012417/http://www.haverford.edu/relg/sells/foca/foca.html.

—— *Mostar*. Community of Bosnia Foundation [Online]. Archived at: http://web.archive.org/web/20050908065424/http://www.haverford.edu/relg/sells/mostar/mostar.html.

—— *Sites of Worship Deliberately Destroyed by Serb Religious Nationalists in Banja Luka*. Community of Bosnia Foundation [Online]. Archived at: http://web.archive.org/web/20050728074832/http://www.haverford.edu/relg/sells/banjaluka/banjaluka.html.

—— *Stolac Pages*. Community of Bosnia Foundation [Online]. Archived at: http://web.archive.org/web/20041214041603/http://www.haverford.edu/relg/sells/stolac/stolac.html.

Ćorić, V. 2005. Princ Čarls želi pomoći obnovu Saborne Crkve. *Nezavisne Novine*, No. 2338, 15 February, 9. Accessed at Infobiro, www.idoconline.info.

Cornwell, R., Savill, A. and Brown, C. 1993. Belgrade to cut supplies to Bosnian Serbs: Rejection of peace angers Milosevic – US demands Europe back quick action. *The Independent*, 7 May. Accessed at NewsBank.

Croatian Radio 1993. Bosnia-Hercegovina: Muslim offensive drives Croats out of 32 villages, 250 killed. *Croatian Radio*, Zagreb 17.00 GMT 8 June 1993. *BBC Summary of World Broadcasts*, EE/1711/C1, 10 June. Accessed at LexisNexis.

Davis, D. 1998. Serbs demand their 'share' of rare Sarajevo Haggadah. *Jewish Telegraphic Agency*, 24 December. Available at: http://archive.jta.org/article/1998/12/25/2892015/serbs-demand-their-share-of-rare-sarajevo-haggadah.

Demick, B.1993. An unceasing terror campaign violence governs the lives of Banja Luka's Muslims. *The Philadelphia Inquirer*, 2 November. Available at: http://articles.philly.com/1993-11-02/news/25948573_1_muslims-bullet-holes-grenades.

Djikic (Đikić), I. 2001. ABC of Croatian crimes in Bosnia-Herzegovina. *Bosnia Report* [online], New Series No. 21/22, January–May 2001. Translation of an article originally published in *Feral Tribune*, Split, 24 February 2001. Available at: http://www.bosnia.org.uk/bosrep/report_format.cfm?articleid=717&reportid=150.

Dnevni Avaz 2003. Posle projekta obnove Starog mosta u Mostaru – nastupice projekat 'Tri naroda' (After Project for Renovation of Old Bridge in Mostar – Project 'Three Nations' to begin), *Dnevni Avaz*, 22 September. Available at: http://www.vibilia.rs/.

—— 2013. Komisija za očuvanje nacionalnih spomenika demantirala Kaplana, *Dnevni Avaz* [online], www.avaz.ba, 11 June. Available at: http://www.avaz.ba/vijesti/iz-minute-u-minutu/komisija-za-ocuvanje-nacionalnih-spomenika-demantirala-kaplana.

Dnevni List 2009. Pravoslavni hram je stup opstanka jednog naroda, *Dnevni List* [online], 9 November. Available at: http://www.dnevni-list.ba/index.php?option=com_content&view=article&id=5942:pravoslavni-hram-je-stup-opstanka-jednog-naroda&catid=3:ivot&Itemid=4

Dodds, J.D. 1998b. Hearts and stones. *Aramco World*, 49:5, September/October 1998, 2–9.

Doder, D. 1993. Warring Bosnia factions practice linguistic cleansing of geographic names, *Baltimore Sun* [online], 4 May. Available at: http://articles.baltimoresun.com/1993-05-04/news/1993124122_1_linguistic-cleansing-serbs-yugoslavia.

Dragić, Z. 2001. Croat scientists appealed for protection of cultural-historical heritage: Their Bosniak colleagues ignored gathering!. *Hrvatska Rijec*, 29 October 2001. Available at: http://www.ex-yupress.com/hrvrijec/hrvrijec47.html.

Drakulic, S. 1993a. 'How is it possible that the people of Bosnia agonise over the loss of a bridge while they become indifferent to the loss of a human life?'. *The Observer* (London), 21 November, News, 21. Accessed at Lexis Nexis.

—— 1993b. Falling down: A Mostar Bridge elegy. *The New Republic*, 209, 13 December.

Dugalić, L. 2008. At the table on the bridge: Fifteenth international painters' gathering in Višegrad. *Serbia National Review* [online], No. 9. Available at: http://www.nacionalnarevija.com/en/tekstovi/br9/010%20Preko%20reke%20-%20Visegrad.html.

Duka, V. 2006. Imam Hodžić: Skrnavljenje spomenika; Vladika Grigorije: Obezvređivanje ideje, *Nezavisne Novine* [online], 3 May. Available at: http://www.nezavisne.com/dnevne/drustvo/dru05042006-03.php.

Edwards, B. 1993. Bosnia loses five-century-old bridge in fighting. *National Public Radio*, Morning Edition, 12 November. Accessed at LexisNexis.

El Pais 1993. Destruido el viejo puente de Mostar, una joya del XVI. *El Pais*, 10 November. Available at: http://elpais.com/diario/1993/11/10/internacional/752886005_850215.html.

Eminović, H. 2006. Pred obnovu Obradovića džamije u Bileći: Čovjek i džamija, *Bileća Online*. Originally published on *Preporod*, 1 May 2006. Available at: http://www.bileca.org/modules.php?name=Sections&op=viewarticle&artid=72.

Fesperman, D. 1995. 'Culturecide' taking heavy toll on art in Balkans war zone. *Baltimore Sun*, 13 April, C–6. Available at: http://www.sfgate.com/cgi-bin/article.cgi?f=/c/a/1995/04/13/MN4695.DTL.

Firestone, D. 1992. Peace rebuff: 'This is not a simple war,' Wiesel is told in Bosnia. *Newsday*, 30 November, 7. Accessed at Lexis Nexis.

Fisk, R. 1992. Travnik trusts in Allah as Serbs advance. *The Independent*, 8 December. Available at: http://www.independent.co.uk/news/world/europe/travnik-trusts-in-allah-as-serbs-advance-robert-fisk-reports-from-the-embattled-bosnian-town-where-the-british-united-nations-troops-do-little-more-than-reassure-people-1562237.html.

—— 1994. Waging war on history. *The Independent*, 20 June. Available at: http://www.independent.co.uk/opinion/waging-war-on-history-in-former-yugoslavia-writes-robert-fisk-whole-cultures-are-being-obliterated-1423832.html.

Ford, P. 2001. Mosques face new danger. *The Christian Science Monitor*, 25 July. Available at: http://www.bosnia.org.uk/bosrep/report_format.cfm?articleID=797&reportid=151.

Đogić, G. 2003. Na početku bi crkva na kraju – knjiga. *Glas Javnosti* [online] 1 April. Available at: www.glas-javnosti.co.yu.

FENA 2005. Zamjenik direktora UNESCO-a u Žitomisliću: Najavljena obnova tri poznata vjerska objekta. *Oslobođenje*, No. 21.158, 20 December, 6. Accessed at Infobiro, www.idoconline.info.

FIRMA 2011. 'Sarajevo City Hall revisited' exhibition opens in Vijecnica.' Press release, Sarajevo: *FIRMA Project*, 14 July. Available at: http://firmaproject.ba/en/news/news.aspx?id=576.

Franjevačka Teologija 2009. Sudbina knjižnice u zadnjem ratu, *www.franjevackateologija.ba*, November 2009. *Available at:* http://www.franjevackateologija.ba/index.php?option=com_content&view=article&id=114:sudbinaknjiznice-clanak&catid=50:knjiznica-category&Itemid=90

Gentile, L. 1994. In Banja Luka, terror seems uncannily normal. *The New York Times*, Opinion, New York, 14 January. Available at: http://www.nytimes.com/1994/01/14/opinion/l-in-banja-luka-terror-seems-uncannily-normal-870200.html.

Gienger, V.G. 2003. Out of sight. *Balkan Reconstruction Report*, *Transitions Online*, 20 February. Available at: http://www.tol.org/client/article/8908-out-of-sight.html.

Glas Srpske 2007. Obnova u ratu srušenih pravoslavnih manastira i crkav. *Glas Srpske*, 14 August. Available at: http://www.glassrpske.com/vijest/11/novosti/2704/lat/Obnova-u-ratu-srusenih-pravoslavnih-manastira-i-crkav.html.

Gordon, M.R. 1993. US memo reveals dispute on Bosnia. *The New York Times*, 25 June. Available at: http://www.nytimes.com/1993/06/25/world/us-memo-reveals-dispute-on-bosnia.html.

Gušić, B. 1995. Život i smrt u okupiranoj banjaluci, *Oslobođenje*, 15–23 March. Accessed at Infobiro, www.idoconline.info. English translations of excerpts formerly online at http://www.haverford.edu/relg/sells, and as *BanjaLuka under Republika Srpska Terror/ Testimony/1 and #2*, https://groups.google.com/forum/, alt.current-events.bosnia, posted by Michael A. Sells, 2 August 1997.

Gušić, B. 2012. In memoriam: 'Banjalučka Ferhadija – ljepotica koju su ubili' (7 maj 1993–7 maj 2012) Zločin bez kazne, *http://bedrudingusic.wordpress.com*, 7 May.

Gutman, R. 1992. Unholy War; Serbs target culture, heritage of Bosnia's Muslims. *Newsday*, 2 September. Accessed at LexisNexis.

Gutman, R./Newsday 1995. Bosnia priest, nun killed, church says – Serbs blamed for terror against Catholics, Muslims. *The Seattle Times*, 18 May. Accessed on NewsBank.

Hadžić, H. 2002. Bijeljina: A bastion of apartheid. *Dani*, Sarajevo, 6 December. Available at: http://www.bosnia.org.uk/bosrep/report_format.cfm?articleID=935&reportid=157.

Hajdarhodžić, S. 2005. Bosnia: Sarajevo's missing tomes. *IWPR Institute for War & Peace Reporting, Balkan Crisis Report*, Iss. 284 [online], 6 September 2005. Available at: http://iwpr.net/report-news/bosnia-sarajevos-missing-tomes.

Hall, B. 1996. Rebecca West's war, *The New Yorker*, 15 April, 74–83.

HAPSBURG 1998. *Banja Luka Mayor Umicevic on the Ferhadija Mosque*. HAPSBURG Discussion Group, 9 May 1998. Available at: http://h-net.msu.edu.

Hedl, D. 2000a. Tapes nail Tudjman cronies. *Institute for War and Peace Reporting* [online], 8 April. Available at: http://iwpr.net/report-news/tapes-nail-tudjman-cronies.

—— 2000b. Bosnian trade off. *Institute for War and Peace Reporting* [online], 4 September. Available at: http://iwpr.net/report-news/bosnian-trade.

Herscher, A. 1998. Remembering and rebuilding in Bosnia. *Transitions*, 15 March 1998, 76–81.

Higham, N. 2001. Croatian holocaust still stirs controversy. *BBC* [online], 29 November. Available at: http://news.bbc.co.uk/1/hi/programmes/from_our_own_correspondent/1673249.stm.

Hoare, M.A. 2008. What do the figures for the Bosnian war dead tell us?. *Greater Surbiton* [online]. 4 January. Available at: http://greatersurbiton.wordpress.com/2008/01/04/what-do-the-figures-for-the-bosnian-war-dead-tell-us.

Husarska, A. 1998. Without a prayer? Rebuilding a Muslim temple is a test of Bosnia's peace. *The Washington Post*, 10 May, C01.

Husedžinović, S. 2002. Zlocinacko unistavanje: spomenika islamske arhitekture u Banjoj Luci. Available at: http:*//bhmedia.se/* [online], http://bhmedia.se/banjaluka/tekstovi/unistavanje.htm.

Huseinovic, S. and Arbutina, Z. 2012. Burned library symbolizes multiethnic Sarajevo, *Deutsche Welle* [online], 25 August. Available at: http://www.dw.de/burned-library-symbolizes-multiethnic-sarajevo/a-16192965

Iglesias Kuntz, L. 2002a. Whose library is it?. *The New Courier*, UNESCO, May 2002, 16–17.

—— 2002b. Mostar, A bridge to peace. *The New Courier*, UNESCO, May 2002, 16–17.

Igrić, G. 1993. On the spot: Zvornik, monster town. *Vreme News Digest Agency*, No. 112, 15 November. Available at: http://www.scc.rutgers.edu/serbian_digest/

—— 1995. Rendering of Accounts. *AIM Press* [online], *Alternativna Informativna Mreža (Alternative Information Network)* 9, 8 February. Available at: http://www.aimpress.ch/dyn/trae/archive/data/199502/50208-005-trae-beo.htm.

IKA. 1999. U Jajce sonešen sarkofag s posmrtnim ostacima posljednjeg Bosanskog kralja Stjepana Tomaševića, *Informativna katolička agencija (IKA)*, 14 September.

Ilić, J. 1993. 'Weeping over a bridge or over Mostar.' *Vjesnik*, 10 November 1993. Cited in Kurspahic 2003.

Independent 1993. Croats questioned over destruction of Mostar bridge, *The Independent* (London) 23 December. Accessed at NewsBank.

Irwin, R. and Šarić, V. 2010. Calls for war memorials divide Bosnia, *Institute for War & Peace Reporting* (*iwpr.net*), 6 December. Available at: http://iwpr.net/report-news/calls-war-memorials-divide-bosnia.

Jelin-Dizdar, T. 2008. UNESCO pomaže obnovu Crkve Svete Trojice, *Radio Slobodna Evropa* [online], 26 March. Available at: http://www.slobodnaevropa.org/articleprintview/1049417.html.

—— 2010. Počela obnova Saborne crkve u Mostaru. *Radio Slobodna Evropa* [online], 22 October, http://www.slobodnaevropa.org/content/mostar_crkva_svete_trojice_obnova/2198656.htm.

JTA 2002. Sarajevo Haggadah goes on display. *JTA (Jewish Telegraphic Agency)*, 3 December. Available at: http://www.jta.org/news/article/2002/12/03/9657/Priceless14thcentu

Judah, T. 1993a. Razing of mosques gives new resolve to Muslims. *The Times*, 14 May.

—— 1993b. Banja Luka loses last mosque. *The Times*, 10 September.

Jukic, E. 2012. Bosnian Muslims appeal over demolished mosques. *Balkan Insight* [online], 16 February. Available at: http://www.balkaninsight.com/en/article/bosnia-islamic-community-to-sue-serb-entity-again.

Jurišin, P. 1995. Bosanski kralj u Splitu. *Oslobođenje* [online], No. 17011, 18 October, 12. Accessed at Infobiro, www.idoconline.info.

Kalamujić, A. 2006. Spor oko provodjenja odluke o nacionalnom spomeniku u Trebinju, *Oslobođenje*, 29 November. Available at: http://www.trebinje.se/index.php?option=com_content&task=view&id=104&Itemid=42

Kamenica, E. 2006. Amerikanci hoće koncesiju na Počitelj, *Oslobođenje* [online], No. 21.472, 4, 30 October. Accessed at Infobiro, www.idoconline.info.

Karadžić, R. 1992. Interview with Radovan Karadžić, *New Perspectives Quarterly*, 9:4 (Fall), 50–51.

Karup-Druško, D. 2000. Ram za sliku opljačkog Sarajeva. *Dani*, 23 June, 50–53.

Kinzer, S. 1993a. Conflict in the Balkans; Serbs reject Bosnia pact, defying friends and foes, and insist on referendum. *The New York Times*, 6 May. Available at: http://www.nytimes.com/1993/05/06/world/conflict-balkans-serbs-reject-bosnia-pact-defying-friends-foes-insist-referendum.html.

—— 1993b. Conflict in the Balkans; The Serbs; Belgrade vows to cut Bosnia aid, rompting skepticism and hope. *The New York Times*, 7 May. Available at: http://www.nytimes.com/1993/05/07/world/conflict-balkans-serbs-belgrade-vows-cut-bosnia-aid-prompting-skepticism-hope.html.

—— 1993c. Conflict in the Balkans; 2 Major mosques blown up by Serbs. *The New York Times*, 8 May. Available at: http://www.nytimes.com/1993/05/08/world/conflict-in-the-balkans-2-major-mosques-blown-up-by-serbs.html.

Klix.ba 2008. Otvoren obnovljeni muzej 2. zasjedanja AVNOJ-a. *www.klix.ba/*, 30 November 2008. Available at: http://www.klix.ba/vijesti/bih/otvoren-obnovljeni-muzej-2-zasjedanja-avnoj-a/081130004

Kolovrat, I. 1999. Arhiv i muzejska zbirka Spomen-područja Jasenovac od 1991 us stanu bivšega kustosa Sime Brdara u Bosanskog Dubici!. *Globus*, 10 September 1999, 30–32.

Komlenović, U. 1994. On the spot: Banja Luka; rolling stones on the Vrbas River. *Vreme News Digest Agency*, No. 156, 19 September. Available at: http://www.scc.rutgers.edu/serbian_digest/

Kovac, E. 2009. Controversial church relocated. *Southeast European Times* [online], 8 June. Available at: http://www.setimes.com/cocoon/setimes/xhtml/en_GB/features/setimes/features/2009/06/08/feature-02.

Kovac, V. 2001. After decade abroad, World War II archive returns home to Croatia. *Jewish Telegraphic Agency* [online], 6 December. Available at: http://archive.jta.org/article/2001/12/06/2906255/after-decade-abroad-world-war-ii-archive-returns-home-to-croatia.

Kovačević, Ž. 2010. Josip Pejaković zaštitio dio blaga BiH, *Radio Slobodna Bosna* [online], 13 December. Available at: http://www.slobodnaevropa.org/content/plp_josip_pejakovic_zastitio_kulturno_blago_bih/2245964.html.

Le Monde 1993. Le Vieux Pont de Mostar n'est plus. *Le Monde*, 11 November. Available at: http://www.lemonde.fr/web/recherche_breve/1,13-0,37-467780,0.html.

Lovrenović, D. 1999. U susret boljoj proslosti, *Dani*, 15 October.

Lovrenović, I. 1994. The hatred of memory: In Sarajevo, burned books and murdered. pictures. *The New York Times*, 28 May, A15.

—— 1999. Bosanski Pompeji. *Feral Tribune*, No. 731, 20 September, 34. Accessed at Infobiro, www.idoconline.info.

Lucić, P. 2000. Shit of the year 1999. *Feral Tribune*, Split, 3 January. Translated version by Ex-Yu Press. Available at: http://www.ex-yupress.com/feral/feralgs99.html.

Luxmoore, J. 2007. Bosnian Serbs make overtures to Church. *The Tablet*, 3 March. Available at: http://m.thetablet.co.uk/article/9434.

Mackic, E. 2009. Historic decisions by Banja Luka court. Justice Report. *Balkan Investigative Reporting Network (BIRN)* [online], 13 November. Available at: http://www.bim.ba/en/192/10/23630/

Malcolm, N. 2001. Britain's fatal foreign policy. *The Sunday Telegraph*, London, 9 December 2001. Available at: http://www.bosnia.org.uk/bosrep/report_format.cfm?articleid=857&reportid=153

Malić, G. 1994. Aladin više ne stanuje ovdje: Stolac grad bez prošlosti, *Feral Tribune*, 2 May, 14. Accessed at Infobiro, www.idoconline.info.

Matvejević, P. 1993. Most. *Feral Tribune*, 16 November. Accessed at Infobiro, www.idoconline.info.

—— 1993. La virtud de un Puente. *El Pais*, Tribuna, 14 December http://elpais.com/diario/1993/12/14/opinion/755823607_850215.html.

Medžlis IZ Trebinje 2011. Obavijest iz Medžlisa IZ Trebinje, *www.medzlis-trebinje.ba*, 15 April 2011. Available at: http://www.medzlis-trebinje.ba/web/index.php?option=com_content&view=article&id=175:obavijest-iz-medlisa-iz-trebinje&catid=1:latest&Itemid=131

Medzlis Trebinje 2011. Trebaju nam primjeri živuće tolerancije a ne slatkorječive i puke retorike, *www.medzlis-trebinje.ba*, 15 September, http://www.medzlis-trebinje.ba/web/index.php?option=com_content&view=article&id=193:trebaju-nam-primjeri-ivue-tolerancije-a-ne-slatkorjeive-i-puke-retorike&catid=1:latest&Itemid=131.

Milliyet 1993. Koprüyü sonunda yıktılar, *Milliyet* (Istanbul), 11 November, 15.

Mlakar, M. 1993. Bijeljina dynamiters: The night the minarets fell. *Vreme News Digest Agency*, No. 78, 22 March. Available at: http://www.scc.rutgers.edu/serbian_digest.

M.Z. 2001. RS Authority cancels permits for construction of mosques in Bijeljina. *Dnevni Avaz*, 18 August 2001, Bosnian Institute News and Analysis [online], 9 October 2001. Available at: http://www.bosnia.org.uk/news/news_body.cfm?newsid=1557.

Mulic-Busatlija, S. 2001. Miracle in the Republic of Srpska: Kozarac. *Dani* [translation online], Sarajevo, 7 December. Available at: http://www.ex-yupress.com/dani/dani76.html.

Musli, E. 2009. Vladika Kačavenda širi laži o Atik džamiji. *Dnevni Avaz*, Sarajevo, 2 February. Available at: http://bosnjaci.net/prilog.php?pid=31715.

Naegele, J. 2000. Saudi aid workers bulldoze Balkan monuments, *Radio Free Europe/ Radio Liberty* [online], Newsline, 8 August. Available at: http://www.rferl.org/content/article/1142212.html.

Naudet, J-B. 1993. La destruction du pont de Mostar est un sacrilège irréparable pour les Musulmans de Bosnie. *Le Monde*, 16 November.

Neier, A. 2011. Helping the Balkans survive a decade of war. *www.opensocietyfoundations.org*, 18 October. Available at: http://www.opensocietyfoundations.org/voices/helping-balkans-survive-decade-war.

New Straits Times 1994. The culture warriors. *New Straits Times* [online], 17 January. Available at: http://news.google.com/newspapers.

Novine 2001. Jasenovac: Muzej holokausta vratice arhivu iz Jasenovca u Hrvatsku. *Novine* [online], Toronto, 23 October. Available at: http://www.novine.ca/arhiva/2001/23_10_01/bivsa_yu.html.

O'Kane, M. 1993. Eyewitness: Peace eludes the people of St Francis. *The Guardian*, 10 June, 24. Accessed at LexisNexis.

ONASA 2003. Bosnian Serb groups warn against rebuilding mosque destroyed during war. *ONASA News Agency* [online], 20 January. From BBC Monitoring International Reports)

—— 2005a. Outgoing British Ambassador to BiH visits Mostar. *ONASA News Agency*, 14 February. BBC Monitoring International Reports.

—— 2005b. Islam is a religion that respects all peoples: UK Ambassador Cliff. *ONASA News Agency*, 24 February. BBC Monitoring International Reports.

—— 2005c. UNESCO to finance reconstruction of Ferhadija mosque. *ONASA News Agency*, 10 October. BBC Monitoring International Reports.

Oslobođenje 2005. Oštećene knjige i slike Franjevačkog samostana Plehan. *Oslobođenje*, No. 21.047, 31 August, 24. Accessed at Infobiro, www.idoconline.info.

Österreichische Nationalbibliothek 2004. Wiederaufbau der Nationalbibliothek Sarajewo. ÖNB *Newsletter*. Vienna: Österreichische Nationalbibliothek, No. 3, September 2004, 14. Available at: http://www.onb.ac.at/files/newsletter04-3.pdf, accessed 23/08/2011.

Palm Beach Post 1993. Serbs reject peace plan – Parliament ignores Western military threats, sets referendum. *The Palm Beach Post, Palm Beach Post Wire Services*, 6 May. Accessed at NewsBank.

Pandurević, T. 1995. Euforični postupak. *Oslobođenje*, No. 17007, 14 October, 12. Accessed at Infobiro, www.idoconline.info.

Pequeux, G. 2003. Rencontre avec Gilles Péqueux, ingénieur responsable de la reconstruction du pont de Mostar. *Courrier des Balkans* [online], transcript by C.H. Groenneberg, 1 March. Available at: http://balkans.courriers.info/article2783.html.

Perlez, J. 1997. Serbian media is a one-man show. *The New York Times*, 10 August 1997

Pettifer, J. 2002. How the Serbs were appeased. *Times Literary Supplement*, 10 May, 29. www.ProfessorJamesPettifer.com.

Phillips, Col. W.R. 1998. Civil-Military Cooperation: Vital to peace implementation in Bosnia. *NATO Review* [online], Vol. 46, No. 1, Spring 1998, 22–25. http://www.nato.int/docu/review/1998/9801-07.htm.

Plazonic, M. 2005. A culture served cold. *Transitions Online*, 9 February. Available at: http://www.tol.org/client/article/13535-a-culture-served-cold.html.

Pomfret, J. 1992a. Battles for Sarajevo intensify as Bosnian peace conference opens. *Associated Press*, 26 August.

Pomfret, J. 1992b. Battling blazes and bullets in Sarajevo. *The Philadelphia Inquirer*, 27 August, A05. Accessed at NewsBank.

—— 1993. As Mostar Bridge crumbles, so does Bosnian dream of ethnic unity. *The Washington Post*, 30 August. Accessed at LexisNexis.

Poskok.info 2007. Livanjski franjevci sačuvali ikone srpske pravoslavne crkve. *www.poskok.info* [online], 19 September.

Radio BH 1993a. Bosnia-Hercegovina; Sarajevo radio reports Serb/Croat attacks in Northern and Central Bosnia, Mostar. *Radio Bosnia-Hercegovina*, Sarajevo, 6 July, *BBC Summary of World Broadcasts*, EE/1735/C1, 8 July 1993. Accessed at Lexis Nexis.

Radio BH 1993b. Mayor of Mostar appeals for international help to save Old Bridge. *Radio Bosnia-Hercegovina*, Sarajevo, 21.00 GMT, 8 November, *BBC Summary of World Broadcasts*, EE/1842/C, 10 November 1993. Accessed at Lexis Nexis.

—— 1993c. Bosnian authorities protest over destruction of Mostar Old Bridge. *Radio Bosnia-Hercegovina*, Sarajevo, 21.00 GMT, 9 November 1993, in Serbo-Croat. *BBC Summary of World Broadcasts*, EE/1843/C, 11 November 1993. Accessed at Lexis Nexis.

Ratkušić, E. 2000. Multietničnosti Mostara nema bez munara i zvonik. *Ljiljan* Sarajevo: 4–11 December, 32.

Reljić, S. and Gojgić, L. 2001. *Hidden crime* [Skriveni zloin]. *NIN*, Belgrade, 15 March. [Serbian] http://www.nin.co.rs/2001-03/15/17048.html and [English] http://www.ex-yupress.com/nin/nin85.html.

Remiković, D. 2012. Islamska zajednica u BiH žalit će se na sudsku presudu u RS po pitanju džamija. *Southeast European Times* [online], 16 February. Available at: http://setimes.com/cocoon/setimes/xhtml/bs/features/setimes/features/2012/02/16/feature-02

RFERL 2002. Bosnian Serbs sentenced over mosque incident. *Radio Free Europe/Radio Liberty Newsline* [online] 29 April 200. Available at: http://www.rferl.org/content/article/1142665.html.

—— 2001. Herzegovinian Croats vandalize mosque site. *Radio Free Europe/Radio Liberty Newsline* [online], 5 December 2001. Available at: http://www.rferl.org/content/article/1142573.html.

Riedlmayer, A. 2008. Banja Luka's Ferhadija mosque rises again. *Bosnian Institute News* [online] 3 March. Available at: http://www.bosnia.org.uk/news/news_body.cfm?newsid=2373.

Riedlmayer, A. and Sells, M.A. *Zitomislici (1566–1992): Meaning, History and Tragic End* [Online], Community of Bosnia Foundation. Archived at: http://web.archive.org/web/20050908065233/http://www.haverford.edu/relg/sells/zitomislici/zitomislici.html.

Roberts, D. 1996. Bosnia/Croats. *Voice of America*, transcript, 28 August. Available at: http://www.hri.org/news/usa/voa/1996/96-08-28.voa.html#02.

Rossini, A. 2003. Ricostruire il ponte di Mostar seguendone lo 'stato dello spirito'. *Osservatorio Balcani e Caucaso* [online], 11 April. Available at: http://www.balcanicaucaso.

org/aree/Bosnia-Erzegovina/Ricostruire-il-ponte-di-Mostar-seguendone-lo-stato-dello-spirito.

RTRS 2001a. Transcript of an interview given to RT RS on 8th May by the Reis-ul-Ulema, efendi Mustafa Ceric. *OHR Afternoon Media Reports, Radio Televizija Republike Srpske (RTRS),* 24 May. Available at: http://www.ohr.int/ohr-dept/presso/bh-media-rep/round-ups/print/?content_id=5707.

RTRS 2001b. Bosnian Serbs request international arbitration on Jasenovac memorial. *Radio Televizija Republike Srpske (RTRS),* 23 February. *BBC Monitoring*. Accessed on NewsBank.

Saborna-Crkva-Mostar.com 2010. Trojičindan u Mostaru. *www.saborna-crkva-mostar.com*, 23 May. Available at: http://www.saborna-crkva-mostar.com/index.php/component/content/article/7-2011-04-08-19-49-41.

Šagolj, Z. 2004. 'Jesam li kriv što nisam Bog?' [Is it my fault I'm not God?]. *Slobodna Dalmacija*, 18 March. Available at: http://www.antegotovina.com/default.aspx?clanak=2319.

Sanchez, L. 2001. SFOR helps to preserve national monuments. *SFOR Informer Online*, 115, 13 June. Available at: http://www.nato.int/sfor/indexinf/115/content.htm.

Santic, N. 2005. Thousand pages about division. *Novi List*, 26 February. Available at: http://www.ex-yupress.com/novi/novilist39.html.

Sarajevo-x.com 2009. Potpisan ugovor o rekonstrukciji Vijećnice. *Sarajevo-x.com* [online], 24 September. Available at: http://www.sarajevo-x.com/bih/sarajevo/clanak/090924093.

Šarić, Ž. 2007. Zbirka vraćena nakon 15 Godina. *Nezavisne Novine*, No. 3067, 22 February, 23. Accessed at Infobiro, www.idoconline.info.

Savić, D. 2008. Pravoslavni hram koji je ujedina Mostar al i cijeli svijet. *Slobodna Bosna*, 1 May, 36–38.

Schiller, B. 1993. Bosnian artists save heritage treasures, *The Toronto Star*, A10, 15 May. Accessed at LexisNexis.

Schork, K. 1992. Jewel of the city destroyed by fire. *The Times*, 27 August. Accessed at NewsBank.

Schwartz, A.E. 1993. What outsiders can't grasp. *The Washington Post*, 9 September. Accessed at LexisNexis.

Schwartz, S. *The Rabbi of Stolac*. Community of Bosnia Foundation [online]. Available at: http://web.archive.org/web/20050507101020/http://www.haverford.edu/relg/sells/stolac/stolacrab.htm.

Sells, M.A. 2001. Erasing culture: Wahhabism, Buddhism, Balkan mosques. First posted on List-Serve Discussion Group, Adabiyat, 14 March 2001, revised and expanded 2 April 2001. Available at: http://www.haverford.edu/relg/sells/reports/WahhabismBuddhasBegova.htm.

Sense Agency 2007. Destruction of Old Bridge seen from two angles. *Sense Tribunal* [online], 17 May 2007. Available at: http://www.sense-agency.com/icty/destruction-of-old-bridge-seen-from-two-angles.29.html?cat_id=1&news_id=10291#

—— 2008. How the Old Bridge was destroyed. *Sense Tribunal* [online], 30 June. Available at: http://www.sense-agency.com/icty/how-the-old-bridge-was-destroyed.29.html?cat_id=1&news_id=8362

—— 2008. Memorable solidarity or ethnic cleansing. *Sense Tribunal* [online], 26 August. Available at: http://www.sense-agency.com/icty/memorable-solidarity-or-ethnic-cleansing.29.html?cat_id=1&news_id=10900#.

—— 2010. Destroying all traces of mosques and churches. *Sense Tribunal* [online], 2 June. Available at: http://www.sense-agency.com/icty/destroying-all-traces-of-mosques-and-churches.29.html.

Serotta, E. 1995. From the ravages of rar, Sarajevo Haggadah appears. *Jewish Telegraphic Agency*, 18 April. Available at: http://www.jta.org/1995/04/18/archive/from-the-ravages-of-war-sarajevo-haggadah-appears.

SFOR Online *Operation Westar*. SFOR Online. Available at: http://www.nato.int/sfor/sfor-at-work/opwestar/indexsaw.htm.

Shanker, T. 1995. Sarajevo postcard: Missing pages. *New Republic*, 13 February.

Sito-Sucic, D. 2004. Postwar squabbles threaten Bosnia's museums, *Reuters* [online], 12 November. Available at: http://www.alertnet.org/thenews/newsdesk/L01684611.htm.

Sito-Sucic, D. 2013. Bosnian activists erect 'guerrilla memorials' to war crimes victims, *Chicago Tribune* [online], 26 October. Available at: http://articles.chicagotribune.com/2013-10-26/news/sns-rt-us-bosnia-memorials-20131026_1_bosnian-serbs-croats-war-crimes.

Slobodna Bosna 2013. Lagumdžija primio vladiku Grigorija, *Slobodna Bosna* [online], 12 December. Available at: http://www.slobodna-bosna.ba/vijest/11712/lagumdzija_primio_vladiku_grigorija.html.

SRNA 2000. Bosnian Serb premier accused of 'giving away' WWII collection to US museum. *SRNA News Agency*, Bijeljina, 27 December, *BBC Monitoring International Reports*. Accessed at NewsBank.

SRNA 2001a. Bosnian Serbs ask Holocaust Museum to keep Jasenovac camp documents. *SRNA news agency*, Bijeljina, 22 October. *BBC Worldwide Monitoring*, 22 October. Accessed on LexisNexis.

SRNA 2001b. Bosnian Serbs urge central government to get back concentration camp documents. *SRNA news agency*, Bijeljina, 27 November. *BBC Worldwide Monitoring*, 27 November. Accessed on LexisNexis.

S.Š. 1998. Nasilje nad spomenicima: kulturni vandalizam, *Svijet*, 06/12/98, 12–13.

Stefanović, N.Lj. 1992. The destruction of cities: We built them – we can demolish them. *Vreme News Digest Agency* (Belgrade), No. 34, 18 May. Available at: http://www.scc.rutgers.edu/serbian_digest.

Stephen, C. 1993. Croat refugees flee Muslim vengeance. *IPS-Inter Press Service*, 10 June 1993. Accessed at LexisNexis.

Stitkovac, E. 1992. Destruction of heritage: More than hatred. *Vreme News Digest Agency*, No. 61, 23 November. Available at: http://www.scc.rutgers.edu/serbian_digest.

Sudetic, C. 1992. Serbs' gains in Bosnia create chaotic patchwork. *The New York Times*, 21 August. Available at: http://www.nytimes.com/1992/08/21/world/conflict-in-the-balkans-serbs- gains-in-bosnia-create-chaotic-patchwork.html.

—— 1993a. Croatian forces face Muslim army. *The New York Times*, 19 June. Available at: http://www.nytimes.com/1993/06/09/world/croatian-forces-face-muslim-army.html

—— 1993b. Mostar's Old Bridge battered to death. *The New York Times*, 10 November. Available at: http://www.nytimes.com/1993/11/10/world/mostar-s-old-bridge-battered-to-death.html.

Taboković, E. 1999. Ostalo je malo pravih Banjalucana koji vole svoj grad. *Dnevni Avaz*, 24 October. Available at: http://www.lablavanda.eu/strana.htm.

Tadic, T. 2000. A report from Kozarusa: Of mosques and politics. *AIM Press* [online]. *Alternativna Informativna Mreža (Alternative Information Network)*, 31 October. Available at: http://www.aimpress.ch/dyn/trae/archive/data/200011/01116-006-trae-sar.htm.

Tanjug 1992a. Yugoslav government sends memorandum to UN and others on 'genocide' in Croatia, *Tanjug*, 31 January. *BBC Summary of World Broadcasts*, 3 February. Accessed on LexisNexis.

Tanjug 1992b. Yugoslav government memorandum on 'genocide' in Croatia delivered to UNESCO. *Yugoslav News Agency (Tanjug)*, 5 February. BBC Summary of World Broadcasts, 7 February. Accessed on LexisNexis.

Tanjug 2011. State broadcaster 'sorry' for 1990s. *B92* [online], 24 May. Available at: http://www.b92.net/eng/news/politics-article.php?yyyy=2011&mm=05&dd=24&nav_id=74507.

TOL 1993. Renaming Fashion, *Transitions Online*, 12 April. Available at: http://www.tol.org/client/article/15051-renaming-fashion.html.

Traynor, I. 1993a. Shells destroy Mostar Bridge. *The Guardian*, 10 November, Foreign page, 11. Accessed at LexisNexis.

—— 1993b. Culture warriors. *The Guardian*, 25 November, 10. Accessed from LexisNexis.

Trebinje Danas 2011. U Pridvorcima otvorena obnovljena džamija, Trebinje Danas [online], 23 July. Available at: http://trebinjedanas.com/node/4745.

Trebinje.se 2006. Peticija za obnovu Begove kuće u Trebinju/Petition for the renovation of the cultural-historical monument Bey's House, Trebinje, 29 April, *www.trebinje.se*. Available at: http://web.archive.org/web/20060430203900fw_/http://medlem.spray.se/trebinje089/experiences7.html.

Vele, F. 2011. Presuda Jadranku Prliću i ostalima potvrdit će da je Hrvatska bila agresor na BiH!. *Dnevni Avaz* [online], 23 April. Available at: http://www.avaz.ba/vijesti/teme/32157-presuda-jadranku-prlicu-i-ostalima-potvrdit-ce-da-je-hrvatska-bila-agresor-na-bih.html.

VOA 2001. Jasenovačfkoj zbirci mjesto je u Jasenovcu. *Glas Amerike* [*Voice of America*], 28 November.

Von Flottau, R., Ihlau, O., Szandar, A. and Ulrich, A. 2004. Deutsche Soldaten: Die hasen vom Amselfeld, *Der Spiegel* [online], 3 May. Available at: http://www.spiegel.de/spiegel/print/d-30748386.html.

V.S. 2007. Ambasador Rajkroft obišao gradilište Princ Čarls spreman pomoći obnovu Ferhadija džamije. *Dnevni Avaz*, No. 4391, 12 December 2007, 7. Accessed at Infobiro, www.idoconline.info.

Vujnović, B. 1995. The Fetish of Nationalism is Unworthy of Man. *AIM Press* [online], Alternativna Informativna Mreža (Alternative Information Network), 3 April.

Ward, G. (ed.) 1996. Revitalisation Research Award. *Revival*, Iss. 6, June 1996. York: Post-war Reconstruction & Development Unit (PRDU), University of York.

—— 1997a. ODA grant for Bosnia project. *Revival*, Iss. 7, January 1997. York: Post-war Reconstruction and Development Unit (PRDU), University of York.

—— 1997b. Analysis of cultural heritage damaged by war in Bosnia-Herzegovina. Synopsis of PRDU research and training project, Bosnia-Herzegovina January–March 1994. *Revival*, Iss. 8, August 1997. York: Post-war Reconstruction & Development Unit (PRDU), University of York.

—— 1997c. The Revitalisation of Počitelj a war-damaged historic settlement in Bosnia-Herzegovina. *Revival*, Iss. 8, August 1997. York: Post-war Reconstruction and Development Unit (PRDU), University of York.

—— 2001. Reconstruction of Pocitelj to go ahead. *Revival*, Iss. 15, August 2001. York: Post-war Reconstruction and Development Unit (PRDU), University of York.

Washington Post Foreign Service 1993. Croat forces evict Muslims, shell famed Mostar bridge. *The Washington Post*, 15 July, A24. Accessed at LexisNexis.

Wenzel, M. 1993b. Eyewitness in Sarajevo, our brother and sister curators and art historians: how do ordinary people live in extraordinary and terrible times?. *The Art Newspaper*, No. 32, November.

Wiesel, E. 2008. Nobel laureate Elie Wiesel remembers Radovan Karadzic, the Serbian murderer. *NY Daily News*, 26 July. Available at: http://www.nydailynews.com/opinions/2008/07/26/2008-07-26_nobel_laureate_elie_wiesel_remembers_rad.html#ixzz185JmM72B.

Williams, C.J. 1993a. Serbs stay their ground on Muslim lands. *Los Angeles Times*, 28 March. Available at: http://articles.latimes.com/1993-03-28/news/mn-16253_1_bosnian-serb.

—— 1993b. Bosnia's hopes fall with historic bridge in Mostar. *Los Angeles Times*, 12 November. Available at: http://articles.latimes.com/1993-11-12/news/mn-56142_1_neretva-river.

—— 1993b. On Bosnia's chessboard, refugees feel like pawns. *Los Angeles Times*, 2 August. Available at: http://articles.latimes.com/1993-08-02/news/mn-19353_1_bosnian-croat-leader.

Zlatović, M. 1999. Sudbina Bosne zrcali se u Jajcu, *Slobodna Dalmacija*, 16 September.

Zubčević, A. 2006. Islamic sites in Bosnia: 10 years after the war. *Islamica Magazine*, Iss. 15, 88.

Žurnal 2010. Grigorijevo Rastrošenije: Vladika u bankrot, crkva na doboš, *www.zurnal.info*, 15 August 2010, http://www.zurnal.info/novost/2520/grigorijevo-rastrosenije-vladika-u-bankrot-crkva-na-dobos-

Zuvela, M. 2012. Bosnia feud threatens end to shared Balkan heritage. *Reuters* [online], 9 January. Available at: http://www.reuters.com/article/2012/01/09/uk-bosnia-feud-threatens-end-to-shared-b-idUSLNE80801L20120109.

Broadcast Media

ITN 1993. Bosnia: British rescue civilians. *ITN News at Ten*, 9 June 1993, Terry Lloyd reporting. *Independent Television News* (ITN). Television news footage; ref: ITN BSP090693016 0. Accessed at ITN Source, http://www.itnsource.com.

Lalic, L. and Sekulic, I. 1999. *Images and Words of Hate: Year Three*, film, ARHITEL, Belgrade.

Reuters 1993. Bosnia-Herzegovina/Croatia: British U.N. troops evacuate 181 Croatians from monastery near Travnik/Refugees from fighting in the town cross the border into Croatia. *Reuters*, 9 June 1993. Television news footage; Reuters ref: 604260160. Accessed at ITN Source, http://www.itnsource.com.

Sky News 1993. Martin Brunt reporting from Central Bosnia. *Sky Television*, 8–9 June 1993. Accessed at YouTube: http://www.youtube.com.

Official Press Releases and Bulletins

British Council BH 2007. *UK Supports Restoration of a National Monument in Bosnia and Herzegovina: From 'Royal Trail' to UNESCO World Heritage Site*. The British Council, press release. Sarajevo, 22 March.

CPIC 2003. *Bijeljina Atik Mosque Reconstruction*. Transcript of Transcript of the International Agency's Joint Press Conference at CPIC. Sarajevo: Coalition Press Information Center, 11 March, 11.00AM.

DELBIH 2012. *EU and BiH Sign Protocol for the Reconstruction of the City Hall in Sarajevo and City Hall in Novi Grad*. Delegation of the European Union to Bosnia and Herzegovina, press release. Sarajevo, 7 March.

IFOR 1996. IFOR AFSOUTH transcript of press briefings, *17–19 October 1996*. http://www.nato.int/ifor/trans/trans.htm.

OHR 1996a. *Destruction of Property on the Site of the Mosque Ferhadija in Banja Luka*. OHR Press Release. Office of the High Representative, Banja Luka:, 17 October.

—— 1996b. *OHR Bulletin 22, Human Rights*. Office of the High Representative, Sarajevo, 24 October.

—— 2000. *BiH TV News Summary, 13 March 2000*. Office of the High Representative, Sarajevo, 13 March.

—— 2001. *Joint Statement on Stolac*. Office of the High Representative, press release, Sarajevo, 1 June.

—— 2002. *High Representative imposes Entity Laws Supporting the Reconstruction of National Monuments*. Office of the High Representative, press release, Sarajevo, 9 February.

—— 2003. *Construction of Stolac Mosque, BiH Media Round-up, 9 January 2003*. Office of the High Representative. Summary of an article in *Večernji List*.

UNDP 2010. *Reconstruction of Three Religious Monuments – Symbols of Multiculturalism Supported by MDGF Culture and Development Programme*. Press release. Sarajevo: UNDP, 20 September 2010. Available at: http://www.undp.ba/index.aspx?PID=7&RID=623.

UNESCO 1995. *UNISIST Newsletter*, General Information Programme, UNESCO, 23:1.

—— 1998. *UNESCO and World Bank to Assist City of Mostar in Rebuilding its Old Bridge*. Press Release No. 98–163. Paris: UNESCO, 30 July 1998.

UNMIBH 2000. *Simultaneous Reconstruction of Religious Sites in BiH*. United Nations Mission in Bosnia-Herzegovina, press release, Sarajevo, 23 May.

USHMM 2001. *Croatian WWII Concentration Camp Records Made Available for First Time by United States Holocaust Memorial Museum*. United States Holocaust Memorial Museum, press release, 13 November.

—— 2002. *Museum Preserves Jasenovac Collection*. United States Holocaust Memorial Museum, press release, winter 2002.

Official Documents

Austrian Development Agency 2007. *Thematic Evaluation: The Relevance of Culture and Cultural Heritage in ADC, Field Report – Bosnia and Herzegovina*, Vienna, January 2007.

CBH (Court of Bosnia and Herzegovina) 2006–2009 Sarajevo. War crimes cases; http://www.sudbih.gov.ba/:

—— *Case No. X-KR-06/241 – Ljubičić, Paško.*

—— *Case No. X-KRŽ-06/299 – Kurtović, Zijad.*

—— *Case No. X-KRŽ-07/442 – Kujundžić, Predrag. First Instance Verdict*, 30 October 2009, 320–21.

CCBiH (Constitutional Court of Bosnia and Herzegovina) 2006. *Sessions. Thirty-Sixth Session of the Grand Chamber, 13 June 2006*. Press release. Sarajevo. Available at: http://www.ccbh.ba.

CIA 1995a. *Sanitized [ ] Version of Ethnic Cleansing Paper*. Central Intelligence Agency,
—— 1995b. *Ethnic Cleansing and Atrocities in Bosnia*. Statement by CIA Deputy Director for Intelligence John Gannon, Joint SSCI SFRC Open Hearing, 9 August. Central Intelligence Agency.
COE (The Council of Europe) 1993–1997. *Information Reports on War Damage to the Cultural Heritage in Croatia and Bosnia-Herzegovina 1–10*. Strasbourg: Parliamentary Assembly of the Council of Europe, presented by the Committee on Culture and Education:
—— *Information Report on War Damage to the Cultural Heritage in Croatia and Bosnia-Herzegovina*. Doc. 6756, 2 February 1993.
—— *Second Information Report on War Damage to the Cultural Heritage in Croatia and Bosnia-Herzegovina*. Doc. 6869, 17 June 1993.
—— *Third Information Report on War Damage to the Cultural Heritage in Croatia and Bosnia-Herzegovina*. Doc. 6904, 20 September 1993.
—— *Fourth Information Report on War Damage to the Cultural Heritage in Croatia and Bosnia-Herzegovina*. Doc 6999, 19 January 1994.
—— *Fifth Information Report on War Damage to the Cultural Heritage in Croatia and Bosnia-Herzegovina*. Doc. 7070, 12 April 1994.
—— *Sixth Information Report on War Damage to the Cultural Heritage in Croatia and Bosnia-Herzegovina*. Doc. 7133, 31 August 1994.
—— *Seventh Information Report on War Damage to the Cultural Heritage in Croatia and Bosnia-Herzegovina*. Doc. 7308, 15 May 1995.
—— *Eighth Information Report on War Damage to the Cultural Heritage in Croatia and Bosnia-Herzegovina*. Doc. 7341, 28 June 1995.
—— *Ninth Information Report on War Damage to the Cultural Heritage in Croatia and Bosnia-Herzegovina*. Doc. 7464, 19 January 1996.
—— *Tenth Information Report on War Damage to the Cultural Heritage in Croatia and Bosnia-Herzegovina*. Doc. 7740, 24 January 1997.
—— 1992. *Minutes of the 90th Session of the Committee of Ministers, Final Communique, Appendix 2, 15 Situation on the Territory of the Former SFR of Yugoslavia*, CM(92)PV1-2-3E, Strasbourg: Council of Europe (COE), 7 May 1993.
—— 1992. *476th Meeting, Decisions Adopted by the Committee of Ministers, Written Question No. 343 by MM Baumel and Tummers on the Cultural Heritage and Former Territory of Yugoslavia*, 7d, CM/Del/Dec(92)476, Strasbourg: Council of Europe, 22 May 1993.
—— 1993. *3rd European Conference of Ministers Responsible for the Cultural Heritage: Proceedings, Malta, 16–17 January 1992, Resolution No. 4 on Conservation in Situations of Conflict*. Strasbourg: Council of Europe (COE), 1993, 60.
—— 1994. *Cultural Situation in the Former Yugoslavia*, Recommendation 1239 (1994), Strasbourg: Parliamentary Assembly of the Council of Europe, 14 April 1994.
—— 1996. *IVth European Conference of Ministers Responsible for the Cultural Heritage Helsinki, 30–31 May 1996 – Report by the Secretary General*, CM(96)97, 9 July 1996, Appendix III: *Appeal in Aid of the Cultural Heritage of Bosnia and Herzegovina and Croatia*, Appendix 1: *Specific Action Plan for Bosnia and Herzegovina.*
—— 1999. *Specific Action Plan for Bosnia-Herzegovina, Preliminary Phase, Final Report*. Technical Cooperation and Consultancy Programme related to the integrated conservation of the cultural heritage, Strasbourg: COE Cultural Heritage Division, March 1999.

—— 2010. *Bosnia-Herzegovina Intermediary Impact Report 2010*. Integrated Rehabilitation Project Plan/Survey of the Architectural and Archaeological Heritage (IRPP/SAAH).
CSCE (Commission on Security and Cooperation in Europe / US Helsinki Commission) 1992. *The Referendum on Independence in Bosnia-Herzegovina February 29–March 1, 1992*. Washington DC. Available at: http://www.csce.gov.
—— 1996. *Banja Luka: Ethnic Cleansing Paradigm or Counterpoint to a Radical Future*. Briefing of the Commission on Security and Cooperation in Europe (US Helsinki Commission), Washington DC. Available at: http://www.csce.gov.
EC/COE 2008. *The Ljubljana Process – Funding Heritage Rehabilitation in South-East Europe, A Joint Action of the European Commission and the Council of Europe*. Available at: http://ec.europa.eu/culture/documents/ce_precatalogue_ljubljana_e.pdf.
EC (European Commission) 2010. *IPA National Programme 2010 – Bosnia and Herzegovina, Fiche 3 'Cultural Heritage'*. Available at: http://ec.europa.eu/enlargement/pdf/bosnia_and_herzegovina/ipa/2010/part-1/pf_03_ipa_2010_cultural_heritage_final_eud_en.pdf.
—— 2010–2012. *Reconstruction works on the Sarajevo and Novi Grad City Halls*. European Commission, Development and Cooperation – EuropeAid. https://webgate.ec.europa.eu/europeaid/online-services/index.cfm?do=publi.welcome&nbPubliList=15&orderby=upd&orderbyad=Desc&searchtype=RS&aofr=130758.
ECMM (European Community Monitor Mission) 1994–1997. *Cultural Heritage Reports on the on the Situation in Croatia and Bosnia and Herzegovina*. Zagreb: ECMM, Humanitarian Section:
—— *Cultural Heritage Report on the on the Situation in Croatia and Bosnia and Herzegovina, No. 1*, 30 December 1994.
—— *Cultural Heritage Report on the on the Situation in Croatia and Bosnia and Herzegovina, No. 2*, 30 April 1995.
—— *Cultural Heritage Report on the on the Situation in Croatia and Bosnia and Herzegovina, No. 3*, 21 July 1995.
—— *Cultural Heritage Report on the on the Situation in Croatia and Bosnia and Herzegovina, No. 4*, 15 February 1996.
—— *Cultural Heritage Report on the on the Situation in Croatia and Bosnia and Herzegovina, No. 5*, 17 December 1996.
—— *Cultural Heritage Report on the on the Situation in Croatia, No. 6*, 21 April 1997.
FSRY 1992. *Memorandum of the Government of Yugoslavia on the Crime of Genocide in Croatia in 1991 and 1992 and the Vandalising of the Memorial of Jasenovac*. Federal Socialist Republic of Yugoslavia, Belgrade, 31 January 1992. Presented to the United Nations Commission on Human Rights, forty-eighth session, 18 February 1992, E/CN.4/1992/71.
HRC (Human Rights Chamber for Bosnia and Herzegovina) 1999–2000. Sarajevo. *Islamic Community in Bosnia and Herzegovina against The Republika Srpska*. Legal rulings:
—— *The Islamic Community in Bosnia and Herzegovina v. The Republika Srpska (Banja Luka Mosques), Decision on the Admissibility and Merits, (Delivered on 11 June 1999)*. Case No. CH/96/29.
—— *Islamic Community in Bosnia and Herzegovina against The Republika Srpska (Zvornik Mosques), Decision on the Admissibility and Merits (Delivered on 9 November 2000)*. Case No. CH/98/1062.

—— *The Islamic Community in Bosnia and Herzegovina v. The Republika Srpska (Bijeljina Mosques), Decision on the Admissibility and Merits (Delivered on 6 December 2000)*. Case No. CH/99/2656.

—— *Islamic Community in Bosnia and Herzegovina (Mrkonjić Grad) against The Republika Srpska, Decision on the Admissibility and Merits (Delivered on 22 December 2003)*. Case No. CH/98/1062.

—— 2000. *Annual Report 1999*, Annex G 1 *Summaries of the Decisions on Merits*, Case No. CH/96/29.

ICJ (International Court of Justice) 1993–2006. The Hague. *Application of the Convention on the Prevention and Punishment of the Crime of Genocide (Bosnia and Herzegovina v. Yugoslavia (Serbia and Montenegro)*:

—— *Application of the Convention on the Prevention and Punishment of the Crime of Genocide (Bosnia and Herzegovina v. Yugoslavia (Serbia and Montenegro)*. 20 March 1993.

—— *Application of the Convention on the Prevention and Punishment of the Crime of Genocide (Bosnia and Herzegovina v. Serbia and Montenegro) Public Sitting Held on Friday 1 March 2006, at 10 a.m., at the Peace Palace, President Higgins Presiding*. Verbatim Record, CR 2006/5. (Laura Dauban pleading on *Destruction of Cultural Heritage in Bosnia and Herzegovina.*)

—— *Application of the Convention on the Prevention and Punishment of the Crime of Genocide (Bosnia and Herzegovina v. Serbia and Montenegro) Public Sitting Held on Friday 17 March 2006, at 10 a.m., at the Peace Palace, President Higgins Presiding*. Verbatim Record, CR 2006/22. (András Riedlmayer as expert witness on *Destruction of Cultural Heritage in Bosnia and Herzegovina*)

ICOM (International Council of Museums) 1995. *18th General Assembly, Resolution No. 2: Protection of the Cultural Heritage during Armed Conflict*, Stavanger, 7 July 1995.

ICOMOS (International Council on Monuments and Sites) 1993. *Resolutions of the 1993 General Assembly, No. 12*, Colombo: 30 July–4 August 1993.

—— 1994. ICOMOS' Risk Preparedness Scheme, *ICOMOS Newsletter, Second Edition 1994*, July 1994, 18.

ICOMOS Sweden 1994. *Information as an Instrument for Protection against War Damages to the Cultural Heritage. Report from a Seminar, June 1994*. ICOMOS Sweden, Central Board of National Antiquities, Swedish National Commission for UNESCO. Stockholm: Svenska Unescoradets skriftserie, 4/1994.

ICRC (International Committee of the Red Cross) 2001. *Resolution 11. Protection of Cultural Property in the Event of Armed Conflict*, Resolutions *of the Council of Delegates – 2001, Adopted at its Session of 11–14 November 2001*. Geneva: Council of Delegates of the International Red Cross and Red Crescent Movement.

ICTY (International Criminal Tribunal for the former Yugoslavia) 2009. *Rules of Procedure and Evidence*. IT/32/Rev. 43, 24 July 2009. The Hague.

—— 2009. *Updated Statute of the International Criminal Tribunal for the Former Yugoslavia*, September 2009.

ICTY (International Criminal Tribunal for the Former Yugoslavia), The Hague. Indictments and other legal rulings 1995–2009:

—— *Prosecutor v. Radovan Karadžić and Ratko Mladić*. Case No. IT-95-18/1-I. *Indictment*, 24 July 1995.

—— *Prosecutor v. Tihomir Blaškić (Lašva Valley)*. Case No. IT-1995-14-T. *Judgment*, 3 March 2000.

—— *Prosecutor v. Dragoljub Kunarac, Radomir Kovač, Zoran Vuković (Foča)*. Case No. IT-96-23-T& IT-96-23/1-T. *Judgement*, 22 February 2001.

—— *Prosecutor v. Dario Kordić and Mario Čerkez (Lašva Valley)*. Case No. IT-95-14/2-T. *Judgment*, 26 February 2001.

—— *Prosecutor v. Radislav Krstić*. Case No. IT-98-33. *Trial Judgment*, 2 August 2001.

—— *Prosecutor v. Momčilo Krajišnik and Biljana Plavšić*. Case No. IT-00-39 & 40-PT. *Amended Consolidated Indictment*, 4 March 2002.

—— *Prosecutor v. Milorad Krnojelac (Foča)*. Case No. IT-97-25-T. *Judgement*, 15 March 2002.

—— *Prosecutor v. Vojislav Šešelj*. Case No. IT-03-67. *Initial Indictment*, 15 January 2003.

—— *Prosecutor v. Miodrag Jokić (Dubrovnik)*. Case No. IT-01-42/1-S. *Judgment*, 18 March 2004.

—— *Prosecutor v. Pavle Strugar (Dubrovnik)*. Case No. IT-01-42-T. *Judgment*, 31 January 2005.

—— *Prosecutor v. Enver Hadžihasanović and Amir Kubura (Central Bosnia)*. Case No. IT-01-47. *Judgement*, 22 April 2008.

—— *Prosecutor v. Radovan Karadžić*. Case No. IT-95-5/18/PT. *Prosecution's Second Amended Indictment*, 18 February 2009.

—— *Prosecutor v. Jadranko Prlić, Bruno Stojić, Slobodan Praljak, Milivoj Petković, Valentin Ćorić & Berislav Pušić (Prlić et al)*. Case No. IT-04-74-T. *Initial Indictment*, 2 March 2004. *Judgement*, 29 May 2013.

IFLA (International Federation of Library Associations) /UNESCO 1995. Call for Solidarity with the Librarians in Sarajevo, January 1995.

MHRR 2005. Uporedna analiza pristupa pravima izbjeglica i raseljenih osoba. Sarajevo: Ministarstvo za ljudska prava i izbjeglice BiH, December. *www.mhrr.gov.ba/PDF/?id=282.*

OHR (Office of the High Representative) 1999–2002 Sarajevo. Decisions:

—— *Decision Removing Mr. Djordje Umicevic from his Position of President of the Municipal Assembly of Banja Luka*. Sarajevo: Office of the High Representative, 29 November 1999.

—— *Decision Imposing the Federation Law on Implementation of Decisions of the Commission to Preserve National Monuments Established under Annex 8 of the Dayton Peace Agreement*. Sarajevo: Office of the High Representative, 8 February 2002.

—— *Decision Imposing the RS Law on Implementation of Decisions of the Commission to Preserve National Monuments Established under Annex 8 of the Dayton Peace Agreement*. Sarajevo: Office of the High Representative, 8 February 2002.

—— *Decision Enacting the Law on Amendments to the RS Law on Urban Planning*. Sarajevo: Office of the High Representative, 14 August 2002.

—— 1998. *Human Rights Report*. Office of the High Representative, Sarajevo, 15 May 1998.

—— 2001. *Agreement on Succession Issues*, agreed in Brussels, 10 April 2001, adopted at the Conference on Succession Issues, Vienna, 29 June 2001. United Nations, *Treaty Series*, Vol. 2262, Chap. XXIX Miscellaneous, 251. Entered into force 2 June 2004, United Nations No. 40296.

OIC (Organisation of the Islamic Conference) 1992–1993. Resolutions and Communiques:

—— *Resolutions adopted by the Fifth Extraordinary Session of the Islamic Conference of Foreign Ministers, Resolution No. 1/5-EX, Situation on Bosnia-Herzegovina*, No. 8, Istanbul: Organisation of the Islamic Conference (OIC), 17–18 June 1992.

—— *Final Communique of the Fifth Extraordinary Session of the Islamic Conference of Foreign Ministers*, Istanbul, Turkey, 17–18 June 1992, Ex-ICFM/5-92/FC (Final), 12.

—— *Resolution No. 19/21-C, Destruction of Mosques and Islamic Monuments in Bosnia-Herzegovina, The Twenty-first Islamic Conference of Foreign Ministers*. Karachi: Organisation of the Islamic Conference (OIC), 25–29 April 1993.

Okružni Sud, Banja Luka 2009. *Ferhadija*. Appeal Judgement (in Serbian). Banja Luka: Okružni Sud u Banjaluci. No. 011-0-Gž-09-000328, 30 September 2009. Available at: http://www.okruznisud-bl.com/

State Commission for Gathering Facts on War Crimes in the Republic of Bosnia and Herzegovina 1992–1993 *Bulletins No. 1–No. 5*.

—— *Bulletin No. 1 (Destruction of Religious Objects)*. Sarajevo: October 1992.

—— *Bulletin No. 2 (Documents and Facts: Destruction of Sacred Complexes)*. Sarajevo: February 1993.

—— *Bulletin No. 3 (Documents and Facts: Destruction of Sacred Complexes)*. Sarajevo: March 1993.

—— *Bulletin No. 4 (Destruction of Sacred Complexes: Serbian Orthodox Church)*. Sarajevo: April 1993.

—— *Bulletin No. 5 (Destruction of Sacred Complexes III)*. Sarajevo: August 1993.

UN 1948a. *United Nations Convention on the Prevention and Punishment of the Crime of Genocide*, Paris: United Nations, 9 December 1948.

—— 1948b. *The Universal Declaration of Human Rights*. United Nations, 10 December 1948.

—— 1966. *International Covenant on Economic, Social and Cultural Rights*. United Nations, 16 December 1966.

—— 1992a. *United Nations Security Council Resolution 764 (1992.) Adopted by the Security Council at its 3093rd Meeting, on 13 July 1992*, 13 July 1992, S/RES/764 (1992) UN.

—— 1992b. *United Nations Security Council Resolution 771 (1992) Adopted by the Security Council at its 3106th Meeting, on 13 August 1992*, S/RES/771 (1992) UN.

—— 1992c. *United Nations Security Council Resolution 780 (1992) Adopted by the Security Council at its 3119th Meeting, on 6 October 1992*, S/RES/780 (1992),

—— 1992d. *The Situation in Bosnia and Herzegovina*. 91st Plenary Meeting of the General Assembly, AG/Res./47/121 of 18 December 1992.

—— 1992e. *Situation of Human Rights in the Territory of the Former Yugoslavia: Resolution/Adopted by the General Assembly*, A/RES/47/147, New York: UN General Assembly, 18 December 1992.

—— 1993a. *Interim Report of the Commission of Experts Established Pursuant to Security Council Resolution 780 (1992)* S/25274, 9 February 1993.

—— 1993b. *Situation of Human Rights in the Territory of the Former Yugoslavia, Report on the Situation of Human Rights in the Territory of the Former Yugoslavia Submitted by Mr. Tadeusz Mazowiecki, Special Rapporteur of the Commission on Human Rights, Pursuant to Commission Resolution 1992/S-1/1 of 14 August 1992*. UN Doc. E/CN.4/1993/50, 10 February 1993.

—— 1993c. *Report of the Secretary-General Pursuant to Paragraph 2 of Security Council Resolution 808 (1993) [Contains text of the Statute of the International Tribunal for the Prosecution of Persons Responsible for Serious Violations of International Humanitarian Law Committed in the Territory of the Former Yugoslavia Since 1991] Resolution 820*

(1993) Adopted by the Security Council at its 3200th Meeting, on 17 April 1993, 7 April 1993, S/RES/820 (1993).

—— 1993d. *United Nations Security Council Resolution 827 (1993) Adopted by the Security Council at its 3217th Meeting, on 25 May 1993*, 25 May 1993, S/RES/827 (1993).

—— 1994a. *Final Report of the United Nations Commission of Experts Established Pursuant to Security Council Resolution 780 (1992)*, S/1994/674, 27 May 1994.

—— 1994b. *Final Report of the United Nations Commission of Experts Established Pursuant to Security Council Resolution 780 (1992), Annex Summaries and Conclusions.* S/1994/674/Add.2 (Vol. I), 28 December 1994.

—— 1999. *Report on the Situation of Human Rights in Bosnia and Herzegovina, the Republic of Croatia and the Federal Republic of Yugoslavia (Serbia and Montenegro) Submitted by the Special Rapporteur of the Commission on Human Rights*, to the General Assembly of the United Nations, 24 October 1999, A/54/396, S/1999/1000.

—— 2005. *Resolution adopted by the General Assembly: The United Nations Basic Principles and Guidelines on the Right to a Remedy and Reparation for Victims of Gross Violations of International Human Rights Law and Serious Violations of International Humanitarian Law, A/RES/60/147*. New York: United Nations, 16 December 2005.

UNESCO 1989. *Information of the Implementation of the Convention for the Protection of Cultural Property in the Event of Armed Conflict, The Hague 1954, 1989 Reports*, CC/MD/11, December 1989.

—— 1992a. *Decision 7.5 The Situation of the Cultural and Architectural Heritage as well as of Educational and Cultural Institutions in Bosnia-Herzegovina* (139 EX/33), *Decisions Adopted by the Executive Board at its 139th Session*, EX139/Decisions, Paris: UNESCO, 18–27 May 1992.

—— 1992b. *Review of the application of the Convention for the Protection of Cultural Property in the Event of Armed Conflict (The Hague, 1954)*, UNESCO doc. 140 EX/26, Paris, 11 September 1992.

—— 1992c. *Address by Mr Federico Mayor, at the First International Symposium on Human Rights La Laguna (Tenerife)*, Doc. DG/92/44, 1 November 1992.

—— 1993a. *Decisions Adopted by the Executive Board at its 141st Session*. 141 EX/Decisions, 9.3, (Paris, 10–28 May 1993). Paris: UNESCO, 17 June 1993.

—— 1993b. *Records of the General Conference, Twenty-seventh Session Paris, 25 October to 6 November 1993, Vol. 1: Resolutions, 4.8 The Situation of the Cultural and Architectural Heritage and of Educational and Cultural Institutions in Bosnia and Herzegovina (Sarajevo National and University Library).*

—— 1994a. *Appeal by Federico Mayor, Director-General of UNESCO for the Reconstruction of the National and University Library of Bosnia and Herzegovina in Sarajevo*. Paris: UNESCO, 13 April 1994.

—— 1994b. *The Situation of the Cultural and Architectural Heritage and of Educational Institutions in Bosnia and Herzegovina*. 145th Session of the General Conference. Provisional Agenda, Doc. 145 EX/36, Paris, 7 October 1994, 8.12.

—— 1995a. *The Situation of the Cultural and Architectural Heritage and of Educational Institutions in Bosnia and Herzegovina. 146th Session of the General Conference. Provisional Agenda*, Doc. 146 EX/46, Paris, 3 May 1995, 9.3.

—— 1995b. *Information on the Implementation of the Convention for the Protection of Cultural Property in the Event of Armed Conflict, The Hague 1954, 1995 Reports*, Doc. CLT-95/WS/13. Paris: UNESCO, December 1995.

—— 1997. *Mostar: Urban Heritage Map and Rehabilitation Plan of Stari Grad/ Mostar: mappa del patrimonio urbanistico e piano di ricostruzione di Stari Grad; Mapa urbanistickog nasljedja i plan rehabilitacije Starog Grada.* Florence: Angelo Pontecorboli.

—— 2005. *Report on the Implementation of the 1954 Hague Convention for the Protection of Cultural Property in the Event of Armed Conflict and its Two 1954 and 1999 Protocols, Report on the Activities from 1995 to 2004*, UNESCO, (Clt-2005/WS/6)

—— 2008. *Address by Mr Koïchiro Matsuura, Director-General of UNESCO, for the Promotion of the Reconstruction of the 3 Monuments: The Ferhadija Mosque in Banja Luka, the Orthodox Cathedral of Mostar and the Franciscan Monastery in Plehan*, Mostar, 26 March 2008.

—— 2008. Mission Report, Old Bridge Area of the Old City of Mostar (Bosnia and Herzegovina) (C 946rev), 26–28 May 2008. Paris: UNESCO, 2008. whc.unesco.org/ document/100757.

US Embassy Sarajevo 2006. *Bosnia: Little Progress in Post-sanction Trebinje*, 16 May. Wikileaks: 06SARAJEVO1075_a. Available at: http://www.wikileaks.org/plusd/ cables/06SARAJEVO1075_a.html.

USCRI 1998. *U.S. Committee for Refugees World Refugee Survey 1998 – Bosnia and Herzegovina*, US Committee for Refugees and Immigrants 1 January 1998.

USDS (US Department of State, Bureau of Public Affairs) 1993. *Dispatch*, 26 July 1993, Vol. 4, No. 30. Bureau of Public Affairs.

—— 2007. *The US Ambassadors Fund for Cultural Preservation, Annual Report 06/07*, United States Department of State.

—— 2008. *International Religious Freedom Report – Bosnia and Herzegovina.* US Department of State, 19 September 2008.

USGAO (US General Accounting Office) 1998. *Bosnia Peace Operation: Pace of Implementing Dayton Accelerated as International Involvement Increased.* GAO/ NSIAD-98-138.

World Bank 1999. *Project Appraisal Document on a Proposed Credit in the Amount of SDR 3.0 Million (US$4.0 million Equivalent to Bosnia and Herzegovina for a Pilot Cultural Heritage Project.* Project Appraisal Document, Report No. 19115-BiH, Infrastructure Sector Unit, Europe and Central Asia Regional Office, 14 July 1999.

—— 2005. *Implementation Completion Report on a Credit in the Amount of US$4.0 Million Equivalent to the Bosnia and Herzegovina for a Pilot Cultural Heritage Project.* Implementation Completion Report, Report No. 32713, Infrastructure Sector Unit, Europe and Central Asia Regional Office, 22 June 2005.

WHC (World Heritage Committee) 1993. *17th Session of the World Heritage Committee*, WHC-93/CONF.002, *Annex V World Heritage Committee Declaration on Bosnia*, Cartagena: 6–11 December 1993;

—— 2005. *Decisions of the 29th Session of the World Heritage Committee (Durban, 2005)*, WHC-05/29.COM/22, Paris: UNESCO, World Heritage Committee, 9 September 2005.

—— 2007. *Decisions of the 31st Session of the World Heritage Committee (Christchurch, 2007)*, WHC.07 /31.COM /24, Paris: UNESCO, World Heritage Committee, 31 July 2007.

Authored Official Reports

Boylan, P.J. 1993. *Review of the Convention for the Protection of Cultural Property in the Event of Armed Conflict (The Hague Convention of 1954)*, (1993). UNESCO: Doc. CLT-93/WS/12.

Čelić-Čemerlić, V. (ed.) 2000. *Katalog obnove kulturno-historijskog i prirodnog. naslijedja Kantona Sarajevo, 1996–2000*. Sarajevo: Kantonalni zavod za zastitu kulturno-historijskog i prirodnog naslijedja Sarajevo.

Čukur, M., Magnusson, K., Molander, J. and Skotte, H. 2005. *Returning Home: An Evaluation of Sida's Integrated Area Programmes in Bosnia and Herzegovina*. Sida Evaluation 05/18, Stockholm: Swedish International Development Cooperation Agency.

de la Brosse, R. 2003. *Political Propaganda and the Plan to Create 'A State For All Serbs' Consequences of Using Media for Ultra-nationalist Ends*. Expert report prepared for the Office of the Prosecutor, the International Criminal Tribunal for the former Yugoslavia (ICTY), Exhibit D01645, January 2003.

Fischer, L. 1994a. *Report on the Cultural Situation in the Former Yugoslavia*, Doc. 6989, Strasbourg: Parliamentary Assembly of the Council of Europe, 14 January 1994.

—— 1994b. *Addendum to the Report on the Cultural Situation in the Former Yugoslavia*, Doc. 6989 Addendum, Strasbourg: Parliamentary Assembly of the Council of Europe, 13 April 1994.

Gelbard, R.S. 1998. *Remarks at the Donors' Conference for Bosnia and Herzegovina*. Ambassador Robert S. Gelbard, Special Representative of the President and Secretary of State for the Implementation of the Dayton Peace Accords, 7 May 1998.

Goodey, B. and Desimpelaere, W. 2003. *External Evaluation of UNESCO's Action in the Preservation of Cultural Heritage Damaged by Conflict: Bosnia-Herzegovina 1996–2003*, UNESCO Doc. IOS/EVS/PI/24.

Hamidović, M., Mićić, L. and Petrović, A. (eds) 2010. *Zaštita kulturne i graditeljske baštine: 2003–2009*, Sarajevo: Federalno ministarstvo kulture i sporta. www.fmks.gov.ba/download/monografija.pdf.

Hanson, D. 2004. *Bosnian Serb Crisis Staffs*. Expert report (updated) prepared for the Office of the Prosecutor, the International Criminal Tribunal for the former Yugoslavia (ICTY), Exhibit P528, 26 November 2004.

Husedžinović, S. 1998. *Cultural Heritage of Special Value Destroyed in the War 1992–1995*. Sarajevo: Federation of Bosnia and Herzegovina, Ministry of Education, Science, Culture and Sports.

Kaiser, C. 2002. (*Report on Destruction of Cultural Property 09-Jul-02, Exhibit P906, Krajišnik* Case No. IT-00-39) *Report on the Damaging and Destruction of Islamic and Roman Catholic Sacral Buildings in the Municipalities of Bosanski Novi, Donji Vakuf, Kljuc, Kotor Varos, Prijedor and Sanski Most in 1992 (April to December) and Preliminary Remarks on Damaging and Destruction of Islamic and Roman Catholic Sacral Buildings in the Municipalities of Bijeljina, Brcko, Bratunac, Cajnice, Doboj, Foca, Ilijas, Rogatica, Visegrad and Zvornik*, 9 July 2002. Prepared for the Office of the Prosecutor, International Criminal Tribunal for the former Yugoslavia (ICTY).

Mulalić Handan, M. et al. 2005. *Preliminary Technical Assessment of the Architectural and Archaeological Heritage in South East Europe: Historic Building, Seranic's House in Banja Luka*. European Commission/Council of Europe Joint Programme on the (IRPP/SAAH), Regional Programme on Cultural and Natural Heritage in South East Europe 2003–2006, Sarajevo: Commission to Preserve National Monuments.

—— 2010. *Report on Assessment of the Architectural and Archaeological Heritage, Bosnia and Herzegovina*. Sarajevo: Commission to Preserve National Monuments of Bosnia and Herzegovina. Regional Programme on Cultural and Natural Heritage in South East Europe, IRPP/SAAH. Available at: http://kons.gov.ba/main.php?mod=vijesti&extra=1075460534&action=view&id_vijesti=669&lang=4/

Pusilo, N. et al. 1995. A report on the devastation of cultural, historical and natural heritage of the Republic of Bosnia and Herzegovina (from April 5, 1992 until September 5, 1995). *Naše Starine*, Vanredni broj XXI. Sarajevo: Institute for the Protection of Cultural, Historical and Natural Heritage of the Republic of Bosnia-Herzegovina [Zavod za zaštitu kulturno-historijskog i prirodnog naslijeđa BiH].

—— 1997. *Izvjestaj o devastaciji kulturno-historijskog i prirodnog naslijedja Bosne i Hercegovine (1992–1995)*. Sarajevo: Institute for the Protection of Cultural, Historical and Natural Heritage of Bosnia-Herzegovina [Zavod za zastitu kulturno-historijskog i prirodnog naslijedja BiH], Centar za naslijedje BiH. Revised edition of 1995 edition.

Riedlmayer, A. 1995. *Killing Memory: The Targeting of Bosnia's Cultural Heritage*, Testimony presented at a hearing of the Commission on Security and Cooperation in Europe, US Congress, 4 April 1995.

—— 2002a. *Destruction of Cultural Heritage in Bosnia-Herzegovina, 1992–1996: A Post-war Survey of Selected Municipalities*. Prepared for the Office of the Prosecutor, International Criminal Tribunal for the former Yugoslavia (ICTY). *Milošević* Case No. IT-02-54, Exhibit P486, Date: 08/07/2003, and *Krajišnik* Case No. IT-00-39, Exhibit P732, Date: 23/05/2005.

—— 2005. *Expert Report: Statistical Addendum: Sacral Sites of the non-Serb Communities in the Surveyed Municipalities Damaged or Destroyed in 1992*, ICTY Exh. P732.B, 22 May 2005.

—— 2009. *Destruction of Cultural Heritage in Bosnia and Herzegovina: A Post-war Survey of Non-Serb Cultural Heritage in the Municipalities of Banja Luka, etc, Prosecutor v. Radovan Karadžić*, ICTY Case No. IT-95-5/18-I.

Shrimplin, R. 1993. *Note on a Visit to Sarajevo to Review Damage to the Cultural Heritage and to Assess Prospects for a Repair/Restoration Programme June 1993*, COE *Third Information Report*, Doc. 6904, *20 September 1993*.

Travers, R. 2012. *MDGF Final Evaluation: Improving Cultural Understanding in Bosnia and Herzegovina*. Sarajevo: United Nations Office of the Resident Coordinator, June 2012.

Walasek, H. and Wenzel, M. 1996. Report on museums in Bosnia-Herzegovina during the present war, in *Ninth Information Report on War Damage to the Cultural Heritage in Croatia and Bosnia-Herzegovina*. Doc. 7464, 19 January 1996. Strasbourg: Parliamentary Assembly of the Council of Europe, presented by the Committee on Culture and Education.

Walasek, H. 1997. Museums in Bosnia-Herzegovina II, A report on museums in Republika Srpska with an update on museums in Sarajevo, in *Tenth Information Report on War Damage to the Cultural Heritage in Croatia and Bosnia-Herzegovina*. Doc. 7740, 24 January 1997. Strasbourg: Parliamentary Assembly of the Council of Europe, presented by the Committee on Culture and Education.

NGO Reports

AKTC/WMF 2004. *Conservation and Revitalisation of Historic Mostar*, Historic Cities Support Programme. Geneva: Aga Khan Trust for Culture/World Monuments Fund.

CHwB 2007. *The Restoration Process in Eight Cities, Towns and Villages in Bosnia-Herzegovina, Impact Assessment Study*, carried out by market research/polling agency Puls. Sarajevo: Cultural Heritage without Borders (CHwB), January 2007.

HRW 1993. *War Crimes in Bosnia-Hercegovina Vol. II*. Helsinki Watch report. New York, Washington, Los Angeles, London: Human Rights Watch, 1 April 1993.

—— 1994. *War Crimes in Bosnia-Hercegovina: U.N. Cease-Fire Won't Help Banja Luka*. Human Rights Watch, Vol. 6, Iss. 4, D608, 1 June 1994.

—— *1995. The Croatian Army Offensive in Western Slavonia and its Aftermath, Ensuing Abuses in Bosnian Serb-Held Areas*, 7:11 (July), 13–15.

—— 1998. *'A dark and closed place': Past & Present Human Rights Abuses in Foca*. Human Rights Watch, Vol. 10, No. 6 (D), 1 July 1998.

—— 2000. *Bosnia and Hercegovina, Unfinished Business: The Return of Refugees and Displaced Persons to Bijeljina. Abuses Against Minorities After the War*. Human Rights Watch, May 2000, 12:7 (D).

—— 2006. *Genocide, War Crimes, and Crimes Against Humanity: A Topical Digest of the Case Law the International Criminal Tribunal for the Former Yugoslavia*. New York: Human Rights Watch.

ICG 1998. *Return of Bosnian Serb Displaced Persons to Drvar, Bosansko Grahovo and Glamoc*, ICG Bosnia Project, International Crisis Group, Report No. 29, 19 January 1998.

—— 1999. *Is Dayton Failing?: Bosnia Four Years After the Peace Agreement*. International Crisis Group, Report, 28 October 1999.

—— 2003. *Building Bridges in Mostar*, International Crisis Group, Europe Report No. 150, Sarajevo/Brussels, 20 November 2003.

Soul of Europe 2004. Newsletter, December 2004. Available at: http://soulofeurope.blogspot.co.uk.

http://soulofeurope.blogspot.co.uk/2004_12_01_archive.html.

NGO Authored Reports

Biörnstad, M. and Johansson, B.O.H. 2002. *Heritage in Distress: Experiences from International Emergency Rescue Aid and Development Co-operation 1995–2002*. Stockholm: Cultural Heritage without Borders (CHwB).

Kälvemark, T. 2007. *Cultural Heritage for Peace and Reconciliation, An Evaluation of Cultural Heritage without Borders (CHwB)*. Stockholm: Cultural Heritage without Borders.

Johansson, B.O.H. et al. 2005. *Cultural Heritage: Reconciliation, Restoration, Hopes for the Future*. Stockholm: Cultural Heritage without Borders (CHwB).

Ljungman, C.M. and Taboroff, J. 2011. *Evaluation of Cultural Heritage without Borders (2008–2011)*. Carried out on behalf of Sida. Stockholm: Indevelop, September 2011.

Matarasso, F. 2005. *Living Heritage Bosnia and Herzegovina: Community Development Through Cultural Heritage Resources, Programme Report 2001–2005*. Brussels: King Badouin Foundation/Mozaik.

Index

For Product Safety Concerns and Information please contact our EU representative GPSR@taylorandfrancis.com
Taylor & Francis Verlag GmbH, Kaufingerstraße 24, 80331 München, Germany

www.ingramcontent.com/pod-product-compliance
Ingram Content Group UK Ltd.
Pitfield, Milton Keynes, MK11 3LW, UK
UKHW020024200926
475702UK00014B/2040

* 9 7 8 1 1 3 8 3 0 8 4 2 8 *